ORDINATION IN THE STUDY OF MORPHOLOGY, EVOLUTION AND SYSTEMATICS OF INSECTS:

APPLICATIONS AND QUANTITATIVE GENETIC RATIONALS

ORDINATION IN THE STUDY OF MORPHOLOGY, EVOLUTION AND SYSTEMATICS OF INSECTS:

APPLICATIONS AND QUANTITATIVE GENETIC RATIONALS

Edited by

J.T. SORENSEN
Department of Food and Agriculture
Insect Taxonomy Laboratory
Sacramento, CA - USA

and

R. FOOTTIT
Agriculture Canada
Research Branch
Ottawa, Ontario - Canada

ELSEVIER

Amsterdam – London – New York – Tokyo 1992

ELSEVIER SCIENCE PUBLISHERS B.V.,
Sara Burgerhartstraat 25
P.O. Box 211, 1000 AE Amsterdam, The Netherlands

ISBN 0-444-89801-8

This book is printed on acid-free paper.

Printed in The Netherlands

Preface

The idea for this volume originated while we were organizing a workshop on "Ordination Methods in Entomological Systematics . . ." for the XVIII International Congress of Entomology, held at the University of British Columbia, Vancouver, in June 1988. Contributions to that workshop, and ordination-based research that was presented at another congress symposium on "The Structural Basis and Analysis of Allometric Growth in Insects," formed the initial basis for this volume.

This book, however, goes well beyond those Vancouver sessions and attempts to portray many current, distinct aspects of ordination analyses that have not been covered before. It attempts to convey a wealth of possibilities for the biological and evolutionary interpretations that these techniques allow. Many of these interpretations are new and controversial. Several of the aspects presented here are still evolving in theory and implementation. Disagreement and varied usage is presented in the book, so that readers can see and understand the philosophies and arguments involved.

In producing a work such as this, we hoped to provide examples for inexperienced workers. We also wanted to provide insights and inspirations for those already experienced in morphometric research, to promote their exploration of the possibilities and new directions in biosystematic research that ordinations can provide. We hope we have adequately included examples and arguments for required cautions and the limitations of the techniques as well. Although there are no magic solutions, there is a wide range of methods from which to choose, each with its own characteristics and potential applications.

Ordinations, perhaps more so than any other group of numerical analyses used in biosystematic research, have a plethora of possible options and variations. These vary from the initial "size-adjustment" of data by shearing or log transformation, to various criteria for the calculation of rotations and plots. Similarly, because any given "class" of ordination has "evolved" during its historic usage, many different names may be applied by various workers to the same, or nearly the same, type of analysis (for example, discriminant function analysis has also been rightfully called discriminant analysis, canonical variates analysis, canonical analysis of discriminance, etc., depending upon a given use). Because of the resulting myriad of possible options and nomenclature for these analyses, we, as editors, have usually and purposely taken no stand on recommending any particular option that is possible, or enforcing in the book any name that should be used; individual authors, in contrast, often have and present points of view.

Unfortunately, we have necessarily had to restrict the book to only principal components and discriminant function analyses, which are the most widely used ordination applications in biology. Both the size of the volume, and a lack of biological research using other types of ordinations, determined this limit. We hope future workers can provide coverage of other important ordination techniques, including principal coordinates analysis and nonmetric multidimensional scaling, that we have not been able to cover, and which have decided advantages when discrete and nonparametric data are available. We also regret restricting the examples of applications to only insects, but because they are the most numerous and diverse of creatures, it should be evident that the techniques can be used with any other biological groups.

The emphasis of this book is thus on how researchers use and interpret ordinations to solve systematic problems in their particular taxonomic groups. We encouraged the contributors to give indications of their philosophies and methodologies for these applications in their biosystematic work. The approaches taken were, of course, influenced by the biology and morphology of the given groups, which in turn reflect their unique evolutionary and ecological processes. The directions and interpretations of these studies are evident: that is, how each researcher views the various matrices, vectors, and plots of specimens, samples and taxa; which characteristics they chose to emphasize (or ignore); and where problems remain unsolved.

We have included chapters: that discuss what ordinations are, that survey options for data gathering and treatment, that assess characters and their allometry, and that demonstrate the circumscription of biological groups. We also present coverage that deals with the genetic foundations and rationales of ordinations, and that explore their potential in phylogenetic and evolutionary inference. We have attempted to group the chapters in the table of contents under subheadings that reflect these thrusts, but readers should be aware that some chapters cover a wider ground than their subheading indicates.

We believe that ordinations have tremendous potential in evolutionary research. We feel they should not be viewed as techniques that are limited to the resolution of patterns alone, but can be used also to reflect processes to a degree. They should be viewed as the morphologically-based methods of choice for taxa with characters that are continuous or hard to define into states, and for taxonomic groups with nebulous or overlapping entities. We hope, this book will serve as a catalyst for the exploration of the potential of ordinations in systematic and evolutionary research. We suspect that ordinations, more than any other type of biosystematic analyses, permit, and in fact require, the researcher to interpret and carefully evaluate their results, yet the results reflect patterns of underlying biological correlation and covariance, whatever their cause, rather than simply a procedural output. Therefore, for more methodologically oriented researchers who are prone to philosophical rationalization, ordinations permit a "journey" that is "its own reward."

We thank the many authors of works on ordinations that have come before us for defining the field. We especially thank Russell Lande and Joe Felsenstein for past discussions, and for their genetic research, which has formed the basis of many of our opinions on, and evolutionary interpretations for, ordinations. We thank the contributors to this volume for their patience, and the high standards of their work. During production of the book, we were aided by: D. Barbe in graphics scanning; J. LeMasters, D. Dethlefsen and L. McCall in text scanning; R.J. Gill in text transfer; R. Fortuner in bibliographic verifications; S.M. Sawyer in bibliographic verifications, indexing and proofing; and by K.H. Sorensen in proofing. We also gratefully acknowledge the assistance and patience of the staff at Elsevier Science Publishers B.V., Amsterdam.

John T. Sorensen, Sacramento
Robert G. Foottit, Ottawa
May, 1992

TABLE OF CONTENTS

ORDINATION METHODS: THEIR CONTRAST TO CLUSTERING AND CLADISTIC TECHNIQUES

Robert G. Foottit[1] and John T. Sorensen[2]

[1] Biological Resources Division, Centre for Land and Biological Resources Research, Agriculture Canada, Research Branch, C.E.F., Ottawa, Ontario K1A 0C6, Canada; [2] Insect Taxonomy Laboratory, California Department of Food and Agriculture, Sacramento, California 95814, U.S.A.

Abstract

Ordinations are multivariate statistical techniques that represent multidimensional patterns of variation in a reduced number of dimensions, and that also reflect the contributions of all attributes under analysis. Basic differences between ordinations versus clustering and cladistic techniques are considered. Although ordinations have historically been considered "phenetic" techniques, they are multidimensional and are, therefore, unlike unidimensional portrayals of variance, such as cluster analysis. Because of this difference, the distances derived from ordination vectors are nonphenetic because they do not represent measures of "overall similarity," as do clusterings. The vectors indicate partitioned variance, which allows directionality (polarity), as do cladistic techniques. A number of types of ordinations, particularly principal components analysis (PCA) and discriminant function analysis (DFA), also known as discriminant analysis and canonical variates analysis, are described and discussed in general terms, including their implications and use in systematics.

1 Introduction

Due to the great diversity of arthropod species and their morphological variability, taxonomists are faced with a daunting task when they attempt to array insects into manageable systematic groups. Often the response of an organism to selective forces will manifest itself as the adaptation of a number of features to many interdependent biological and environmental factors. Morphologically, responses may occur in a multidimensional fashion, rather than as a change in a single character (Blackith, 1960; Blackith and Reyment, 1971; Gould and Johnston, 1972; Sokal, 1986). Populations and species may overlap when characters are studied individually, but they may become distinct entities when many characters are considered jointly.

Numerical analytical techniques provide an objective, operational approach for the examination of data for systematic patterns. Morphometrics, the "measurement and analysis of form" (Daly, 1985) has developed into a major research approach for systematists concerned with this task (Blackith and Reyment, 1971; Pimentel, 1979; Reyment et al., 1984; Bookstein, et al., 1985; Rohlf and Bookstein, 1990). The tremendous development of computer technology now allows the entomological systematist to automate the measurement process (Rohlf, 1990a; Meacham, this volume) and to simultaneously analyze many characters (Daly, 1985). In addition, computerization allows for precision and repeatability in the analysis of taxonomic patterns and the development of classifications. For these reasons, computer-based taxonomic analysis has itself led to a better understanding of the taxonomic process.

A major methodological approach in morphometrics is the use of ordinations, which can be defined as the representation of OTUs (Operational Taxonomic Units) in a reduced character space that is defined by the major axes of variation, each ranked by the amount of variation it explains (Rummell, 1970; Sneath and Sokal, 1973; Moss, 1979; Neff and Marcus, 1980). Ordinations are mainly used in an exploratory way for dimensional reduction of data sets, determination of character relations, calculation of the major patterns of variation, and for ordering OTUs within an newly defined space. In this paper, we compare ordinations with other methodological approaches, namely, cluster analysis and cladistics, and we provide a brief overview of the available ordination methods and their operational aspects.

2 Overview of Ordination Methods

When attempting to solve taxonomic problems, systematists must deal with complex patterns of multidimensional morphometric variation. Choice of the appropriate ordination method can also be a complex problem. Ordination methods can be grouped into two categories, those for the analysis of single samples (principal components analysis, for example) and those for the analysis of two or more samples (such as discriminant function analysis). In addition to reducing the dimensionality of the data, these methods plot the OTUs under study in the reduced space by using two or three of the components of variation; as a result, finding interpretable clusters among the OTUs and the variables is also possible. Therefore, the choice of an appropriate ordination method depends on the characteristics of the data set to be analyzed, and on the objectives of the study.

In this section, the various ordination methods are briefly described, emphasizing the assumptions of each and their uses. Mathematical and computational details are not considered here, but are available in such works as those by Seal (1964), Rummell (1970), Tatsuoka (1971), Van de Geer (1971), Gnanadesikan (1977), Morrison (1976), Harris (1985), and Gower (1987). The taxonomic applications of ordination methods are reviewed with differing levels of detail in works by Blackith and Reyment (1971), Sneath and Sokal (1973), Clifford and Stephenson (1975), Pimentel (1979), Neff and Marcus (1980), Dunn and Everitt (1982), Reyment et al. (1984), and Rohlf and Bookstein (1990).

2.1 Principal Components Analysis

Principal components analysis (PCA) has been widely used as a dimension-reducing and data exploration technique. PCA has been the most commonly used ordination technique for the examination of relationships among taxa in systematics studies. Because it assumes no a priori division of the OTUs into separate groups, PCA concentrates on the relationships among variables and among individuals within a single sample. It reduces the dimensionality of the original set of variables, with as little loss of information as possible, by transforming the original variables into a new set of axes or components. These components are simply linear combinations of the original variables.

The principal components, or axes, are computed in a way such that they have a number of useful properties. The components are orthogonal (independent), that is, for example, scores on the first component are uncorrelated with scores on the second component. The components are arrayed in order of decreasing sample variance, that is, the first or largest component will maximally discriminate among the sample OTUs. Often, the data set will be reduced to a few components that account for the major, independent, patterns of variation that produce the greatest differentiation among the OTUs in the data set. The principal components can be visualized geometrically as a new set of axes with all axes at right angles to each other, such that the original array of the OTUs is most closely estimated.

Because these vectors are each a linear combination of the original variables, each can be described in terms of the original variables. Each variable has a weight or loading that is a scaled value, and which is unique on each component. These loadings can be visualized geometrically, as a function of the angle between the variable and the component. The rotational angles of the variables to the component axes produce an optimum position in the multidimensional hyperspace. The summation of total variance among the variables that is explained by each component is expressed as an eigenvalue, which is associated with an eigenvector corresponding to that principal component. The total of all of the eigenvalues equals the summed total of all of the variances of the original variables. The score of a specimen on a particular component axis can be found by solving the equation for that component, using the variable measurements for that specimen.

A useful review of the statistical and biological assumptions of PCA along with comments on possible applications is provided by Neff and Marcus (1980).

2.2 Discriminant Function Analysis

Unlike PCA, where there is no a priori grouping of OTUs, discriminant function analysis (DFA) involves grouping the samples before analysis; these groups may be chosen so as to represent known taxonomic entities or have some other feature in common. The main objective of DFA is to calculate a linear function that is based on the original variables as measured on each original sample. This linear function must best discriminate among the a priori groups, and can be used to assign individuals to these a priori groups. Unlike PCA, DFA places emphasis on the gaps, or divergence, between the grouped samples.

Fisher (1936) originally developed the discriminant function to maximize the difference between two groups, but the method was later generalized for the analysis of situations involving many groups. When this method is used to reduce the dimensions of data, it is termed canonical variates analysis. Most aspects of DFA are dealt with by Lachenbruch (1975), Nie et al. (1975), Morrison (1976), and Pimentel (1979, this volume). Albrecht (1980) and Campbell and Atchley (1981) provide overviews of DFA, including geometrical interpretations.

In calculating the linear combination of the variables to produce the discriminant functions, emphasis is placed on those variables that maximize the among-groups (OTUs) variance relative to the within-group variance. As a result, the original set of variables is transformed into a set of functions, each independent of the other, of which the first discriminant function (DF) accounts for the largest amount of total, independent variation and the second DF accounts for the next largest proportion of the remaining variation. This process is carried out until all of the variation has been accounted for among groups, to a maximum of DFs = groups - 1. Usually most of the variation and potential discrimination can be obtained with less than the maximum possible number of discriminant functions.

Two kinds of variation matrices, containing measures of the amount of absolute variation within and among the variables, are manipulated in order to calculate the DFs. After the variances and covariances between the variables in each group have been calculated, they are averaged over all of the groups to produce a pooled within-groups variance-covariance matrix. The sample means are then used to calculate an among-groups variance-covariance matrix. Next, various discriminant equations (see Pimentel, this volume: equations 14-18) are solved, the roots of one of which represents the DFs. Each DF is orthogonal, and represents a pattern of variation that is uncorrelated with that of other DFs. The DFs contain coefficients that are weighted according to the variance structure of the original variables; they are calculated so that differences between the group means are maximized. The mathematical derivation of these coefficients is described in detail in Seal (1964) and in Cooley and Lohnes (1971).

DFA can be compared geometrically with PCA (Albrecht, 1980; Campbell and Atchley, 1981; see Pimentel, this volume). As a variance-maximizing technique, PCA can be seen as a rotation of the original axes system so that the first principal axis orients to the maximum amount of the variation in the data. Then, rotation around this fixed axis is carried out to produce the second principal axis. This procedure is carried out further if more variables were involved. Multiple discriminant analysis also performs a rotation to maximize variance, where it rotates the original coordinate axes so that they become parallel with the major axes of within-groups dispersion. It then adjusts for the effect of within-group variability in the characters by scaling the distances between groups with a value equal to one standard deviation of the within-groups variation. This has the effect of changing the distances between group centroids. The last step is to perform another variance-maximizing rotation along the major axes of between-groups dispersion. The rescaled, standardized axes are rotated so as to be parallel with the major and minor axes of between-groups dispersion.

Thus, each DF can be interpreted as an axis in geometric space and has an eigenvalue associated with it that represents the variance of the sample means on that axis. A measure of the total variance in the discriminating variables is provided by the sum of the eigenvalues. Therefore, a measure of the relative discriminatory ability, and the amount of variability accounted for by each discriminant function, is given when each eigenvalue is expressed as a percentage of the sum of the eigenvalues.

The DF coefficients can be standardized by scaling each coefficient for an independent variable by multiplying it by the pooled standard deviation of that variable. The standardized DF coefficients can then be used as a measure of the importance of each variable to a particular DF. Irrespective of its sign, the magnitude of a standardized coefficient indicates the contribution of a particular variable to the optimized discrimination among groups along that particular axis.

In DFA, single specimens can be allocated into a number of a priori groups on an objective basis by using discriminant functions in order to provide a quantitative check on the discriminatory power of particular sets of variables. This is carried out by computing a set of separate identity functions for each a priori group. Probabilities of membership in each group are determined from the function scores; an individual specimen is allocated into the group with the closest sample centroid in multivariate space. Because the actual group membership of each specimen to be allocated is known beforehand, it can be compared with the results of the above procedure and summarized in the form of an identification table.

There is an upward bias in the calculated percentage of correct allocations when the allocation procedure is based on the use of specimens that were also used in the calculation of the identity functions of the a priori groups. The amount of bias can be checked by allocation of a number of specimens that were not used to calculate the identity functions of the groups. If the amount of bias is not too great, allocation procedures provide a means of characterizing the extent of overlap between groups (Foottit and Mackauer, 1980).

2.3 Additional Ordination Methods

Principal coordinates analysis (PCORD) (Neff and Marcus, 1980) is a method of calculating the coordinates of OTUs along principal components based on distance or similarity measures. A more general procedure than PCA, PCORD is done without any reference to the values of the original variables. PCORD can be applied to similarity coefficients and, therefore, it can be used on any type of data, including qualitative or mixed data. The procedure is limited in the sense that the results depend on the distance measure that is chosen and that the trends cannot be related to combinations of the original variables.

Factor analysis (FA) (Neff and Marcus, 1980; Harris, 1985; Johnson and Wichern, 1988) is similar to PCA in that its purpose is to find structure in a data set by defining the linear combinations of the several observed variables as a new set of fewer factors. However, FA differs from PCA in that FA emphasizes the relationships among the variables.

Canonical correlation analysis (CANCOR) (Neff and Marcus, 1980; Harris, 1985; Johnson and Wichern, 1988) is an extension of multiple regression procedures to cases where there are several dependent variables rather than the usual case of only one dependent variable. CANCOR is used for the analysis of the correlation between two sets of variables measured for the same set of OTUs. This procedure reduces the dimensions of the two sets of variables by determining that linear combination of the variables in one set that is most strongly correlated with a linear combination of the variables in the second set. The procedure does not work well on categorical data or other types of data that are not summarized well by correlation or linear variable combinations.

Correspondence analysis (COA) (Pimentel, 1979; James and McCulloch, 1990) is used to determine new variable combinations from tables consisting of counts by reducing the number of dimensions. COA is considered to be inefficient for data other than counts.

Nonmetric multidimensional scaling (NMDS) (Kruskal, 1964; Kruskal and Wish, 1978; Neff and Marcus, 1980; Schiffman et al., 1981; Davidson, 1983) is designed to produce the best representation of OTU fit in a restricted set of reduced dimensions (usually two), which are based on a matrix of their total multidimensional distances or similarities. For example, three or more vector dimensions can be scaled (compressed) into only two, so that the amount of stress (a dimension compression artifact) is minimized in the distances between OTUs shown in the two-dimensional portrayal. Thus, NMDS differs from PCORD in that it attempts to preserve the original, close and distant-order distances between the OTUs. Usually, OTU coordinates that have been derived from PCORD are used as a starting point for NMDS compressions.

A limitation of NMDS is that it uses only the ranks of the distances among the OTUs (rank order data), rather than their actual distances. NMDS differs from cluster analysis of the same matrix in that it produces a spatial representation of the distance matrix, while cluster analysis results in a tree (one-dimensional) representation. The local-global distortion artifacts produced by both these techniques (stress for NMDS, and the cophenetic correlation coefficient for clustering) must be assessed to determine the relative confidence that can be placed in their respective portrayals of relationships.

3 Contrasting Ordinations, Cluster Analysis and Cladistics

Ordinations are not simply techniques, they have a philosophical underpinning that emphasizes the exploration and portrayal of taxonomic data (Sorensen, 1987; Sorensen and Foottit, this volume). The interpretation of the results of an ordination analysis are dependent upon an awareness by the user of the particular context of the original analysis (Sorensen, this volume).

Cluster analysis, which encompasses those methods that identify sets of similar entities from a previously unpartitioned set of entities (Sneath and Sokal, 1973; Everitt, 1980), can be compared and contrasted with ordination methods. As mentioned above, ordination methods emphasize the reduction of data sets, the determination of trends of variation and the placement of OTUs within a newly defined character space. Clustering methods, by classifying similar OTUs into groups reflecting their commonality, place an emphasis on discontinuities and on a hierarchical structure rather than on continua and exploration for trends. Clusterings are also based upon the overall similarity among the data, rather than on its inherent covariance patterns, as are ordinations.

Cluster analysis of OTUs will result in groupings, but these will not necessarily reflect natural groups and the method will only develop a one-dimensional view of the relationships. The results of cluster analysis procedures depend largely on the choice of similarity measure and clustering algorithm. In spite of these characteristics, cluster analysis can be used for data reduction and exploration in cases where there are large matrices of similarity measures (Sneath and Sokal, 1973). In cases such as these, visual interpretation of ordinations may be difficult.

Conventionally, cladograms are identified with phylogenetic hypotheses, while ordinations are associated with hypotheses about phenetic (similarity) relationships among the OTUs being studied (Sneath and Sokal, 1973). Cladistic relationships as depicted by cladograms are usually temporal relationships, while ordinations are representative of spatial relationships.

Ordination techniques have been used most often for infraspecific analysis, where the portrayal of continuous variation is particularly important. Morphometric methods can be used to define homologous characters and to quantify character states for cladistic analysis. In addition, ordinations (particularly DFA) can be used as phylogenetic approaches. OTUs may be positioned in an ordination space on the basis of OTU group definition, or the definition of regions of ordination space on the basis of shared derived character states (Faith, 1989; Sorensen, this volume). In this sense, ordinations in general, and DFA in particular, are not really "phenetic" in the restricted sense of overall similarity analysis (Sneath and Sokal, 1973). The partitioning of data variance multidimensionally, among vectors, as all ordinations do, or into divergent versus common variance, as does DFA, indicate that the portrayals created by these techniques capture the direction of character change among groups or OTUs, rather than their overall similarity.

Although phenetic relationships are those due primarily to similarity (Sneath and Sokal, 1973), they should not be viewed as being antagonistic to cladistic and evolutionary approaches, but rather complementary. Phylogenetic reconstruction cannot take place without an assessment of similarity. In fact, all three methodological approaches have different, but equally important, contributions to make to the study of biosystematics.

4 Ordinations: Practical Considerations

Specialized software is available but many of the ordination methods can be carried out using programs available in such general use packages as, SPSS (Nie, et al., 1975), BMDP (Dixon, 1983), and SAS (SAS Institute, 1988). Programs for computing ordinations are available in some excellent packages that have been developed specifically for taxonomic applications, such as BIOΣTAT II (Pimentel and Smith, 1986), Michigan Morphometrics Workshop (Rohlf and Bookstein, 1990) and NTSYS (Rohlf, 1990b).

Although ordination methods are an important part of the tools available to systematists, like every set of approaches they have their individual advantages and disadvantages in different situations. The taxonomist has to choose the most appropriate methods for each situation. Ordinations should not be used as a "black box," but rather their use requires thought and experience in order to obtain reliable results (Gower, 1987).

There is not one best ordination method. The choice of a method will depend on such conditions as the nature of the data to be analyzed and the sample sizes of the OTUs and the characters being studied. The researcher's experience and the complexity of the taxonomic structure to be developed will have a bearing on the choice of method.

There are few published comparative studies of ordination techniques. Those available (Rohlf, 1972; Thorpe, 1980; Pimentel, 1981) have come to different conclusions as to which approach is better. In fact, each approach has its own advantages and disadvantages and can be affected by the type of data being analyzed. It is a good strategy to use more than one method, given that the results of different approaches are often similar.

James and McCulloch (1988) provide a useful review of the statistical assumptions and problems in interpretation in the use of multivariate morphometric techniques. They caution against the over-interpretation of data, particularly in the absence of experimentation. For example, in PCA the orthogonal components do not necessarily represent independent patterns in nature, therefore, interpretations of the components should be made carefully. As methods are refined, future work in ordination analysis will involve a more deductive approach and an emphasis on the development of a genetic, developmental and ecological framework for the assessment of morphometric pattern.

5 References

Albrecht, G.H., 1980. Multivariate analysis and the study of form, with special reference to canonical variate analysis. *American Zoologist*, 20: 679-693.

Blackith, R.E., 1960. A synthesis of multivariate techniques to distinguish patterns of growth in grasshoppers. *Biometrics*, 16: 28-40.

Blackith, R.E. and Reyment, R.A., 1971. Multivariate Morphometrics. Academic Press, New York, 412 pp.

Bookstein, F.L., Chernoff, B., Elder, R.L., Humphries, J.M., Smith, G.R. and Strauss, R.E., 1985. Morphometrics in Evolutionary Biology. Special Publication, 15. Academy of Natural Sciences, Philadelphia, 277 pp.

Campbell, N.A. and Atchley, W.R., 1981. The geometry of canonical variates analysis. *Systematic Zoology*, 30: 268-280.

Clifford, H.T. and Stephenson, W., 1975. An Introduction to Numerical Classification. Academic Press, New York, 229 pp.

Cooley, W.W. and Lohnes, P.R., 1971. Multivariate Data Analysis. John Wiley and Sons, New York, 364 pp.

Daly, H.V., 1985. Insect morphometrics. *Annual Review of Entomology*, 30: 415-438.

Davidson, M.L., 1983. Multidimensional Scaling. John Wiley and Sons, New York, 242 pp.

Dixon, W.J. (Editor), 1983. BMDP Statistical Software, (1988 Revised Edition). University of California Press, Berkeley, 725 pp.

Dunn, G. and Everitt, B.S., 1982. An Introduction to Mathematical Taxonomy. Cambridge University Press, Cambridge, U.K., 152 pp.

Everitt, B., 1980. Cluster Analysis, (Second Edition). Heinemann Educational Books Ltd., London, 136 pp.

Faith, D.P., 1989. Homoplasy as pattern: Multivariate analysis of morphological convergence in Anseriformes. *Cladistics*, 5: 235-258.

Fisher, R.A., 1936. The use of multiple measurements in taxonomic problems. *Annals of Eugenics*, 7: 179-188.

Foottit, R.G. and Mackauer, M., 1980. Morphometric variation between populations of the balsam woolly aphid, *Adelges piceae* (Ratzeburg) (Homoptera: Adelgidae), in North America. *Canadian Journal of Zoology*, 58: 1494-1503.

Gnanadesikan, R., 1977. Methods for Statistical Data Analysis of Multivariate Observations. John Wiley and Sons, New York, 311 pp.

Gould, S.J. and Johnston, R.F., 1972. Geographic variation. *Annual Review of Ecology and Systematics*, 3: 457-498.

Gower, J.C., 1987. An introduction to ordination techniques. In: P. Legendre and L. Legendre (Editors), Developments in Numerical Ecology, NATO Advanced Studies Institute, Series G, Number 14. Springer-Verlag, New York, pp. 3-64.

Harris, R.J., 1985. A Primer of Multivariate Statistics, (Second Edition). Academic Press, New York, 332 pp.

James, F.C. and McCulloch, C.E., 1990. Multivariate analysis in ecology and systematics: panacea or Pandora's box? *Annual Review of Ecology and Systematics*, 21: 129-166.

Johnson, R.A. and Wichern, D.W., 1988. Applied Multivariate Statistical Analysis, (Second Edition). Prentice-Hall Publishers, Englewood Cliffs, New Jersey, 607 pp.

Kruskal, J.B., 1964. Multidimensional scaling by optimizing goodness of fit to a nonmetric hypothesis. *Psychometrica*, 34: 287-326.

Kruskal, J.B. and Wish, M., 1978. Multidimensional Scaling. Series: Quantitative Applications in the Social Sciences, 11. Sage Publications, Beverly Hills, California, 93 pp.

Lachenbruch, P.A., 1975. Discriminant Analysis. Hafner Press, New York, 128 pp.

Meacham, C.A., (this volume). Practical machine vision for morphometric data acquisition.

Morrison, D.F., 1976. Multivariate Statistical Methods. McGraw-Hill Book Company, New York, 338 pp.

Moss, W.W., 1979. Phenetic approaches to classification. *American Zoologist*, 19: 1217-1223.

Neff, N.A. and Marcus, L.F., 1980. A Survey of Multivariate Methods for Systematics. American Museum of Natural History, New York, 221 pp.

Nie, H.H., Hull, C.G., Jenkins, J.G., Steinbrenner, K. and Bent, D.H., 1975. SPSS, Statistical Package for the Social Sciences, (Second Edition). McGraw-Hill Book Company, New York, 675 pp.

Pimentel, R.A., 1979. Morphometrics, the Multivariate Analysis of Biological Data. Kendall/Hunt Publishing Company, Dubuque, Iowa, 276 pp.

Pimentel, R.A., 1981. A comparative study of data and ordination techniques based on a hybrid swarm of sand verbenas (*Abronia* Juss.). *Systematic Zoology*, 30: 250-267.

Pimentel, R.A., (this volume). An introduction to ordination, principal components analysis and discriminant analysis.

Pimentel, R.A. and Smith, J.D., 1986. BIOΣTAT II, a Multivariate Statistical Toolbox. Sigma Soft, Placenta, California.

Reyment, R.A., Blackith, R.E. and Campbell, N.A., 1984. Multivariate Morphometrics, (Second Edition). Academic Press, New York, 233 pp.

Rohlf, F.J., 1972. An empirical comparison of three ordination techniques in numerical taxonomy. *Systematic Zoology*, 21: 271-280.

Rohlf, F.J., 1990a. Morphometrics. *Annual Review of Ecology and Systematics*, 21: 299-316.

Rohlf, F.J., 1990b. NTSYS-PC. Numerical Taxonomy and Multivariate Analysis System, (Version 1.60). Exeter Publishing Ltd., Setauket, New York.

Rohlf, F.J. and Bookstein, F.L., 1990. Proceedings of the Michigan Morphometrics Workshop. Special Publication, 2. The University of Michigan Museum of Zoology, Ann Arbor, 380 pp.

Rummell, R.J., 1970. Applied Factor Analysis. Northwestern University Press, Evanston, Illinois, 617 pp.

SAS Institute Inc., 1988. SAS Language Guide for Personal Computers, (Release 6.03 Edition). SAS Institute Inc., Cary, North Carolina, 558 pp.

Schiffman, S.S., Reynolds, M.L. and Young, F.W., 1981. Introduction to Multidimensional Scaling, Theory, Methods and Applications. Academic Press, New York, 413 pp.

Seal, H., 1964. Multivariate Statistical Analysis for Biologists. Methuen, London, 209 pp.

Sneath, P.H.A. and Sokal, R.R., 1973. Numerical Taxonomy, the Principles and Practice of Numerical Classification. W.H. Freeman and Company, New York, 573 pp.

Sokal, R.R., 1986. Phenetic taxonomy: theory and methods. *Annual Review of Ecology and Systematics*, 17: 423-442.

Sorensen, J.T., 1987. The multivariate evolution and taxonomic analysis of leafhopper biotypes and species complexes: use of character correlations and quantitative genetics methods. In: M.R. Wilson and L.R. Nault (Editors), Proceedings of 2nd International Workshop on Leafhoppers and Planthoppers of Economic Importance, Brigham Young University, Provo, Utah USA, 28th July - 1st August 1986. CAB International Institute of Entomology, London, pp. 217-234.

Sorensen, J.T., (this volume). The use of discriminant function analysis for estimation of phylogeny: partitioning, perspective and problems.

Sorensen, J.T. and Foottit, R.G., (this volume). The evolutionary quantitative genetic rationales for use of ordination analyses in systematics: phylogenetic implications.

Tatsuoka, M.M., 1971. Multivariate Analysis: Techniques for Educational and Psychological Research. John Wiley and Sons, New York, 310 pp.

Thorpe, R.S., 1980. Microevolution and taxonomy of European reptiles with particular reference to the grass snake *Natrix natrix* and the wall lizards *Podarcis sicula* and *P. melisellensis*. *Biological Journal of the Linnean Society*, 14: 214-234.

Van de Geer, J.P., 1971. Introduction to Multivariate Analysis for the Social Sciences. W.H. Freeman and Company, New York, 293 pp.

AN INTRODUCTION TO ORDINATION, PRINCIPAL COMPONENTS ANALYSIS AND DISCRIMINANT ANALYSIS

Richard A. Pimentel

Biological Sciences Department, California State Polytechnic University, San Luis Obispo, California 93407, U.S.A.

Abstract

This article provides a basic discussion of some ordination techniques that are widely used in biological investigations. The ordinations that are covered are principal components analysis (PCA) and discriminant analysis; the latter is also known by several other terms: canonical analysis of discriminance (CAD), which is preferred here, canonical variates analysis, canonical analysis, and discriminant function analysis. Other ordination methods, such as metric multidimensional scaling, nonmetric multidimensional scaling, nonlinear mapping and principal coordinates analysis, are not discussed. PCA and CAD are classification methods that reduce the dimensionality of data, plotting n individuals in r-space. The importance of the operational assumptions of these methods, and their biological interpretations are discussed. The differences between PCA and CAD are examined in a step-by-step manner, using a geometric approach and a minimum of mathematics.

1 Introduction

Ordination encompasses methods that attempt to summarize, in a reduced r-space, the maximum possible information from a total t-space. Generally, r is restricted to two or sometimes three dimensions, a space that facilitates interpretation from plots. Naturally, the r-space summary is the basis for these methods being termed data reduction techniques.

As is true for many of the terms of multivariate analysis, there is some disagreement in the application of the term ordination, and hence ordination methods. In the broad sense of the above definition, ordination addresses methods of classification and discrimination. In classification methods, individuals start as members of a single conceptual group, with the goal of the analysis being the assignment of individuals to groups. In discriminant analysis, individuals start as, and remain, members of samples from recognized populations, with the goal of the analysis being the disclosure of uniqueness and examination of relationships among the different populations.

I shall confine the use of ordination to one of two kinds of classification methods, the other being cluster analysis. Examples of such ordination methods are principal components analysis (Hotelling, 1933, 1936), metric multidimensional scaling (Torgerson, 1952) and the very similar principal coordinates analysis (Gower, 1966), nonmetric multiple dimensional scaling (Kruskal, 1964a, 1964b; Shepard, 1962a, 1962b, 1966), and nonlinear mapping (Sammon, 1969). To be more specific, classification methods operate on *n* individuals. Regularly, the *n* individuals represent several populations (e.g., species), but, conceptually at least, they are treated as a single group. The purpose of analysis is to segregate the individuals into groups. The criteria for defining groups and assigning individuals to groups are purely arbitrary. As a matter of course, interpretation is based mostly upon the interpretation of an ordination graph, a plot of *n* individuals in *r*-space. In cluster analysis, definition of groups, or clusters, and assignment of individuals to groups is based upon branching, tree-like graphs termed dendrograms. Thus, interpretation of classification is more a matter of heuristics rather than of statistical analysis.

Discriminant analysis, specifically canonical analysis of discriminance, CAD, is a true data reduction technique. Like classification methods such as principal components analysis, PCA, CAD summarizes the maximum information from a *t*-space in a *r*-space. In both analyses, plots of individuals are involved; however, there are two fundamental differences between the two methods. First, CAD operates on *g* samples that may or may not be of equal size, rather than on a single conceptual group. Second, the purpose of discriminant analysis is to evaluate the degrees of similarity and difference among the *g* groups and the uniqueness of each group, rather than to define groups. Below, we shall see that discrimination readily extends into a detailed evaluation of groups, a vastly more detailed analysis than that of a classification method.

Although there are many data reduction methods, this book is virtually restricted to principal components analysis and to canonical analysis of discriminance. This chapter is intended for those not familiar with these methods. It represents a serious attempt to provide insight into the utility and shortcomings of the methods. The primary avenues for indicating the nature of these methods are diagrams showing what PCA and CAD do with data. Although several formulas are provided, their mathematical basis is ignored. Emphasis remains upon the diagrams indicating what formulas do with data, and hence the following treatment is largely nonmathematical. Those readers desiring more formal treatment are left to their own devices.

2 Principal Components Analysis

Originally PCA was proposed as a method to study variation in single samples from single populations. Its application, as an ordination method for defining groups from a single conceptual group of *n* individuals, came later. Here, discussion is restricted to PCA as an ordination method. Ordination via PCA is regularly applied as an exploratory procedure that starts from a set of *p* measurements made on each of *n* individuals. An individual can represent any biological unit (e.g., a single organism, a quadrat containing plants) or any physical unit of the environment. In any case, the data must conform to a $p \times n$ data matrix, and consist of *p* rows, each pertaining to a measurement, and *n* columns, each pertaining to an individual.

If data are applied in the classical sense, the character by individual matrix is used to calculate a variable by variable matrix, either a variance-covariance matrix or a correlation matrix. Both of the latter are $p \times p$ matrices that differ only in one containing variances and covariances and the other containing correlation coefficients. A PCA derived from a variance-covariance matrix, **S**, can be said to examine absolute variation, and a PCA based on a correlation matrix, **R**, to examine relative or standardized variation. The variation examined in **S** is absolute in the sense that it contains variances and covariances that measure variation in the original variables. Thus, absolute variation is no more than variation measured in terms of the original data. The variation examined in **R** is relative or standardized in the sense that a correlation matrix is also a variance-covariance of z-scores of the original variables, the perfect intracorrelation of the ith variable (i.e., $r_{ii} = 1$) is the z-score variance for that variable, and the bivariate correlation between the ith and jth variables (i.e., r_{ij}) is their z-score covariance. We see that the term relative or standardized refers to the variances of any set of z-scores being equal, specifically to unity. It follows that the latter is a case of standardization and can be considered to involve relative variation. Thus, both **S** and **R** are variance-covariance matrices, also termed dispersion matrices.

Next, we will consider a PCA derived from a variance-covariance matrix of the original data. First, we will get somewhat mathematical. Several of the formulas involved in PCA will be introduced. However, as stated above, the primary discussion of methodology will be delayed until presentation of a pictorial indication of what these formulas actually do with data.

Discussion shall first stress the use of the matrix **S** to create new variables. The end result is that the original p variables, $\boldsymbol{x}$, are transformed to new p variables, $\boldsymbol{y}$, via a so called characteristic equation and eigenvector function. Later, it will be seen that each of the new y-variables represents a combination of fractions of each of the original x-variables. As a consequence, each y-variable is said to describe a pattern of variation (i.e., of the original x-variables. Each new y-variable provides a set of n component scores, which are plotted in reference to a component axis. To summarize, analysis results in p sets of n component scores and, conceptually at least, each set is plotted in reference to one of p axes.

The first step in analysis involves the characteristic equation

$$|\mathbf{S} - \lambda \mathbf{I}| = \mathbf{0}\,, \tag{1}$$

which provides p values of λ, so called eigenvalues. Thus, λ is a variable. To generalize, any characteristic equation involving a $p \times p$ dispersion matrix, such as **S,** will provide p values of λ. It is important to know that, as applied here, the characteristic equation serves to transform a $p \times p$ matrix **S** to a $p \times p$ matrix **L**, where

$$\mathbf{L} = \lambda \mathbf{I}\,. \tag{2}$$

The $p \times p$ matrix **I** is an identity matrix. Identity matrices have all diagonal elements of unity and all off-diagonal elements of zero. The purpose of including **I** is to enable the product $\lambda\mathbf{I}$ to form **L**, which differs from **I** only in that its diagonal elements contain eigenvalues.

Three features of **L** and its included eigenvalues are of paramount importance:

(1) **L** is a diagonal matrix, specifically one in which the diagonal contains the p values of λ. Any diagonal matrix has all off-diagonal elements of zero, but the diagonal elements of such matrices can differ from unity. Thus, in transforming **S** to **L**, an original data dispersion matrix of x-variables is transformed to a dispersion matrix of the y-variables. The consequence is that the eigenvalues of **L** are successive y-variable variances, and the off-diagonal elements of **L** are covariances of zero. Furthermore, because all covariances are equal to zero, and covariances are involved in calculating the correlation coefficients, the correlation between any pair of y-variables must be zero.

(2) Successive eigenvalues represent maximum possible variances (although successive y_i are restricted to being uncorrelated). Otherwise they have the maximum variance possible for their given dimension. Thus, the first two axes present the maximum possible variation among uncorrelated variables, or individuals, that can be shown in two-dimensional space. Naturally, the latter constitutes a basis for the interest in PCA as an ordination method.

(3) The sum of the eigenvalues equals the sum of the original data variances. For example, for a PCA of **S**,

$$\sum_{i=1}^{p} s_i^2 = \sum_{i=1}^{p} \lambda_i \, . \tag{3}$$

The latter translates into the variation of original X-variables and transformed y-variables being identical (i.e., PCA neither adds nor subtracts the magnitude of original variation). Thus, **L**, like **S**, represents a valid and complete measure of the variation in the data. However, because **L** is a diagonal matrix that leads to zero correlation between any pair of y-variables, **L** provides a simpler representation of the variation than does **S**.

When the p λ values are obtained, solution of the eigenvector function becomes possible as

$$(\mathbf{S} - \lambda_i \mathbf{I})\, \mathbf{a}_i = \mathbf{0} \, , \tag{4}$$

where each of the p $\mathbf{a}_i$ are eigenvectors of p elements. Because each of the p eigenvectors contains p elements, the total set of eigenvectors can be summarized in a $p \times p$ matrix **A**.

The sum of the squares of the p elements of any $\mathbf{a}_i$ can be scaled to any value. For example, it can be set to the value of the corresponding eigenvalue, as

$$\sum_{i=1}^{p} \mathbf{a}_i = \lambda_i \, . \tag{5}$$

However, it is often more meaningful to set the value of any such sum to unity, as in

$$\sum_{i=1}^{p} \mathbf{a}_i = 1 . \tag{6}$$

Then, the coefficients of the ith eigenvector define the rotation from each of the original variable axes to the ith component axis. This is the case because, then, the value of each coefficient of the ith eigenvector is the cosine of the angle between the axis of the corresponding original X-variable axis and the ith new y-variable axis. In addition, such eigenvectors are readily translated into a particular combination of the original variables. Because each p such combinations is independent of all others, each is often evaluated as an independent pattern of variation.

Patterns of variation are recognized as general or bipolar. A general component is an eigenvector having all coefficients of the same sign. Often such a component can be interpreted as a size increase in all variables, and the magnitudes of its coefficients as the extent of increase in size of the variables. If all coefficients are of the same magnitude, size increase is assumed to be equal in all variables. For example, if five coefficients are equal (each would equal $1/\sqrt{5} = 0.447$), all features might be hypothesized to increase 5 mm in length at the same later date. Such an equal magnitude of increase is said to characterize isometric growth. On the other hand, cases of coefficients of the same sign, but of unequal magnitude, characterize allometric growth (i.e., growth resulting in unequal increases in features). Frequently, allometric growth is referenced as shape changes that are dependent upon size changes.

A bipolar component contains coefficients of mixed sign. In morphological studies, bipolar components indicate shape changes because they denote increases in some features and decreases in others. However, they differ from allometric growth in that the changes are more-or-less independent of size changes. Again, the magnitude of coefficients can be changed into convenient units to indicate the degree of change in each of the variables. Interest in patterns of variation usually is restricted to cases where the n individuals represent a sample from a single population, or to cases where ordination and a component score graph is of minor interest.

In spite of the utility of the latter scaling of eigenvectors, some users prefer the λ_i scaling: for example,

$$\sum_{i=1}^{p} \mathbf{a}_i = \lambda_i . \tag{7}$$

Such $\mathbf{a}_i$ are often called factor loadings. From one point of view, these $\mathbf{a}_i$ are regression coefficients and are termed factor pattern coefficients. From another point of view, they are correlation coefficients between original variables and corresponding eigenvectors and are termed factor structure coefficients.

In the last step of a PCA, the new y-variables are obtained by

$$\boldsymbol{y} = \mathbf{A}' (\boldsymbol{X} - \bar{\boldsymbol{X}}) = \mathbf{A}'\boldsymbol{x} , \tag{8}$$

where $\bar{X}$ is a p-element centroid (vector of all X-value means), X is a p-element vector representing any one of the n sets of measurements for an individual, and $\mathbf{A}$ is a $p \times p$ matrix containing p eigenvectors, each having p-elements. Thus, y represents any one of p possible sets of n component scores, one score for each individual. It might be helpful to consider the above formula in the form

$$y_i = \mathbf{a}_i'(X - \bar{X}) = \mathbf{a}_i' x \,, \tag{9}$$

where $\mathbf{a}_i$ is the ith of the p eigenvectors, y_i is the vector of n component scores for the ith component axis, and x is a vector of p deviation scores.

Fig. 1 represents an attempt to portray the relationships discussed above. Particularly, it indicates the consequences of PCA analysis for the bivariate case; however, it also indicates the case for any number of variables. In the various diagrams, data are indicated by an ellipse, a geometric form that can be expected to circumscribe the data. In Fig. 1A, the original bivariate configuration of coordinates for variables X_1 and X_2 are characterized by an ellipse. The reason for using deviation scores is convenience (i.e., so the zero means of x_1 and x_2 occur at the origin of their axes). This is indicated in Fig. 1B. Fig. 1C represents the endproduct of PCA. Note that the y_1 axis corresponds to the major axis of the ellipse of points. The consequence is that the y_1 axis is so placed that the variance of y_1, λ_1, is the largest possible. Note that the y_2 axis is perpendicular to the y_1 axis and conforms to the minor axis of the ellipse. If there were more than two axes, y_2 would be the axis representing the maximum variance from all possible axes at right angles to y_1. The zero correlation between the y_1 and y_2 scores can be appreciated from Fig. 1C. Because y_1 traverses the longest possible path of points, it is analogous to a line of regression with zero slope, and hence it would necessitate a zero correlation between y_1 and y_2.

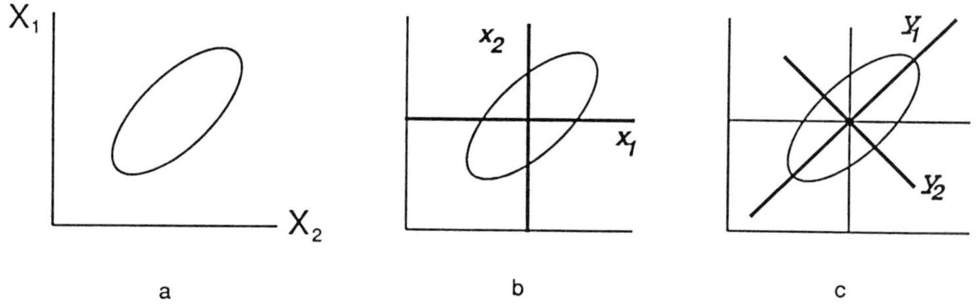

Figure 1. Steps in principal components analysis. (A) An ellipse indicating a swarm of coordinates pertaining to axes X_1 and X_2. (B) The result of transforming variables X_1 and X_2 to deviation scores, x_1 and x_2. The origin of axes x_1 and x_2 are at the means of those variables. (C) The result of rotation from axes x_1 and x_2 to component axes y_1 and y_2, the major and minor axes of the conceptual ellipse of coordinates.

The foregoing discussion of a PCA derived from **S** is directly applicable to a PCA of **R**. Aside from involving **R** rather than **S**, there are no differences whatsoever in the calculations. The identity of calculations follow **R** being a variance-covariance matrix of z-scores of the original data. However, from a given set of data, PCAs of **S** and **R** will give different values for eigenvalues, eigenvectors, and component scores. This is indicated by convention in notation. For example, the parallel to the above formulas involved in a PCA from **S** could be

$$|\mathbf{R} - \lambda \mathbf{I}| = 0 \,, \tag{10}$$

$$(\mathbf{R} - \lambda_i \mathbf{I})\, \boldsymbol{v}_i = 0 \,, \tag{11}$$

$$\boldsymbol{y} = \mathbf{V}' \boldsymbol{z} \,, \tag{12}$$

and

$$\mathbf{y}_i = \mathbf{v}_i' \mathbf{z} \,. \tag{13}$$

It follows that the various calculations serve to bring about a transformation of the n sets of p original correlated X-variables to n sets of p uncorrelated y-variables. Furthermore, the transformation has absolutely no influence on the juxtaposition or distances among the n original coordinates. The coordinates are merely redefined in terms of a new set of component axes. However, from the point of view of ordination, the most important property of the y-variables is the amount of the original variation explained by each (i.e., successive axes pertain to variables having the maximum possible variance while remaining uncorrelated with all other variables). Consequently, the first r axes summarize the largest possible amount of the original variable variation in r-space. Furthermore, the sum of all eigenvalues, λ_i, is equal to the sum of the variances of the original X-variables. Hence, the amount of variation in the y-variables is identical to that in original variables. Again, in the case of a PCA of **R**, the original variables are z-scores.

In addition to examining absolute or relative variation, it is possible to examine variation in proportions. This is accomplished by a transformation of raw data to logarithms. Once the transformation is made, all further calculations are as for **S** or **R**. Regularly, aside from a logarithmic transformation being noted, there are no special symbols indicated in the mathematical formulas of the characteristic equation, etc. It should be appreciated that in multivariate analysis, transformations can have little influence on satisfying statistical assumptions, such as normality. Rather, their primary function is to establish the variation being studied. Although contrary statements can be found, when original data do not satisfy one or more assumptions of a particular multivariate analysis, it is unlikely that any transformation will alter the situation.

PCA can be done in two modes, Q or R. The above discussion was of R-mode. R-mode analyses obtain eigenvalues, eigenvectors, and component scores from a variable x variable matrix, generally a variance-covariance or correlation matrix as discussed above. In such cases, each eigenvalue provides the variance of the corresponding eigenvector, which describes an independent pattern of variation. Any interest in individuals is derived from their graph of component scores, because these pertain to individuals.

On the other hand, Q-mode analyses obtain eigenvalues and eigenvectors from an individual x individual matrix. The latter matrix can be a variance-covariance or correlation matrix, but one pertaining to individuals rather than to variables. However, it is often a similarity matrix of some sort. For example, the matrix might contain the compliments of dissimilarity coefficients (e.g., Euclidean or some other distance), or the matrix might contain a similarity coefficient, such as the simple matching coefficient. Each eigenvalue measures the amount of variation between individuals that is associated with a particular axis, and each eigenvector provides coordinates of individuals for a particular axis. The latter follows from eigenvalues and eigenvectors being extracted from an individual x individual matrix.

Although Q-mode component scores would pertain to variables, they normally are omitted from analysis. This follows from Q-mode component scores being relatively uninformative about variables. For example, the information about variables obtained from Q-mode component scores falls far short of the patterns of variation described by R-mode eigenvectors. For this reason, a Q-mode is usually limited to cases where interest is more-or-less restricted to individuals. Also, where ordination of individuals is the primary interest, the analysis of choice usually is Q-mode.

The limitations of PCA, as an ordination method, reflect the assumptions. As is the case for any statistical method, each individual must be representative of its population. This is true whether each individual of the conceptual single group is a member of an unique population, or the total sample size contains replicates of all populations included. Regularly, this assumption is stated in terms of random samples. The only other mandatory assumption is linearity. As is the case in linear regression, it is assumed that the functional relationship among the variables exists in the form of a straight line. Unfortunately, this is often not the case. For example, when one taxon has a body part longer than wide, another the opposite, then the relationship among the taxa is nonlinear. Also, if the fundamental difference among taxa is size, size differences are regularly allometric, and hence nonlinear.

Naturally, one should try to satisfy assumptions, but the techniques are somewhat robust. Obviously, individuals must be representative of their taxon of membership. In many cases, the only way to verify this is by testing for outliers in a sample of each population to be represented in an ordination. The absence of serious nonlinearity can be verified by examining variable relationships via cross tabulation, plotting, or regression. Computer programs that provide three-dimensional plotting and allow rotation of coordinates can be especially helpful and simplify the process of examining linearity. Often most revealing is a two-dimensional plot of component scores. Such plots should display individual scores that approximate a straight line. Coordinate plots should be serial, traversing the graph in a continuous linear pattern, or nodal, conforming to a similar pattern, but interrupted by breaks in the points. The latter results in "clusters" (sensu spatial arrays) that are recognized as groups of related individuals. Nonlinearity is recognized by an arch-like or horseshoe-shaped pattern reflecting a polynomial function among the points, and hence among the original variables. Although simple arches place no serious restriction on interpretation, complex polynomial patterns can prevent any meaningful evaluation of results. It is possible that transformation of data to logarithms or a z-score model will allow meaningful patterns to result, but in many instances a valid PCA is not possible. Then, nonlinear methods, such as nonmetric multidimensional scaling or nonlinear mapping, should be applied.

It is often stated that PCA necessitates (assumes) that variables approximate the multivariate normal distribution. This is not true. Normality is mandatory only if tests of significance (e.g., of successive eigenvalues, hence component axes, explaining significant variation) are to be applied. Regularly, the two or three component axes used in an ordination can be expected to describe significant variation. For example, it is extremely unlikely that the first axis, in the case of a two-dimensional ordination, or even the first two, in a three-dimensional ordination, will summarize all significant variation. On the other hand, the presence of multivariate normality is useful. In addition to allowing parametric tests, it facilitates interpretation of the components as independent patterns of the total variation that reflect stimuli, such as selection forces (see Sorensen and Foottit, this volume). If normality is absent, the components merely represent successive orthogonal axes that summarize maximum variation among the individuals. This may, or may not, be all that is desired.

It is unlikely that the assumption for normality can be justified for the vast majority of ordination studies. This follows from ordinations regularly involving individuals from several different populations, and normality being restricted to a single population. Furthermore, there is no basis to presume that multivariate normality is mimicked by individuals from different populations. Thus, as a practical matter, interpretation of ordinations is restricted to evaluating the groupings of individuals that is obtained.

An additional difficulty that is inherent in any ordination of individuals from different populations is the nature of the variation being examined. The variation portrayed in a plot of a single conceptual group is of two types, shared and unshared. Shared variation is simply the variation common to several biological populations. In an analysis of variance (ANOVA), it is the within-group variation. Unshared variation is that which is unique to single populations. In the sense of ANOVA, it is included in the among-group variation. Collectively, shared and unshared variation are the counterpart of the total mean square of ANOVA. Unfortunately, classification methods make no distinction between these two types of variation. This stems from these methods operating upon shared and unshared variation. Thus, different placement of points in the plots of such analyses (coordinates on an ordination graph) can reflect either minor variation that is consistent with shared variation, or major variation that is consistent with unshared variation. Normally, it is difficult to know which. Below it will be seen that discriminant analysis focuses upon unshared variation.

If practical, a multigroup, R-mode PCA should be considered. Here, adequate samples are taken from each of the involved (separate) populations, a PCA is performed on the within-group dispersion matrix, perhaps after logarithmic transformation of the original data, and individuals (component scores) are plotted relative to the resulting eigenvectors. The method can be extended to test equality of group dispersion matrices and to comparing eigenvectors from groups having unequal dispersion matrices (Pimentel, 1979). In such cases, it is often possible for normality to be assumed (i.e., for each sample), and hence there can be benefit from the extend interpretation of results. The program MPCA of the BIOΣTAT II statistical package (Pimentel and Smith, 1986) performs such an analysis.

3 Discriminant Analysis

The term discriminant analysis is frequently applied erroneously. It can describe analyses using no more than a univariate or multivariate analysis of variance, or even those featuring no more than visual examination of data. True discriminant analysis is of two types. Both have been referred to as discriminant function analysis. I shall follow other terminology conventions in calling them classification function analysis and canonical analysis of discriminance. Although applying the term classification to the former might cause confusion with the above classification methods, there is no suitable substitute. Classification analysis entails the calculation of so called classification functions, one for each sample being studied. The primary purpose of these functions conforms to their designation (i.e., allocation of each individual to the group that it most closely resembles). Originally, this was simply termed classification, but the widespread use of classification to encompass methods of ordination and cluster analysis of single conceptual groups makes it difficult to retain the original designation. For this reason, I will stress the application of "allocation."

Allocation does not actually involve reassigning individuals to new groups. Original group membership is retained, in spite of the fact that each individual might be "allocated" to another group that it most closely resembles. This is indicated by emphasis being placed on allocation of individuals to their own group, termed "hits," or to any other group, termed "misses." The percentage of hits and misses indicate the uniqueness of each group and the distribution of misses indicates the similarity of groups.

In classification analysis, for each individual, g scores are calculated, one score for each group. Each of these scores represents a measure of the distance between the individual and a group centroid. In its simplest form, the scores might be an unmodified distance between the individual and each group centroid. For the kth individual of the jth group, the distance from the ith group is the square root of

$$d_{ijk}^2 = (\boldsymbol{X}_{jk} - \bar{\boldsymbol{X}}_i)' \, \mathbf{S}^{-1} \, (\boldsymbol{X}_{jk} - \bar{\boldsymbol{X}}_i) \,, \qquad (14)$$

which translates into the products of the transpose of the deviation of the kth individual of the jth group from the ith group centroid, the inverse of the within-group dispersion matrix, and the deviation of the kth individual of the jth group from the ith group centroid. The effect of including $\mathbf{S}^{-1}$ is to cause d_{ijk} to be the distance of the individual from the ith centroid in what is termed discriminant space (see below).

The individual is allocated to the group for which it has the shortest d_{ijk}. Consequently, allocation of an individual is to the group whose centroid most closely approximates the location of that individual in the defined space. Of additional interest is the fact that each of these d^2_{ijk} approximates the chi-square distribution. For this reason, for each, a probability can be assigned for its allocation to each group.

The application of classification functions to allocation, their calculation for each sample, and the absence of successive axes that summarize the maximum possible differences among groups, all negate these functions being a classification ordination method. Again, obtaining successive axes that provide the maximum possible differences among groups necessitates application of the characteristic equation. The nature of the simplification, by any application of the characteristic equation, is termed reduction to canonical form, and serves as a basis for the present method being termed canonical analysis of discriminance, CAD. Canonical analysis of discriminance is also termed canonical variates analysis, canonical analysis, discriminant function analysis, or simply discriminant analysis. Although discrimination can involve several statistical methods, further discussion of discriminant analysis shall stress CAD.

CAD is an operational extension of PCA, but there are several fundamental differences:

(1) The CAD experimental design necessitates that each individual is a member of a sample drawn from a single population. The concept of a single conceptual group is neither assumed, nor pertinent.

(2) In CAD, the focus of analysis is upon the degree of difference among populations. In ordination via PCA, the focus of analysis is upon assigning individuals to groups that are created as a function of the analysis, instead of erected before hand.

(3) CAD portrays contrasting variation above and beyond that common to the groups or samples (i.e., within-group variation). Measures of differences, so called generalized distances, are measures of variation above and beyond shared, within-group variation. (The distances calculated above for classification functions are generalized distances.) PCA extracts patterns of variation from the total variation (within plus among) of the involved individuals.

(4) CAD rotates original variable axes through angles whose cosines duplicate the bivariate correlations of the original variables to create a discriminant space. Thus, original variable axes become oblique. The consequence is to define a space of total form. Then, coordinate axes are defined in the latter non-Euclidean space. PCA merely replaces one set of Cartesian coordinates with another set. The consequence is to define new variables from the original ones.

(5) Although CAD eliminates the orthogonality of original variable axes, orthogonal axes are applied to the discriminant space. These axes, canonical axes, provide the coordinates for the scores of individuals, canonical variates, in discriminant space. On the other hand, PCA maintains orthogonal axes while creating uncorrelated, orthogonal variables, as the component scores.

Although CAD involves the characteristic equation, as expected, equation elements differ from those of PCA. Rather than eigenvalues and eigenvectors being extracted from **S** or **R**, they are extracted from $\mathbf{BW}^{-1}$, both being $p \times p$ variable x variable matrices, so that

$$|\mathbf{BW}^{-1} - \lambda\, \mathbf{I}| = \mathbf{0}\,, \tag{15}$$

where **B** is the between-group cross products matrix and **W** is the within-groups cross products matrix. If each were divided by its degrees of freedom, **B** would become the between-group dispersion matrix and **W** the within-groups dispersion matrix (i.e., the **S** involved in the above calculation of generalized distances). The dispersion matrices formed from **B** and **W** are, respectively, the multivariate counterparts of the among-group mean square and within or error mean square of a univariate analysis of variance.

$\mathbf{W}^{-1}$ is the inverse of **W**, and an inverse is used to allow multiplication to perform the equivalent of division, because division is impossible in matrix algebra. Thus, the product of $\mathbf{BW}^{-1}$ is mathematically equivalent to dividing the between cross products matrix by the within cross products matrix. It likewise has a univariate ANOVA parallel. If the product were of dispersion matrices, it would be analogous to a one-way ANOVA *F*-ratio. Hence, it can be appreciated that calculations stress variation among groups.

In the present case, successive eigenvalues represent the maximum possible portions of the among-groups variation that can be extracted in successive dimensions. This maximizing of intergroup differences is often implied to magnify among-group differences. This is not the case. The true situation is as for PCA. Both methods merely serve to extract successive dimensions providing maximum possible information from a total t-space.

If we consider p variables for each individual and g samples or groups, the number of total dimensions, and hence the possible number of eigenvalues that can be extracted, is a minimum of p and g - 1. Once the eigenvalues are obtained, the eigenvector function is solved by

$$(\mathbf{BW}^{-1} - \lambda_i\, \mathbf{I})\, \mathbf{a}_i = \mathbf{0}\,, \tag{16}$$

where each eigenvector, $\mathbf{a}_i$, is scaled so that the variance along its axis is $\mathbf{a}_i'\mathbf{S}\mathbf{a}_i = 1$. The effect is to cause the defined discriminant space to have units of standard deviations which are expressed as Mahalanobis', or generalized, distances. The eigenvectors, here termed canonical vectors, define the axes of this space and the contrasting patterns of variation that describe intergroup differences. In the sense that the first r of the p axes are used in the analysis, the r-space is sometimes referred to as a canonical space. The scores, termed canonical variates, for the k individuals of the j groups are obtained from

$$\boldsymbol{y}_{jk} = \mathbf{A}'(\boldsymbol{X}_{jk} - \bar{\boldsymbol{X}}) = \mathbf{A}'\boldsymbol{x}_{jk}\,, \tag{17}$$

which for the kth individual of the jth group on the ith axis is

$$y_{ijk} = \mathbf{a}_i'(\boldsymbol{X}_{jk} - \bar{\boldsymbol{X}}) = \mathbf{a}_i'\boldsymbol{x}_{jk}\,, \tag{18}$$

where $\bar{\boldsymbol{X}}$ is the grand centroid, a vector containing the means of all variables (i.e., it is calculated from all observations of all groups, the total sample size). As a consequence of using $\boldsymbol{x}_{jk}$ deviations of observations from the grand centroid, the grand centroid of the deviation scores is a null vector (i.e., contains all zeros and is located at the origin of the canonical vectors).

The transformation of data can be appreciated as a three step process. In Fig. 2A, four ellipses represent the bivariate group dispersions of the deviation scores of four samples and a fifth ellipse, centered at the origin of the deviation scores, represents the within-group variation. Because all ellipses are of the same size, shape, and orientation, it follows that all populations have identical dispersion matrices.

In the first step, Fig. 2B, the principal components of the within-group dispersion are calculated. As must be the case, the figure indicates that the component axes, y_1 and y_2, correspond to the major and minor axes of the ellipse.

In the second step, Fig. 2C, the total space of the four group and the within ellipses are so modified that the within-group ellipse, and hence the group ellipses as well, are transformed to circles with a radius of one standard deviation. Generally, this standardization to unit variance entails a contraction of the original space because the original variable variances often are greater than unity. Naturally, if original variances are less than unity, an expansion is involved. This establishment of a unit radius is the basis the units of discriminant space being standard deviations. At this point, discriminant space is formed. All that is lacking is the definition of coordinate axes for this space. Although it might not be obvious from the differences in Figs. 2A and 2C, the standardization from ellipses to circles entails the above mentioned rotation of the original deviation score axes, x_1 and x_2, through an angle equal to their correlation coefficient. The end result of the rotation is indicated in Fig. 2C.

In the final step, Fig. 2D, a PCA is performed on the among-group variation; this is conceptually on the variation among individual group centroids. The resulting two component axes are $\mathbf{a}_1$ and $\mathbf{a}_2$. The reason for two axes follows from the minimum of the number of groups minus 1 (=3) and of number of variables (=2). These axes are the canonical axes, which serve as the coordinate axes of discriminant or canonical space.

We see that the transformation from the original variable space to discriminant space is very different from that for PCA. Here, transformation goes well beyond a simple redefinition of the coordinates of a Cartesian space. The standardization causes changes in the distances among coordinates as defined by the original variables. However, there is no change whatsoever in the juxtaposition of points. This occurs because all changes result strictly from rotating the original variable axes, so that the cosine of the angle between any original pair of variable axes is equal to their correlation coefficient.

The assumptions of discriminant analysis correspond to those of the analysis of variance. Thus, assumptions encompass random sampling, normality, additivity, independence of errors, and equality of population dispersion matrices. Additivity translates into linearity. The only critical assumption is equality of population dispersion matrices, homoscedasticity (Burnaby, 1966; Olson, 1974; Geisser, 1977). Nonnormality is known to have little, if any, influence on analysis. Nonlinearity is unlikely to be of sufficient magnitude to create a problem. This is because the only departure from linearity that is of any concern is that within samples. If adults are used, the potential nonlinearity is minimized. This follows from the strong linearity during total growth being virtually absent in the minor growth in adults.

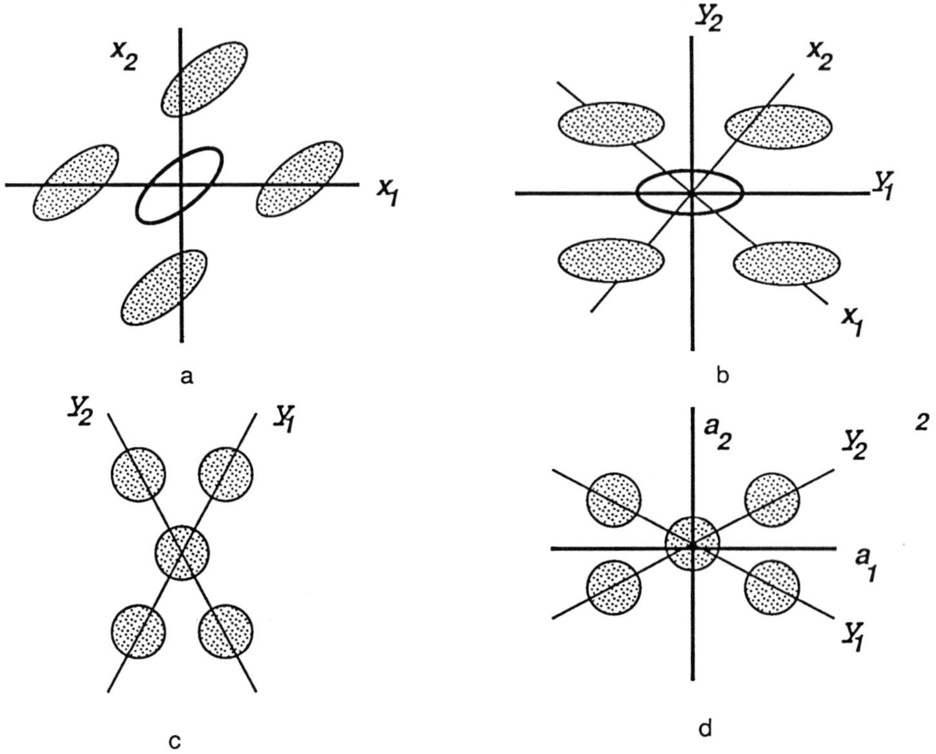

Figure 2. Steps in canonical analysis of discriminance. (A) Four stippled ellipses, each indicating the swarm of coordinates pertaining to deviation scores axes, x_1 and x_2, for one of four samples and a bold line ellipse centered at the grand centroid and indicating the common dispersion of the four groups. (B) The result of rotation from deviation scores axes, x_1 and x_2, to component axes, y_1 and y_2. (C) The result of standardization of component scores to unit variance. In the process of transforming ellipses of group coordinates to circles, component axes, y_1 and y_2, are rotated and no longer orthogonal. This step defines discriminant space, but not the coordinate axes of that space.

Even inequality of group dispersions is not disastrous. The effect of heteroscedasticity is to alter significance levels of the test of equality of group centroids and to modify values of generalized distances among group centroids. If sample sizes are equal and not small, the influence on tests and the modification of distances is minimal (Geisser, 1977).

The factors that contribute to discrimination are well known. These factors are portrayed for the bivariate case in Figs. 3A-D. In each of the figures, two groups are shown. In all figures, the distances between group centroids are identical and the sum of the variances of X_1 and X_2 are likewise constant. Naturally, those groups represented by an ellipse have a nonzero within-group correlation between X_1 and X_2 and those represented by a circle have zero within-group correlation. Maximum discrimination occurs when within- and between-group correlations are high and of opposite sign. One such case is

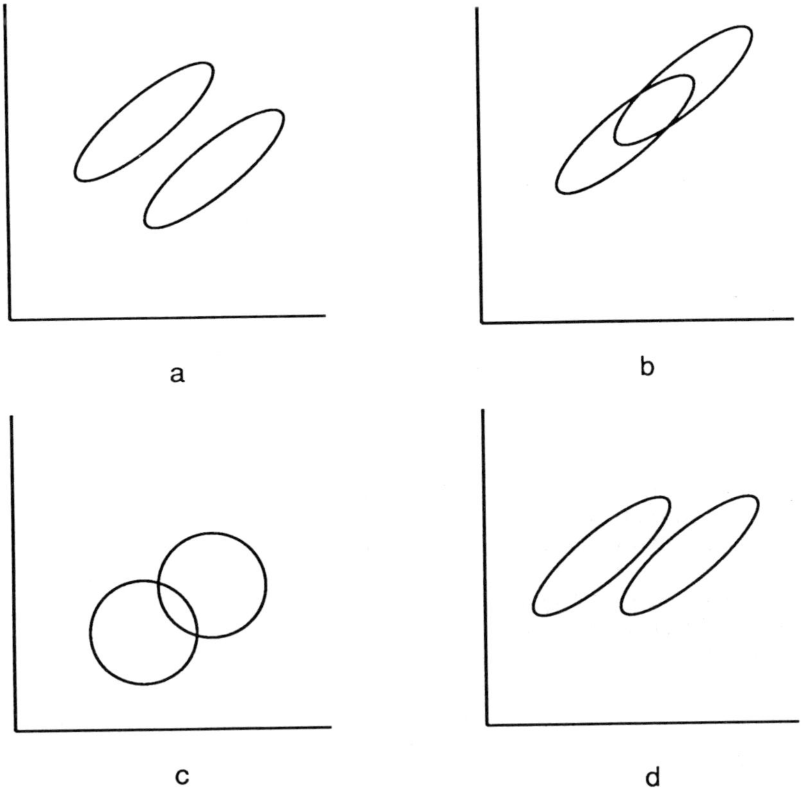

Figure 3. Factors influencing discrimination. The ellipses in A, B and D, and the circle in C, are of equal area, implying identical variation in all populations. Also, the distances between the centers of ellipse pairs and the circle pair are identical, implying identical between population differences. It is intended that the orientation of ellipse pairs, and difference between ellipses and circles, indicate the factors influencing discrimination. (A) The condition of maximum discrimination, within-group and between-group correlations are high and of opposite sign. In the diagram, within-group correlation is positive and between-group correlation is negative. (B) The condition of minimum discrimination, within and between-group correlations are likewise high, but of the same sign, here positive. (C) The condition of intermediate discrimination, within-group correlation is zero. This condition is not influenced by the between-group correlation being negative, positive, or zero. (D) The condition where a variable, here X_2, has the same mean for two groups but serves to aid in discrimination. Note that X_1 alone allows considerable overlap between the two populations.

implied in Fig. 3A, where the within-group correlations are positive and the between-group correlations are negative. In such situations, maximum discrimination prevails because there is maximum separation among individuals of the two groups. Minimum discrimination again occurs in a situation of maximum correlation, but in the situation where both within and between-groups are of the same sign. Fig. 3B indicates the condition where both correlations are positive. Minimum discrimination follows from some overlap occurring between the two groups. An intermediate degree of discrimination occurs when there is zero within-group correlation and is implied by Fig. 3C. Of necessity, the conformation of individuals in each group must approximate a circle.

Fig. 3D portrays a property of variables that denies the application of methods for selecting the best discriminating variables. Although the intergroup distance is maintained in the means of X_1, there is no difference whatsoever in the group means for variable X_2. However, it is clear that X_2 is mandatory for complete discrimination between the two groups (i.e., X_1 alone allows considerable overlap between the two groups). This exemplifies the condition where two populations have unlike form from, say, different lengths of a structure, but identical widths. It further serves to stress that discriminant analysis contrasts the total form of populations, rather than individual variables. Equally important, it shows why stepwise discriminant analysis should be avoided. The stepwise method supposedly selects "best discriminators" from a suite of variables. Regularly, such an approach would consider X_2, or any other variable not differing between populations, to be a poor discriminator and purge it from the data set.

CAD can be applied as only one of five methods that provides a detailed discrimination among groups. Collectively, the methods provide a detailed description of the degree of differences and relationships among the populations sampled. Program MDA of the BIOΣTAT II statistical package (Pimentel and Smith, 1986) performs such an analysis. The methods are as follows:

(1) The first method is Bartlett's test of equality of group dispersions or homoscedasticity. Although unusual, accepting equality clearly satisfies the single critical assumption of discriminant analysis. (It is important to realize that multivariate normality is necessary for the test of homoscedasticity, because significance can indicate no more than nonnormality.) Results of this test can be applied to decide the nature of an application of multigroup principal components analysis. If homoscedasticity is accepted, a PCA of the within-group dispersion matrix serves to examine patterns of variation that are common to all groups. If homoscedasticity is rejected, a PCA of each group can be applied to contrast patterns of variation among the groups, and thus engage a relatively simple Euclidean avenue for examining shared, modified, and perhaps unique patterns of variation.

(2) The second method is the analysis of variance, both univariate and multivariate. Use of univariate ANOVAs is useful in justifying significance in the multivariate MANOVA where significance of the prior test indicates heteroscedasticity. Then, known robustness of the univariate ANOVA can verify MANOVA significance (i.e., inequality of group centroids) even if population dispersions are deemed unequal. It is necessary to reject centroid equality to justify the applications of the detailed tests and measures of discrimination among groups.

(3) The third method is generalized distance analysis. Here, distances between pairs of group centroids can be examined for individual and all variables with a view towards gaining some appreciation of the discrimination power of each variable and the magnitude of intergroup differences. If no more than four populations are being compared, the distances are confined to a three-dimensional space that can be readily appreciated. Otherwise, it

becomes necessary to consider these distances in a reduced r-space.

(4) The fourth method is allocation. Here, the shortest generalized distance from a group centroid, d_{ijk} as indicated above or a related statistic is applied to allocate each individual to one of the groups. A feature of d_{ijk}^2 that is noteworthy is that it approximates the chi-square distribution, so that each of the g d_{ijk}^2 values can be assigned a probability (i.e., of membership to one of the g groups).

(5) The final method is CAD. In addition to the statistics involved, the canonical graph can serve as an r-space pictorial summary of what was learned from prior methods, especially generalized distance analysis and allocation.

Only two forms of data, original data and logarithms of the original data, are usually examined via discriminant analysis. As one might appreciate from the discussion of PCA, original data lead to discrimination on the basis of absolute differences among variables, whereas logarithms lead to discrimination on the basis of proportional differences. An additional benefit that can arise from applying logs is to cause sample dispersion matrices (i.e., those based on logs) to more closely approximate equality than do those for original data; if interpretation of results from original and log data agree, there is support for the analytical results.

4 References

Burnaby, T.P., 1966. Distribution-free quadratic discriminant functions in paleontology. *Computer Contributions, State Geological Survey, Kansas*, 7: 70-76.

Geisser, S., 1977. Discrimination, allocatory and separatory, linear aspects. In: J. Van Ryzin (Editor), Classification and Clustering. Academic Press, New York, pp. 301-330.

Gower, J.C., 1966. Some distance properties of latent root and vector methods used in multivariate analysis. *Biometrika*, 53: 325-338.

Hotelling, H., 1933. Analysis of a complex of statistical variables into principal components. *Journal of Education Psychology*, 24: 417-441, 498-520.

Hotelling, H., 1936. Simple calculation of principal components. *Psychometrika*, 1: 27-35.

Kruskal, J.B., 1964a. Multidimensional scaling by optimizing goodness of fit to a nonmetric hypothesis. *Psychometrika*, 29: 1-27.

Kruskal, J.B., 1964b. Nonmetric multidimensional scaling: a numeric method. *Psychometrika*, 29: 115-129.

Olson, C.L., 1974. Comparative robustness of six tests of multivariate analysis of variance. *Journal of the American Statistical Association*, 69: 894-908.

Pimentel, R.A., 1979. Morphometrics, the Multivariate Analysis of Biological Data. Kendall/Hunt Publishing Company, Dubuque, Iowa, 276 pp.

Pimentel, R.A. and Smith, J.D., 1986. BIOΣTAT II, a Multivariate Statistical Toolbox, (Version 2). Sigma Soft, Placentia, California.

Sammon, J.W. Jr., 1969. A nonlinear mapping algorithm for data structure analysis. *I.E.E.E. Transactions on Computers*, C18: 401-409.

Shepard, R.N., 1962a. The analysis of proximities: multidimensional scaling with an unknown distance function. I. *Psychometrika*, 27: 125-140.

Shepard, R.N., 1962b. The analysis of proximities: multidimensional scaling with an unknown distance function. II. *Psychometrika*, 27: 219-246.
Shepard, R.N., 1966. Metric structures in ordinal data. *Journal of Mathematical Psychology*, 3: 287-315.
Sorensen, J.T. and Foottit, R.G., (this volume). The evolutionary quantitative genetic rationales for use of ordination analyses in systematics: phylogenetic implications.
Torgerson, W.S., 1952. Multidimensional scaling. I. Theory and method. *Psychometrika*, 17: 401-419.

THE EVOLUTIONARY QUANTITATIVE GENETIC RATIONALES FOR THE USE OF ORDINATION ANALYSES IN SYSTEMATICS: PHYLOGENETIC IMPLICATIONS

John T. Sorensen[1] and Robert G. Foottit[2]

[1] Insect Taxonomy Laboratory, California Department of Food and Agriculture, Sacramento, California 95814, U.S.A.; [2] Biological Resources Division, Centre for Land and Biological Resources Research, Agriculture Canada, Research Branch, C.E.F., Ottawa, Ontario K1A 0C6, Canada.

Abstract

Evolutionary quantitative genetics is discussed in relationship to systematics, with respect to the nature and use of correlated phenotypic traits as estimators of genotypic differentiation between groups. Ordinations are discussed as vehicles for the assessment of such differentiation, as is their potential use in phylogenetic inference using Lande's (1979) phenotypic model for multivariate evolution. Under this model, the mean phenotype of a group moves across an adaptive topology of fitness, along orthogonal net selection gradients, as a function of the historically imposed selection and the genetic covariance matrix of the population(s). The character responses necessary to account for group differentiation along the selection vectors can be calculated, and a measure of intergroup anagenic distance can be derived. Intergroup anagenic distance based on divergence (apomorphy) can be calculated using Landean minimum selective mortality indices. Using discriminant function analysis, an apomorphically based phyletic pathway between groups can be derived. The techniques provide a way to overcome the limitations imposed by the use of correlated, continuous characters in phylogenetic inference. Arguments and evidence for and against the model and such procedures are discussed.

1 Introduction

To understand the implications of multivariate ordination techniques for systematics, it is necessary to portray their relationship to quantitative genetics (QG) with respect to evolutionary mechanisms. Quantitative genetics deals with correlated, continuous phenotypic attributes that are assumed to be predominantly genetic. It differs from conventional genetics by its concern with the integrated aspects of a multitude of correlated and linked genes that covary together, as suites, yielding continuous phenotypic variation among individuals. In these suites, each individual gene contributes to phenotypic variance but has a relatively minor influence in the change created by the summation effect of the suite as a whole. The field of quantitative genetics has historically been divided into applied quantitative genetics, concerned largely with breeding improvements in plant crops and animals, and evolutionary quantitative genetics, concerned with the theoretical evolutionary mechanics of population genetics and natural selection on groups and traits. The latter expressly differs from relatively simple theories of phylogenetic deductions, such as cladistics, which employ discrete state characters without thought to the selection pressures or trait correlations involved in character changes during evolution.

In contrast to quantitative genetics, systematics deals with the taxonomic division and placement of organisms. The desirable information to be derived in any systematic treatment is usually the evolutionary affiliation or relationship among groups, although not necessarily in a cladistic sense. Systematists (sensu Mayr, Simpson) have classically deduced evolutionary relationships through the mental analysis of the multistate traits occurring among groups. The estimated relationships arrived at usually reflect the correlation among assemblages of these traits among the groups. If the traits being analyzed have derived (apomorphic) states that are largely correlated, the greater the correlated occurrence of the derived character states among a series of taxa, the closer the systematist usually considers the taxa to be related. This conclusion would be expected also in the most parsimonious solution for a phylogenetic tree (but not necessarily a classification) of a cladistic treatment when many correlated traits are used. The conclusion is justified because the phenotypic correlation of apomorphic character states among traits for closely related, derived taxa is due to their common phylogenetic lineage; it is imparted by shared genetic covariance in the synapomorphic traits that are displayed by the taxa. Usually, among "more advanced" forms, the greater the phenotypic covariance of apomorphic character states for traits between entities, the more common their evolutionary lineage has been and the closer is their phyletic relationship. Note, we have discussed only derived states of traits, not plesiomorphic attributes.

In both evolutionary quantitative genetics and systematics, phenotypic traits are usually used to be operational and estimate relationships for convenience, although very few systematists (and certainly no quantitative geneticists) would not suggest that strictly genotypic information undoubtedly provides a more proper and better estimate. Because a phenotypic trait is the product of a genotype, whose expression may have been environmentally modified, and because the environmental component involved with the trait *usually* provides little or no significant modification of the genotype, it is commonly assumed, in the absence of better data, that evolutionary estimates that are based upon phenotype will provide a reasonable approximation of the actual evolutionary relationships between groups. There are, however, certainly exceptional animals, where environmental influences do cause significant variance in phenotype (e.g., Claridge and Gillham, this volume; Blackman, this volume), which can potentially confound the interpretation of results; this suggests that studies that are based on phenotype for such animals must be cautious, and should, perhaps, include an extrinsic, but closely related, species as a reference indicator for environmental effects.

Recent evidence suggests that genotypic estimates may be able to be operationally based upon phenotype, in as much as large sample sizes apparently allow a reasonable approximation of the genetic correlations for groups by using their phenotypic counterparts. In attempting to estimate whether phenotype could successfully be used to estimate genotype, Cheverud (1988) analyzed 41 pairs of matrices from the literature and found that squared genetic correlations were, on average, much higher than squared phenotypic correlations, and that genotypic and phenotypic correlations were only broadly similar; the differences were possibly due to multiple factors causing imprecise estimates of the genetic correlations. He further found that when sample sizes were high (effectively, $n = 40+$ individuals), the phenotypic correlations were likely to be fair estimates of their genetic counterparts, and that the genetic and environmental causes of phenotypic variation tend to act on growth and development in a similar manner. Willis et al. (1991), however, dispute Cheverud's (1988) assertion, and conclude that phenotypic variance has little predictive value for assessing past selection among groups. In any event, use of phenotype has been an operational limitation to most systematists in the past and probably will continue as such for the bulk of taxonomic work until molecular genetic techniques are common place.

Although they usually prefer to work with discrete state phenotypic attributes, systematists have for decades used measured characters in ordinations to produce nonhierarchical scattergrams of the relationships among groups of organisms. This is especially true where few or no discrete state characters are apparent, such as between sibling species or closely related species that show only gradient characters within problematic genera. For the experienced systematist, the scattergrams usually seem to provide a biologically meaningful portrayal of the relationships among the groups or individuals in the analyses. Yet, the actual (evolutionary) mechanism for the results of ordinations that indicate such a portrayal is often evasive to the systematist. This is because biologists usually understand ordinations only as a mysterious "black box" that data goes into, and from which meaningful results (usually) emerge.

When statisticians without (or sometimes with) biological knowledge are questioned as to why these techniques seem to process data to yield an output that is biologically meaningful, they may often state that the techniques are just "statistical processes" that have "no inherent biological meaning." Even many statistically sophisticated systematists, who understand the manipulations within ordination analyses, often fail to see how or why they empirically mimic any theoretical evolutionary mechanism. Alternatively, they may hesitate to consider using models of evolution that are based upon optimization, stating there is no reason why groups should evolve in such a way, even though cladists using discrete state characters overwhelmingly accept such a parsimony criterion as a philosophically necessary operational "limit" when assessing phylogenetic relationships. Some may not like to "divide out" information that is common among all taxa, although this is exactly what plesiomorphy is, or to "reduce" the information in a set of characters to fewer, orthogonal vectors, an operational necessity for incorporating all that information and yet provide uncorrelated variables with which to work. Others may not like the "instability" of output from ordinations, which can change to some degree depending upon the characters and individuals used to construct the matrix, even though other types of analyses are even more unstable with regard to the character choice and coding.

Nevertheless, there must be a reason why the output of ordinations usually provides results that are meaningful in systematics. In fact there is, as tentative and imprecise as the evolutionary quantitative genetic models that serve as the rationales may be. Because correlation and covariation among continuous morphometric attributes results from genetic linkages (Falconer, 1981), multivariate statistical treatments, such as principal components analysis (PCA) and discriminant function analysis (DFA), which sort statistical differences among parametric variables based upon covariance or correlation, are particularly suited as taxonomically oriented evolutionary quantitative genetics probes. Actually, basic statistical concepts that deal with regression, correlation and analysis of variance were initially devised to analyze inheritance and the evolution of continuous traits in applied quantitative genetics, and it is entirely appropriate that evolutionary biologists should employ multivariate statistical techniques and adopt the methods of quantitative genetics (Lande, 1988). Techniques such as PCA and DFA create covariance or correlation matrices from morphometric data and construct eigenvectors, upon which each original morphometric variable has a input weight coefficient proportional to its covariance with the derived vector. Such techniques are powerful discriminators that have served to classify intraspecific variants (Dupraw, 1964, 1965; Daly and Balling, 1978), to provide insight into the phylogenetic lineage (Schluter, 1984; Sorensen, 1987a; Wood and Pesek, this volume) or character evolution (Sorensen, 1991) within closely related groups, and to determine phenological changes between sibling species (Sorensen, 1987b; Sorensen and Sawyer, 1989).

Thus, ordinations basically work by portraying covariant relationships among the data analyzed in reduced dimensional space that is defined by the covariance. From the data, an overall covariance matrix is deduced and covariance is partitioned (at least initially) among orthogonal, independent vectors, as described by Pimentel (this volume). Using the inferences of evolutionary quantitative genetics, the relationships of groups that are shown in ordinations reflect the evolutionary proximity of the groups with respect to a set of dimensionalized, historical gradients of selection pressure that the groups have been exposed to during their common evolutionary history. The levels of phenotypic covariance among groups allow this portrayal, and also determination of the past relative responses of the groups to these differing, previously imposed selection regimes.

This use of ordinations allows an estimation of the evolutionary anagenic distance between the groups. The time frame in which these selection pressures have worked upon the groups analyzed is that between the (current) observation time, and the time of divergence, from a common ancestor, of all considered monophyletic groups. The mechanism involved is the placement of groups or individuals within a space defined by their character information with regard to orthogonal vectors of the phenotypic covariance matrix for the traits and individuals analyzed. Groups that are most phenotypically similar for the traits analyzed share many characters in common; this is expressed as a relatively high level of phenotypic covariance among those attributes. Inversely, the more phenotypically distinct that related groups are, with reference to many characters, the less phenotypic covariance of traits that they will share (assuming that homoplasy does not dominate the covariance matrix). A species or infraspecific grouping has a relatively high level of covariance among its member individuals or populations. Whereas a genus has less common covariance among its components, but still serves as a relatively covariant level for grouping in comparison to more distant, extrageneric relationships among taxa.

As an analytical instrument, ordinations are necessarily relative (rather than absolute) inasmuch as the outcome of an analysis is dependent upon the individuals and characters chosen, so that the classification results or the scattergrams produced may change if the input matrix changes. The scope of an analysis depends upon the determined vectors among the groups analyzed, which in turn depend upon the incorporation of any and all groups to be considered in a given (local) perspective (Sorensen, this volume). By failing to include a particular group in an ordination analysis, that group's value in adjusting the vectors among all groups will not be accounted for, just as for a parsimony analysis in numerical cladistics, where the failure to incorporate any member of a monophyletic set may change the most parsimonious interpretation for the minimum state-change network that is arrived at. Such a relative viewpoint in local perspective thus is not unique to ordinations; all analytical procedures have, to some degree, local restrictions or limits in their attempted estimation of global "truth."

2 Genetic Implications of Correlation Among Characters

Because the character information that is being analyzed is taken from phenotypic measurements, the phenotypic covariance is considered to be derived from characters that are under polygenic control. Such polygenic traits usually are expressed as a Gaussian distribution that yields a bell-shaped, normal curve. An example of a polygenic trait is human height, an attribute that is expressed in a normal distribution within a population. In contrast, if height were controlled by a single gene with two allelic states, "short" and "tall," then humans would basically come in these two sizes. The individuals with these short and tall states would obviously not be absolutely identical in height because environmental influence would cause a spread around the means of both allelic states. On the other hand, if height were caused by a single gene with five or 10 allelic states, the resulting population would similarly be arrayed along a rough five or 10 "stair-step" distribution in height. Height is, however, controlled by very many genes that act in a correlated fashion. If only genotype were involved, this theoretically would cause a distribution that has a multitude of discrete, but infinitesimally small, stair-steps that would be barely (if at all) discernable. The actual, expressed distribution of height for a very large and random population is a smoothed curve because each very minute, discrete step, expressed by the genetic component alone, is blurred by environmental influences.

Consider, however, that height is not a single trait, rather it is a summation of the lengths of several body components (i.e., tibia, femur, pelvis, spinal vertebra, etc.). Each component is controlled by different genes that act in a correlated fashion, expressible together allometrically as a general-size factor. These separate components are also correlated with others (i.e., parts of the arm), which are not themselves involved directly with height, but that respond to the variances in height in a coordinated fashion. The components of the body are correlated allometrically, so that selection for a particular length component, such as a tibia, would result in a correlated response to that selection in the form of a generally larger individual with correspondingly larger sizes for other components. Other components may be negatively correlated and subject to restriction by physical parameters. For example, avian clutch size usually has a negative correlation with egg size; birds can be bred with relative ease that lay many small eggs or a few large ones, but it is less easy to breed birds that lay many large eggs because of energy restrictions. In essence, all measurable traits usually show either positive or negative correlation with all others, and selection for a given trait will also produce positive correlation responses in those traits with positive correlation, but negative correlation responses in those having a negative correlation with the selected trait.

In any assessment of phylogenetic relationships among groups, the effects of character correlation would confound a true estimation of phyletic lineage. Nevertheless, numerical cladistic analyses that use discrete state characters assume all such characters evolve independently, thereby attempting to negate the concerns of character correlation. Although parsimony often is used as a justification, such clearly false assumptions render most such cladistic algorithms logically void by violating their own operational assumptions. In deed, Felsenstein (1982: 386) has suggested that the development of methods to remove the effects of correlation among traits is "perhaps the most important unsolved problem facing phylogenetic inference." Because the traits of organisms evolve in a correlated fashion, however, it seems appropriate to make use of their covariance information in assessments of relationships, rather than try to completely remove it or simply pretend that it does not exist.

3 Multivariate Evolution of the Mean Phenotype of Populations

How then can information on the correlation among traits be useful in determining the phyletic relationships among groups, the response of the groups to historical selection regimes? A phenotypic model for the multivariate evolution of correlated suites of morphological characters has been developed by Lande (1979), and permits a reconstruction of the net genetic gradients of natural selection that have acted among evolving groups. (Although only one of several possible QG based models, Lande's was chosen here because of its *relative* simplicity and additive variance structure; for others see Felsenstein [1988].)

Lande's (1979) model provides a relative estimation of the degree to which historical net selective pressures have influenced correlated phenotypic traits that covary genetically (Lande, 1980). This multivariate evolution for a population is expressed as a changing of the phenotypic mean for the group upon the mathematical vectors of an adaptive topology (for geometrically based examples see Phillips and Arnold, 1989). The model can be used (a) if each individual has traits that collectively comprise a phenotype, which is predominantly composed of an underlying genotype whose expression has been modified by minor additive environmental factors; and (b) if the genetic covariance matrices for all groups being analyzed are functionally similar.

Lande's model, therefore, can be used to determine evolutionarily covarying suites of traits among related groups, and so can approximate the historical net natural selection forces that have acted upon them. It also permits quantitative assessment of the evolutionary divergence among groups and, in conjunction with phylogenetic networking algorithms, permits analysis of their phylogenetic relationships (Sorensen, 1987a, this volume).

Under the above model, the phenotype of an individual is represented by the summation of a vector of additive genetic effects, its breeding values, and an independent vector of environmental effects, broadly defined to include developmental noise, dominance and epistatic genetic variance as well as the interaction of the genotype with its microenvironment (Lande, 1988). Multivariate evolution works on the mean values of the phenotypic traits of an individual, representable as a columnar vector $z = (z_1, z_2, z_3, \ldots, z_n)^T$, where superscript T indicates a matrix transposition.

The model requires these assumptions (Lande, 1979) before generalized natural selection can impart expected fitness, $W(z)$, to individuals with phenotype z: (1) the phenotypic, **P**, genotypic, **G**, and environmental, **E**, matrices are nonsingular and are related as $\mathbf{P} = \mathbf{G} + \mathbf{E}$; (2) the vectors of additive genetic effects, x, and environmental effects, e, including nonadditive genetic effects, are independent and have Gaussian distributions, while the additive phenotypic effects, z, equals $x + e$; (3) the mean of additive environmental effects is 0, so that the mean group phenotype, $\bar{z}$, equals mean group genotype, $\bar{x}$.

Denoting the phenotype distribution as $p(z)$, and the (average) fitness of phenotype z as $w(z)$, the mean fitness before selection in any generation of a population with a mean phenotype $\bar{z}$ can be expressed as $\bar{W} = \int p(z)\, W(z)\, d(z)$ (Lande, 1979: eq 4). In populations with discrete, nonoverlapping generations and panmictic breeding, the change that can be measured before selection on each generation along the mean phenotypic vector, can be expressed by (Lande, 1988: eq 1a) as

$$\Delta \bar{z} = \mathbf{G}\, \nabla \ln \bar{W} \ , \tag{1}$$

where $\nabla \ln \bar{W}$ represents the changes in mean phenotype along the gradient of the surface of log mean fitness and ∇ is defined as the gradient operator (Lande, 1979, 1988). This is equivalent to (Lande, 1979: eq 6)

$$\mathbf{G}^{-1}\, \Delta \bar{z} = \mathbf{P}^{-1} S \,, \tag{2}$$

where both expressions are equal to the selection vectors, $\nabla \ln \bar{W}$, and S represents the total selection differential on prereproductive adults. Therefore, this form can be rewritten as

$$\Delta \bar{z} = \mathbf{G}\mathbf{P}^{-1} S \,, \tag{3}$$

which is recognizable as the customary expression form in quantitative genetics (Young and Weiler, 1960; Harvey and Bearden, 1962; Lande, 1988).

This multivariate model for the evolution of the mean phenotype in a population recognizes both Wright's (1931, 1932, 1940, 1977) adaptive topologies for gene frequency evolution and Simpson's (1944, 1953) concept of phenotypic adaptive zones. Using this concept, the evolving population is represented as its mean, $\bar{z}$, which is a point on the surface of mean fitness, $\bar{W}$; the latter is a function of the mean phenotypic values of the traits in the population (Lande, 1988). Evolution in this system optimizes the phenotypic mean of the group and is not concerned with the distribution of genotypes or phenotypes for individuals within the population (Lande, 1979). Because the speed and direction of evolution of the populational mean is partly determined by **G**, $\bar{z}$ does not evolve in the steepest uphill course indicated by the selection gradient. Rather, the direction and speed of evolution of $\bar{z}$ are determined by the concerted action of **G** and $\nabla \ln \bar{W}$, which act jointly upon correlated character suites to provide impetus for modifying the mean group phenotype through time and over the evolutionary adaptive topology.

Under the model, if **G** remains stable during evolution among the group(s) under analysis, the net selection product is expressible as the summation of net selection gradients acting upon $\bar{z}$ during the episode, over generations 0 to t - 1. This can be determined (Lande, 1979: eq 9) as

$$\sum_{0}^{t-1} \nabla \ln \bar{W} = \mathbf{G}^{-1} [\bar{z}(t) - \bar{z}(0)] , \tag{4}$$

which is equivalent phylogenetically, across all evolved generations, to

$$\sum_{0}^{t-1} \beta(z) = t\bar{\beta} , \tag{5}$$

where $\beta = \nabla \ln \bar{W}$, as the net selection gradient over a single generation, and $t\bar{\beta}$ is its summation over all generations during the phylogenetic period that is addressed. The resulting measure of intergroup anagenic distance, β of Schluter (1984), deals only with the directional forces of selection, not stabilizing or disruptive selections, which do not alter mean group phenotype.

The summation $t\bar{\beta}$ estimates the net evolutionary distance transversed by the mean group phenotype along the selection vectors between the initial and contemporaneous times. The time frame considered is implicitly defined by the period of common ancestry among the (nonpolyphyletic) groups that are analyzed (Sorensen, this volume). Hence, a determination of the anagenic distances between groups is possible if the genetic covariance matrix (**G**) remains proportionally stable during selective change through evolutionary time. Then, the evolutionary vectors of relative selective forces can be calculated as orthogonal, additive sets of the genetic and environmental factors that result in the evolution of the group mean phenotype. The time dependence of this system incorporates both natural and sexual selection effects into the fitness changes that are experienced by the mean phenotype(s) during evolution. Therefore, $t\bar{\beta}$ approximates only the net forces and evolutionary speeds encountered during evolution, which during the actual evolutionary episode of divergence may vary in both intensity and duration in real time (Lande, 1988).

4 Interaction of Correlated Characters During Evolution

In this model, the behavior of characters during evolution differs from what would be expected in nonmultivariate models of evolution, where the effects of character correlation are not considered and thus neither is any protagonistic and antagonistic influence from other genetically linked traits. It is this accountability for genetic linkages, via correlation and covariance, that makes multivariately based character information superior to that which is estimated univariately, for the description of biological variance (Willig et al., 1986; Lofsvold, 1988). During evolution for any single trait, the directional selection can be expressed as the covariance or regression of relative fitness upon that character (Price, 1970). In a multivariate generalization of this, the forces of directional selection on correlated character suites are best described by a vector of partial regression coefficients of individual relative fitnesses on the characters in a suite (Lande and Arnold, 1983), although estimating the multivariate fitness of, or selection pressures on, traits has been addressed by several methods recently (i.e., Schluter, 1988; Phillips and Arnold, 1989; Wade and Kalisz, 1990; Crespi, this volume).

To illustrate the difference between univariate and multivariate effects on characters during evolution, consider the simple univariate selection of a single phenotypic trait (z_b) and the effects of selection on a correlated trait (z_i) within a population. If the measured traits are scaled and show phenotypic variances among groups that are roughly constant and independent of their means, $\bar{z}_b$ and $\bar{z}_i$, and their additive values on the phenotype of trait z_b have a linear regression with a Gaussian distribution, then, assuming no initial genotypic-environmental correlation exists (Lande, 1979: eq 1a),

$$\Delta \bar{z}_b = (\mathbf{G}_{bb} / \sigma_b^2) S_b = h_b^2 S_b \tag{6}$$

represents the change during evolution in trait z_b for a group during one generation under selection, and (Lande 1979: eq 1b)

$$\Delta \bar{z}_i = (\mathbf{G}_{ib} / \sigma_b^2) S_b = h_i h_b \gamma_{ib} (\sigma_i / \sigma_b) S_b \tag{7}$$

represents the correlated response of trait z_i, to selection on z_b. In both equations the heritability of z_b (as h_b^2) equals the trait's additive genetic variance ($\mathbf{G}_{bb}$) divided by its total phenotypic variance (σ_b^2); S_b is the selection differential, and $\mathbf{G}_{ib}$ and γ_{ib} are the covariation and correlation, respectively, of the additive genetic values of z_b and z_i within the group (Lande, 1979).

The behavior of trait z_i differs considerably under multivariate evolution, where changes to z_i depend not only upon selection that acts directly upon it, but also upon the indirect selection acting upon all genetically correlated traits, to create a response (Lande, 1988: eq 1b) as

$$\Delta \bar{z}_i = \sum_{j=1}^{n} \mathbf{G}_{ij} (\partial \ln \bar{W} / \partial \bar{z}_j) = \sum_{j=1}^{n} \mathbf{G}_{ij} \beta_j , \tag{8}$$

where $\beta_j = \partial \ln \bar{W} / \partial \bar{z}_j$ is the force of directional selection acting directly upon z_j, and $\mathbf{G}_{ij}$ represents the additive genetic covariance between z_i and z_j, (Lande, 1988). The latter is the difference in Malthusian mean fitness in a population that is produced by a selective change in the mean of trait z_j, while holding the variance of that trait and the distribution of all other traits constant (Lande, 1988). This reaction then accounts for both the positive and negative correlated effects of other traits on z_j. The adaptive effect of the latter on the mean phenotype of a population is that more selective pressure must be exerted on trait z_j to produce that same evolutionary response that would occur in a population with genetically independent traits (Lande, 1979). Phylogenetic models (e.g., discrete state numerical cladistic procedures), which ignore character correlation by (erroneously) assuming character independence (in the absence of testing) would necessarily misestimate the degree of selection that must be applied to modify the mean phenotype during evolution and, therefore, the calculated distance transgressed; however, models are being developed to recognize this problem (i.e., Maddison, 1990), by testing for correlation between binary traits on the terminal branches of phylogenetic trees.

Lande and Arnold (1983) indicate that the selection gradient acting upon characters can be thought of as a vector of the partial regression coefficients in a linear multiple regression of relative fitness, w (which equals $W/\bar{W}$, so that $\bar{w} = 1$), on phenotypes, z, for individuals (for an alternative interpretation, however, see Crespi, this volume). This vector is expressible (Lande, 1988: eq 2a) as

$$\beta = \mathbf{P}^{-1} s = \mathbf{P}^{-1} [z^* - z] = \mathbf{P}^{-1} \operatorname{cov} [w, z] , \tag{9}$$

where s, the observed selection differential equals the selection within a generation, $[\bar{z}^* - \bar{z}]$, with the asterisk denoting a quality after selection. Lande (1988) notes that the components in this selection gradient are equivalent to the selection index, I, used by animal breeders (Dickerson et al., 1954, 1974; Magee, 1965; Yamada, 1977) in applied quantitative genetics, where

$$I = \beta^{\mathrm{T}} z = s^{\mathrm{T}} \mathbf{P}^{-1} z . \tag{10}$$

If the intraspecific allometry between two traits during evolution at any given time is considered, the allometric relationship between the traits should follow the common generalized equation (Fleagle, 1985; Lande, 1985)

$$\log(z_j) = \log(\mathrm{d}) + b\log(z_i)\,, \tag{11}$$

where d is a constant, and b is the allometric coefficient for the pair of traits, expressed as

$$b = \gamma_p\, p_y / p_x\,, \tag{12}$$

in which x and y are the respective logs of traits z_i and z_j, p_x and p_y are their respective phenotypic standard deviations, and γ_p is the correlation between the traits (Zeng, 1988).

When evolution proceeds to morphological divergence between two closely related species, however, the interspecific differentiation between ancestral and progeny species follows an allometric response between a direct selection on z_i and a correlated response on z_j (Zeng, 1988: eq 25) as

$$R_y / R_x = \gamma_g\, g_y / g_x\,, \tag{13}$$

where γ_g is the genetic correlation, and g_x and g_y are the genetic standard deviations, respectively; assuming trait z_j to be selection neutral and that stabilizing selection does not occur on z_i. If strict neutrality, in terms of selection, does not occur for z_j during evolution, then a higher interspecific regression slope is realized than is predicted by the last equation (Lande, 1979; Zeng, 1988).

Stabilizing selection on either trait during evolution, however, greatly complicates these regression relationships. Zeng (1988: eq 27) estimates the complicating effect of stabilizing selection on traits during their evolutionary responses. Zeng further presents a model of long-term correlated evolution among multiple correlated traits that partitions selection into stabilizing and directional components, assuming the former to be less variable than the latter. Under those assumptions, character responses to selection within a population differ, in that short-term responses are solely dependant upon genetic correlation between traits. Character responses in the long-term under Zeng's model are dependent only upon the shape of the fitness function for directional and stabilizing selection, and are independent of phenotypic and genotypic correlations, assuming that **G** is not singular or nearly singular (i.e., that there is no combination of characters for which substantial genetic variance does not exist). Nevertheless, under the assumption of constant **G**, it is valid to estimate the net selection gradient as a measure of cumulative forces of directional selection acting to differentiate taxa (R. Lande, personal communication). (Also, see Felsenstein [1988: 452, eq 5] for the implications of "selective covariance" concerning Lande's model.)

5 Determination of Historical Net Selection Gradients

Character analysis under the rather simple multivariate evolution model allows the differing responses between characters to be quantified through multivariate analyses of covariance and correlation. It also permits those covariant suites of attributes that react similarly to gradients of evolutionary selective pressure to be distinguished. PCA is particularly suited as a probe for determining the covariant relationships among attributes that indicate a trait's sequestered participation among several independent genetic suites of characters, because the correlation and covariation among morphometric variables result predominantly from pleiotropy and other genetic linkages (Lande, 1980; Falconer, 1981).

Quantification of character interaction during evolution can be accomplished by inverting an individual x character data matrix prior to PCA, so that a R- instead of an Q-type analysis is conducted (Sneath and Sokal, 1973). Normally in Q-analysis, PCA creates a correlation matrix from morphometric character data and constructs additive linear eigenvectors upon which each original morphometric variable has an input coefficient weight that is proportional to its covariance with the derived vector. The relative positions of the individuals are calculated along each orthogonal vector, using the eigenvector coefficients derived from the attributes, and these positions are then plotted in unrotated A-(*attribute*)-space (Williams and Dale, 1965). If, however, the initial data matrix is inverted beforehand to allow an R-analysis, then each individual provides the input coefficient along the orthogonal vectors, and it is the positions of the original characters that are plotted along the unrotated vectors in I-(*individual*)-space (Williams and Dale, 1965).

Thus, an R-analysis depicts two important patterns with reference to the multivariate model of evolution. First, it shows the degree of association and interaction of each trait with each orthogonal vector, the latter having been generated by the covariance(s) derived from the (largely) genetic correlation patterns (due to linkages, sensu Sorensen, 1991) among the attributes and representing a covariant character suite that responds to net evolutionary selective pressures in an integrated fashion (sensu Lande, 1979). All traits belong in varying degrees to all covariant character suites, as the PCA vectors, in that they have a contributing (weighted) input for each vector that varies among the vectors. Dominating traits ("drivers") for each covariant character suite are those traits with the highest absolute loading coefficients (and correlations) on the vector representing their suite. Traits with loading coefficients nearer to zero on any vector are more subordinate for that vector, and more respectively neutral in response to the net selection pressures engaging that vector. Therefore, when selection is imposed upon any suite (vector) it acts largely upon the dominating "driver" traits for the vector, causing modification of all traits in proportion to their loading input and correlation for the vector (but see estimations for the effects of stabilizing selection for covarying traits in Zeng, 1988).

The second pattern that R-analysis depicts is that when two or more dimensions (as orthogonal vectors) in PCA are plotted, spatial arrays (constellations) of traits are often apparent, as "virtual groups" of characters. Because of their proximity in the dimensional space that is so defined, the traits in these virtual groupings tend to react in a relatively similar fashion to the ("activated") combination of independent selective pressure gradients, which are represented by only the plotted orthogonal vectors. Within such a virtual group, the orchestrated response among its traits is actually their similar, but independent, individual summations of responses to each of the orthogonal selection gradients that are portrayed. Such virtual groupings may exist (at least) temporarily, whenever only their respective component selection gradients are activated in unison (Sorensen, 1991). Such virtual groupings may exist over any or all of the g - 1 vectors, or any partial subset combinations thereof, depending upon which vectors are activated and the degree to which the selection is imposed.

It is important to remember that although these virtual groups of characters may react similarly across multiple, orthogonal vectors during evolution, their association is transient and their traits may change affiliation in the sense that the response of individual traits is dictated by their interaction along the selective pressure gradient of each vector separately and orthogonally. Further, it should be stressed that assessments of single vector based character suites, and of transient, multiple vector based, virtual groups of traits, are operational conveniences for employing ordinations under the evolutionary model; they serve to empirically quantify and describe character reactions to imposed selection, rather than to imply a true, biologically based condition among the traits and the functional architecture (sensu Houle, 1991) of their loci. Although it may be doubtful that all loci actually effect all traits via strong aquisition-allocation pleiotropy (Wagner, 1989; Houle, 1991), parameters with similar input for new mutational variance (e.g., Charlesworth, 1990) suggest the net effects for character reaction to selection under such optimization models would be substantially the same under an assumption of no aquisition-allocation pleiotropy at all.

Using the multivariate evolution model, Sorensen (1991) has attempted to estimate the relative degree of genetic linkage (assuming $\mathbf{E} = 0$) among traits in a phenotypic matrix for a species, as a function of the communality (sensu Sneath and Sokal, 1973) among the character variables in the matrix; this information, from a mother-taxon, was then used to predict character interaction during the multivariate evolutionary episode to its putative daughter-taxon. Many studies (see Lande, 1988) have analyzed natural selection on subsets of traits for populations across a variety of organisms; these generally suggest that small subset of traits may be focused upon by natural selection during intermittent periods of abnormal environmental conditions.

The relative nature of ordinations, in local perspectives, is not problematic under the multivariate evolution system. Although any given analysis has the restriction that the viewing of contributions and interactions of the traits being analyzed is relative with respect to the matrix employed locally, the model is probabilistic (sensu Felsenstein, 1983, 1984) in its assessment of the true, global phenotypic mean (sensu μ) for a group. Thus, results may change to some degree if either characters or individuals are removed or added to the local analysis (Sorensen, 1987b, 1991). This effect may occur even though the overall genetic covariance matrix for a species or population undergoes selection for locally maximized mean fitness, because various, arbitrarily chosen subsets of traits that are analyzed may not reflect the mean maximized fitness for the (global) genetic covariance matrix as a whole. Such deviation in local perspective can be caused by particular traits that may evolve maladaptively in reaction to genetically correlated responses to selection on other traits. Adaptive change for a species, therefore, results in an evolutionary compromise between genetically correlated traits, at least in the short term (Zeng, 1988). These compromises occur, in part, as a response to their relative importance, as expressed by their loading coefficients on the orthogonal fitness vectors. Thus, a change to the input matrix of samples of different characters or individuals functionally modifies the sampled effects of genetic positive and negative correlation during the estimation of the true, overall genetic covariance matrix for the total group. In this respect, the system must cope with estimation errors similar to those anticipated when trying to approximate the true (global) mean (μ) of a population by using subsets that provide slightly differing (local) sample means. Despite this, however, traits that are analyzed multivariately provide a significantly better estimate of their interaction and evolutionary importance, than when considered univariately (Willig et al., 1986; Lofsvold, 1988).

6 Apomorphic Character Information and the Estimation of Phylogeny

Systematists generally accept only estimations of evolutionary distance from derived traits (apomorphic information). It is important to consider only such information in determining anagenic distances between groups when estimating a phylogeny. Under the multivariate evolution model of phenotype, if the actual evolutionary time involved in genetic divergence among the groups is known, then both the magnitude and direction of the net selective forces involved in their history can be estimated directly. Unfortunately, the evolutionary time is usually not accurately known; however, the past magnitude of selection can be estimated indirectly and conservatively from the minimum selective mortality that is necessary to account for the observed rate and direction of evolution. In this extension of the model, the mortality estimate considers only divergence (derived) information and is equivalent to a selection index (Hazel, 1943) or discriminant function (Smith, 1936), where each original character is weighted by the directional selection acting upon it to cause evolution of the mean phenotype (Lande, 1979: eq 10) as

$$I = (\nabla \ln \bar{W})^{T} z = \Delta \bar{z}^{T} \mathbf{G}^{-1} z . \tag{14}$$

Here, I is the selection index that changes the direction of evolution of the group mean by altering the phenotypic responses of characters in proportion to their additive genetic covariance with I. Therefore, calculation of I, as in DFA, allows the optimized (parsimonious), apomorphic evolutionary distance between groups to be determined, because only that portion of variation for characters that maximally segregates groups (*unshared* variance sensu Pimentel, this volume) is employed, while maximum intragroup cohesion is also maintained. This "divergence" portion of **G** corresponds to cladistic apomorphic information, whereas "nondivergent" variation, which is equivalent to plesiomorphic variance (under an assumption of "common equals primitive"), is not employed and is discarded as information on group communality in the DFA calculation (Sorensen, 1987a, 1987b). A discriminant vector is thus analogous to an apomorphic covariant character suite, where the degree of selection on each character equals the absolute value of that character's factor loading coefficient for the discriminant vector, and where opposite signs for character loading coefficients indicate membership in genetically antagonistic covariance suites (Sorensen, 1987a, 1987b, 1991; Sorensen and Sawyer, 1989).

According to the above, if discriminant vectors are employed for measured characters, and their dimensional coordinate, as Mahalanobis' distances, are summarized, a nonhierarchical estimation of apomorphic anagenic distance for $\Delta \bar{z}$ among closely related groups can be ascertained. Using DFA, analysis of anagenic divergence during differentiation among several (g) groups creates g - 1 net selection vectors, as minimum selective mortality indices, whose Mahalanobis' distance is apomorphic and is derived from **G**, as represented by **P**. Each DFA vector thus acts as an optimized, transformed suite of covarying phenotypic traits, and represents a parsimonious net selection gradient that has acted upon the groups during their evolution from a common ancestor. These optimized evolutionary vectors each have an input component from each trait that is weighted in proportion to the trait's share in the explanation of the observed divergent distance that occurs between the groups. The relative strengths of these vectors are proportional to their eigenvector values within **G**.

Character correlation in this minimum selective mortality based model is dealt with by summarizing the divergence information among the measured characters and by sequestering this information to the vectors, which function as covariant character suites where loading coefficients represent the absolute value of apomorphic information used to move the groups centroids along the vector gradients. The orthogonal vectors thus represent newly constructed and optimized, independent (uncorrelated) characters, and the group centroid scores upon each vector represent the "state" value for a given group on the particular discriminant function.

The "new" characters, as apomorphic covariant trait suites, are orthogonal and uncorrelated; as a result the problems of character correlation are eliminated among them. Further, because the DFA has scaled the variance among the groups, and is weighted to optimally and parsimoniously separate the groups in the discriminant space, the problems of character scaling, weighting and coding are eliminated. The problem of homoplasy among vectors has also been eliminated by the optimal separation of groups on each of the discriminant vectors; because each original character may contribute a portion of its divergence information to each discriminant vector, and each vector represents only apomorphic information, the contributions of each original trait to multiple vectors (as new characters) represents only orthogonal and nonredundant information. (Homoplasy may exist, however, along a given vector in the sense that the evolving mean phenotype can undergo directional reversals on it; this is accounted for in the *net* nature of multivariate evolution, and is analogous to character-state reversals on a given branch of a discrete state cladistic tree.)

Although the dimensional space created by the DFA represents only a nonhierarchical portrayal of the apomorphic information in evolution, the centroid scores on each vector can be analyzed using standard "distance" based numerical cladistic algorithms to generate a phyletic pathway representing a phylogenetic network. Under this model, the evolution of groups through the g - 1 dimensional space occurs as net probabilistic movements (see Felsenstein, 1979, 1981, 1984) among the group means, as DFA centroids, which move between the temporal positions $\Delta \bar{z}(0)$ and $\Delta \bar{z}(t)$ among all vectors. This movement (evolution) occurs through the discriminant space in a Euclidean fashion (Lande, 1979) among the vectors (not in Manhattan distance fashion, directly along each vector), although the space involved is based in Mahalanobis' distance, which is non-Euclidean (Pimentel, this volume). The apomorphic phyletic pathway, $t\bar{I}$, between centroids is expressible as

$$\sum_{0}^{t-1} \beta^{\mathrm{T}} z = t\bar{I}\,, \tag{15}$$

and can be generated with a cladistic networking algorithm, but rooting is problematic (Sorensen, 1987a, 1990, this volume).

Examples of such phylogenetic estimation include Sorensen (1987a), where centroids were connected using a maximum-likelihood algorithm, and Wood and Pesek (this volume), where a Euclidean algorithm was similarly used. Sorensen (in press) went further in extracting temporally based evolutionary information. He examined, among three taxa in a terminal clade within a genus (*Essigella*: Sorensen, 1987a), their relative, nonhierarchical distances along DFA vectors as selection gradients in a local perspective (sensu Sorensen, this volume). He then compared these local nonhierarchical distances to the phyletic distances between these taxa that were separately generated on a $t\bar{I}$ pathway (Sorensen, 1987a, as "β"), and that incorporated all generic representatives plus the immediate generic out-group (a global perspective). He concluded that changes in the relative, apomorphic, anagenic distances shown among the three taxa in local (Sorensen, in press) versus global (Sorensen, 1987a) views, probably represent a form of relatively recent multivariate character displacement among them. This displacement necessarily must have occurred since the divergence of the three taxa from their most recent common ancestor (Sorensen, this volume) in response to competition for their pine resources (Sorensen, in press, unpublished data).

Such usages should be considered truly cladistic analyses in the sense that only apomorphic information was used to generate phylogenetic relationships. Further, they represent a parsimony criterion, in that the divergence information employed was optimized to maximize intergroup differences (distances), while maintaining maximal intragroup cohesion.

7 Refinement of the Rationale

The multivariate evolutionary model does have several shortcomings in its admittedly simplistic approach to deducing evolutionary history, although this simplicity is not the naivety found in discrete state cladistic methods that ignore character correlation (Sorensen, 1990). The multivariate model depends to large extent upon the stability of the genetic covariance matrices among the groups analyzed over the period considered in their evolution. This stability is probably adequately true for intraspecific populations, for closely related species within species complexes, and is also possibly true for "tight" and problematic genera (e.g., Sorensen, 1987a), although it may not adequately be true for analyses above the generic level (but see Lofsvold, 1988; Wood and Pesek, this volume).

Lande (1979) argues on theoretical grounds, that the stability of **G** is maintained for a group by a balance between selection and mutation, which keeps genetic variances and covariances roughly constant. He suggests the stability of **G** can be inferred indirectly when the phenotypic patterns that are displayed are consistent between current related groups, because constancy in the phenotypic covariance matrix (**P**) is not likely without a stable genetic matrix if the latter dominates the additive components (Lande, 1979). If this were not true, and **G** changed during evolution, it would indicate that the selective pressures that have been exerted upon the groups during evolution have favored mutations with different pleiotropic and linked allelic combinations for various traits (Lande, 1979; Hegmann and Possidente, 1981).

Nevertheless, Turelli (1984) indicates that Lande's (1979) calculation of the genetic variance that is maintained by mutation rates is high. Lande (1988), however, states that a simple assumption of effective constancy for **G** and **P** during evolution is justified by many artificial selection experiments, where a constant intensity of selection creates a steady response that changes the mean phenotype by three to five phenotypic standard deviation units. Experiments where reversed artificial selection is imposed (Robertson, 1980; Falconer, 1981) suggest that countervailing natural selection, and not exhaustion of additive genetic variance, is usually responsible for the limits or plateaus to selection shown in populations that are large enough so that genetic drift does not cause a loss of effective, heritable genetic variance. Therefore, Lande (1988) suggests that for populations near evolutionary equilibrium, it is reasonable to conclude that heritable variance is effectively maintained by a balance of mutation, migration, recombination, selection and population mating systems.

The temporal stability of **G** is thus quite important in deciding to what extent the multivariate evolution model can be used to imply actual genetic divergence, rather than simply phenotypic differentiation (confounded by environmental influence). Systematists have generally assumed, perhaps unknowingly, that **G** retains approximate proportionality within genera (see Lemen and Freeman, 1984). Arnold (1983) and Ayres and Arnold (1983) show examples that support the constancy of **G** among groups within lineages. At the tribal level, Cheverud (1989) found similar patterns (but also minor differences) in the morphological variance of the cranium among species of Papionins, indicating that their patterns of variance and correlation in facial morphology have tended to remain stable in the clade, and that selection gradients could be reconstructed for the group.

In contrast, Lofsvold (1986) estimated **G** among *Peromyscus* species and subspecies under controlled conditions, stating that **G** is proportional within the tested species, but is less so between them; he suggested Lande's model should be modified to allow for fluctuations of **G** through evolutionary time. However, Kohn and Atchley (1988) reanalyzed Lofsvold's (1986) data, and state that the estimated **G** between differing genera, represented by laboratory rats and mice, show no statistically significant association indicating that an assumption of equal genetic variance-covariance across divergent taxa may be problematic. They demonstrate that both the phenotypic and genotypic covariation and correlation are similar between laboratory rats and mice, and that the phenotypic and genotypic correlation is also similar between them. Further, in another reanalysis of his 1986 data, Lofsvold (1988) compared the among-locality covariance matrix with the additive genetic covariance matrix and found no significance in the difference of proportionality. He interpreted this to mean that random genetic drift was not responsible for the evolutionary divergence shown between his populations, and went on to reconstruct the net selection gradients that were necessary to account for the observed interpopulational differentiation. Lofsvold (1988) concluded, using these generated minimum selective mortality measures, that on any time frame selected, the required populational truncation necessary to account for selection was quite low, not exceeding levels that are shown among extant populations and, therefore, he suggested that it is entirely credible that the imposed directional selection can produce species-level differences in a relatively short time.

If **G** retains stability, as the model assumes, then β reflects the changing net selective pressures acting upon the groups under divergence. However, if, as Lofsvold (1986) suggests, **G** varies significantly among groups, then the groups would diverge even under overall, constant, net selection pressures. In either case, the multivariate evolution model provides a practical assessment of phenotypic divergence within and among closely related groups, and employing DFA to yield an apomorphically based phyletic pathway appears to be operational possible; however, before recommended employment for groups above the generic level, the model requires more theoretical development and testing.

To date, the model has been used with empirically credible results several times. Schluter (1984) has produced a functional phylogeny of finches (by recalculating **P** among divergent groups) that mirrors electrophoretic evidence. Sorensen (1987a) has produced an operational phylogeny for a problematic aphid genus that appears substantiated by both (the few available) independent, qualitative apomorphies, and by genetic interbreeding relationships among their hosts. Wood and Pesek (this volume) has similarly produced a reasonable phylogenetic assessment for treehoppers. Sorensen and Sawyer (1989) have demonstrated biologically inherent and differing attributes in the seasonal phenological response of morphological traits between sibling species of leafhoppers. Simon (this volume) states that two populations of 13-year periodical cicadas, which were originally thought to be misclassified by a (nonhierarchical) DFA as 17-year cicadas (Simon, 1983), have now been shown, using analysis of mitochondrial DNA (Martin and Simon, 1988), to be actually 17-year cicadas that have secondarily adapted a 13-year brood cycle. Therefore, the DFA based model appears functional for analyses of intraspecific populations or closely related species complexes, and it can be operational for phylogenetic and evolutionary inference within, at least, less diverse genera if **G** is relatively similar among the species. Note, however, that in instances where PCA, rather than DFA, is used for attempted phylogenetic inference (e.g., Roskam and Zandee, this volume), there are problems that are, no doubt, due to the incorporation of both plesiomorphic and apomorphic information (*shared* and *unshared* variance, sensu Pimentel, this volume) into the PCA vectors, rather than only apomorphy, as occurs with DFA.

In general, the multivariate evolution approach to the analysis of phylogeny and trait interaction is criticized by Crespi and Bookstein (1989) and Crespi (this volume) as being too assumptive. Admittedly, many of Crespi's contentions are valid; not all important characters upon which selection acts are usually measured (or known) in any given analysis and thus potential errors in interpretation are possible. However, there is no guarantee, or even reasonable expectation, that any phylogenetic procedure involving any data source (i.e., discrete or continuous phenotypic traits, or nucleotides) will ever, or can ever, meet such rigorous expectations. Robust estimates, nevertheless, are possible (e.g., Price et al., 1984; Price and Grant, 1985), which appear operational for the use of the model with caution (Turelli, 1988); perhaps, after appropriate tests of **G** among populations (Shaw, 1991).

Many of Crespi's criticisms appear to be due often to the relative, rather than absolute, nature of ordinations, and they might be minimized or mitigated by running several, separate analyses using differing matrices that contain varied characters and individuals, and then resolving any differences by consensus or averaging. With regard to his criticism based upon the methods used to determine selection, Lande (1988) has suggested in defense that a partial resolution of the effects of direct and indirect selection is better that no resolution, and that at least the estimated directional and stabilizing (or disruptive) selection coefficients for a particular trait cannot be attributed to the indirect effects of selection on any other traits included in the analysis.

Criticism of the model suggesting that orthogonal vectors, or those subsequent to the first eigenvector, are biologically arbitrary (not necessarily a fact, see Blackman [this volume] for an example) misses an important point: that they must be employed to yield an *objective*, *operational* procedure that is based upon the *parsimony* criterion, and which employs *uncorrelated* traits (as new characters represented by the transformed orthogonal vectors). (There have been apparently no successful attempts to use continuous character data in phylogenetic estimation when employing the data from individual traits as separate characters, regardless of various coding schemes; see Felsenstein [1988] for a critical review of character coding.)

Likewise, although organisms may not (necessarily) evolve parsimoniously, or in a fashion that optimizes intergroup distance or differences, the employment of such criteria in phylogenetic analyses is philosophically necessary as an operational *limit* for choice when a multitude of possible analytical outcomes (trees) is possible; this philosophical premise has been accepted in phylogenetic inference for over a decade. Quite simply, DFA generated evolutionary space gives a parsimonious portrayal of the divergence between treated groups, while minimizing the problems of homoplasy and avoiding the problems of correlated character usage. When an appropriate networking algorithm (see Sorensen, this volume) is run through that space, the tree created must also necessarily have the desirable attributes, and lack of undesirable attributes, of the "characters" (DFA vectors) that were used to generate it.

Crespi's recommended divorce of general-size and shape, with elimination of the former, is philosophically problematic in that there is no reason to suspect that size is not important in evolution (see Daly, this volume), especially at the intrageneric level. Size may well be the most important component in the final evolution of closely related species (e.g., within a species complex). This is because changes along a general-size regression line between "entities" that are diverging evolutionarily should be more energy efficient than changes in shape, which necessarily require disassociations or rearrangements of genetic linkages (Sorensen, 1991). Nevertheless, size is often dismissed because of the ease with which it is changed during evolution (Felsenstein, 1988). To automatically ignore size is unwise, and to remove it correctly and without a degree of aberration may indeed not be totally feasible.

8 Conclusions

Alas, no analytical technique for phylogenetic investigation is without at least one of the following: problematic assumptions, operational problems, philosophical detractions, or a requirement for subjective (and thus potentially misleading) decision input; but these are not reasons why they should not be used and contrasted, if their shortcomings are understood and stated. What is important is that the use of ordinations, particularly DFA, for evolutionary inferences allows analysis and incorporation of the covariance that is encoded within the groups being analyzed, and that this covariance is most probably a genetic result of common phyletic lineage. Denying that it exists, or failing to attempt to employ it as a major evolutionary clue, is often a fault in phylogenetic analyses. Modern molecular techniques will no doubt prove to be a better estimate of phylogeny (although they also have problems, such as selection of the appropriate nucleotide segments for the relative taxonomic levels being dealt with) than any technique that is operationally based on phenotype (including isozymes), but at least one case has occurred where after running both methods, the results of a nonhierarchical DFA were found to largely corroborate those of DNA techniques (see Simon, this volume).

If reasons why something might not work are searched for in science, they are seldom hard to find in any context or pursuit. There are a great many reasons why phylogenies "cannot" be successfully inferred with any sense of realism by using any morphologically based character scheme (see Felsenstein [1988] for a pessimistic review of these). Yet reasonable phylogenetic models of many groups, made with preDNA data, exist. No single model can be expected to account for even a fraction of the plethora of possible evolutionary mechanisms, most of which we do not yet know. As evolutionary conditions are realized and modelled, however, they can be added to various models in extended refinements. Similarly, "If under the simplifying assumptions of quantitative genetics we are able to show that there is insufficient information to infer phylogenies from quantitative characters, then when those assumptions are removed the situation can only get worse." (Felsenstein, 1988: 468).

The alternative to pursuit of morphologically based phylogenetic modeling and estimation, of course, is to "throw up one's hands" in frustration at the multitude of possible inabilities that might be perceived; ultimately, at its extreme, this school of thought requires the exclusion of any consideration of relationships that are based on morphological grounds in any degree, as unpalatable to conventional systematics as that may sound. "Nevertheless it is always necessary to keep trying to improve the realism of the models and not surrender to despair. ... As heuristic devices they are always subject to skepticism based on their oversimplified nature, but they are nevertheless useful in honing our understanding and training our intuition." (Felsenstein, 1988: 468). This is particularly true when "In morphometrics, few workers are taking advantage of some precise mathematical methods for the definition of size and shape and their covariation." (James and McCulloch, 1990: 159).

We hope there will be enough room in systematics for all evolutionary models of phylogenetic assessment until sufficient evidence exists to suggest that a particular morphological approach is entirely unsuitable, as determined by testing against real molecular results. Clearly no morphological approach will be a panacea (Felsenstein, 1988; Barton and Turelli, 1989).

9 Acknowledgments

We thank Russell Lande for discussions on multivariate evolution, and for his earlier review of, and suggestions for, this chapter. We also thank Joseph Felsenstein and Richard Pimentel for previous discussions and insight into the problems of ordinations, correlated characters and phylogenetic inference.

10 References

Arnold, J.S., 1983. Morphology, performance and fitness. *American Zoologist*, 23: 347-361.

Ayres, F.A. and Arnold, S.J., 1983. Behavioral variation in natural populations. IV. Mendelian models and heritability of a feeding response in the Garter snake, *Thamnophis eleganes. Heredity*, 51: 405-413.

Barton, N.H. and Turelli, M., 1989. Evolutionary quantitative genetics: how little do we know? *Annual Review of Genetics*, 23: 337-370.

Blackman, R.L., (this volume). The use of ordination techniques to discriminate within pest aphid species complexes.

Charlesworth, B., 1990. Optimization models, quantitative genetics, and mutation. *Evolution*, 44: 520-538.

Cheverud, J.M., 1988. A comparison of genetic and phenotypic correlations. *Evolution*, 42: 958-968.

Cheverud, J.M., 1989. A comparative analysis of morphological variation patterns in the Papionins. *Evolution*, 43: 1737-1747.

Claridge, M.F. and Gillham, M.C., (this volume). Variation in populations of leafhoppers and planthoppers (Auchenorrhyncha): biotypes and biological species.

Crespi, B.J., (this volume). Natural selection and morphometrics.

Crespi, B.J. and Bookstein, F.L., 1989. A path-analytic model for the measurement of selection on morphology. *Evolution*, 43: 18-28.

Daly, H.V., (this volume). A statistical and empirical evaluation of some morphological variables of honey bee classification.

Daly, H.V. and Balling, S.S., 1978. Identification of Africanized honeybees in the Western Hemisphere by discriminant analysis. *Journal of the Kansas Entomological Society*, 51: 857-869.

Dickerson, G,E., Blunn, C.T., Chapman, A.B., Kottman, R.M., Krider, J.L., Warwick, E.J. and Watley, J.A. Jr., 1954. Evaluation of selection in developing inbred lines of swine. *North Central Regional Publication* 38, *Missouri Agricultural Experiment Station Research Bulletin*, 551.

Dickerson, G.E., Hetzer, H.O., Krehbiel, E.V. and Flower, A.E., 1974. Effectiveness of reciprocal selection for performance of crosses between Montana No. 1 and Yorkshire swine. III. Expected and actual response.

Journal of Animal Science, 39: 24-41.
Dupraw, E.J., 1964. Non-Linnean taxonomy. *Nature*, 202: 859-852.
Dupraw, E.J., 1965. Non-Linnean Taxonomy and the systematics of honey bees. *Systematic Zoology*, 14: 1-24.
Falconer, D.S., 1981. Introduction to Quantitative Genetics, (Second Edition). Longman, London, 340 pp.
Felsenstein, J., 1979. Alternative methods of phylogenetic inference and their interrelationship. *Systematic Zoology*, 28: 49-62.
Felsenstein, J., 1981. Evolutionary trees from gene frequencies and quantitative characters: finding maximum likelihood estimates. *Evolution*, 35: 1229-1242.
Felsenstein, J., 1982. Numerical methods for inferring evolutionary trees. *Quarterly Review of Biology*, 57: 379-404.
Felsenstein, J., 1983. Statistical inference of phylogenies. *Journal of the Royal Statistical Society*, 146: 246-272.
Felsenstein, J., 1984. Distance methods for inferring phylogenies: a justification. *Evolution*, 38: 16-24.
Felsenstein, J., 1988. Phylogenies and quantitative characters. *Annual Review of Ecology and Systematics*, 19: 445-471.
Fleagle, J.S., 1985. Size and adaptation in primates. In: W.L. Junger (Editor), Size and Scaling in Primate Biology. Plenum Press, New York, pp. 1-20.
Harvey, W.R. and Bearden, G.D., 1962. Tables of Expected Genetic Progress in Each of Two Selected Traits. *United States Department of Agriculture, Agricultural Research Service, Publication* 20-12.
Hazel, L.N., 1943. The genetic basis for constructing selection indexes. *Genetics*, 28: 476-490.
Hegmann, J.P. and Possidente, B., 1981. Estimating genetic correlations from inbred strains. *Behavior Genetics*, 11: 103-114.
Houle, D., 1991. Genetic covariance of fitness correlates: what genetic correlations are made of and why it matters. *Evolution*, 45: 630-648.
James, F.C. and McCulloch, C.E., 1990. Multivariate analysis in ecology and systematics: panacea or Pandora's box? *Annual Review of Ecology and Systematics*, 21: 129-166.
Kohn, L.A.P. and Atchley, W.R., 1988. How similar are genetic correlation structures? Data from mice and rats. *Evolution*, 42: 467-481.
Lande, R., 1979. Quantitative genetic analysis of multivariate evolution applied to brain:body size allometry. *Evolution*, 33: 402-416.
Lande, R., 1980. The genetic covariance between characters maintained by pleiotropic mutations. *Genetics*, 94: 203-215.
Lande, R., 1985. Genetic and evolutionary aspects of allometry. In: W.L. Junger (Editor), Size and Scaling in Primate Biology. Plenum Press, New York, pp. 21-32.
Lande, R., 1988. Quantitative genetics and evolutionary theory. In: B.S. Weir, E.J. Eisen, M.M. Goodman and G. Namkoong (Editors), Proceedings of the Second International Conference on Quantitative Genetics. Sinauer Associates Inc., Sunderland, Massachusetts, pp. 71-84.
Lande, R. and Arnold, S.J., 1983. The measurement of selection on correlated characters. *Evolution*, 37: 1210-1226.
Lemen, C.A. and Freeman, P.W., 1984. The genus: a macroevolutionary problem. *Evolution*, 38: 1219-1237.

Lofsvold, D., 1986. Quantitative genetics of morphological differentiation in *Peromyscus*. I. Tests of the homogeneity of genetic covariance structure among species and subspecies. *Evolution* 40: 559-573.

Lofsvold, D., 1988. Quantitative genetics of morphological differentiation in *Peromyscus*. II. Analysis of selection and drift. *Evolution*, 42: 54-67.

Maddison, W.P., 1990. A method for testing the correlated evolution of two binary characters: are gains or losses concentrated on certain branches of a phylogenetic tree? *Evolution*, 44: 539-557.

Magee, W.T., 1965. Estimating response to selection. *Journal of Animal Science*, 24: 242-247.

Martin, A. and Simon, C., 1988. Anomalous distribution of nuclear and mitochondrial DNA markers in periodical cicadas. *Nature*, 336: 237-239.

Phillips, P.C. and Arnold, S.J., 1989. Visualizing multivariate selection. *Evolution*. 43: 1209-1222.

Pimentel, R.A., (this volume). An introduction to ordination, principal components analysis and discriminant analysis.

Price, G.R., 1970. Selection and covariance. *Nature*, 227: 520-521.

Price, T.D. and Grant, P.R., 1985. The evolution of ontogeny in Darwin's finches: a quantitative genetic approach. *The American Naturalist*, 42: 54-67.

Price, T.D., Grant, P.R. and Boag, P.T., 1984. Genetic changes in the morphological differentiation of Darwin's ground finches. In: K. Wohrmonn and V. Loeschcke (Editors), Population Biology and Evolution. Springer-Verlag, New York, pp. 49-66.

Robertson, A. (Editor), 1980. Selection Experiments in Laboratory and Domestic Animals. Commonwealth Agricultural Bureau, Farnham Royal Slough, U.K.

Roskam, J.C. and Zandee, M., (this volume). Phylogenetic analysis of quantitative and qualitative characters of gall-inducing midges and historical relation to their hosts: *Dasineura* (Diptera: Cecidomyiidae) on Rosaceae and monocots.

Schluter, D., 1984. Morphological and phylogenetic relationships among the Darwin's Finches. *Evolution*, 38: 921-930.

Schluter, D., 1988. Estimating the form of natural selection on a quantitative trait. *Evolution*, 42: 849-861.

Shaw, R.G., 1991. The comparison of quantitative genetic parameters between populations. *Evolution*, 45: 143-151.

Simon, C., 1983. Morphological differentiation in wing venation among broods of 13- and 17-year periodical cicadas. *Evolution*, 37: 104-115.

Simon, C., (this volume). Discriminant analysis of year classes of periodical cicada based on wing morphometric data enhanced by molecular information.

Simpson, G.G., 1944. Tempo and Mode in Evolution. Columbia University Press, New York, 237 pp.

Simpson, G.G., 1953. The Major Features of Evolution. Columbia University Press, New York, 434 pp.

Smith, H.F., 1936. A discriminant function for plant selection. Annals of *Eugenics*, 7: 240-250.

Sneath, P.H.A. and Sokal, R.R., 1973. Numerical Taxonomy, the Principles and Practice of Numerical Classification. W.H. Freeman and Company, New York, 573 pp.

Sorensen, J.T., 1987a. Multivariate statistical approach to deduction of phylogeny within *Essigiella* [sic] (Aphididae: Lachninae). In: J. Holman, J. Pelikan, A.G.F. Dixon and L. Weismann (Editors), Population Structure, Genetics and Taxonomy of Aphids and Thysanoptera, Proceedings of International Symposia, held at Smolenice, Czechoslovakia, September 9-14, 1985. SPB Academic Publishing, The Hague, pp. 243-260.

Sorensen, J.T. 1987b. The multivariate evolution and taxonomic analysis of leafhopper biotypes and species complexes: use of character correlations and quantitative genetics methods. In: M.R. Wilson and L.R. Nault (Editors), Proceedings of 2nd International Workshop on Leafhoppers and Planthoppers of Economic Importance, Brigham Young University, Provo, Utah, USA, 28th July - 1st August 1986. CAB International Institute of Entomology, London, pp. 217-234.

Sorensen, J.T., 1990. Taxonomic partitioning of discrete-state phylogenies: relationships of the aphid subtribes Eulachnina and Schizolachnina (Homoptera: Aphididae: Lachninae). *Annals of the Entomological Society of America*, 83: 394-408.

Sorensen, J.T., 1991. Phylogenetic character responses for shape component variance during the multivariate evolution of eulachnine aphids: redescription of *Pseudessigella* Hille Ris Lambers (Homoptera: Aphididae: Lachninae). *Pan-Pacific Entomologist*, 67: 28-54.

Sorensen, J.T., (this volume). The use of discriminant function analysis for estimation of phylogeny: partitioning, perspective and problems.

Sorensen, J.T., (in press). Deciphering biological groupings in the *Essigella knowltoni* complex (Homoptera: Aphidinea: Lachnidae), and comparison to variation in *Pinus contorta* (Coniferae: Pinaceae). *Entomologica Generalis*, 17 (1).

Sorensen, J.T. and Sawyer, S.M., 1989. Assessing the multivariate evolutionary responses of phenological differentiation for sibling species: biosystematics in the *Macrosteles fascifrons* complex (Homoptera: Cicadellidae). *Annals of the Entomological Society of America*, 82: 250-261.

Turelli, M., 1984. Heritable genetic variation via mutation-selection balance: Lerch's Zeta meets the abdominal bristle. *Theoretical Population Biology*, 25: 138-193.

Turelli, M., 1988. Phenotypic evolution, constant covariances, and the maintenance of additive variance. *Evolution*, 42: 1342-1347.

Wade, M.J. and Kalisz, S., 1990. The causes of natural selection. *Evolution*, 44: 1947-1955.

Wagner, G.P., 1989. Multivariate mutation-selection balance with constrained pleiotropic effects. *Genetics*, 122: 223-234.

Williams, W.T. and Dale, M.B, 1965. Fundamental problems in numerical taxonomy. *Advances in Botanical Research*, 2: 35-62.

Willig, M.R., Owen, R.D. and Colbert, R.L., 1986. Assessment of morphometric variation in natural populations: the inadequacy of the univariate approach. *Systematic Zoology*, 35: 195-203.

Willis, J.H., Coryne, J.A. and Kirkpatrick, M., 1991. Can one predict the evolution of quantitative characters without genetics? *Evolution*, 45: 441-444.

Wood, T.K. and Pesek, J.D., (this volume). Pronotal shape: a source of confusion or panacea in systematic studies of treehoppers (Homoptera: Membracidae)?

Wright, S., 1931. Evolution in Mendelian populations. *Genetics*, 16: 197-195.

Wright, S., 1932. The roles of mutation, inbreeding, crossbreeding, and selection in evolution. *Proceedings of the First International Congress of Genetics*, 1: 356-366.

Wright, S., 1940. Breeding structure of populations in relation to speciation. *The American Naturalist*, 74: 232-248.

Wright, S., 1977. Evolution and the Genetics of Populations, Volume 3. Experimental Results and Evolutionary Deductions. University of Chicago Press, Chicago, 622 pp.

Yamada, Y., 1977. Evaluation of the culling variate used by breeders in actual selection. *Genetics*, 86: 885-899.

Young, S.S.Y. and Weiler, H., 1960. Selection for two correlated traits by independent culling levels. *Journal of Genetics*, 57: 329-338.

Zeng, Zhao-Bang, 1988. Long-term correlated response, interpopulation covariation, and interspecific allometry. *Evolution*, 42: 363-374.

NATURAL SELECTION AND MORPHOMETRICS

Bernard J. Crespi

Department of Biological Sciences, Simon Fraser University, Burnaby, British Columbia V5A 1S6, Canada

Abstract

Character correlations plague the study of natural selection, systematics, and morphometrics. Path analysis, Wright's method for integrating statistical parameters and biological insight into hypotheses of causation, compels explicit analyses of within-group covariance structure and thereby facilitates discovery of the phenotypic variables interacting among themselves and interacting with the environment. Multivariate techniques such as multiple regression, principal components and discriminant function usually fail to treat character correlation in a biologically realistic manner. As a result, these methods are useful primarily for optimal prediction, recognition of the presence of previously unknown groups, and sorting of phenotypes into already known groups. Conflict between interdependence of phenotypic characters and the independence assumptions of linear statistics seriously impedes causal analyses of morphology using any sets of one-dimensional characters. The only logical escape from this dilemma is the use of new methods that are explicitly and directly two or three dimensional.

1 Introduction

"How to develop methods to remove the effects of character correlation is perhaps the most important unsolved problem facing phylogenetic inference" (Felsenstein, 1982: 386). The phenotypic variation in morphology that we observe among and within species is caused by the interaction of genetics with environments over time. The interaction comprises two parts: first, phenotypes and environments interact to cause variation in reproductive success by the process of natural selection; and second, genotypes and environments interact to cause development. Three disciplines, morphological systematics, morphometrics and quantitative genetics, seek to measure and explain the patterns that these processes generate. The goal of this paper is to discuss the importance of natural selection analysis in drawing these fields together, and to consider the disparate methodologies that may thwart such a synthesis. I suggest here that Sewall Wright's method of path analysis may be useful in translating between the languages and aims of these fields, by prompting each to explain rather than describe morphology.

The paper defines morphology, discusses methods for the analysis of natural selection on continuous morphological characters, and analyzes the relevance of natural selection measurement in systematics.

2 Morphology

I define a morphological phenotype as a group of cells, caused by development, in one temporal and three spatial dimensions. Phenotypes are, therefore, purely biological entities that are, in principle, unmeasurable for mathematical representation. Faced with such hopeless complexity, we measure "characters," which I define as measured distances between two points that are homologous across a series of individuals. Characters are "designed" as input for the parametric machinery of statistics. In measuring characters, we have gained in mathematical tractability but have lost in biological relevance, because character choice is necessarily based on a more or less arbitrary compromise between measurability and a priori expectations of importance for any given investigator's purpose. Moreover, morphological characters are usually highly correlated among themselves, as a result of their joint developmental causes (e.g., Riska, 1986), and characters may often interact in their effects on fitness. Morphological "factors" (sensu Bookstein et al., 1985) mediate between unmeasurable phenotypes and overly simplistic, correlated characters. Factors are linear combinations of characters that are intended to represent and explain the causes of character correlation. Factors such as "size" and "shape" are hybrids between biology and statistics, and depend for their existence and usefulness upon the integration of causal biological knowledge with consistent patterns or homogeneity of statistical coefficients.

Analyzing morphological phenotypes and their evolution requires admitting both the vagaries of measuring phenotypes linearly, and the subjectivity of both characters and factors. The primary symptom of the unresolved tension between biological geometry and linear statistics is the presence of character correlation. On a proximate level, character correlation in a phenotype, **P**, is caused by genetic correlation (due to pleiotropy and linkage disequilibrium), as expressed by the genetic variance-covariance matrix, **G**, environments, expressed by the matrix **E**, and genotype-environment interaction. These causes are usually inaccessible in natural populations, although their effects are not. Ultimately, characters and their correlation are caused by natural selection, involving varying degrees of optimization for varying aspects of phenotypes, and by constraints, defined as the aspects of developmental systems that bias the range of phenotypes that selection can produce (Maynard Smith et al., 1985). Of these causes, only natural selection can be measured simply and fairly easily for many organisms. If we detect selection for certain aspects of phenotypes and their functional relationships, then we may be able to explain phenotypic change or variation as a more or less optimal expression of certain aspects of environmental variation. By contrast, consistent failure to measure selection leads directly into the labyrinths of developmental genetics and genetic drift. Finally, to the extent that the varying processes of selection within species are expressed as among-species differences, analyzing selection may be useful in developing models for phylogenetic inference.

3 Measuring Selection on Morphology

There are two approaches to the measurement of selection on morphology. Lande and Arnold (1983) assume that the measured characters included in a selection analysis affect fitness independently and completely. They model selection by multiple linear regression, whereby the partial regression coefficients (selection "gradients") represent expected changes in fitness caused by unit changes in each character, with all of the other characters remaining constant. Whenever characters are correlated for biological reasons, and whenever characters that are subject to selection are not included in the analysis, the coefficients are more or less incorrect (see Mitchell-Olds and Shaw, 1987; Crespi and Bookstein, 1989; Crespi, 1990). Lande and Arnold (1983) suggest the use of principal components analysis for highly correlated (multicollinear) data sets and argue that partial resolution of selection processes is better than none at all. They further justify the use of multiple regression analysis by appeal to one of its ends: usefulness in quantitative genetic models for the prediction of multivariate evolutionary change.

Crespi and Bookstein (1989) argue that, for most morphometric data sets, violation of the assumption of multiple regression vitiates the analysis. They assume that selection occurs on unmeasured (but estimable) size and shape factors, and that characters are correlated usually as a result of joint dependence on a factor for general-size. Thus, rather than holding constant the full covariance matrix of the characters to compute selection coefficients using multiple regression, Crespi and Bookstein's method holds constant only general-size, the explicable aspect of character correlation. A path diagram depicting their model is presented in Fig. 1B. Adjusting the characters only for general-size avoids the problem of multicollinearity by recognizing its usual cause. The problem of "missing" characters is mitigated in two ways. First, because only general-size is adjusted for in computing the selection coefficients, the coefficients are much more robust than multiple regression coefficients to the inclusion and exclusion of characters and a very large number of characters can be included in the analysis (contra, see Kingsolver and Schemske, 1991). In the temporary relative absence of information about the causes and objects of selection, measuring a multitude of characters increases the likelihood of having measured aspects of the phenotype relevant to selection. Second, the selection coefficients from this model are considered not as the "force" of selection on each character, but as imperfect indicators of what the true objects of selection may be. Statistical information in the sign and magnitude of the selection coefficients is integrated with biological insight concerning possible causes of selection, to derive hypotheses of selection on size and shape factors indicated by subsets of the characters with similar coefficients and an explanation for their "selective covariance" (Felsenstein, 1988). By constructing such factors and searching for measured characters that approximate them closely, the model is continually improved by reducing violation of the assumptions of independent and missing character effects.

Crespi and Bookstein's (1989) model uses the philosophy and statistics of path analysis, Wright's (1934, 1954) method for integrating biological and statistical information, to explain causal relations among sets of measured and estimated variables. Path analysis involves developing a hypothesis of causation for each set of data. This hypothesis is depicted in a path diagram,

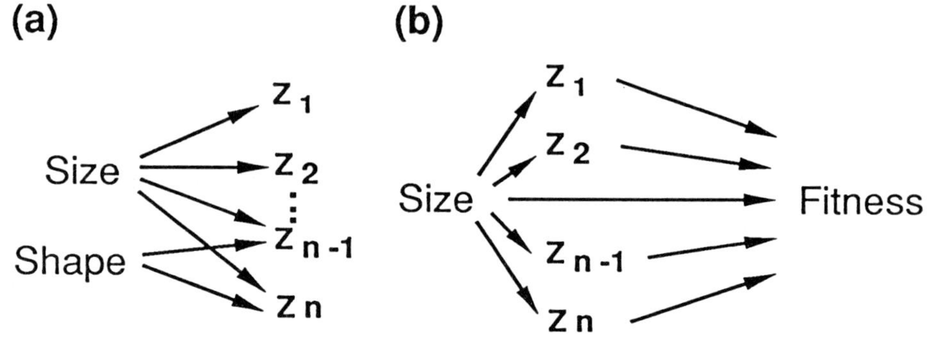

Figure 1. Path models for (A) explaining character covariances within groups by Wright's (1954) method, and (B) measuring selection on morphology by the method of Crespi and Bookstein (1989). Lande and Arnold's (1983) approach to within-group analysis is to consider characters as the column vector z, and thus ignore the causes of character covariance; their approach to selection analysis is to adjust for the full covariance matrix of the characters themselves, rather than for the only usual explicable aspect of character covariances, general-size.

which is comprised of two types of variable, measured characters and unmeasured factors, each standardized to mean zero and unit variance. Pairs of variables are connected by two types of relation, directional paths, which represent hypothesized causes whose importance (relative to other causes) is evaluated by the attached regression (path) coefficient, and symmetrical paths, double-headed arrows of correlation which connect pairs of ultimate determining variables and admit our current ignorance of their joint causes (see Li, 1975).

A path-analytic approach seeks neither to optimize nor maximize any particular quantity, in contrast to the standardized procedures of such methods as multiple regression, discriminant function, and principal components. Instead, path modeling seeks a compromise between statistical information, revealed in patterns of bivariate and multivariate covariance, and biological knowledge of causes, derived from the history of experimentation and observation across the levels of genetics, development, physiology and behavior.

The primary advantage of path modeling for selection studies is that it provides a robust interactive method for determining the objects of selection. Such robustness comes, however, with a cost in subjectivity: each model for each data set must be justified step by step. The primary advantage of a single fixed model of selection, such as multiple regression, is that it serves as a touchstone for comparing results of diverse studies. The cost of such generality is the extremely restrictive assumptions of completeness and independence, whose violation leads to results that are not simply partial, but also incorrect.

The multiple regression and path analysis approaches to measuring selection represent an extension of the divergent genetic views of Fisher and Wright to the level of phenotypes. Thus, Fisher's deterministic, predictive models of change in independent genes correspond closely to Lande and Arnold's predictive models of independent character evolution. Conversely, Wright's view of genetics stresses the importance of gene interaction, and the uncertainty of prediction in a shifting landscape of genotypes and environments. Similarly, path-analytic modeling of selection seeks to uncover character interaction, and to uncover and explain causal processes of change rather than assuming and extrapolating from their form (Crespi, 1990). In this context, the two approaches are complementary. Path modeling of selection is designed to continually reduce violation of the assumptions of character independence and completeness in revealing the objects of selection. Such variables can then be used in the predictive quantitative genetic framework of Lande and Arnold. This type of combined analysis within species may, in time, lead to a thorough understanding of the processes of selection which, along with drift, constraints and vicariance, have produced the differences among taxa that are the focus of systematics.

4 Systematics and Selection

I consider here two of the primary purposes of systematics: identifying groups, and describing and quantifying differences among groups with respect to branch order and length in a phylogenetic hypothesis. Many systematic studies using continuous morphological characters focus at the species level and below. I present here a caricature of the use of "conventional" morphometrics (e.g., Reyment et al., 1984) in such among-group identification, description, and quantification. The caricature is intended to highlight the usual absence of methodology incorporating models of hypotheses of biological causation from morphometric analysis in systematics.

Groups, be they putative species, subspecies, ecological types or collection sites, are either recognized a priori with more or less confidence or sought. In either case, we measure a variable number of homologous linear distances, often choosing such characters for their desired high levels of variation among groups rather than within. The characters are then ordinated. Without a priori groups, we may conduct a principal components analysis and see if definable groups appear on any of our new axes. With a priori groups, we may calculate a discriminant function separating them "optimally," using either the original characters or some useful combination or subset. In any case, we determine if and how groups exist and differ by optimizing or maximizing some quantity derived from a chosen set of characters. We may then interpret the results of these statistical gymnastics with some hypothesis of biological cause, such as those leading to genetic or ecological differentiation.

I argue here that such analyses have only two logical purposes: methods such as principal components analysis may be useful in recognizing the existence of previously unknown groups, and methods such as discriminant function analysis may be useful for classifying unknown individuals into one of several known groups. In each case, I contend that nearly all analyses and interpretations beyond these purposes, such as interpretation of coefficients or axes in biological terms or generation of evolutionary "distances" between groups, are invalid.

There are two reasons for such a nihilistic view of morphometrics in systematics. First, no causal information has entered into the procedures; we have used no causal model based in genetics or development for "analyzing" the variation and covariation either within or among groups. Moreover, there is as yet little or no evidence in favor of any model that we might use. Second, we have treated character correlation in one or two hopelessly artificial ways, each diametrically opposed to the truism that phenotypes comprise sets of functionally related traits. Using methods such as principal components analysis, we have collapsed character correlation onto a small set of orthogonal axes that sequentially maximize variance. Although the first (largest) principal component often behaves as a general-size factor, subsequent principal components are biologically arbitrary and may usually be quite meaningless; simply put, phenotypes cannot be modeled as a set of orthogonal axes (Bookstein et al., 1985; Mitchell-Olds and Shaw, 1987). Using methods such as discriminant function analysis, we have divided out character correlation, by computing a linear combination of characters whereby the hypothesized groups vary by the greatest multiple of their within-group variation. But we know, of course, that discrete groups of phenotypes have not evolved in the context of maximizing their differences for the set of characters that we happen to measure. To circumvent these two dilemmas, we need to understand rather than collapse or divide out the within-group covariances of characters. A path-analytic approach is well suited for this goal.

Analyzing within-group covariance structure involves devising a path model hypothesis depicting the covariance among measured characters and factors in causal terms. The path diagram is a model of the phenotype. Morphometric phenotypes usually comprise three types of variation: joint variation in all characters (i.e., general-size variation), joint variation in subsets of the characters which is independent of general-size ("shape" variation), and variation in each character uniquely (here, a form of error variation). These levels of global, local and unpartitioned variation correspond to Wright's (1932) terminology of general, group and special size factors, and we can use his method of morphometric factor analysis to model phenotypes at these levels. This method involves: (1) computing a general-size factor from the correlation matrix of the characters with self-correlations omitted; (2) computing the residual correlations between each pair of variables, as the simple correlations minus the product of their factor loadings on size; (3) inspecting the matrix of residual correlations to see if they form any patterns with respect to the presence of shape variation (e.g., a positive correlation for two or more functionally-related characters or two or more characters in geometric proximity on the form; and (4) sequestering these explicable correlations from the matrix, recomputing general-size, and deriving a path model for each particular set of data (see Wright, 1954; Bookstein et al., 1985; Crespi, 1990).

The result of the modeling is thus a set of size and shape factors that represent our best approximations to "independent" phenotypic variables. Of what use is such a model of phenotypes in systematics? First, to the extent that **G** and **P** are similar (see Cheverud, 1988; Willis et al., 1991), recognition of independent phenotypic variables will help in recognizing genetically independent variables. (Genetic correlation matrices may be factored in the same way as those of phenotypes, but the correlations may be less likely to range only between the values of zero and one which permit the analysis). Genetically independent characters are expected to evolve independently across lineages (assuming a constant **G**), and use of such variables reduces the problem of character correlation and "too many degrees of freedom" whatever our model of phenotypic change (see also Schaffer, 1986). Second, path models of phenotypic covariation may be useful in comparing closely related taxa to assess how they differ. For such groups, we can compute a pooled within-group size factor and observe the pattern of sign and magnitude for the size-adjusted mean differences between groups, in relation to independently-derived models of the phenotypes (cf., Humphries et al., 1981; Bookstein et al., 1985). The groups may differ in size, shape, or both, and each group may or may not vary for these same factors; interpreting the correspondence of within and among-group covariation patterns requires causal information from the study of genetics, development, and selection. Third, we can measure selection in one or more groups of a taxonomic assemblage, and relate the pattern of selection coefficients derived from the Crespi-Bookstein (1989) path model to the patterns of within and among-group size and shape factors. This information can then be merged. For example, if selection occurs on one aspect of shape, and if such selection is responsible for an among-species difference, then we might expect: (1) evidence, preferably both genetic and phenotypic, for the homologous shape factor within each species; (2) evidence for selection on this factor within species, recognized by a consistent pattern in the selection coefficients for the characters comprising the factor; and (3) evidence for a species or population difference in the magnitude of the factor.

To the extent that these lines of evidence converge, morphological evolutionary change may be predictable from incomplete data. For example, Bumpus's (1899) female house sparrows exhibit shape factors for the leg and head, and the selection coefficients are similar for each set of characters comprising each shape factor (Crespi, 1990); the possibility of subspecific differentiation for these factors awaits analysis. To the extent that the lines of evidence noted above diverge, we are forced to recognize the unpredictability of morphological change, and turn research effort toward the less spectacular goal of explanation.

The methodology of path analysis and selection measurement thus has several important uses in morphological systematics. First, path analysis of within-group covariance facilitates the recognition of characters or factors that may evolve independently. Path modeling, therefore, serves as a tool for choice of characters that fit the assumptions of phylogenetic models of change. Second, path analysis to determine the presence and objects of selection may tell us which phylogenetic model is appropriate for which type of character; that is, whether selection or drift is most credible explanation for explaining among-group variation in different types of phenotypic variable in any given clade. For example, Lande's (1979) model for "selective transformation" among taxa assumes a primacy of selection, in addition to assuming constant **G**, no **G** by **E** interaction, and no effects of novel environments on change in **G** or **P**.

Theoretical and empirical evidence for these assumptions is equivocal at best (see Lofsvold, 1986; Mitchell-Olds and Rutledge, 1986; Mitchell-Olds and Shaw, 1987; Turelli, 1988; Zeng, 1988; Gillespie and Turelli, 1989). If, however, we are willing to build inferences upon such weak assumptions, we can compute "selective differences" between taxa and infer past selection (e.g., Schluter, 1984; Sorensen, 1987, this volume; Sorensen and Sawyer, 1989; Sorensen and Foottit, this volume). Conversely, under a model of character change by drift (e.g., Bookstein, 1988), analogous measures of distance can be computed given enough independently-evolving characters and used these to infer phylogeny (Lynch, 1989). If, however, we are unwilling to accept approximation to either extreme, quantitative morphological characters are not useful for statistical phylogenetic inference, because there is an infinity of models for their evolution.

5 Conclusions

There are three possible escapes from the problem of character correlation that follows from ignorance of character causes. The first is to escape from the problem, using nucleotide data for which explicit statistical models of change are available and robust (e.g., Lake, 1987; Lewontin, 1989). Evolutionary lineages are, after all, lineages of DNA only, painstakingly encoding aspects of the environment. Arguments that such analyses are impractical with many taxa lose force with each advance in DNA technology. The second is to escape the characters of morphometrics, by using cladistic methods with morphology and other types of phenotypic variables. Although we gain here in assessing polarity and having a general model of discrete differences based in parsimony or compatibility, ignorance of causes of character correlation may still bias the results (reviewed in Felsenstein, 1982, 1983). The third escape is the aforementioned attempts to avoid correlation by using maximally independent variables, though some realistic model of their evolution must then be assumed.

Regardless of philosophy or method, it is impossible to determine how morphology evolves using phylogenies derived from morphological data (Felsenstein, 1988). Given phylogenies inferred from molecular variation, and using causal data from the analysis of natural selection and the processes underlying quantitative genetic variation (cf., Atchley, 1983), we can assess the causes of morphological change. The problem of character correlation can be solved, in part, by the use of path-analytic methods, and, perhaps, nearly in its entirety by developing an explicitly geometrical morphometrics amenable to parametric statistics. Bookstein's methods of analyzing size and shape directly in two dimensions by two-point registration of triangles (Bookstein et al., 1985) and principal warps (Bookstein, 1989, in press) are first leaps in this direction.

6 References

Atchley, W.R., 1983. Some genetic aspects of morphometric variations. In: J. Felsenstein (Editor), Numerical Taxonomy, NATO Advanced Study Institute, Series G, Number 1. Springer-Verlag, New York, pp. 346-363.

Bookstein, F.L., 1988. Random walks and the biometrics of morphological characters. *Evolutionary Biology*, 23: 369-398.

Bookstein, F.L., 1989. Principal warps: thinplate splines and the decomposition of deformations. *I.E.E.E. Transactions of Pattern Analysis*

and Machine Intelligence, 11: 567-585.
Bookstein, F.L., (in press). Uniform factors and relative warps: a feature space of shape variation for landmark data. *Annals of the Institute for Statistics and Mathematics.*
Bookstein, F.L., Chernoff, B., Elder, R.L., Humphries, J.M., Smith, G.R., and Strauss, R.E., 1985. Morphometrics in Evolutionary Biology. Special Publication, 15. Academy of Natural Sciences, Philadelphia, 277 pp.
Bumpus, H., 1899. The elimination of the unfit as illustrated by the introduced house sparrow, *Passer domesticus*. *Biological Lectures, Woods Hole Marine Biological Station*, 6: 209-226.
Cheverud, J.M., 1988. A comparison of genetic and phenotypic correlations. *Evolution*, 42: 958-968.
Crespi, B.J., 1990. The measurement of selection on phenotypic interaction systems. *The American Naturalist*, 135: 32-47.
Crespi, B.J. and Bookstein, F.L., 1989. A path-analytic model for the measurement of selection on morphology. *Evolution*, 43: 18-28.
Felsenstein, J., 1982. Numerical methods for inferring evolutionary trees. *Quarterly Review of Biology*, 57: 379-404.
Felsenstein, J., 1983. Methods for inferring phylogenies: a statistical view. In: J. Felsenstein (Editor), NATO Advanced Study Institute, Series G, Number 1. Springer-Verlag, New York, pp. 315-334.
Felsenstein, J., 1988. Phylogenies and quantitative characters. *Annual Review of Ecology and Systematics*, 19: 445-471.
Gillespie, J.H. and Turelli, M., 1989. Genotype-environment interactions and the maintenance of polygenic variation. *Genetics*, 121: 129-138.
Humphries, J.M., Bookstein, F.L., Chernoff, B., Smith, G.R., Elder, R.L. and Poss, S.G., 1981. Multivariate discrimination by shape in relation to size. *Systematic Zoology*, 30: 291-308.
Kingsolver, J. and Schemske, D., 1991. Path analyses of selection. *Trends in Ecology and Evolution*, 6: 276-280.
Lake, J.A., 1987. Determining evolutionary distances from highly diverged nucleic acid sequences: operator metrics. *Journal of Molecular Evolution*, 26: 59-73.
Lande, R., 1979. Quantitative genetic analysis of multivariate evolution, applied to brain:body size allometry. *Evolution*, 33: 402-416.
Lande, R. and Arnold, S.J., 1983. The measurement of selection on correlated characters. *Evolution*, 37: 1210-1226.
Lewontin, R.C., 1989. Inferring the number of evolutionary events from DNA coding sequence differences. *Molecular Biology and Evolution*, 6: 15-32.
Li, C.C., 1975. Path Analysis - A Primer. Boxwood Press, Pacific Grove, California, 347 pp.
Lofsvold, D., 1986. Quantitative genetics of morphological differentiation in *Peromyscus*. I. Tests of the homogeneity of genetic covariance structure among species and subspecies. *Evolution*, 40: 599-573.
Lynch, M., 1989. Phylogenetic hypotheses under the assumption of neutral quantitative-genetic variation. *Evolution*, 43: 1-17.
Maynard Smith, J., Burian, R., Kaufmann, S., Alberch, P., Campbell, J., Goodwin, B., Lande, R., Raup, D. and Wolpert L., 1985. Developmental constraints and evolution. *Quartery Review of Biology*, 60: 265-187.
Mitchell-Olds, T. and Rutledge, J.J., 1986. Quantitative genetics in natural plant populations: a review of the theory. *The American Naturalist*,

127: 379-402.
Mitchell-Olds, T., and Shaw, R.G., 1987. Regression analysis of natural selection: statistical inference and biological interpretation. *Evolution*, 41: 1149-1161.
Reyment, R.A., Blackith, R.E. and Campbell, N.A., 1984. Multivariate Morphometrics, (Second Edition). Academic Press, New York, 233 pp.
Riska, B., 1986. Some models for development, growth, and morphometric correlation. *Evolution*, 40: 1303-1311.
Schaffer, H.B., 1986. Utility of quantitative genetic parameters in character weighting. *Systematic Zoology*, 35: 124-134.
Schluter, D., 1984. Morphological and phylogenetic relations among the Darwin's finches. *Evolution*, 38: 921-930.
Sorensen, J.T., 1987. Multivariate statistical approach to deduction of phylogeny within *Essigiella* (sic) (Aphididae: Lachninae). In: J. Holman, J. Pelikan, A.G.F. Dixon and L. Weismann (Editors), Population Structure, Genetics and Taxonomy of Aphids and Thysanoptera, Proceedings of International Symposia, held at Smolenice Czechoslovakia, September 9-14, 1985. SPB Academic Publishing, The Hague, pp. 243-260.
Sorensen, J.T., (this volume). The use of discriminant function analysis for estimation of phylogeny: partitioning, perspective and problems.
Sorensen, J.T. and Foottit, R.G., (this volume). The evolutionary quantitative genetic rationales for use of ordination analyses in systematics: phylogenetic implications.
Sorensen, J.T. and Sawyer, S.M., 1989. Assessing the multivariate evolutionary responses of phenological differentiation for species: biosystematics in the *Macrosteles fascifrons* complex (Homoptera: Cicadellidae). *Annals of the Entomological Society of America*, 82: 250-261.
Turelli, M., 1988. Phenotypic evolution, constant covariances, and the maintenance of additive variation. *Evolution*, 421: 1342-1347.
Willis, J.H., Coryne, J.A. and Kirkpatrick, M., 1991. Can one predict the evolution of quantitative characters without genetics. *Evolution*, 45: 441-444.
Wright, S., 1932. General, group and special size factors. *Genetics*, 17: 603-619.
Wright, S., 1934. The method of path coefficients. *Annals of Mathematics and Statistics*, 5: 161-215.
Wright, S., 1954. The interpretation of multivariate systems. In: O. Kempthorne, T.A. Bancroft, J.W. Gowen, J.L. Lush (Editors), Statistics and Mathematics in Biology. Iowa State University Press, Ames, Iowa, pp. 11-33.
Zeng, Zhao-Bang, 1988. Long-term correlated response, interpopulation covariation, and interspecific allometry. *Evolution*, 42: 363-374.

THE USE OF DISCRIMINANT FUNCTION ANALYSIS FOR ESTIMATION OF PHYLOGENY: PARTITIONING, PERSPECTIVE AND PROBLEMS

John T. Sorensen

Insect Taxonomy Laboratory, California Department of Food and Agriculture, Sacramento, California 95814, U.S.A.

Abstract

This article, as a continuation of Sorensen and Foottit (this volume), discusses how discriminant function analysis can be implemented for phylogenetic estimation. The partitioning of variance within continuous characters is discussed, as are the problems of relative perspective in multivariate evolution. Procedures for handling distortion problems between global and local perspectives are suggested. Problems that are inherent in the use of morphological data for phylogenetic estimation are noted. For a given data set, and its subsets, trees that were created using both Manhattan and Maximum-likelihood networking algorithms are reviewed and contrasted. Discriminant function analysis is concluded to be useful for phylogenetic estimation with certain precautions. Recommendations for this use are provided.

1 Introduction

Given that the assumptions and rationales of the model of multivariate evolution (sensu Lande, 1979, 1988) that are discussed in Sorensen and Foottit (this volume) are operationally reasonable, this chapter briefly attempts to portray, in a geometric sense, how variance can be partitioned from the distribution of a continuous character, and how discriminant function analysis (DFA) can be used for phylogenetic estimation. Please read the chapter on quantitative genetic rationales (Sorensen and Foottit, this volume) before this chapter, which is written as a continuation; I will not attempt to redefine items concerning the multivariate evolution model that are discussed adequately there.

2 Partitioning the Variance of Continuous Characters

The variance in a given morphometric trait of Gaussian distribution, can be partitioned in a number of ways. The most frequently mentioned variance partitioning in the literature involves many controversial statistical criteria and methods to completely partition multivariate "shape" from "size" in data matrices (e.g., Burnaby 1966a; Humphries et al. 1981; Bookstein et al. 1985; Bookstein 1986, 1989; Somers 1986, 1989; Rohlf and Bookstein 1987; Sundberg 1989; Strauss, this volume). Most methods appear to adulterate the generated evolutionary space to some degree, such as shearing or otherwise rotating the generated vectors from orthogonality. If orthogonality is to be maintained, however, the task is difficult. Frequently, the first generated vector of a principal components analysis (PCA) can be regarded as chiefly reflecting general-size, if its loadings and correlations are high and positive (Sneath and Sokal, 1973); subordinate vectors are then taken as shape portrayals, although the separation is usually incomplete, at least among the first few vectors. To maintain orthogonality in accord with Lande's (1979) model, Sorensen (1991) conservatively chose this PCA approach, but necessarily at the cost of incomplete separation of size. In any event, once orthogonality is breached, it may be difficult to establish or conceptualize exactly how size and shape are partitioned, because these are abstract items at best, and are, therefore, relatively poor examples for visualization.

Other types of partitions are possible, however, which might be easier to conceptualize for demonstration. For example, consider the length of the tibia in an adult human individual. Because $\mathbf{P} = \mathbf{G} + \mathbf{E}$ (as the phenotypic, genotypic and environmental covariance matrices, respectively), the measured length of the bone has both a genotypic and an environmental component; the latter determined by environmental factors that have modified the genotypic component to express the given phenotype (e.g., abundant food or good health, or starvation or disease, during juvenile development). This is readily apparent if genetically identical twins fare differently during development, and show different tibial lengths at maturity; no matter how minor these differences might be, they are due to **E**. Operationally, of course, it is very difficult, if not impossible, to perfectly partition out **E**.

Similarly, at a populational level, variance in a morphometric attribute also can be partitioned into various components. Consider the tibia again, but this time in a population of 100 adult humans: 50 males and 50 females. If we take the distribution of tibial length for all 100 adults, we get a range of variance for the overall population that equals the spread between the extremes of its largest and smallest members (for the sake of the example, a male and female, respectively). Part of this variance is sexual in nature, for if we consider separately the 50 males and the 50 females, each sex will probably have a different mean and distribution, with the mean for males being usually larger than that for females. In essence, we have a sexual dimorphism (of variance distribution and mean) at the populational level for this morphometric trait, although this dimorphism is not as apparent as it would be for a more conventional, sexually dimorphic trait, such as a beard or bald head.

In the example, if the spread of variances between the sexes in such a sexually differing trait were similar, but displaced in their ranges and means, then that part of variance in the female population that was below the minimum distribution tail of the male curve could be considered to be "female

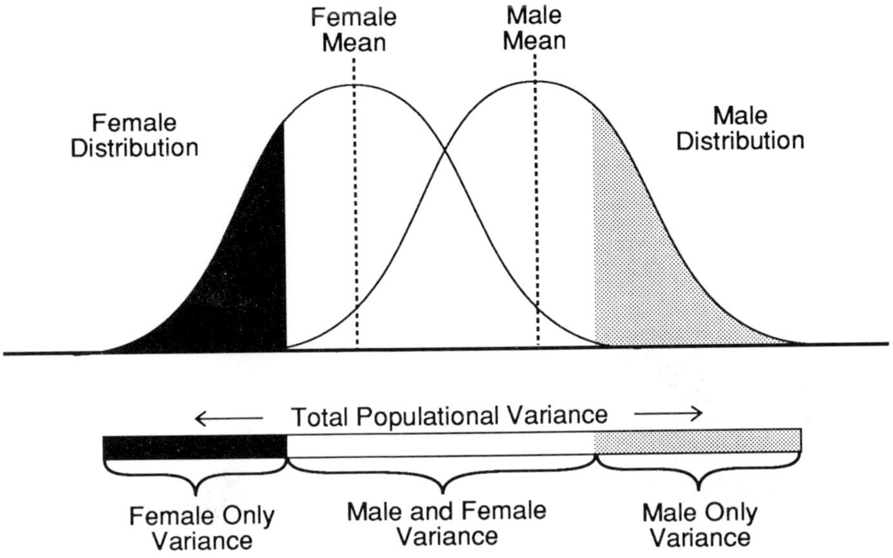

Figure 1. An example of how sexual variance can be partially partitioned from a continuous trait that is thought to be nonsexually dimorphic in the conventional sense. The trait (e.g., tibial length) varies in mean and (overlapping) variance distribution between males and females of a population. That portion of the trait's distribution for each sex that is beyond the distribution for the opposite sex can be though of as being sexually unique, and thus dimorphic. The overlapping variance is attributable to both sexes.

only" in origin, and similarly that part of the variance in the male population that was above the maximum distribution tail of the female curve would be "male only." The range of overlapping variance for both sexes that occurs between these extremes would be "male and female variance" (Fig. 1).

Although the above example was univariate, extrapolation to a multivariate situation, with population centroids representing the mean populational phenotype (sensu Lande, 1979), can easily occur. Using DFA, Sorensen and Sawyer (1989) employed this concept of sexually partitioned variance to separate the sexes (as well as species and phenological differences) between sibling species of leafhoppers based upon continuous morphometric traits of the legs and wings that showed no apparent sexual dimorphism in a conventional sense.

Now, quite aside from this example of sexual variance, consider how apomorphic variance can be partitioned from either a univariate, continuous attribute or from multivariately based population variance. Fig. 2 shows the univariate means (or multivariate centroids) and Gaussian distributions of three populations, along an axis (or evolutionary vector, as $\nabla \ln \bar{W}$). The center population, P (plesiomorph), represents the ancestral group from which daughter populations A1 and A2 (apomorphs) have evolved along the axis (vector). As such, the distribution of variance for P can be considered to represent the plesiomorphic state (as white in Fig. 2) for populations A1 and A2. The derived conditions for variance in both of the daughter populations, is the apomorphic (black) portion of their variance that is beyond (in either direction) the distribution tails of the P population. The evolution away from P could, of course, also be sequential and occur unidirectionally along the axis (i.e., populations P and A1 could be transposed, with evolution along the vector occurring to the right).

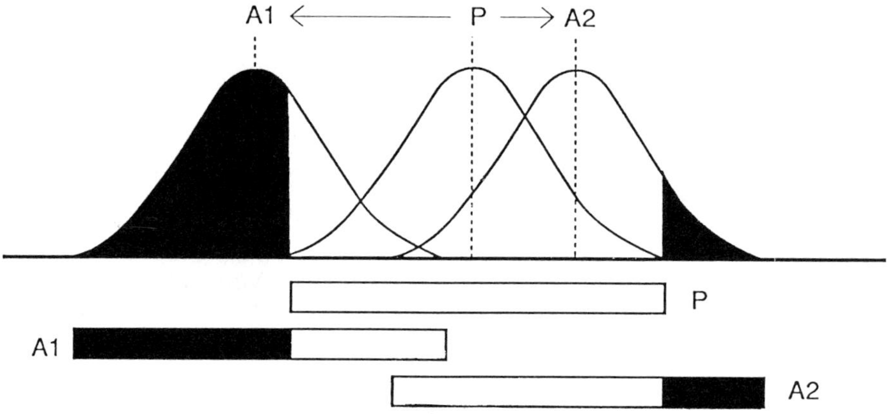

Figure 2. A visualization of how apomorphic variance occurs in, and can be partitioned from, either variance in a univariate, continuous attribute or multivariate variance in a population. The vertical dashed lines represent univariate means (or multivariate centroids) and Gaussian distributions of three traits (or populations), along an axis that acts as an evolutionary vector. P (plesiomorph) represents the ancestral state (or population), which has diverged into states (or daughter populations) A1 and A2 (apomorphs) along the vector. Both the variance of P and the overlapping variance in A1 and A2 represent the plesiomorphy [= white], while the variance in both A1 and A2 that is nonoverlapping with P represents apomorphy [= black]

Plesiomorphy, in this example, is the variance that occurs in common with P. Equally, however, if P were not present in the example, and it was assumed that both A populations were derived from an unseen common ancestor, it would be reasonable (and, in fact, operationally necessary) to assume that the overlapping portion of variance common between both A populations would represent plesiomorphy. This is the "common equals primitive" stance that is sometimes necessary in discrete state cladistics, when lacking other information for character polarization. If this commonly shared information were divided out of a data matrix representing the groups, as DFA does (see Pimentel, this volume), then only apomorphic, divergent variance would remain among the groups involved. Any attempt at constructing a hierarchical portrayal of relationships using such information would necessarily be phylogenetic, especially if cladistic networking algorithms that estimate furcation event points (as ancestors) were used.

3 The Importance of Perspective in Multivariate Evolution

Therefore, when DFA, as a minimum selective mortality index, I, creates a $(\nabla \ln \bar{W})^{T}z$ vector, it is only apomorphic information that is used to determine the evolutionary anagenic distance between groups; communality is divorced in the calculation (unlike PCA). Usually, because no plesiomorphic population is included in a DFA, the distance that is calculated among contemporaneous and monophyletic populations, as members of an in-group, is inferred from the necessarily nearest ancestor common to all of them, as represented by their common, and eliminated, variance. Any depiction of intergroup distances among such in-group members in this "out-group *excluded*" case necessarily occurs in the perspective of local space that is defined by the in-group members alone; only the apomorphic information among them is incorporated into the perspective. The evolutionary space that is so generated is already polarized (devoid of plesiomorphy) to the immediate ancestor (its node if a phyletic network did already exist) for the in-group as a whole, but the "location" of that ancestor, as a historical evolutionary divergence point, in the space is not depicted.

In contrast, inclusion of a plesiomorph representative along with the in-group members, to represent an "out-group *included*" analysis, provides a more-global perspective for viewing the relationships among the in-group centroids; however, this view necessarily imposes separate, out-group based apomorphy on all intergroup distances that are displayed along each vector. This "out-group apomorphy" is the divergence that has been independently evolved by the plesiomorph, since its own evolutionary separation from its own immediate ancestor (or again, that respective node if a network across all groups already existed; that node, of course, is the phyletic furcation event that is immediately "below," in a phylogenetic sense, the common ancestral node for all in-group members). This interpretation of out-group apomorphy does not consider the plesiomorph representative to be a (direct) hypothetical ancestor of the in-group, of the type represented by an array of entirely "O" (plesiomorphic) states in a character set, as in discrete state cladistics, but rather to represent the nearest contemporaneous population. The out-group apomorphy is necessarily defined with reference to the matrix employed and, therefore, the character set chosen to analyze the in-group members. The vectors that are generated in an out-group included analysis are prepolarized to the ancestor (node) common to both the out-group and in-group members.

When several discriminant functions are used to generate an apomorphic phyletic pathway, $t\bar{I}$, between in-group centroids only (out-group excluded case), that pathway is also necessarily created through the local perspective space that is based upon the in-group alone. Unfortunately, in such a case, it is impossible to root the generated network because there is no certainty as to where exactly to place the root (i.e., along which segment, or between which nodes) without a globally based reference.

In contrast, the inclusion of a plesiomorph representative (out-group included case), while necessarily distorting the local evolutionary space (and, therefore, the phyletic pathway) to a degree between the in-group members, has the great advantage that the pathway produced among the in-group centroids is now rooted (the relative location of the ancestral node for the in-group is estimated and inserted upon the network). However, it is the greater phyletic network, which spans between the in-group and out-group, that now has no root. This is not a problem because we know that the root must be placed somewhere along the phyletic segment intervening between the plesiomorph and the node common to all in-group centroids. The actual placement of that root along that segment is somewhat irrelevant: (a) because the relative anagenic distances between the root and plesiomorph terminal, versus the root and nearest nonterminal node (in-group ancestor), will never be resolved (a midpoint placement along the segment seems most appropriate); and (b) because the actual intervening segment is, itself, distorted in the restricted evolutionary space that is defined by the analysis (to be more accurately calculated and rooted, the network would require another next-step-greater analysis, which would include a plesiomorph to all groups contained in the current analysis). (Note that Strauss [in press, this volume] has mapped existing cladograms that have been generated extrinsically, over centroid positions in discriminant space and, if such additional information exists, that approach can also functionally help determine root placement on DFA generated networks.)

4 Problems: Morphological Data Itself?

It should now be noticeable that, using DFA and multivariate evolution rationales, the generation of a phyletic pathway for a set of closely related groups appears to have a problem in trade-offs, depending on whether a more-local (out-group excluded) or more-global (out-group included) perspective is chosen. The local perspective excludes the extraneous information of out-group apomorphy that is contributed by the plesiomorph representative, but at the cost of not knowing where to root the generated local phyletic pathway among the in-group centroids. On the other hand, the more-global perspective roots the in-group pathway, but necessarily distorts, to some degree, the local space (and possibly the pathway) between the in-group members.

This discrepancy should not be perceived as a fault of DFA in this regard, but rather indicates one if its strengths as a model of true evolution; nevertheless, it is operationally troublesome when trying to achieve a most resolute phyletic estimate. The reason for this difference in local versus global portraits is quite simple. When DFA vectors are generated, they encompass the covariance displayed among groups. The greater the distance between centroids in DFA space, the further their evolutionary divergence. DFA functions, as $(\nabla \ln \bar{W})^{T}z$ vectors, simply reflect optimized, minimum selective mortality depictions of the parsimonious evolutionary distance that must be transversed to account for the observed separation of defined groups on PCA vectors, as $\nabla \ln \bar{W}$ gradients of historical *net* natural selection.

Selection gradients change over time during any defined evolutionary episode. Any selection gradient that is imposed upon several, fairly closely related contemporaneous groups seen *today* will necessarily reflect the net selection regime that occurred to separate those groups. The temporal reign of that selection regime starts in the present and goes back to the time of most immediate, common ancestor for those groups, but no further. In DFA this time sequence is definable in the present by the current observations (measurements for the groups), and in the past by a representation of this ancestor, as is calculated by discounting the communality between the groups.

If another group is inserted into a given DFA, as a plesiomorph that is monophyletic with the in-group members, the covariance encompassed by the data matrix obviously changes to include that of the out-group also. The communality among all groups in the analysis is then taken as the most immediate, common ancestor for all treated groups. On a tree, it would be represented by the phyletic node that occurs just prior to the immediate, common ancestral node for the in-group only. The generated net selection gradients now necessarily reflect the reinterpreted net selection regime that spanned all of the additional time that must be accounted for by the more inclusive DFA. This increased evolutionary period is represented on a tree by the phyletic segment that intervenes between the respective common ancestors of the overall groups versus the in-group only, and the segment representing the out-group stem. However, because the more-global DFA only allows generation of a network that is technically unrooted (requires interpreted root placement), both these stem segments are represented on that network by the single segment that occurs between the plesiomorph and the common in-group node.

For the analysis to mean much phylogenetically, three conditions must be met. First, any plesiomorph chosen must, in fact, represent the most immediate sister-group of the in-group. If this condition is not met, the generated net selection gradients must necessarily cover the increased time for the overall net selection regime, back to the period of existence of whatever was the most immediate, common ancestor to all included groups. Second, if a more-global perspective is desired, only one plesiomorph representative should be inserted; the ideal situation is a monotypic sister-group. If the immediate out-group contains more that one "entity," a "single estimate" for the plesiomorph might be treatable in two ways: (a) by choosing only the most primitive member of out-group (the closest to the ancestral node for that group); or, if need be, (b) by calculating some estimate to represent a "mean" for the out-group as a whole, or preferably for a primitive subcomponent of it. The first of these methods is, perhaps, the most appropriate treatment. The objective, of course, is to minimize insertion of out-group apomorphy into the DFA and, thus, minimize distortion of the relationships that are expressed among the in-group centroids.

A third condition requires that the (in-)groups treated under any chosen perspective, must be monophyletic in the sense that missing groups must not occur among them (the estimation for the plesiomorph should be considered to be a single entity under this statement of implied monophyly). Any missing group will change the vectors that are calculated, to a degree reflecting the misestimation of the actual net selection gradients that it causes. This effect could be expected to be relatively minor if the missing group is closer to the common ancestral node for the in-group tree, but could be severe if the omitted group belonged out among the most derived positions of the tree. (In discrete state cladistics, an analogous effect might occur if a taxon were omitted from an analysis, and that taxon showed a character state that differed from the state possessed by all other included taxa on the branch lineage upon which the omitted taxon would have occurred, had it been included; retrospective inclusion of the omitted taxon would affect the needed number of character state changes and, therefore, the length for that branch; its omission incorrectly estimates the anagenic distance that has evolved for at least the given lineage, because this distance is represented as the number of necessary character state changes along the given branch.)

Although the use of DFA as a phylogenetic probe and mimic seems quite reasonable and powerful if properly directed, problems can occur regarding which of the alternative, but all quite valid, evolutionary perspectives should be chosen, and also in how to handle any transitions that occur between the varying perspectives. This necessarily creates an operational dilemma.

This dilemma, somewhat akin to "Heisenbergian" uncertainty, where one "state" is knowable at the expense of another, is a function of the relativity of this (and other) type(s) of analysis. Using DFA and the multivariate evolution model, the vector space and, hence, the pathway that are generated are "relative" to the groups that are included, although the analysis, itself, objectively partitions and treats the variance that is present in the data set. Indeed most, if not all, analyses are similarly subject to the data analyzed. For instance, outcomes in discrete state cladistics also change when the taxa that are included are changed, but in that type of analysis subjectivity is also present in the coding and transformation of characters, especially qualitative attributes that have nonlinear transformations. Such coding is often determined very subjectively by the investigator and, if erroneous, can cause disastrous results in output that may go unrecognized. This subjectivity has even spawned procedures (e.g., transformation series analysis: Mickevich, 1982) that objectively try to overcome the problem by iteratively fitting characters and trees.

The operational dilemma that is present in phylogenetic inference using DFA seems to be comparable to Robin Thompson's example of "confounding" that is cited by Felsenstein (1988: 458), wherein, given character covariances, it is possible to calculate a phyletic tree, and vise versa, but it appears to be impossible to estimate both. According to Felsenstein (1988: 458), if it were not for this Thompsonian confounding problem, the ideal solution to estimating both covariances and trees would be, given a set of characters, to "estimate a provisional tree" and then, "using the contrasts dictated by that tree," . . . "revise our estimate of the covariances of the characters, from those get a new tree, and continue iteratively until the process converges." (This is why iterative procedures that attempt to converge characters and trees and, therefore, overcome the faults of coding problems in discrete state cladistics, should not work).

In fact, Felsenstein, in an important follow-up discussion (1988: 458-459) on this sort of uncertainty, concludes that unknown character correlation (and covariance) prevents *any morphological data* (!) from being employable for phylogenetic inference (although with discrete characters, covariance is not *directly* involved), because both parsimony and likelihood methods of phyletic estimation require, explicitly or implicitly, character independence (zero character correlation and covariance). He notes, however, that our progress in understanding phylogenetic relationships in general over the last century has been based solely upon morphological grounds, but concludes that these estimations have been possible because they have involved a large number of characters that show covariances of character change between pairs that are at least tenuous enough, so that the characters and tree converge.

Felsenstein states that although this morphologically based estimate of the overall phylogeny of living things is credible, our ignorance of character correlations impedes any statistical use of the data. He concludes that we could not make statistically based statements about which of several generated phylogenies to chose, therefore, unless we make either the invalid assumption of character independence, or we can compute a measure of uncertainty for any given tree by using its (unknown) character correlation information. Felsenstein suggests that it is DNA data that can best be used for phylogenetic reconstruction, because of the prior reason to believe that its molecular sites evolve independently, thus freeing the molecular analyses from the problems of character correlation, and allowing statistical statements concerning confidence of choice between competing trees.

Thus, in summary, according to Felsenstein (1988), morphology cannot be used for phylogenetic inference because either its character correlation, or a given procedure's invalid assumption of the lack thereof, prevent any confidence assessments among competing trees that may be generated, in view of his "Thompsonian uncertainty" regarding estimation of both covariances and trees. Put another way, this suggests that either: (a) character data, of no matter what source, must indeed have no correlations; or (b) that such correlations, if they do in fact exist, must be known and accounted for; or (c) that new phylogenetic inference procedures be developed that can cope with this lack of information. Given that the latter solution is probably an unlikely event, either of the former two seems to be a better approach.

It is certainly true that the original, measured variables that are gathered for any morphometric analysis necessarily suffer from the problems of character correlation; this is why any attempt to use them as raw data directly (e.g., see Simon, this volume), whether coded or not, will not succeed. However, when morphometric data is subjected to DFA it is transformed, so that Mahalanobis' distances and orthogonal vectors are generated. (For discussion of Mahalanobis' distance, see Atchley et al., 1982.) These orthogonal vectors, as minimum selective mortality indices (sensu Lande, 1979), serve as new "characters" of which there are g - 1 possible, and where the character state for any given group is represented by its centroid score for the given vector. Because of their orthogonality, these new "characters" (functions) circumvent the problems of correlation because they are independent (J. Felsenstein, personal communication). Operationally, only they, among morphologically based data, and molecular DNA data, seem to fit the working assumptions of character independence that are required, either implicitly or explicitly, by existing phylogenetic inference algorithms. Note that this "side benefit" of DFA transformed data has also been accomplished by taking into account (in fact requiring) the calculated character correlations and covariances in the data matrix, and using that information in its parsimonious estimation of the apomorphic, anagenic distance (character state) that separates the groups along each independent vector (character).

Let us assume that morphometric data that has been transformed by DFA is fit for phylogenetic inference employing existing phyletic algorithms, because such DFA transformed data seems to show uncorrelated "characters" as functions that have been generated by taking the correlation among the original raw traits into account. This still leaves us with the operational dilemma for DFA that was described above. Therefore, the question still seems to be which perspective to chose for the most resolute results?

Why not use both perspectives, in two analytical runs, despite Felsenstein's uncertainty warnings? The first run, with out-group included, should be used to realize where the network among in-group centroids should be rooted. (Again, this should not be thought of as "polarization," in the same sense as a mid-point rooting option in discrete state cladistics, because DFA uses only prepolarized apomorphic information to generate the vector space and network). The second run should be conducted with the out-group excluded, to attach the in-group network root point appropriately in the more restricted local space perspective. These perspective networks could then be appropriately patched together and in so doing create a composite tree. This would accomplish a transition between the perspectives, and could be sequentially repeated if necessary. Cumbersome, perhaps, but potentially operational, nevertheless, because the attempted convergence will necessarily be relatively crude by any standards that might require an absolute knowledge of both stem lengths and furcation patterns.

One criticism of such a composite tree approach might be that because the DFA vectors act as new, transformed "characters," and because different DFAs conducted on more restrictive subsets of groups necessarily generate new vectors, any composite tree that would be generated would necessarily be created using "different characters" to cover its various portions (R. Fortuner, personal communication). Although this is philosophically true at the level of "vectors equal characters," the new vectors would be created using the same original morphometric character set. In this sense, the original morphometric traits, themselves, have not changed (although the new matrix is minus some excluded groups). Rather, it is the weighted input coefficients (as relative "character states" at the "morphometric trait equals character" level) of these original traits on each of the vectors that has been modified. Such changes reflect only the new minimum selective mortality contributions of the original traits to the vectors, as redefined historical net selection gradients that have acted upon the more restrictive subset of groups alone. In this sense, such "new" vector coefficients, as bits of optimized apomorphic information, can only be meaningfully applied to those sets of restricted taxa from which they were generated; such application is logically restrictive, in the same sense that when speaking of fish, hair, as an apomorphic character state, cannot be useful information for the condition of scales, as a character.

The major operational drawback of the suggested approach is the introduced uncertainty, at each transition of perspective, of anagenic distance on either side of the placed rooting node (between it and the adjacent node/terminal) along the more-local pathway. In this respect, the solution proposed for local-to-global transition is unlike the Thompsonian uncertainty example cited by Felsenstein (1988); perhaps the obfuscation of segmental distances associated with root-placements serves as enough of a "buffer" of sorts for the system to converge. Although convergence in such a system should accrue, it would necessarily be at the expense of distance ambiguity for the root-associated branch lengths. Such obfuscated segments would occur throughout the final composite tree, wherever a perspective was realigned. The resulting tree would be less resolute for these instances of muddled anagenic distance, but hopefully the furcation patterns of the tree would be resolute (and, after all, from a cladist's view of phyletic truth, it is the furcation patterns alone that are of importance in phylogenetic inference).

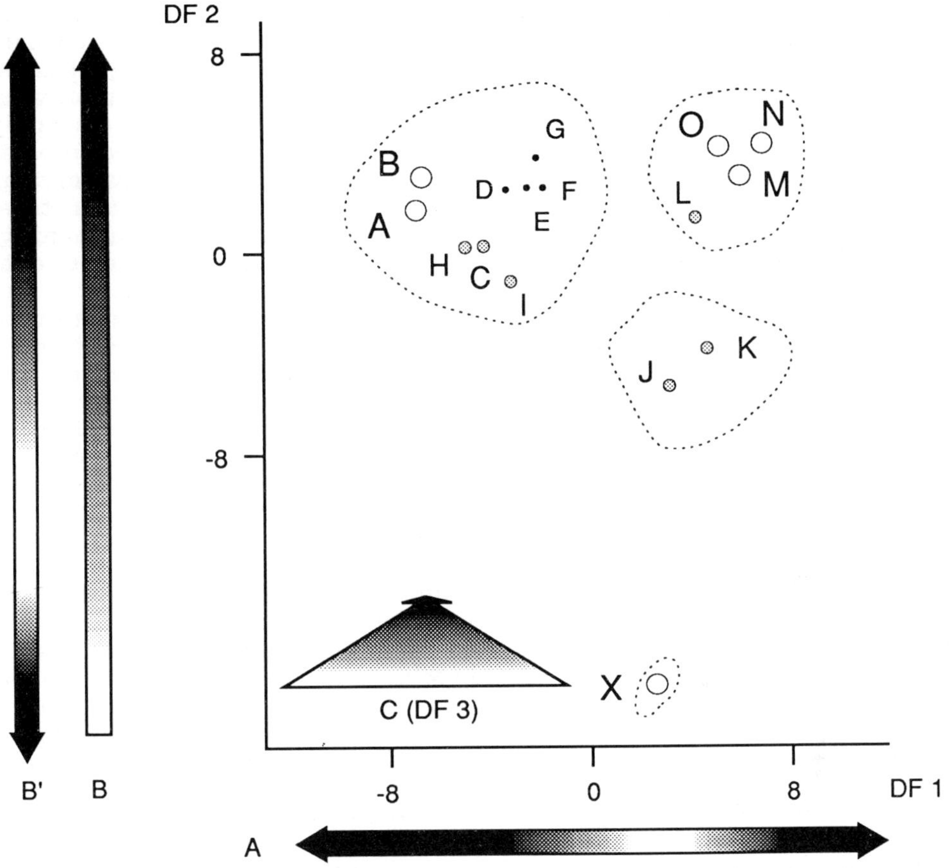

Figure 3. A spatial portrayal of *Essigella* taxa [A - O], as in-group centroids, and *Pseudessigella* [X], as the out-group centroid, in the two-dimensional discriminant space of DFA-3 vectors 1 and 2, with the third dimension (DF 3) implied. Three major subgroups of *Essigella* taxa exist, as circumscribed. The orthogonal third vector is implied by the color/size of the centroid symbols: large, white symbols are nearer than the page; intermediate, gray symbols are on the page; and small, black symbols are more distant that the page. Gradient vectors [A = DF 1; B, B' = DF 2; C = DF 3] indicate the direction of divergence, towards black and away from white (see text).

5 Implementation: The *Essigella* Data

To illustrate the use of DFA in phylogenetic estimation, consider the series of attempts to construct a phyletic pathway within the eulachnine aphid genus *Essigella* (Sorensen, 1983, 1987; present data), using *Pseudessigella* as a plesiomorphic representative (Sorensen, 1990, 1991). *Essigella*, a genus of 15 taxa (13 species, 2 subspecies: Sorensen, 1983) is represented here by the letters "A" through "O" in Fig's. 3-9. (Letters are used here, and in Sorensen [1987], because a generic revision of *Essigella* is pending.) *Pseudessigella*, represented as "X" in those figures, is a monotypic genus and the immediate sister-group of *Essigella*, based upon discrete state cladistics of qualitative traits (Sorensen, 1990). Unfortunately, *Essigella* is quite problematic, showing neotenic evolution (reductionism) of characters, and very few, discretely codeable qualitative attributes that are reliable, beyond those synapomorphies that define the genus. Those traits that do exist within the genus serve as synapomorphies for groups of taxa, or autapomorphies for individual taxa; they provide little or no information on hierarchy within *Essigella*, and chiefly imply multifurcal branching from the basal node of the genus. Therefore, to eliminate the problems involved with subjective interpretations in polarity and character coding, and to circumvent the inadequate information present in the small number of available qualitative traits, Sorensen (1983, 1987) used morphometric attributes alone in constructing a phylogeny for the group.

To conceptualize the spatial array for these groups, Fig. 3 shows their dispersion in a plot of vectors 1 and 2 in the space generated by a DFA (as DFA-3, below), with vector 3 implied. In that figure, three apparent groupings within *Essigella* exist. Each discriminant vector, can be thought of as a transformed character suite showing a minimum selective mortality portrayal of the divergence information between the centroids. In Fig. 3, if the direction of relative divergence were to be hypothesized for each of the vectors with respect to *Pseudessigella* as a starting point (this is functionally equivalent to an increasing degree of apomorphy along a given vector in any direction of divergence, although all DFA generated space is, in effect, apomorphic), it would be different between the depicted vectors. This is shown by the "clinal" arrow gradients for the vectors in Fig. 3, where gradient A represents vector 1, gradients B and B' represent vector 2, and gradient C represents vector 3. Under this assumption, divergence along each of the vectors would begin at centroid X (shown as white on the arrows in the figure), and proceed away from it (towards black). Such a hypothesis is technically erroneous, however, because *Pseudessigella*, is not the actual ancestor of *Essigella*, but rather its immediate sister-group; therefore, *Pseudessigella* contributes out-group apomorphy to the total given vector space. An accurate portrayal of the arrows in Fig. 3 would require the white zone on gradient B for vector 2 to be shifted away from X, as in gradient B', to an exact location that is unknown but is somewhere between X and J. The white zones on gradients A and C for vectors 1 and 3, respectively, *may* also require similar minor shifts, but in these cases the direction of shift is uncertain because the *Essigella* taxa span the vector, and some co-occur with *Pseudessigella* on these vector sections.

5.1 **WAGNER-78 Trees**

To construct an initial phyletic network for *Essigella*, Sorensen (1983) used WAGNER-78 (Farris, 1972) in distance mode and DFA. The discriminant space that was generated by that analysis (DFA-1; Fig. 4A) involved only a local perspective for the genus (out-group excluded case). To insure that the most parsimonious tree was created and chosen, several WAGNER runs were conducted and the data set was shuffled between runs; this was because the WAGNER-78 algorithm is, to a degree, data set order dependant and does not necessarily provide the most parsimonious tree on the initial run.

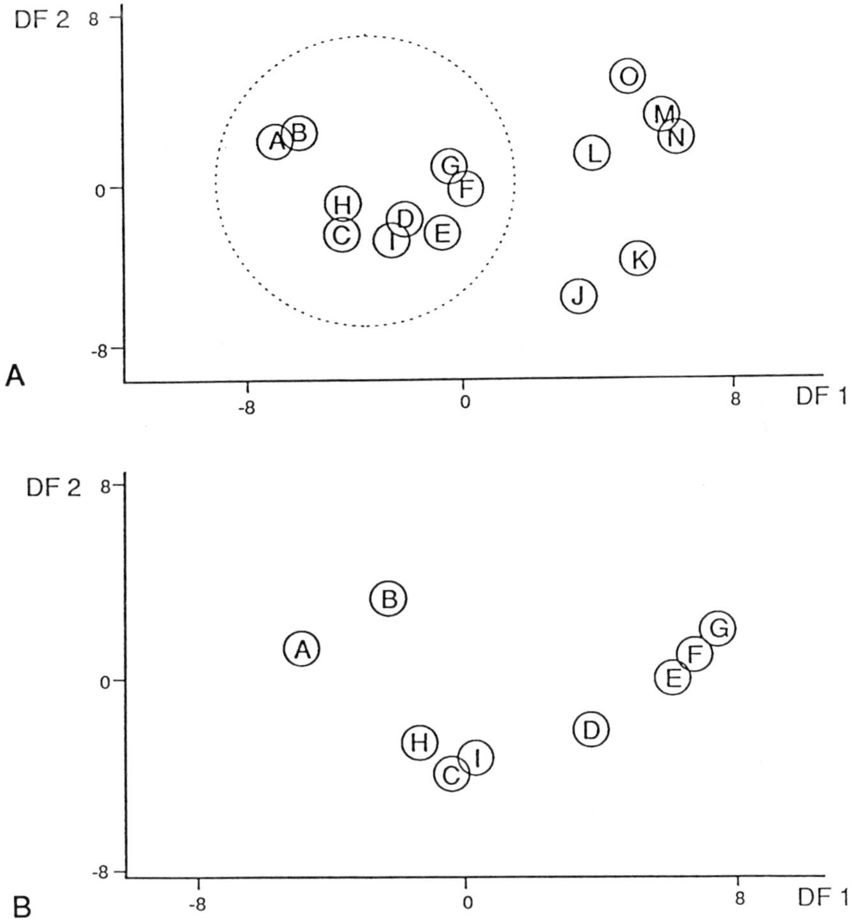

Figure 4. The effect of reorientation of the first two discriminant vectors on the distribution of *Essigella* centroids. (A) The distribution of centroids A - O when X (*Pseudessigella*) is excluded [= DFA-1]; circumscription shows those taxa further analyzed along in Fig. 4B. (B) The distribution of centroids A - I when J - O are excluded [= DFA-2]; note that most of the separation between centroids G versus A that is shown in Fig. 4B along vector 1, occurs in Fig. 4A along vector 3, which is not shown, but which is orthogonal to the page.

The most parsimonious WAGNER network was then subjectively rooted. Lacking convincing evidence for other action at the time, the chosen phyletic network (WAGNER-1) produced from DFA-1 was rooted at the node (Fig. 5A: "a") between the most primitive pair of *Essigella* species (Fig. 5A: "J" and "K"); the relative primitiveness of J and K was assumed because neither showed any qualitative apomorphy that was extrinsic to the analysis, beyond those synapomorphies that defined the genus (Sorensen, 1983). With this rooting, a basal trifurcation was necessarily created on the WAGNER-1 tree,

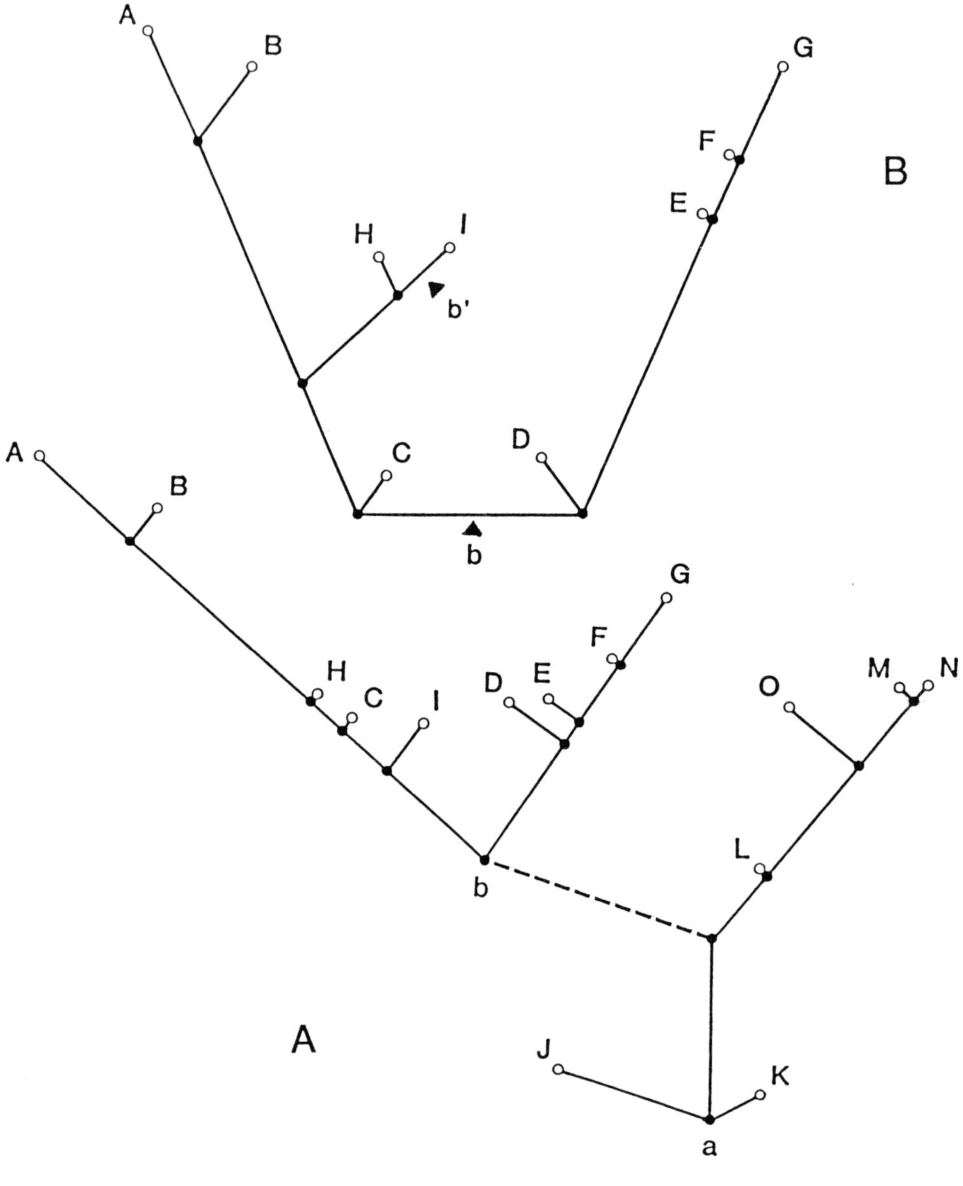

Figure 5. (A) The WAGNER-1 (HENNIG-1]) network, created from DFA-1. This WAGNER network was initially rooted at node "a" between centroids J and K. The dashed phyletic segment below node "b" shows where it would be broken to patch on the WAGNER-2 (HENNIG-2) network (see Fig. 5B) to create composite network WAGNER-1+2 (HENNIG-1+2). (B) The WAGNER-2 (HENNIG-2) network, created from DFA-2. All segmental lengths are proportionate within each network, but not between them, and reflect distance in standard deviation units determined by the appropriate discriminant analysis.

which involved J, K, and the rest of the genus as a single lineage (Fig. 5A). The WAGNER-1 tree is shown in the discriminant space generated by DFA-3 in Fig. 6A; this depiction preserves the furcation event sequence of the tree, but does not represent its appropriate segmental lengths, as shown in Fig. 5A.

Based upon biological information that was extrinsic to the analysis, a distortion of relationships among the taxa in the most derived lineage (Fig. 3: "A" through "I") of the WAGNER-1 tree was perceived; therefore, a more resolute analysis of that lineage was conducted using a second DFA (Sorensen, 1983: DFA-2). DFA-2 excluded the taxa (Fig. 5A: "J" through "O") that were extrinsic to the lineage in question and were judged to be less derived, based on the WAGNER-1 tree. This allowed a vector reorientation (Fig. 4B) to portray the divergence among taxa A through I, as a group, in response to the selective pressure regimes acting on those taxa only (Sorensen and Foottit, this volume; Sorensen, in press).

The vector reorientation for DFA-2 was done to allow a shift from a more-global (DFA-1; Fig. 4A) to a more-local (DFA-2; Fig. 4B) perspective within the genus, thereby minimizing problems for the Manhattan distance based networking algorithm, which was designed to spread or average inherent dimensional distortion over the network produced. This dimensional distortion is caused by projecting (forcing) n-dimensional variance into a unidimensional network portrayal. This "error" is considered to be homoplasy in cladistics, but is analogous to the distortion indicated by the cophenetic correlation coefficient in clustering analyses, or by "stress" in multidimensional scaling.

A second network (WAGNER-2; hereafter all multiple trees produced are given the numerical suffix of their respective DFA) (Fig. 5B) was generated from DFA-2, and was "patched" (appended) onto the WAGNER-1 tree over its respective section to create a composite tree (WAGNER-1+2) for *Essigella* that used information from both the DFA-1 and DFA-2 perspectives. The choice of the exact rooting location for the WAGNER-2 patch was open to interpretation, however. Therefore, an operational rule was devised for rooting such patches: under the "*bracketing taxa*" method, the root on the patch would be inserted somewhere along the patch segment(s) analogous to the base tree segment(s) that separated the most adjacent pair of taxa bracketing the common ancestral node, on the base tree, of the section to be patched.

In this case, the common ancestral node for the WAGNER-2 patch was node b on the WAGNER-1 tree (Fig. 5A: "b"), and its immediate set of bracketing taxa on the WAGNER-1 tree were I and D. Unfortunately, I and D were not also adjacent on the WAGNER-2 network. Therefore, C replaced I as the next most adjacent taxon, and the patch root was inserted between C and D. This choice of a C/D rooting anchorage also sequestered I to a minor monophyletic group (under the imposed rooting assumption) along with H, on the WAGNER-2 pathway (Fig. 5B); monophyly for the resulting H-I lineage seemed to be supported by a qualitative character that was extrinsic to the analysis, and which was originally thought to be a corroborative synapomorphy.

The composite WAGNER-1+2 pathway that was generated by Sorensen (1983) is not illustrated directly here, but can be visualized by substituting the WAGNER-2 tree (Fig. 5B) over its respective section on the WAGNER-1 tree (Fig. 5A: that section of WAGNER1 beyond the dashed line), and functionally rooting node b of Fig. 5A to point b of Fig. 5B.

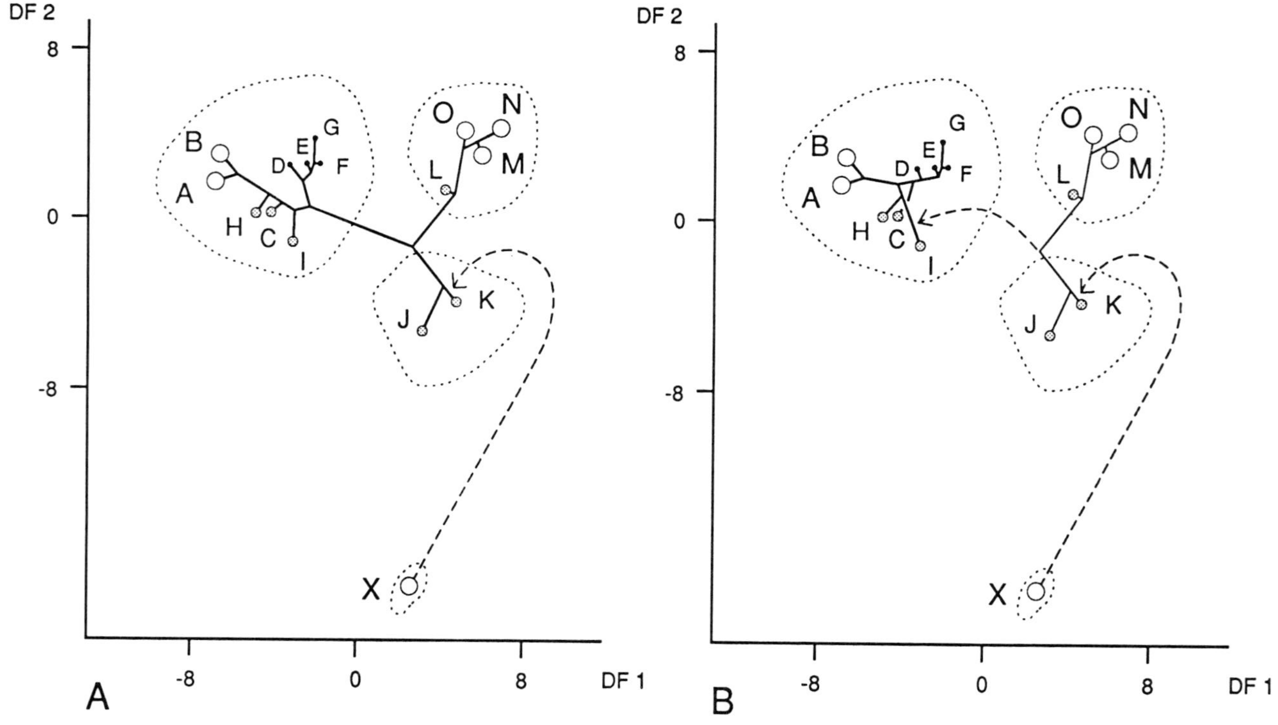

Figure 6. (A) The navigational path of the WAGNER-1 (HENNIG-1) network [= solid segments among centroids] shown in the discriminant space created by DFA-3. The root from the segment connecting X would be placed near the base of the segment common to J and K. (B) The navigational path of the WAGNER-2 (HENNIG-2) network [= thicker solid lines among centroids] shown in the discriminant space created by DFA-3; the thinner solid line section represents the residual basal section of the WAGNER-1 network. The WAGNER-1 section shown assumes rooting at b' (Fig. 5B) along the stem of I (most adjacent taxon method); the entire patched network here [= dashed and solid segments] represents the composite WAGNER-1+2 tree (with b' rooting). Segmental lengths are not proportionate, but the networks reflect the furcation pattern through the defined space. (See Fig. 3 for third dimension information.)

Later, an alternative and more operational rooting scheme for patches was devised to circumvent the problem experienced above, when the taxa bracketing a common ancestral node on a base tree were not immediately adjacent on the patch also. This method, the "*most adjacent taxon*" method, calls for insertion of the patch root along the stem segment of the taxon that, on the basal tree, had the shortest total segmental distance between it and the common ancestral node for the segment to be patched. If that rooting rule had been implemented during the 1983 analysis, the WAGNER-2 root would have been inserted adjacent to I (Fig. 5B: "b'"). This reinterpreted WAGNER-1+2 tree can similarly be visualized by rooting node b of Fig. 5A to point b' of Fig. 5B. It is functionally illustrated in Fig. 6B, as the sectional components of its WAGNER networks.

After some reflection, I later decided that the initial attempt (Sorensen, 1983) at operational objectivity in producing the phyletic tree for *Essigella* was unnecessarily subjective, in that only the dominant three discriminant functions of DFA-1 and DFA-2 were employed in generating both pathways for the WAGNER-1+2 tree. For example, of the total apomorphic variance that was present in the 14 possible functions (as *g* - 1, where *Essigella* groups = 15) of DFA-1, only 78.3% of the variance was present in the first three vectors; similarly, the first three vectors of DFA-2 contained only 83.7% of the variance present in its eight possible functions. Although the first few dominant functions in DFA usually portray intergroup distances adequately, there are exceptions (for example, see Blackman, this volume); despite many attempts to develop various thresholds at which to determine when (or if) subordinate vectors cease to be relevant (see Sneath and Sokal, 1973), there appears to be no such method that is not arbitrary to a degree. Furthermore, because all variance that is used by DFA can be considered to be apomorphic, and because any possible homoplasy in the apomorphic variance that is present in the original morphometric characters is functionally eliminated by DFA, due to the orthogonality of the discriminant vectors, there should be no a priori reason to exclude the information present in any discriminant vector (unless it can be judged to be biologically unsound, or shows very poor canonical correlation coefficients [the latter, at least, requiring a subjective decision]).

Beyond these philosophical reasons, the basal trifurcation of the WAGNER-1+2 tree was unsatisfying; and the corroborative "synapomorphy" for H-I monophyly, was increasingly suspected to be a symplesiomorphy, or at the very least a qualitative character homoplasy. Therefore, to resolve these problems, and to increase operational objectivity, a second phyletic tree was later generated (Sorensen, 1987).

5.2 **CONTML Tree**

Because it was believed that a certain measure of distortion in the phyletic network created earlier was due to a basic "incompatibility" between the Manhattan distance based networking algorithm and the nature of the vector space generated by DFA, the second attempt at phylogenetic tree generation used a maximum-likelihood based algorithm because it more closely approximates a Euclidean solution among the DFA centroids in the vector space. (Although DFA space is based in Mahalanobis' distance and is, therefore, non-Euclidean [Pimentel, this volume], the centroid scores among

DFA vectors can be linked using a Euclidean based, or approximating, algorithm; this is because only the group centroids, as mean group phenotypes [sensu Lande, 1979], are of interest in the multivariate evolutionary model, and not the actual distribution of individual phenotypes.) The networking algorithm that was chosen for the second attempt was the REML algorithm CONTML (PHYLIP, version 2.4: Felsenstein, 1984a), using the continuous characters option (option-c).

To resolve the trifurcation rooting of the WAGNER-1+2 tree, Sorensen (1987) employed a more-global perspective and used 16 taxa: the original 15 *Essigella* taxa matrix plus *Pseudessigella* as a plesiomorphic representative; in addition, all of the 15 possible discriminant vectors that were generated from the new DFA (DFA-3) were used. The resulting CONTML tree (Sorensen, 1987) is shown in Fig 7. The segmental lengths on that CONTML tree are expressed in standardized units of standard deviation, pooled across all groups in the analysis. As such, they show the relative anagenic divergence among all mean group phenotypes (centroids) for the taxa with reference to the averaged standard deviation among all taxa depicted. In this regard, the total anagenic distance between *Pseudessigella* (X) and the most primitive *Essigella* taxa (K) on the generated tree is 18.9 standard deviation units. Therefore, for this tree and data matrix of traits/individuals, the intergeneric distance that separates *Pseudessigella* and *Essigella*, is 18.9 times the standard deviation of variance that is expressed among all biological groups in both genera.

Fig. 8A depicts how the CONTML tree would be projected through the vector space of DFA-3. Note that this depiction preserves the furcation event sequence of the tree, but does not represent its appropriate segmental lengths, as shown on the CONTML tree (Fig. 7) and which are based upon 15, not three, dimensions. In the evolutionary space represented in Fig. 8A, note how the phyletic path travels between the group centroids in the third (approximated) dimension. At X, the path begins out from the page, and proceeds to J and K on the page plane. It then continues on the page plane and furcates, with the branches going to L versus I and H, respectively, which are all still on the plane of the page . The branch from L comes back out from the page to M, N, and O. The alternative branch splits on the page plane, with one line coming out from the page to A and B, and the other going to C on the page plane and then continuing into the page to D, E, F and G. This evolutionary transition through the three-dimensional space shown seems quite reasonable and efficient.

Note several other things about the CONTML tree. First, the global CONTML tree (Fig. 8A) is virtually identical to the patched WAGNER-1+2 tree, if the most adjacent taxon criterion (root b') is used for its WAGNER-2 section. Also, the CONTML tree defines K as the most primitive *Essigella*. This is despite the relatively closer position of J to X in reduced dimensional portrayal of Fig. 8A, and is because the apomorphic information present among subordinate vectors, 3-15, combine to indicate a greater distance between X and J than between X and K. This indicates the dangers of restricting the number of vectors incorporated in tree construction. The CONTML tree also circumvents another problem in subjective decision making: the patch rooting possibilities within the WAGNER-1+2 tree (points b versus b' in Fig. 5B). It eliminates the dangers associated with attaching a rooting node to a more-local perspective phyletic pathway, especially if the taxa sequence around the proposed root insertion point is not the same between the trees.

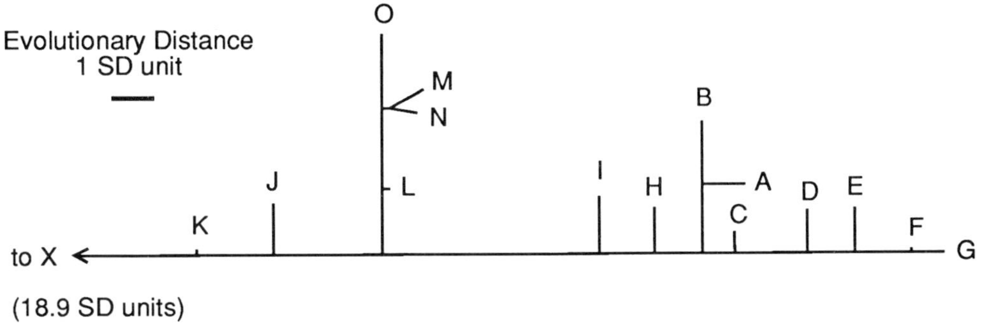

Figure 7. The CONTML network (modified from Sorensen, 1987a: fig. 1.), created from DFA-3. All segmental lengths are proportionate and reflect distance in standard deviation units determined by the appropriate discriminant analysis.

At the time of construction of both the initial WAGNER-1+2 tree, with its point b rooting (Sorensen, 1983), and the CONTML tree (Sorensen, 1987), I attempted to corroborate each of them with overlays of transformations of qualitative traits that were extrinsic to the analyses. Although the WAGNER-1+2 tree was overlaid with 21 traits (Sorensen, 1983) (five subjectively deemed as probably not homoplasious, 10 as possibly homoplasious, and six as probably homoplasious), and the CONTML tree with only nine (Sorensen, 1987) (all subjectively deemed as "reasonable"), the latter set of traits are considered, in retrospect, to be better, more reliable and objective estimators of true synapomorphy. Further corroboration of the CONTML tree (Sorensen, 1987) was based on biogeographic and hybridization data for the host plants of the aphids (chiefly *Pinus*, but also *Pseudotsuga* and *Picea*). Although this extrinsic data supported major trends in the CONTML tree (e.g., in Fig. 5B lineages: "A-B," "D-E-F-G," "L-M-N-O," "A-B-C-D-E-F-G-H-I," "A-B-C-D-E-F-G-H-I-L-M-N-O") it provided little insight into the phyletic pathway through the critical region of vector space surrounding C, H and I. Sorensen (1987) discusses the corroborative evidence supporting the CONTML tree in more detail. In essence, the CONTML tree is favored over the composite WAGNER trees because of its increased operational objectivity; it includes the data from all vectors, lacks requirement for subjective root placements, and is not refuted by overlays of extrinsic traits.

5.3 **HENNIG-86 Trees**

With the advent of the HENNIG-86 procedure (Farris, 1988), a refinement of the WAGNER algorithm, another attempt at tree construction was conducted (present data) to see what effect might have on the *Essigella* data, especially when *Pseudessigella* is initially included in the matrix, in an out-group included case. The DFA-3 data matrix was analyzed using HENNIG-86 to generate another parsimony tree (HENNIG-3) for the genus. The HENNIG-3 tree is shown in Fig. 9, which depicts both the furcation pattern among the

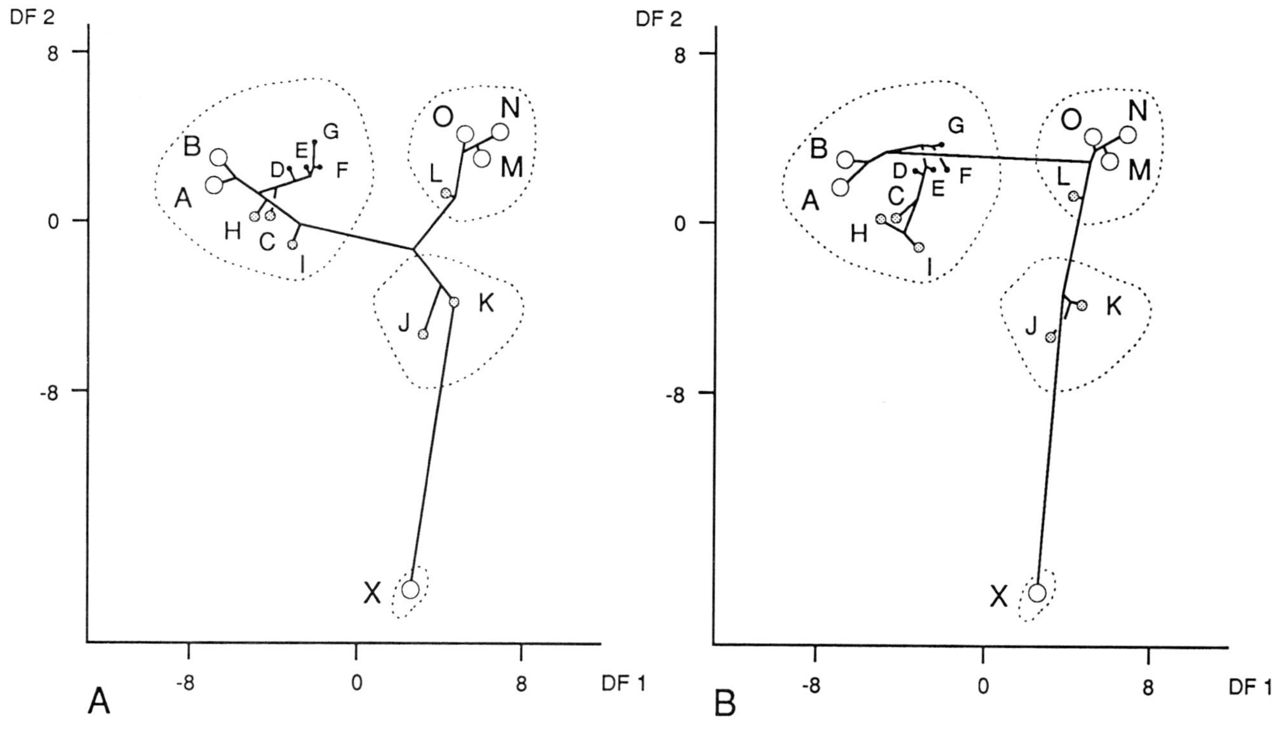

Figure 8. (A) The navigational path of the CONTML network among centroids in the discriminant space created by DFA-3. (B) The navigational path of the HENNIG-3 network among centroids in the discriminant space created by DFA-3. Segmental lengths are not proportionate, but the networks reflect the furcation pattern through the defined space. (See Fig. 3 for third dimension information.)

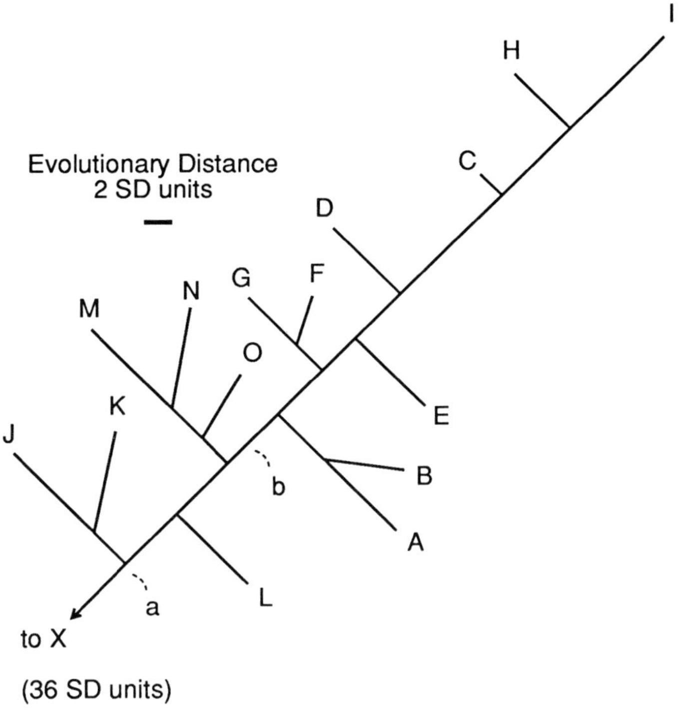

Figure 9. The HENNIG-3 tree, created from DFA-3. Point "a" should be the rooting insertion for the HENNIG-1 network (see WAGNER-1, Fig. 5A), more accurately determined here by the inclusion of X. Point A, here, also suggests the best point to break this network for reanalysis (as HENNIG-1), because of distortion near L versus M, N, and O. Point b, here, seems to be the best point for an even more local reanalysis (as HENNIG-2), because of severe distortion among taxa linkages that are more terminal to it. All segmental lengths are proportionate and reflect distance in standard deviation units determined by the appropriate discriminant analysis.

centroids and the relative anagenic distances (taken as HENNIG-86 stem lengths) for pathway segments.

Notice that despite an identical global input matrix, the HENNIG-3 tree differs substantially from the CONTML tree. The effect of plesiomorph inclusion has "jumbled" the furcation pattern of the HENNIG-3 tree within *Essigella*, producing a network that is quite different from the any of the WAGNER trees or the CONTML tree. The HENNIG algorithm has erroneously spread the dimensional distortion of its Manhattan distance throughout the tree in global perspective only. This occurs at the expense of local relationships, as determined by both CONTML, which calculates the most appropriate (most probable) pathway between centroid pairs in a more-local based fashion, and the more local perspective WAGNER networks that were generated.

Several problems exist with the HENNIG-3 tree: (1) J and K are considered a minor monophyletic group, yet all extrinsic qualitative traits for this pair of taxa represent symplesiomorphies only; (2) L is separated from monophyletic group M-N-O, with which it was previously associated in both the WAGNER-2 and CONTML trees, despite convincing, extrinsic, qualitative synapomorphies indicating that grouping is monophyletic; (3) D, E, F and G have become paraphyletic, again despite convincing, extrinsic, qualitative synapomorphies indicating monophyly; and (4) H, I and C, as a grouping, have become monophyletic due solely to their spatial proximity in intermediate vector space, an event that is not supported by any convincing, extrinsic synapomorphy that is so restricted.

These relationship anomalies not only are not overlaid by the extrinsic synapomorphies that do exist, they require reversals in several of them. The HENNIG-3 tree is also counter-intuitive to the expected evolutionary relationships of the aphids among the hosts, which are supported by the CONTML (Sorensen, 1987) and the b' rooted WAGNER-1+2 solutions. The HENNIG-3 solution would require a serial evolutionary sequence of host plant associations that involves the colonization of a major section of *Pinus* and radiation through it, followed by a host capture (from that lineage) of another section of *Pinus*, repeating the process at least three times. In effect, this is like a "stepping-stone" type of radiation among quite genetically diverse groups of pines, and one that would seem to occur without regard to the implied relative ages of such pine groups (Sorensen, 1987).

In contrast, the CONTML and WAGNER-1+2 solutions also show colonization of sections of *Pinus* and radiation within them, but these colonization events are radial in nature, rather than serial; the end point of each specialized lineage is further specialization of its taxa on particular host species, rather than escape to attain a new taxonomic section of hosts (Sorensen, 1987).

Because the centroid scores on the vector characters of the input matrix were expressed in pooled standard deviation units, as in CONTML, the segment lengths of the HENNIG-3 tree are also expressed in this measure. Interestingly, however, the HENNIG-3 tree shows 36 SD units along the segment of the phyletic pathway connecting *Pseudessigella* and the basal node of *Essigella*, while the CONTML tree shows only 18.6 SD units. A very puzzling result, considering that HENNIG-86 is considered to be a "parsimony" package. This long pathway segment between the genera is due to the Manhattan calculation used in HENNIG-86, which tally its distances *along the vectors* between the centroids, instead of *directly between* the centroids, as in Euclidean based algorithms.

The consequences of the HENNIG-3 pathway in the vector space that is shown in Fig. 8B are severe when compared to Fig. 8A; the network between the major groupings of *Essigella* is substantially inefficient (unparsimonious) in comparison to that of CONTML. (Remember, that the HENNIG-3 network was constructed using 15-dimensional data, as was the CONTML tree; therefore, minor "discrepancies" in the anticipated pathway that is navigated among the centroids in the three-dimensional space could be expected for both cases because of dimensional distortion in the vector space portrayal.) Nevertheless, the distortion of the HENNIG-3 tree in evolutionary space (Fig. 8B) seems unnecessarily large in comparison to that of the CONTML tree (Fig. 8A). The "backbone" of the HENNIG-3 pathway occurs out from the plane of the page, chiefly navigating to connect X versus M, N and O versus A and B, as groups within that plane. Spur routings are sent separately to the page plane to connect J and K, as a group, and L to the main path. The path deviates near A and B, into the page to link D, E, F and G, and then returns to the page plane to link H, I and C. This is a seemingly circuitous route at best.

To see if the HENNIG-86 and WAGNER-78 algorithms performed similarly in more-local perspectives, two additional runs were conducted using the former in restricted analyses. These were composed of the taxa subsets that were encompassed by the two WAGNER-78 analyses (HENNIG-1 [= WAGNER-1] used DFA-1, and HENNIG-2 [= WAGNER-2] used DFA-2); in the HENNIG-86 analyses, however, all of the possible functions that were generated by the respective DFAs were used, instead of only the dominant three. It was expected that these algorithms would perform very similarly, if not identically, and that if their results differed, this would reflect the added information of the subordinate vectors that were present in the HENNIG-86 matrix only.

Both the HENNIG-1 and HENNIG-2 unrooted networks turned out to be identical to their WAGNER-78 counterparts (see Fig. 5) in their furcation patterns, and virtually so in the relative lengths of their network segments as anagenic distances. This indicates several things. First, it seems to confirm that the CONTML algorithm, with its Euclidean navigation between centroids, handles the transition between local-to-global perspectives in vector space more effectively that does either WAGNER-78 or HENNIG-86, as parsimony packages with a Manhattan navigation style. This may be a partial reason why Farris (1981), a supporter of Manhattan calculation, believes that distance data are inappropriate for cladistic analysis, despite Felsenstein's (1984b) justification for the use of distance data in probability based phyletics. Second, it also shows, at least in the case of this data set, that the addition of information present in the vectors subordinate to the first three (approximately 20 percent of the total) had no noticeable effect on the Manhattan distance trees that were generated. That result was not unexpected, but it is philosophically more satisfying nonetheless to know that all available data was used to generate the trees. Third, if the first deduction is true, then obviously phyletic "patching" must be done for the HENNIG-86 output, as it was for the WAGNER-78 output, to allow the Manhattan distance calculations to accurately reflect the local aspects of phyletic relationships among the taxa.

When patching occurs among the HENNIG networks, however, so that a composite tree is created, it is evident that any given basal tree that is chosen to determine the appropriate root location for a patch must, itself, be of a sufficiently local perspective to have undistorted phyletic relationships among its taxa in the vicinity of the node that is common to all taxa in the patch. We don't know if the phyletic relationships among taxa adjacent to that node are correct, or at least reasonable, without reliable corroborative evidence that is extrinsic to the analysis Alternatively, by reanalyzing the data matrix several times and successively eliminating the more plesiomorphic taxa and lineages during this sequence of analyses, we should get an idea of where potential problems are because the rooting choices might become so broad as to make the procedure nonoperational. Because of the better handling of local/global transitions by CONTML, and because of only symplesiomorphic similarity between J and K in all available extrinsic traits, the CONTML solution is still preferred over either Manhattan distance based solution.

6 Conclusions and Recommendations

It seems empirically correct to conclude that a systematist who wishes to estimate the phylogeny of a set of taxa should use the best information that is accessible, and attempt to corroborate the findings with all other available evidence. In this regard, the use of discriminant function analysis appears to be reasonable in estimating phyletic lineages among closely related groups. Such use of DFA is possible when groups are closely related and monophyletic, and is particularly indicated when they display either (a) morphometric traits that are the only available characters or (b) qualitative traits that are few and unreliable due to homoplasy or uncertain transformation. Therefore, phylogenetic use of DFA seems especially appropriate when dealing with troublesome species complexes. In contrast, the use of PCA in phylogenetics appears to be philosophically limited, because it neither employs only information that can be considered apomorphic (noncommunality-based) among groups, nor are its generated distances optimized to any limit as a maxima or minima.

If DFA is used as a phylogenetic procedure, and Lande's model is employed as a justification for the method, then two major assumptions are: (a) that **E** among the groups be reasonably considered to be operationally 0, with **P** = **G** + **E**); and (b) that **G** is functionally similar and stable across the groups to be analyzed. The **E** = 0 concern can prove problematic in "environmentally sensitive" groups, (e.g., Blackman, this volume; Claridge and Gillham, this volume), and where warranted some assessment and operational stabilization of "environment" may be necessary (i.e., Sorensen and Sawyer, 1989) to minimize, or at least equalize, its affect. The concern over a stable and similar **G** should be chiefly addressed by restriction of the analyses to quite closely related groups (but see Wood and Pesek [this volume] for a broader usage).

In addition several operational recommendations appear to be warranted for phylogenetic use of DFA:

(1) That the groups to be analyzed be considered to be monophyletic, and that initially all members be used in the analysis.

(2) That if an out-group is included in an analysis as a plesiomorph representative, it be a single entity (i.e., monotypic genus) and the immediate sister-group of the groups to be analyzed, as was *Pseudessigella* here. If the nearest out-group is not monotypic, it may be necessary to represent the out-group as a pooled estimate that is preferably relatively plesiomorphic within the overall out-group variance. For example, if *Essigella* were to be used as a out-group for another, more derived genus, it might be appropriate to pool the variance in only *Essigella* species J and K for such use.

(3) That all g - 1 generated discriminant vectors be used in constructing the phyletic network to maximize available information and eliminate a subjective decision during the procedure.

(4) That any networking algorithm used be Euclidean in nature, rather than Manhattan. The latter is much less successful in handling the transitions between differing perspectives of DFA generated space.

(5) That if it is necessary to construct a composite tree to accommodate several temporal perspectives, sequentially more-local perspective DFAs should be used and these should be rooted, each in turn, based upon an appropriate "undistorted" nodal area from the preceding tree, preferably by the "most adjacent taxon method" described earlier.

(6) That the phylogenetic estimate that is created be corroborated, if possible, by features and facts that are extrinsic to the matrix used in the analysis.

The justification for the use of ordinations that have been presented here and in Sorensen and Foottit (this volume) have been based upon Lande's (1979) model of multivariate evolution of mean group phenotype. That model was chosen largely because it partitions phenotypic variance into linearly additive, orthogonal vectors, which are similar to the procedures used in PCA and DFA. Of these procedures, the use of DFA seems to be able to provide an objective and operational method for phylogenetic inference among groups where only nondiscrete data are available.

If the phylogenetic DFA procedure outlined can be criticized on any grounds, it might be that the technique uses "pattern" in its spatial portrayals, from which the networks are generated (see Faith, 1989; although his usage cites multidimensional scaling). However, because the discriminant space portrayed in DFA reflects only the parsimonious divergence (apomorphy) among groups after communality is eliminated, and because parsimony networking algorithms can be used to network through that space (although it appears they should not be recommended to do so), even that argument appears to be blunted. Even the argument that DFA space is somehow created by "averaging" its variance should not be problematic. After all, even discrete state, Manhattan-based, cladistic algorithms "average" homoplasy over their entire trees, sometimes causing severe distortion (state reversals) and necessitating a "consistency index" measure for gauging the problem. In DFA space, however, any such averaging occurs on each of the orthogonal vectors, which provides a superior and more locally sensitive refinement in resolution.

Ultimately, whether or not Lande's (1979) particular genetic model, in its strict sense, stands as a reasonable approximation for an evolutionary mechanism seems unimportant. Any somewhat similarly functioning genetic hypothesis could serve equally well to explain how phenotypic data might be processed by ordinations to provide phylogenetically useful information. This should be apparent because most early, crude phylogenetic approximations (e.g. any pre-Hennigian tree), which are still quite meaningful, were based simply on the genetic assumption of phylogenetic response to natural selection as a driving mechanism; when cladistic theory arose, it provided only a "screening" mechanism to employ solely apomorphic information in tree construction. Thus modification, or abandonment, of Lande's model should not preclude the use of DFA in phylogenetic estimation, any more than the absence of a genetically-based evolutionary model has hindered the use of discrete state cladistics to estimate acceptable phylogenies.

The phylogenetic usefulness of DFA as a evolutionary probe has three foundations: (a) the orthogonality of its vectors, which eliminate the problems of character correlations among the original traits (a shortcoming of most other phylogenetic procedures); (b) its divergence-based spatial distances, which represent only apomorphy; and (c) the parsimonious portrayal of relationships among groups in the generated evolutionary space. Those attributes will remain unchanged, even if the current, rather simplistic, quantitative genetic models that can be useful in rationalizing such use are increasingly refined. Further development of this philosophical rationale is possible, and should be explored. For example, although DFA, as a statistical method, in strict sense assumes that all input data are parametric, it nevertheless is certainly robust enough to handle nonparametric data successfully in portraying group relationships (e.g., Hafernik, 1983). Statistically, the assumptions of normality for DFA are required (Pimentel, personal communication) only when one desires to classify "unknowns" into a DFA generated space, so that probability assignments for group affiliation for such individuals are generated, based on defined territories for "known" groups (e.g., Daly and Balling, 1978). When applying a parametrically-based genetic theory, such as the Landean models, as a rationale for the phylogenetic use of DFA, parametric assumptions are also necessary to facilitate the interpretation of the vectors as independent patterns of the total variation that reflect stimuli, such as selection forces (Pimentel, this volume). However, quadratic refinements to DFA (e.g. Burnaby, 1966b) could lessen its philosophical problems with any nonnormal data, and the development of appropriate non-Gaussian phenotypic models (see Karlin, 1988) might provide accommodating rationales for phylogenetic usage in such cases.

The far more troublesome aspect of approximating phylogenies in general remains the problem in estimating changes across transitions between global and local perspectives of selection responses on the time continuum. This occurs in the phylogenetic usage of DFA as network changes during any global-local transition, and is evident as secondary rooting uncertainties. It is implied in discrete state cladistics by erroneously averaging homoplasy over the entire tree; the shortcomings of that approach are evident in various attempts to accommodate character state change to fit the generated tree successfully (e.g., Mickevich, 1982; Strauch, 1984), or to examine restricted lineages for enhanced resolution (see Felsenstein, 1988). Interestingly, such uncertainty phenomena are not restricted to phylogenetic inference alone (i.e., Rohrlich, 1983), and in phylogenetics the problem well may be unresolvable with a confident degree of certainty, as Felsenstein (1988) suggests. That, however, does not mean that attempts at resolution should not be continuously made.

7 Acknowledgments

Numerous people have helped me through discussion (and criticism) of this subject over the years; although no agreement or consensus is implied, these include: D.H. Blocker, B.J. Crespi, H.V. Daly, J.T. Doyen, T.O. Duncan, G.F. Estabrook, J. Felsenstein, E.M. Fisher, R.G. Foottit, R. Fortuner, R.W. Garrison, D. Hille Ris Lambers, R. Lande, L.P. Lefkovitch, J.K. Liebherr, C. Meacham, R.A. Pimentel, J.A. Powell, F.J. Rohlf, C.W. Schaefer, D. Schluter, S. Selvin, J.D. Smith, R. Somerby, D. Swofford, B.W. Triplehorn, M. Turelli, D.J. Voegtlin, L. Watrous, Q.D. Wheeler, and T.K. Wood. Also, of aid have been the discussions and questions with many of the participants at several meetings where this subject has been presented, among them: the eighteenth Numerical Taxonomy Meeting (NT-18), held at Cornell University, Ithaca, New York, October, 1984; the second International Symposium on the Population Structure, Genetics and Taxonomy of Aphids, held at facilities of the Czechoslovakian Academy of Sciences, Smolenice, Czechoslovakia, September, 1985; the Second International Workshop on Leafhoppers and Planthoppers of Economic Importance, held at Brigham Young University, Provo, Utah, July-August, 1986; and the Workshop on Ordination Techniques in Entomological Systematics (IW-1), XVIII International Congress of Entomology, held at the University of British Columbia, Vancouver, British Columbia, July, 1988.

8 References

Atchley, W.R., Nordheim, E.V., Gunsett, F.C. and Crump, P.L., 1982. Geometric and probabilistic aspects of statistical distance functions. *Systematic Zoology*, 31: 445-460.

Blackman, R.L., (this volume). The use of ordination techniques to discriminate within pest aphid species complexes.

Bookstein, F.L., 1986. Size and shape spaces for landmark data in two dimensions (with discussion). *Statistical Science*, 1: 181-242.

Bookstein, F.L., 1989. "Size and shape": a comment on semantics. *Systematic Zoology*, 38: 173-180.

Bookstein, F.L., Chernoff, B., Elder, R.L., Humphries, J.M., Smith, G.R. and Strauss, R.E., 1985. Morphometrics in Evolutionary Biology. Special Publication, 15. Academy of Natural Sciences, Philadelphia, 277 pp.

Burnaby, T.P., 1966a. Growth-invariant discriminant functions and generalized distances. *Biometrics*, 22: 96-110.

Burnaby, T.P., 1966b. Distribution-free quadratic discriminant functions in paleontology. *Computer Contributions, State Geological Survey, Kansas*, 7: 70-76.

Claridge, M.F. and Gillham, M.C., (this volume). Variation in populations of leafhoppers and planthoppers (Auchenorrhyncha): biotypes and biological species.

Daly, H.V. and Balling, S.S., 1978. Identification of Africanized honey bees in the Western Hemisphere by discriminant analysis. *Journal of the Kansas Entomological Society*, 51: 857-869.

Faith, D.P., 1989. Homoplasy as pattern: multivariate analysis of morphological convergence in Anseriformes. *Cladistics*, 5: 235-258.

Farris, J.S., 1972. Inferring phylogenetic trees from distance matrices. *The American Naturalist*, 106: 645-668.

Farris, J.S., 1981. Distance data in phylogenetic analysis. In: V.A. Funk and D.R. Brooks (Editors), Advances in Cladistics. Proceedings of the First Meeting of the Willi Hennig Society. New York Botanical Garden, New York, pp. 3-23.

Farris, J.S., 1988. HENNIG, Hennig86 Reference, (Version 1.5). J.S. Farris, Department of Ecology and Evolution, State University of New York at Stony Brook, Stony Brook, New York.

Felsenstein, J., 1984a. PHYLIP, Phylogeny Inference Package, (Version 2.5). J. Felsenstein, Department of Genetics, University of Washington, Seattle.

Felsenstein, J., 1984b. Distance methods for inferring phylogenies: a justification. *Evolution*, 38: 16-24.

Felsenstein, J., 1988. Phylogenies and quantitative characters. *Annual Review of Ecology and Systematics*, 19: 445-471.

Hafernik, J.E. Jr., 1983. Phenetics and ecology of hybridization in buckeye butterflies (Lepidoptera: Nymphalidae). *University of California Publications in Entomology*, 96, 109 pp.

Humphries, J.M., Bookstein, F.L., Chernoff, B., Smith, G.R., Elder, R.L. and Poss, S.G., 1981. Multivariate discrimination by shape in relation to size. *Systematic Zoology*, 30: 291-308.

Karlin, S., 1988. Non-Gaussian phenotypic models of quantitative traits. In: B.S. Weir, E.J. Eisen, M.M. Goodman and G. Namkoong (Editors), Proceedings of the Second International Conference on Quantitative Genetics. Sinauer Associates Inc., Sunderland, Massachusetts, pp. 123-144.

Lande, R., 1979. Quantitative genetic analysis of multivariate evolution applied to brain: body size allometry. *Evolution*, 33: 402-416.

Lande, R., 1988. Quantitative genetics and evolutionary theory. In: B.S. Weir, E.J. Eisen, M.M. Goodman and G. Namkoong (Editors), Proceedings of the Second International Conference on Quantitative Genetics. Sinauer Associates Inc., Sunderland, Massachusetts, pp. 71-84.

Mickevich, M.F., 1982. Transformation series analysis. *Systematic Zoology*, 31: 461-478.

Pimentel, R.A., (this volume). An introduction to ordination, principal components analysis and discriminant analysis.

Rohlf, F.J. and Bookstein, F.L., 1987. A comment on shearing as a method for "size correction." *Systematic Zoology*, 36: 356-367.

Rohrlich, F., 1983. Facing quantum mechanical reality. *Science,* 221 (4617): 1251-1255.

Simon, C., (this volume). Discriminant analysis of year classes of periodical cicada based on wing morphometric data enhanced by molecular information.

Sneath, P.H.A. and Sokal, R.R., 1973. Numerical Taxonomy, the Principles and Practice of Numerical Classification. W.H. Freeman and Company, New York, 573 pp.

Somers, K.M., 1986. Multivariate allometry and removal of size with principal components analysis. *Systematic Zoology*, 35: 359-368.

Somers, K.M., 1989. Allometry, isometry and shape in principal components analysis. *Systematic Zoology*, 38: 169-173.

Sorensen, J.T., 1983. Cladistic and Phenetic Analysis of *Essigella* Aphids: Systematics and Phylogeny in Relation to their Pinaceae Host Plants (Homoptera: Aphididae, Lachninae). Ph.D. Thesis, University of California, Berkeley, (unpublished for the purposes of taxonomic nomenclature), 606 pp.

Sorensen, J.T., 1987. Multivariate statistical approach to deduction of phylogeny within *Essigiella* [sic] (Aphididae: Lachninae). In: J. Holman, J. Pelikan, A.G.F. Dixon and L. Weismann (Editors), Population Structure, Genetics and Taxonomy of Aphids and Thysanoptera, Proceedings of International Symposia, held at Smolenice, Czechoslovakia, September 9-14, 1985. SPB Academic Publishing, The Hague, pp. 243-260.

Sorensen, J.T., 1990. Taxonomic partitioning of discrete-state phylogenies: relationships of the aphid subtribes Eulachnina and Schizolachnina (Homoptera: Aphididae: Lachninae). *Annals of the Entomological Society of America*, 83: 394-408.

Sorensen, J.T., 1991. Phylogenetic character responses for shape component variance during the multivariate evolution of eulachnine aphids: redescription of *Pseudessigella* Hille Ris Lambers (Homoptera:

Aphididae: Lachninae). *Pan-Pacific Entomologist*, 67: 28-54.

Sorensen, J.T., (in press). Deciphering Biological groupings in the *Essigella knowltoni* complex (Homoptera: Aphidinea: Lachnidae), and comparison to variation in *Pinus contorta* (Coniferae: Pinaceae). *Entomologia Generalis*, 17 (1).

Sorensen, J.T. and Foottit, R.G., (this volume). The evolutionary quantitative genetic rationales for use of ordination analyses in systematics: phylogenetic implications.

Sorensen, J.T. and Sawyer, S.M., 1989 Assessing the multivariate evolutionary responses of phenological differentiation for sibling species: biosystematics in the *Macrosteles fascifrons* complex (Homoptera: Cicadellidae). *Annals of the Entomological Society of America*, 82: 250-261.

Strauch, J.G. Jr., 1984. Use of homoplastic characters in compatibility analysis. *Systematic Zoology*, 33: 167-177.

Strauss, R.E., (in press). Parsimonious mapping of continuous characters onto evolutionary trees for studies of character evolution. *Paleobiology*.

Strauss, R.E., (this volume). Lepidopteran wing morphology: the multivariate analysis of size, shape, and allometric scaling.

Sundberg, P., 1989. Shape and size-constrained principal components analysis. *Systematic Zoology*, 38: 166-168.

Wood, T.K. and Pesek, J.D., (this volume). Pronotal shape: a source of confusion or panacea in systematic studies of treehoppers (Homoptera: Membracidae)?

THE ANALYSIS OF SHAPE VARIATION USING ORDINATIONS OF FITTED FUNCTIONS

F. James Rohlf

Department of Ecology and Evolution, State University of New York, Stony Brook, New York 11794, U.S.A.

Abstract

This paper is concerned with morphometric methods that can be used for the description of shapes of structures with few (if any) landmarks. A useful approach is to fit a function (such as polynomials, splines, Fourier, or "eigenshapes") to the outline shape of the structure, and then to use the parameters of the fitted function as descriptors of the shape. A number of such approaches are described and their usefulness compared for both portions of an outline between two landmarks and for complete outlines. It is shown that many of the different methods result in parameters that can be expressed as linear transformations of the input coordinates. The transformations are not orthogonal for all methods. Thus, ordination analyses, such as principal components analysis of the OTUs (specimens, taxa), are not expected to give the same results. Analyses that standardize the within-group covariance structure (such as canonical variates analysis) are invariant with respect to linear transformations of the variables, whether they are orthogonal or not. Thus, the problem of discrimination between known groups is not effected by this problem.

1 Introduction

Morphometrics is an important area of current research in systematic biology. Most numerical taxonomic analyses (both phenetic and cladistic) are built upon the assumption that one knows how to describe the diversity of variation in size and shape that is found among the organisms under study.

Unfortunately, this is often not true. There has been confusion about the appropriateness of different methods, and there seems to be little consensus on the situations in which different analyses are most appropriate. The problem is complex because structures can vary by being bent, enlarged in one region, stretched, twisted, etc. When the pattern of variation is simple and the mode of growth understood, it may be possible to find a meaningful parameter such as a particular ratio or angle (e.g., Raup, 1966). In most cases it is necessary to use a more brute force empirical approach.

The most appropriate approach depends, in part, on how many morphological landmarks are recognized. This paper is limited to cases where there are few, if any, landmarks. When the shape of a structure is well represented by a collection of landmarks, more powerful approaches are possible which take advantage of this additional information (see, for example, Bookstein, 1989).

One approach used to describe the shapes of complex structures with few landmarks is to digitize coordinates of points along the outline, fit some function to the points, and then to use the parameters of this function as shape descriptors. The most common example is the use of Fourier analysis to fit sums of sine and cosine functions to an outline. Other examples include the use of polynomials, cubic splines, eigenfunctions, and Bezier functions. This paper compares these methods, especially in terms of the application of ordination analyses that are based on these different sets of parameters.

Although this paper contains formulas that are useful in practical applications, its emphasis is on the relationships among the methods. Texts, such as Green and Carroll (1976) or Graybill (1969), can be referred to for proofs of most of the matrix relationships that are used here. Most texts on multivariate analysis (e.g., Krzanowski 1988) also provide summaries of matrix algebra. In order to simplify some of the equations presented here, equally spaced points along an outline will be assumed when relatively simple mathematical formulations can be achieved. This will make the important properties of each method easier to see and compare with those of other methods. Because the properties of linear transformations of vector spaces play an important role in this paper, a summary of the mathematical properties of linear transformations are reviewed below.

Let

$$\mathbf{B} = \mathbf{T}\mathbf{A}, \tag{1}$$

where $\mathbf{A}$, for example, is a p variables by n specimens data matrix which can be interpreted as giving the coordinates of n points in a p-dimensional space, $\mathbf{T}$ is a $p \times p$ transformation matrix, and $\mathbf{B}$ is the resulting $p \times n$ matrix of the n points in a new p-dimensional space. Alternatively, the transformation $\mathbf{B} = \mathbf{A}\mathbf{T}$ can occur, where $\mathbf{A}$ and $\mathbf{B}$ are as before, but $\mathbf{T}$ is now an $n \times n$ matrix. In either case, if $\mathbf{T}$ is an orthonormal matrix ($\mathbf{T}^t\mathbf{T} = \mathbf{T}\mathbf{T}^t = \mathbf{I}$, which implies that $\mathbf{T}^t = \mathbf{T}^{-1}$), then the transformation will not only be linear but also orthogonal, and the distances between the n points in the p-dimensional space will be the same whether computed from the coordinates given in the $\mathbf{A}$ or $\mathbf{B}$ matrices. This can be demonstrated quite easily. Let B_i and B_j be two columns of the $\mathbf{B}$ matrix. This distance between points i and j is

$$d_{ij} = \sqrt{(B_i - B_j)^t (B_i - B_j)} \quad (2)$$

$$= \sqrt{(\mathbf{T}A_i - \mathbf{T}A_j)^t (\mathbf{T}A_i - \mathbf{T}A_j)} \quad (3)$$

$$= \sqrt{(A_i - A_j)^t \mathbf{T}^t \mathbf{T} (A_i - A_j)} \quad (4)$$

$$= \sqrt{(A_i - A_j)^t (A_i - A_j)} . \quad (5)$$

What this means is that the relationships among the points will be shown to be the same by methods based on distances (such as principal components analysis, principal coordinates analysis, and canonical variates analysis), and will also be the same whether based on the **A** or **B** matrices (Anderson, 1984).

If **T** is not an orthonormal matrix, then the distances between points need not be the same in the two spaces. A nonorthogonal transformation implies a uniform "stretching" of the multivariate space in some directions and a "compression" in others. This will change the directions that correspond to maximum variance, and hence will effect the orientation of the principal component axes. Krzanowski (1988) demonstrated that even if **T** were just a diagonal matrix with unequal elements the principal components would not bear a simple relationship to those based on the untransformed data. The properties of orthogonal linear transformations will be used below to demonstrate the equivalence, or lack of equivalence, of different methods.

On the other hand, analyses such as linear discriminant function analysis, canonical variates analysis, and other analyses based on Mahalanobis' generalized distances (Mahalanobis, 1936) are invariant to nonsingular linear transformations, whether orthogonal or not. This can be demonstrated as

$$D_{ij} = \sqrt{(B_i - B_j)^t \mathbf{S}_{\mathrm{B}}^{-1} (B_i - B_j)} \quad (6)$$

$$= \sqrt{(\mathbf{T}A_i - \mathbf{T}A_j)^t \mathbf{T}^{t-1} \mathbf{S}_{\mathrm{A}}^{-1} \mathbf{T}^{-1} (\mathbf{T}A_i - \mathbf{T}A_j)} \quad (7)$$

$$= \sqrt{(A_i - A_j)^t \mathbf{T}^t \mathbf{T}^{t-1} \mathbf{S}_{\mathrm{A}}^{-1} \mathbf{T}^{-1} \mathbf{T} (A_i - A_j)} \quad (8)$$

$$= \sqrt{(A_i - A_j)^t \mathbf{S}_{\mathrm{A}}^{-1} (A_i - A_j)} , \quad (9)$$

where $\mathbf{S}_{\mathrm{A}}$ and $\mathbf{S}_{\mathrm{B}}$ are within-groups variance-covariance matrices based on the original and transformed data, respectively. The above relationships hold as long as $\mathbf{T}^{-1}$ exists (i.e., as long as the transformation does not imply a reduction in dimensionality of the space).

2 Capturing an Outline

One of the simplest methods to fully "capture" the shape of an organism is to record the coordinates of points closely spaced along its outline. This can be done easily with the aid of microcomputer-based coordinate digitizers and video image analysis systems. Different methods can be used to express the coordinates of each point, not just the usual rectilinear coordinate system. For example, polar coordinates are often used in studies using Fourier analysis.

Mathematically, it does not matter which coordinate system is used if the points from one coordinate system can be transformed to another. For example, conversion from the usual rectilinear coordinate system to polar coordinates can be done using the standard equations

$$\rho = \sqrt{(x - x_c)^2 + (y - y_c)^2} \tag{10}$$

$$\theta = \tan^{-1} \frac{y - y_c}{x - x_c},$$

where x_c, y_c are the coordinates of a central reference point which is to be used as the origin of the polar coordinate system. Reconversion, from polar back to rectilinear coordinates, uses the equations

$$x = \rho \cos \theta + x_c \tag{11}$$

$$y = \rho \sin \theta + y_c .$$

The statistical consequences of the use of different coordinate systems can be very important, however. Most statistical procedures are designed to give equivalent results (significance levels, etc.) under different linear transformations of the data, but the same method applied after a nonlinear transformation can give very different results. The use of a particular coordinate system, rather than another, emphasizes different aspects of shape. Unless a theoretical model exists for the expected variation in shape of a collection of organisms, the choice of alternative coordinate systems is usually arbitrary. A polar coordinate system might be appropriate if it is known that a set of points differ mainly in their distance from a central landmark. On the other hand, if variation were mostly due to expansion in one direction, then aligning one of the coordinate axes with that direction would enable such variation to be captured as a single coordinate. These simple approaches are not expected to be sufficient in general, however, because organisms can differ in complex ways, as shown quite clearly by the complexity of most of the transformation grids of Thompson (1948).

3 Parametrization of an Outline

Using coordinates of large numbers of points along an outline is not a very efficient means to describe a shape. Raw coordinates are expected to be highly correlated with each other and, thus, contain large amounts of redundant information. It will be shown below that in many cases the parameters of functions fitted to the outline correspond to a simple transformation of the original set of coordinates to another.

We must distinguish two major classes of outlines because they require different methods. First, there are outlines that correspond to *open contour curves*. Two landmarks are required, one to indicate where the curve begins, and the other where it ends. Second, there are outlines that correspond to *closed contour curves.* These curves represent the complete outline of an organism or a structure. In most applications, there is assumed to be a morphological landmark that is used as a reference point for the start of the contour. In some cases, an additional landmark is used as a reference point for the center of a polar coordinate system, or as a reference to enable each object to be rotated to a standard orientation.

For both classes, there are several approaches that can be used to select the points for the functions to describe the curves:

(1) An a priori defined coordinate system can be used. For example, rather using the simple rectilinear coordinates to describe shell shape, Raup (1966) used parameters that reflected an understanding of the way in which shells grow. The advantage of the use of such coordinates is that it is easier to give direct geometrical and biological interpretation to the measurements.

(2) A family of convenient functions (such as polynomials or trigonometric series) can be used, with the resulting coefficients as parameters.

(3) An "empirical function" (sensu Lohmann, 1983) can be used. The approach is similar to the previous, but the function being fitted is simply a linear combination of the eigenvectors of the data matrix. Relatively few parameters (the parameters are merely the projections of each object onto the eigenvectors) are needed to obtain a certain degree of fit; but, each function is represented by as many numbers as there are data points. An interesting aspect of this method is that an ordination analysis is performed as part of the fitting process.

3.1 Open Curves

The subsections below are concerned with fitting functions to curves representing the shape of the outline of a structure between two landmarks.

3.1.1 Polynomial Functions

When the y-coordinates along the outline of a structure correspond to a monotonically increasing set of x-coordinates, then polynomials such as $y = b_0 + b_1x + b_2x^2 + b_3x^3 + \ldots + b_{p-1}x^{p-1}$ can be fit to the y-coordinates. The parameters, b_j, can be found by first forming a matrix, $\mathbf{Z}$, consisting of a column containing the y-coordinates of p points along the outline of a structure, and then constructing an $\mathbf{X}$ matrix, whose first column (which we will call X_0) is a vector with all elements set equal to 1. The second column, X_1, of the latter contains the vector of x-coordinates. The elements of the other columns are set equal to powers of X_1 (i.e., $X_i = X_1^i$). The vector of parameters, $\mathbf{B}$, of the least-squares best fitting solution can be obtained using the equation $\mathbf{B} = (\mathbf{X}^t\mathbf{X})^{-1}\mathbf{X}^t\mathbf{Z}$. The details of the necessary computations are described in many textbooks on statistics (e.g., Sokal and Rohlf, 1981: section 16.6). Most major statistical packages provide for their computation.

The equation for $\mathbf{B}$ can be interpreted as a linear transformation, $\mathbf{B} = \mathbf{TZ}$, of the original data matrix, where $\mathbf{T} = (\mathbf{X}^t\mathbf{X})^{-1}\mathbf{X}^t$. This does not correspond to an orthogonal transformation. In fact, the matrix $\mathbf{X}^t\mathbf{X}$ will often be nearly singular and, consequently, the computation of $\mathbf{B}$ will present serious computational difficulties. If the higher order terms are ignored, because they usually have very small coefficients, then the transformation corresponds to a nonrigid rotation followed by a projection onto a lower dimensional subspace. The effect of ignoring the higher order terms is to fit a smooth curve to the observed data, which may correspond to an irregular curve. If the x-coordinates are equally spaced, then we can simplify the $\mathbf{X}$ matrix by replacing the ith column with the coefficients of the ith degree orthogonal polynomial (for example, see Sokal and Rohlf, 1981: 677-683). If all of the possible terms are used (up to degree p - 1), then $\mathbf{T}$ will then be an orthogonal matrix.

3.1.2 **Bezier Functions**

Given an arbitrary ordered sequence of p points in a two- or three-dimensional space, it is possible to construct a $q = p - 1$ degree Bezier function (Bezier, 1972) that passes through these points. The parameters of a Bezier curve are the locations of $q + 1$ vertices of an open polygon. The first and last vertices are the first and last data points; the other vertices are constructed. These vertices define a unique polynomial function of degree q that interpolates smoothly between the first and last vertices. Its use implies that two morphological landmarks (the two end points of the curve) exist, and that there is interest in describing the shape of the curve between these two points.

The points need not correspond to a curve that is single-valued because the polynomial is a function of an intrinsic parameter, t, which is the chordal distance along the curve that is scaled to go from 0 to 1 between the first observed point and the last. The vector of coordinates of the Bezier vertices can be constructed as follows. The Bernstein basis function is defined as

$$J_{q,k}(t) = \binom{q}{k} t^k (1 - t)^{q - k} \tag{12}$$

where

$$\binom{q}{k} = \frac{q!}{k!(q - k)!} \, . \tag{13}$$

The degree of the polynomial is q and k is an index to the Bezier vertex to be constructed (numbered from $k + 1 = 1$ to $q + 1$. The observed coordinate of the jth point along the curve can be expressed as

$$Z_j = \sum_{k=0}^{q} X_k \, J_{q,k}(t_j) \, , \tag{14}$$

where t_j is the cumulative chordal Euclidean distance of the ith along the observed curve, normalized to that $0 \le t_j \le 1$. X_k is a vector containing the coordinates of the kth Bezier vertex.

Thus, we may write

$$\mathbf{Z} = \mathbf{JB}, \tag{15}$$

where $\mathbf{J}$ is a $p \times q + 1$ matrix of coefficients, $\mathbf{Z}$ is a matrix with rows corresponding to the coordinates of the p observed points, and $\mathbf{B}$ is a matrix with rows corresponding to $q + 1$ coordinates of the fitted Bexier vertices. With $q + 1 = p$ points, the matrix $\mathbf{J}$ will be square and the equation can be solved for the $\mathbf{B}$ as

$$\mathbf{B} = \mathbf{J}^{-1}\mathbf{Z}. \tag{16}$$

Lower order polynomials can be fitted. This is desirable when many coordinates have been recorded, but the shape of the outline is fairly simple. Engel (1986) shows how to construct least-squares estimates of the vertices. When $q < p - 1$ the estimated location of the $q + 1$ vertices can be obtained as

$$\hat{\mathbf{B}} = (\mathbf{J}^t\mathbf{J})^{-1}\mathbf{J}^t\mathbf{Z}, \tag{17}$$

where the various matrices are as defined above. Points along the estimated curve can be computed as

$$\hat{\mathbf{Z}} = \mathbf{J}\hat{\mathbf{B}}. \tag{18}$$

The $q + 1$ vertices are such that the residual sum of squares,

$$SS = \mathrm{tr}[(\mathbf{Z} - \hat{\mathbf{Z}})(\mathbf{Z} - \hat{\mathbf{Z}})^t], \tag{19}$$

is minimized. Engel (1986) suggests that the degree of the polynomial be taken as low as possible in order to minimize error due to numerical instabilities and rounding off during the matrix inversion. It should be noted that the usual statistical tests cannot be applied to the least squares estimates of the Bezier parameters because the independent variable, t, is computed from the coordinates of the data points that were used as the dependent variable.

This method has several interesting properties. The fitted parameters are easy to interpret. The polygon defined by the coordinates of the vertices forms a convex hull around the observed curve. The position of the ith vertex is mostly a function of nearby data points (but there is also some influence due to the global configuration of points). The parameters, $\hat{\mathbf{B}}$, are linear functions of the original data points because

$$\hat{\mathbf{B}} = \mathbf{TZ}, \tag{20}$$

where

$$\mathbf{T} = (\mathbf{J}^t\mathbf{J})^{-1}\mathbf{J}^t. \tag{21}$$

Because **T** will not usually be an orthogonal matrix, ordination analyses based on these parameters will not give the same results as those based on the original data points, even when the fit is perfect. However, because closely spaced data points along a smooth curve are obviously highly redundant (the coordinates of adjacent points must be highly correlated), the constructed vertices can be viewed as representing a new set of coordinates that are much less redundant, and hence expected to be more efficient descriptors of morphometric variation. Engel (1986) shows examples of fitting Bezier curves to shell valves of *Anodonata* and of human sagittal profiles of male and female skulls of various ages.

3.1.3 **Cubic Splines**

Cubic splines can be fitted to data in which the x-coordinates of the points being fitted are monotonically increasing (the next section will described parametric cubic splines which do not have this limitation). A cubic spline is a series of cubic polynomial functions that are pieced together, so that the fitted function passes through all of the observed data points and its first and second derivatives are continuous throughout the range being fitted. In a practical application, a decision has to be made about the properties of the fitted curve at each end of its range. One choice is to require the second derivatives to be zero at each end point (this results in a "natural spline"). Another choice is to set the first derivatives equal to the actual derivatives of a known, underlying function (these are, of course, usually unknown in morphometrics). Atkinson and Harley (1983), De Boor (1978), and Press et al. (1986) give helpful introductions to cubic splines.

Let y_i, for $i = 1, \ldots, p$, be a set of y-coordinates along an outline for a set of monotonically increasing x-coordinate values. For any interval (e.g, x_i to $x_i + 1$), we can interpolate for y using the cubic polynomial

$$y = Ay_i + (1 - A)y_{i+1} + Cb_i + Db_{i+1}, \qquad (22)$$

where $A = \dfrac{x_{i+1} - x}{\Delta x_i}$, $C = \frac{1}{6}(A^3 - A)(\Delta x_i)^2$, $D = \frac{1}{6}[(1 - A)^3 - (1 - A)](\Delta x_i)^2$, $\Delta x_i = x_{i+1} - x_i$, and the b_i (as the second derivatives of the interpolating polynomial function) are the parameters of the curve. It can be shown that the b_i satisfy

$$b_{i-1}\Delta x_{i-1} + 2b_i(\Delta x_i + \Delta x_{i-1}) + b_{i+1}\Delta x_i = \frac{6}{\Delta x_i}(y_{i+1} - y_i) - \frac{6}{\Delta x_{i-1}}(y_i - y_{i-1}), \qquad (23)$$

for $i = 2, 3, \ldots, p - 1$. This corresponds to $p - 2$ equations with p unknowns. To get a solution, we can add the equations $b_1 = 0$ and $b_p = 0$. The spline equations can be expressed as

$$\mathbf{CB} = \mathbf{A}, \qquad (24)$$

where **B** is the column vector of p parameters,

$$\mathbf{C} = \begin{pmatrix} 1 & 0 & 0 & \dots & 0 & 0 \\ \Delta x_1 & 2(\Delta x_1 + \Delta x_2) & \Delta x_2 & \dots & 0 & 0 \\ 0 & \Delta x_2 & 2(\Delta x_2 + \Delta x_3) & \dots & 0 & 0 \\ \vdots & \vdots & \ddots & \ddots & \vdots & \vdots \\ 0 & 0 & 0 & \dots & 2(\Delta x_p + \Delta x_{p-1}) & \Delta x_p \\ 0 & 0 & 0 & \dots & 0 & 1 \end{pmatrix}, \tag{25}$$

and

$$\mathbf{A} = 2 \begin{pmatrix} 0 \\ \frac{3}{\Delta x_2}(y_3 - y_2) - \frac{3}{\Delta x_1}(y_2 - y_1) \\ \frac{3}{\Delta x_3}(y_4 - y_3) - \frac{3}{\Delta x_2}(y_3 - y_2) \\ \vdots \\ \frac{3}{\Delta x_{p-1}}(y_p - y_{p-1}) - \frac{3}{\Delta x_{p-2}}(y_{p-1} - y_{p-2}) \\ 0 \end{pmatrix}. \tag{26}$$

This tridiagonal system of equations can be easily solved as

$$\mathbf{B} = \mathbf{C}^{-1}\mathbf{A}. \tag{27}$$

The use of cubic splines results in as many parameters as there are data points; this is expected because the interpolating spline function is required to pass through all p points. If, as is usually the case, recorded points exist along a fairly smooth curve, its preferable to fit a curve with many fewer parameters. One procedure is to resample the curve with fewer points and then check that the new interpolated function stays fairly close to the original points.

The matrix **A** can be factored as

$$\mathbf{A} = 6\mathbf{W}(\mathbf{D} - \mathbf{P}^t\mathbf{D}\mathbf{S})\mathbf{Y}, \tag{28}$$

where **W** is a diagonal matrix of $1/\Delta x_i$ values, **P** is a permutation matrix of the form

$$\mathbf{P} = \begin{pmatrix} 0 & 0 & 0 & \dots & 0 & 1 \\ 1 & 0 & 0 & \dots & 0 & 0 \\ 0 & 1 & 0 & \dots & 0 & 0 \\ \vdots & \vdots & \ddots & \ddots & \vdots & \vdots \\ 0 & 0 & 0 & \dots & 0 & 0 \\ 0 & 0 & 0 & \dots & 1 & 0 \end{pmatrix}, \tag{29}$$

and **D** is a matrix that computes first differences

$$\mathbf{D} = \begin{pmatrix} 0 & 0 & 0 & \dots & 0 & 0 \\ 0 & -1 & 1 & \dots & 0 & 0 \\ 0 & 0 & -1 & \dots & 0 & 0 \\ \vdots & \vdots & \ddots & \ddots & \vdots & \vdots \\ 0 & 0 & 0 & \dots & -1 & 1 \\ 0 & 0 & 0 & \dots & 0 & 0 \end{pmatrix}, \tag{30}$$

Matrix **P** is an orthonormal matrix because $\mathbf{PP}^t = \mathbf{I}$. The importance of this factoring is that we can now express the parameters as

$$\mathbf{B} = \mathbf{C}^{-1}6\mathbf{W}(\mathbf{D} - \mathbf{P}^t\mathbf{DS})\mathbf{Z}, \tag{31}$$

were **Z**, is a matrix of y-coordinates. Hence, the parameter matrix, **B**, is a linear (but not orthogonal) transformation of the original y-coordinates.

Atkinson and Harley (1983) provide a straight forward Pascal procedure that can be used to compute cubic splines. Press et al. (1986) give space efficient programs in both Fortran and Pascal.

3.1.4 Parametric Cubic Splines

When a curve is complex, so that the x-coordinates need not be monotonically increasing, parametric cubic splines can be used. The method is as described above, but with x and y-coordinates (or even z-coordinates) separately fitted by splines with the cumulative chordal distance, t_i, of the ith point used as the abscissa. That is, in the equations of the previous section, replace Δx_i with Δt_i, and then generalize the matrix **Z** to a $p \times n$ matrix with the columns corresponding to the x, y, and possibly z-coordinates of the p points along the outline. The solution matrix, **B**, then will also have n columns. Except for the fact that t is a nonlinear function of **Y**, Equation 31 demonstrates that a **B** is linear (but not orthogonal) transformation of **Y** in this case also.

Useful general references include Spath (1974) and De Boor (1978). Gasson (1983) describes its application to computer graphics. Evans et al. (1985) describe the method and furnish examples of its application to microfossil outlines and to rock folds.

3.2 Closed Contours

The most usual application of these methods is to complete closed outline contours, such as the outline of a shell, mosquito wing, etc. The methods used most commonly in morphometrics are described below. Other methods also exist that are beyond the scope of the this paper. One of the more unique approaches is Blum's (1973) medial axis transformation. The resulting parameters correspond to coordinates of loci of points that form a "skeleton" in the interior of the region bounded by a closed curve. Another approach is that of moment invariants (Hu, 1962; Rohlf, 1990). The relationship of parameters to the coordinates of the initial points is complex for both of these methods.

3.2.1 Fourier Functions

This method is used very commonly to describe the shapes of complete outlines of biological structures. Fourier analysis has often been applied to the lengths of equally spaced radii in a polar coordinate representation of points along the outline of an organism. This simple approach, however, cannot be used when an outline is so complex that a radial vector intersects the outline at more than one place (i.e., the x-coordinates do not increase monotonically). Zahn and Roskies (1972) proposed a solution to this problem. They suggested that rather than using lengths of radii, ρ, analysis should involve the function

$$\phi^{*}(t) = \phi(t) - t\,, \tag{32}$$

where t is the cumulative chordal distance along the curve (scaled as if it were an angle ranging from 0 to 2π), $\phi(t)$ is the angular difference in orientation between a tangent at the starting point ($t = 0$) and at a point at position t along the outline. This function is designed so that a circle will have $\phi^{*}(t)$ values equal to 0 for all values of t.

The ρ or ϕ^{*} values for the ith specimen are entered as columns of the **Z** matrix. The last value (corresponding to the radial vector at an angle of 2π or to the last tangent angle) is not entered in the matrix, because it is assumed to be identical to the value in the first row (corresponding to an angle of 0). A Fourier analysis can then be expressed as the matrix multiplication

$$\mathbf{B} = \mathbf{F}^{t}\mathbf{Z}\,, \tag{33}$$

where **Z** is as defined above and **F** is a $p \times p$ Fourier transformation matrix defined as follows. The first column, F_1, is $1/\sqrt{p}$ for all rows, i. Let us assume that p is odd. Then, each harmonic from $k = 1$ to $p/2$ generates a pair of columns (from column 2 to $p - 1$), whose elements are

$$\sqrt{\frac{2}{p}}\cos(2\pi ki/p) \tag{34}$$

and

$$\sqrt{\frac{2}{p}}\sin(2\pi ki/p) \tag{35}$$

for even and odd numbered columns, respectively, and i is the row number. If, on the other hand, p is even, then the last column is $1/\sqrt{p}\cos(\pi i)$.

F is an orthogonal matrix (Tunnicliffe-Wilson, 1984). The product, **B**, is a matrix giving the Fourier coefficients (weights) for the contribution of the harmonics (rows) to the shape of each outline (columns). In practice, however, all $p/2$ harmonics are not used because it is usually found that the higher order harmonics have very small amplitudes and contribute relatively little to the variance among the object outlines. By dropping out the higher order harmonics, information is selectively lost about the fine details of the outline shape. Depending upon the application, this may or may not be desirable. The matrix **F** will then have fewer than p columns.

In some studies, a landmark is not available to serve as a homologous starting point on different specimens. In such cases, it is not appropriate to place ρ or ϕ^* values for different specimens as columns in the same **Z** matrix because the alignment of each outline is arbitrary. One solution is to combine the numbers in the pair of columns into a single harmonic amplitude (the sum of the squares of corresponding elements in the pair of columns). This gives values that are independent of the starting position on each outline. I agree with Lohmann (1983) and Bookstein et al. (1985) that if a reference point can be defined, then the information on phase angles should be retained. When the phase angle information is ignored, rather different looking shapes can have identical sets of descriptors (Bookstein et al., 1982). In addition, the resulting parameters no longer represent a linear transformation of the **Z** matrix.

Another solution is to rotate the specimens so that they align as well as possible according to some criterion. Lohmann (1983) rotated each specimen so as to maximize its correlation with a standard reference specimen. Ferson et al. (1985) aligned specimens according to the orientation of the major axis of the best fitting ellipse. Full and Ehrlich (1986) refer to the alignment of two outlines, so that they have maximal agreement, as "mathematical homology."

3.2.2 **Elliptic Fourier Analysis**

The Fourier method that Rohlf and Archie (1984) found to be most effective and convenient in their study was elliptic Fourier analysis (Kuhl and Giardina, 1982). An example of its use in conjunction with a variety of multivariate analyses is given in Ferson et al. (1985). The elliptic Fourier method is based on Fourier decompositions of first differences of the x and y-coordinates (Δx_i and Δy_i) as functions of the cumulative chordal distance of points around the outline. Thus, it is a Fourier analysis of the *change* in coordinates between adjacent points along an outline, as a parametric function of the cumulative chordal distance along the outline. The Fourier coefficients for the kth harmonic of the x-projection of the outline are (Kuhl and Giardina, 1982)

$$A_k = \frac{T}{2p^2\pi^2} \Sigma^p \frac{\Delta x_i}{\Delta t_i} \left(\cos \frac{2\pi k t_i}{T} - \cos \frac{2\pi k t_{i-1}}{T}\right), \qquad (36)$$

$$B_k = \frac{T}{2p^2\pi^2} \Sigma^p \frac{\Delta x_i}{\Delta t_i} \left(\sin \frac{2\pi k t_i}{T} - \sin \frac{2\pi k t_{i-1}}{T}\right),$$

where p is the number of steps in the trace, $\Delta x_i = x_i - x_{i-1}$, Δt_i is the length of the chordal distance between steps i - 1 and i, t_i is the accumulated length of such segments at step i, and T (= t_p) is the total length of the contour as approximated by the trace polygon. Because a complete outline is assumed, $\Delta x_1 = x_1 - x_n$. The coefficients for the y-projection, $\mathbf{C}_i$ and D_i, (and even z-projections) are found in the same way.

For simplicity, assume that the points are equally spaced along the contour (this is not required for the actual application of the method) and that the specimens have been aligned and scaled to a common size. Then, let **Z** be a matrix with p rows and the columns partitioned in the form $\mathbf{Z} = (\mathbf{Z}_x | \mathbf{Z}_y)$, where $\mathbf{Z}_x$ is a matrix of x-coordinates with one column for each of the n objects to be analyzed, and $\mathbf{Z}_y$ a corresponding matrix of y-coordinates. If the number of steps are unequal, then coordinates for different objects cannot be placed together in the same matrix.

For the case of equal steps, Rohlf (1986) showed that the $p \times 2n$ Fourier coefficient matrix is

$$\mathbf{B} = \mathbf{F}^t(\mathbf{I} - \mathbf{P})^t \frac{1}{\Delta t} (\mathbf{I} - \mathbf{P})^t \mathbf{Z} \tag{37}$$

$$= \frac{1}{\Delta t} \mathbf{F}^t(2\mathbf{I} - \mathbf{P} - \mathbf{P}^t)\mathbf{Z} \tag{38}$$

$$= \mathbf{TZ}, \tag{39}$$

where **I** is a $p \times p$ identity matrix, **P** is a permutation matrix of the form defined in Equation 29, and **Z** is the $p \times 2n$ matrix of the x and y-coordinates of each point for each of the n outlines. Equation 38 takes advantage of the fact that $\mathbf{P}^t\mathbf{P} = \mathbf{I}$. The **F** matrix is as defined in a previous section. If the lengths of the t_i's are unequal, then the scaler $1/\Delta t$ has to be replaced with the inverse of a diagonal matrix of segment lengths. The **F** matrix would also have to be adjusted in that case.

Although the **F** matrix corresponds to an orthogonal transformation, **T** does not. Thus, elliptic Fourier analysis corresponds to a linear, but not an orthogonal, transformation of the x, y-coordinates.

3.2.3 **Parametric Cubic Splines**

Evans et al. (1985) present an example in which they fit a parametric cubic spline to a closed curve corresponding to the outline of an ostracode. No special modification of the method for parametric open curves was made. The fact that the first point, x_1y_1, was the same as x_py_p was not taken into account explicitly. Thus, the parameters for this method also represent a linear, but not an orthogonal, transformation of the x, y-coordinates.

3.2.4 **Eigenshape Analysis**

Eigenshape analysis (Lohmann, 1983) represents what seems like a very different approach from those described above. In this case, we do not fit some simple nonlinear curve to the data; instead the function is constructed empirically, as a linear function of the observed data across one or more specimens.

The method begins with x, y-coordinates of p points along the outlines of each of n objects. The following steps are then performed:

(1) The x, y-coordinates of each outline are transformed into normalized net

angular changes (Zahn and Roskies, 1972), as in Equation 32. This representation is invariant to size, translation, and orientation of the object in the x, y-plane, but it is sensitive to the location of the starting point and to whether the outline is traced in a clockwise or counterclockwise direction (hence, it is sensitive to reflection of the object in the x, y-plane).

(2) Each outline must be rotated to a standard orientation. Lohmann (1983) rotated each outline so as to maximize its correlation, or cross covariance, with a standard reference outline.

(3) Lohmann (1983) and Lohmann and Malmgren (1983) then standardized the vector of $\phi^*(t)$ values for each object, but that is not an essential (or perhaps even desirable) step.

(4) Assemble the $\phi^*(t)$ vectors into a $p \times n$ matrix $\mathbf{Z}$.

(5) Perform a singular value decomposition, SVD, of the $\mathbf{Z}$ matrix (Golub and Reinsch, 1970; Eckhart and Young, 1936), with

$$\mathbf{Z} = \mathbf{USV}^t , \tag{40}$$

where the k columns of $\mathbf{V}$ are the normalized eigenvectors of the $n \times n$ matrix $\mathbf{Z}^t\mathbf{Z}$, $\mathbf{S}$ is a $k \times k$ diagonal matrix of the square roots of the eigenvalues of $\mathbf{Z}^t\mathbf{Z}$, the k columns of $\mathbf{U}$ are the normalized eigenvectors of the $p \times p$ matrix $\mathbf{ZZ}^t$, and k is less than, or equal to, the rank of the $\mathbf{Z}$ matrix (usually n because there usually are more points than specimens).

The columns of $\mathbf{U}$ are empirical shape functions. The matrix $\mathbf{SV}^t$ represents the matrix of parameters for each outline. This is because the ith column of $\mathbf{Z}$ (which describes the ith outline of the object) can be expressed as

$$\mathbf{Z}_i = \mathbf{U}(s_i \mathbf{V}_i^t) , \tag{41}$$

where s_i is the ith element of $\mathbf{S}$ and $\mathbf{V}_i{}^t$ is the ith column of $\mathbf{V}^t$. Because $\mathbf{SV}^t = \mathbf{U}^t\mathbf{Z}$ the parameters represent an orthogonal transformation of the $\mathbf{Z}_i$ (but not of the original x, y-coordinates because the $\phi^*(t)$ function is being analyzed). The ith column of $\mathbf{SV}^t$ gives the projections (weights) of the ith object onto the k empirical shape functions.

4 Ordination Analyses

Although there are other types of ordination analyses, this paper is concerned only with linear methods, such as principal components analysis, PCA, and canonical variates analysis, CVA (for a discussion of factor analysis, see Mulaik, 1972). A PCA is usually performed by projecting the data for each specimen onto the eigenvectors that are computed from a correlation or variance-covariance matrix among the variables. It is convenient to note, however, that a PCA can be expressed in terms of a singular value decomposition. Let $\mathbf{Z}$ be the p variables by n objects data matrix (standardized by rows if a PCA based on a correlation matrix is desired). Then,

$$\mathbf{Z} = \mathbf{U \Lambda V}^t , \tag{42}$$

where $\mathbf{L}$ is a diagonal matrix of eigenvalues. The columns of $\mathbf{V}\Lambda^{1/2}$ give the eigenvectors that would be obtained from the variance-covariance matrix (or correlation matrix if $\mathbf{Z}$ were standardized). The ith row of $\mathbf{U}\Lambda^{1/2}$ gives the projections of the ith object onto each of the principal component axes. From the decomposition given above,

$$\mathbf{B} = \mathbf{T}^t\mathbf{U}\Lambda\mathbf{V}^t\,. \tag{43}$$

If it were not for the fact that the correlation (or variance-covariance) matrix is itself of interest, the SVD approach would be used more routinely.

It was shown in previous sections that various coefficients (e.g., elliptic Fourier, Bezier, cubic splines, etc.) represented linear transformations of the input data. What are the expected effects of using these coefficients instead of the raw data? If an open curve is fit using orthogonal polynomials, for example, the matrix of coefficients is $\mathbf{B} = \mathbf{T}^t\mathbf{Z}$. By substituting $\mathbf{Z}$, with its singular value decomposition, this becomes

$$\mathbf{B} = \mathbf{T}^t\mathbf{U}\Lambda\mathbf{V}^t\,. \tag{44}$$

Because $\mathbf{T}$ is an orthonormal matrix, $\mathbf{T}^t\,\mathbf{U}$ is the left eigenvector matrix of $\mathbf{B}$. Thus, the effect of using orthogonal polynomial coefficients, rather than raw coordinates, is merely to perform a rigid rotation of the eigenvectors, relative to the original variables (x and y-coordinates). The ordination of the objects (projections onto the principal component axes), given by $\mathbf{V}$ is unchanged. However, if the linear transformation does not correspond to an orthonormal matrix, then there is no such simple relationship, and Equation 44 will no longer correspond to a SVD of $\mathbf{Z}$.

When there are a priori groups, canonical variates analysis is often used both to test for group differences and to create a plot that shows the best view of the among-group differences relative to the within-group variation. This technique is a generalization of the usual linear discriminant function analysis to allow for more than two groups. The distance between points i, j in the canonical variates space is proportional to the Mahalanobis' generalized distance, D_{ij}, defined in Equation 6. Because that distance is invariant to nonsingular linear transformation (whether orthogonal or not), the configuration of points, group means, etc., in the canonical variates space is also invariant to these transformations (see also, Rao, 1973).

5 Discussion

Note that the function t_i was treated as a "given" in the accounts describing parametric representations of curves. That function is, of course, a nonlinear function of the x and y-coordinates. Thus, methods used on parametric descriptions of curves and those based solely on a single coordinate (such as ρ or y) cannot be related exactly to the parameters by a simple linear transformation. The error that this introduces is expected to be relatively small in most practical applications (where most curves are similar in shape).

There are several important results that should not be lost in the details of the matrix algebra presented above. First, most methods for describing outlines do not yield parameters that imply the same relationships among the objects under study when analyzed by methods based on Euclidean distances; even though the parameters fitted by the various methods differ (at least approximately) by only a linear transformation. This means that it is necessary to justify the choice of descriptors that are used (e.g., why one chooses Elliptic Fourier coefficients rather than Bezier points). Practical computational considerations alone cannot dictate which method should be used because any curve can be fit by many different types of functions. This problem is quite fundamental, both for morphometrics and for numerical taxonomy. In what space should the similarities or difference among organisms be measured? The solution is not clear unless a model specifying the most appropriate metric to measure similarities can be determined.

A second important point is that ordinations based on generalized distances (such as canonical variates analysis) are invariant to linear transformations of the variables. Thus, alternative methods of description, that differ only by a linear transformation, will yield the same results. In this case, the decision of which method should be used can be based solely on practical considerations, such as the availability of software or computational expense. This does not "solve the problem" for morphometrics and numerical taxonomy, however. Generalized distances express differences among groups, in terms of the probability of a specimen from one group being identified (incorrectly) as belonging to another. Although this represents a solution to the problem of distinguishing different groups (however defined), that problem is quite different from the one involving the construction of groups based on their overall similarity and differences from other groups in the study.

6 Acknowledgments

This work was supported, in part, by the National Science Foundation under grant BSR-86-14382. Comments and suggestions by members of the Numerical Taxonomy Discussion group, at the State University of New York at Stony Brook, contributed to this study and are gratefully acknowledged. This is contribution number 793 from the Graduate Studies in Ecology and Evolution, State University of New York at Stony Brook.

7 References

Anderson, T.W., 1984. An Introduction to Multivariate Statistical Analysis, (Second Edition). John Wiley and Sons, New York, 675 pp.

Atkinson, L.V. and Harley, P.J., 1983. An Introduction to Numerical Methods with Pascal. Addison-Wesley, Reading, Massachusetts, 300 pp.

Bezier, P., 1972. Numerical Control; Mathematics and Applications. A.R. Forrest and A.F. Pankhurst (Translators). John Wiley and Sons, New York, 240 pp.

Blum, H., 1973. Biological shape and visual science (part I). *Journal of Theoretical Biology*, 38: 205-287.

Bookstein, F.L., 1989. Principal warps: thin-plate splines and the

decomposition of deformations. *I.E.E.E. Transactions on Pattern Analysis and Machine Intelligence*, 11: 567-585.
Bookstein, F.L., Chernoff, B., Elder, R.L., Humphries, J.M., Smith, G.R. and Strauss, R.E., 1985. Morphometrics in Evolutionary Biology. Special Publication, 15. Academy of Natural Sciences, Philadelphia, 277 pp.
Bookstein, F.L., Strauss, R.E., Humphries, J.M., Chernoff, B., Elder, R.L. and Smith, G.R., 1982. A comment upon the uses of Fourier methods in systematics. *Systematic Zoology*, 31: 85-92.
De Boor, C., 1978. A Practical Guide to Splines. Springer-Verlag, New York, 392 pp.
Eckhart, C. and Young, G., 1936. The approximation of one matrix by another of lower rank. *Psychometrika*, 1: 211-318.
Engel, H., 1986. At least squares method for estimation of Bezier-curves and surfaces and its applicability to multivariate analysis. *Mathematical Biosciences*, 79: 155-170.
Evans, D.G., Schweitzer, P.N. and Hanna, M.S., 1985. Parametric cubic splines and geological shape descriptions. *Mathematical Geology*, 17: 611-624.
Ferson, S., Rohlf, F.J. and Koehn, R.K., 1985. Measuring shape variation of two-dimensional outlines. *Systematic Zoology*, 34: 59-68.
Full, W.E. and Ehrlich, R., 1986. Fundamental problems associated with "eigenshape analysis" and similar "factor" analysis procedures. *Mathematical Geology*, 18: 451-463.
Gasson, P.C., 1983. Geometry of Spatial Forms. Ellis Horwood Limited, Chichester, U.K., 601 pp.
Golub, G.H. and Reinsch, C., 1970. Singular value decomposition and least squares solutions. *Numerische Mathematik*, 14: 403-420.
Graybill, F.A., 1969. Introduction to Matrices with Applications in Statistics. Wadsworth, Belmont, California, 372 pp.
Green, P.E. and Carroll, J.D., 1976. Mathematical Tools for Applied Multivariate Analysis. Academic Press, New York, 376 pp.
Hu, M.K., 1962. Visual pattern recognition by moment invariants. *I.R.E. Transactions on Information Theory*, 8: 179-187.
Krzanowski, W.J., 1988. Principles of Multivariate Analysis: a User's Perspective. Clarendon Press, Oxford, U.K., 563 pp.
Kuhl, F.P. and Giardina, C.R., 1982. Elliptic Fourier features of a closed contour. *Computer Graphics and Image Processing*, 18: 236-258.
Lohmann, G.P., 1983. Eigenshape analysis of microfossils: a general morphometric procedure for describing changes in shape. *Mathematical Geology*, 15: 659-672.
Lohmann, G.P. and Malmgren, B.A., 1983. Equatorward migration of *Globorotalia truncatulinoides*. *Paleobiology*, 9: 414-421.
Mahalanobis, P.C., 1936. On the generalized distance in statistics. *Proceedings of the National Institute of Science, India*, 2: 49-55.
Mulaik, S.A., 1972. The Foundations of Factor Analysis. McGraw-Hill Book Company, New York, 453 pp.
Press, W.H., Flannery, B.P., Teukolsky, S.A. and Vetterling, W.T., 1986. Numerical Recipes: the Art of Scientific Computing, (Fortran and Pascal Version). Cambridge, New York, 818 pp.
Rao, C.R., 1973. Linear Statistical Inference and Its Applications, (Second Edition). John Wiley and Sons, New York, 625 pp.

Raup, D.M., 1966. Geometric analysis of shell coiling: general problems. *Journal of Paleontology*, 40: 1178-1190.

Rohlf, F.J., 1986. The relationships among eigenshape analysis, Fourier analysis, and the analysis of coordinates. *Mathematical Geology*, 18: 845-854.

Rohlf, F.J., 1990. An overview of image processing and analysis techniques for morphometrics. In: F.J. Rohlf and F.L. Bookstein (Editors), Proceedings of the Michigan Morphometrics Workshop. Special publication number 2, Michigan Museum of Zoology, Ann Arbor, 37-60 pp.

Rohlf, F.J. and Archie, J.W., 1984. A comparison of Fourier methods for the description of wing shape in mosquitoes (Diptera: Culicidae). *Systematic Zoology*, 33: 302-317.

Sokal, R.R. and Rohlf, J.F., 1981. Biometry, (Second Edition). W.H. Freeman and Company, New York, 859 pp.

Spath, H., 1974. Spline Algorithms for Curves and Surfaces. Utilitas Mathematica Publishing, Winnipeg, Canada, 198 pp.

Thompson, D.W., 1948. On Growth and Form. Cambridge University Press, Cambridge, U.K., 1116 pp.

Tunnicliffe-Wilson, G., 1984. Time series. In: E. Lloyd (Editor), Handbook of Applicable Mathematics, Volume VI, Part B. John Wiley and Sons, New York, 783-846 pp.

Zahn, C.T. and Roskies, R.Z., 1972. Fourier descriptors for plane closed curves. *I.E.E.E Transactions on Computers*, C-21: 269-281.

PRACTICAL MACHINE VISION FOR MORPHOMETRIC DATA ACQUISITION

Christopher A. Meacham

Department of Integrative Biology and Office of Information Systems and Technology, University of California, Berkeley, California 94720, U.S.A.

Abstract

Machine vision techniques permit the capture of morphometric data with greatly increased speed and accuracy over manual methods. Machine vision software can be designed so that it contributes to effective morphometric data acquisition by allowing direct interactive intervention in the feature extraction process, by permitting easy modification and editing of extracted features, and by providing a command structure to implement a variety of measurement schemes. It is important to consider many elements in setting up a machine vision workstation for morphometric data acquisition. Chief among these elements are illumination, optics and camera. The present goal of machine vision research in systematic and evolutionary biology is to extend present techniques to allow automated data acquisition across a broader range of circumstances. The issues of how effectively to use machine vision and how best to develop machine vision methods deserve attention from systematic and evolutionary biologists.

1 Introduction

Much research in systematic and evolutionary biology is limited by the difficulties of data acquisition, either in terms of limited personal time or limited funds for research assistants. Empirical sciences depend fundamentally on the quality and quantity of empirical data that are captured. Data capture can be accomplished by machine vision techniques at lower cost and with better accuracy than by other means. When appropriately implemented in an interactive computer program, so that the investigator has control over the process, even the basic image analysis techniques described below typically yield a ten-fold increase in data capture speed (e.g., Lindberg, 1990). Some users attain an increase of as much as 100 times that of manual measurement, capturing data in a few hours rather than several weeks. Such a drastic reduction in data capture effort demonstrates the importance and potential of machine vision in systematic and evolutionary biology. However, I will not focus on the future of machine vision or theoretical issues that are of interest to workers who will develop new machine vision techniques. Instead, I will describe hardware and software techniques that have been made to work successfully, and that can be used now to capture morphometric data. I will also discuss practical issues involved in setting up and using a machine vision workstation.

The techniques described here have been incorporated in a microcomputer program, MorphoSys (Meacham and Duncan, 1987), which runs in an MS-DOS™ environment. MorphoSys was developed in collaboration with Thomas Duncan at the University Herbarium and has been supported by the University of California and by the IBM Corporation.

2 Machine Vision in a Microcomputer Environment

The advent of the microcomputer has made machine vision techniques accessible to systematic and evolutionary biologists. Microcomputers have greatly reduced the cost of computing power, and have allowed manufacturers to develop inexpensive specialized hardware in the form of expansion boards, usually called "frame grabbers," for image capture and display. (A single video image is called a "frame.") These frame grabbers are independent processors that occupy a single circuit board and plug into the main bus, or internal communication path, of a microcomputer. Over the past decade, frame grabbers have become increasingly sophisticated. MorphoSys uses the PCVISION*plus*® frame grabber that is manufactured by Imaging Technology Incorporated. I will describe the functions of such a state-of-the-art board for microcomputers rather than older or less capable technology, which is mostly of historical interest now. A modern frame grabber board performs three main functions: image capture, image display, and image access.

Capture of the image is accomplished by receiving from a video camera a continuous waveform that describes the varying brightness as the light sensor in the video camera scans the image. The frame grabber converts ("digitizes") the waveform into a matrix of numbers that resides in memory on the board. This special memory is called the "frame buffer." Each number in the frame buffer represents the brightness of the image at a single point and is called a "pixel." The size of this matrix determines the spatial resolution of the frame grabber and the maximum precision for dimensional measurements taken with the board. Typical values are 640 columns by 480 rows of pixels. The number of different brightness levels that can be represented by a pixel determines the gray scale resolution. Current boards usually have eight bits (one byte) per pixel and can, thus, distinguish 256 distinct gray scale values. This is adequate for morphometric purposes, although systems designed to investigate the limits of the human visual system need at least 12 bits, or 4096 gray levels.

The second function is image display. The frame grabber is also responsible for converting the image that is stored in the frame buffer into a standard television signal that can be displayed on an ordinary television monitor for viewing. These first two functions, image capture and display, are performed by the board automatically when initiated by software running on the host microcomputer. These functions, once initiated, require no further intervention by the host.

The third function is to allow access to the image by software running on the host microcomputer. A program running on the microcomputer can retrieve specific pixels from the frame buffer and can change the contents of pixels. This allows software to modify the image, trace the edges of objects, or display a cursor. Software also controls the overall activity of the board. For example, software can instruct the board to cease capture of frames from the camera. This will freeze the current frame in the frame buffer for special processing or analysis.

Many frame grabbers can perform some very elementary image enhancement functions as an image is digitized. Often an input look-up table (LUT) is used that will map any input pixel value (any number from 0 to 255 for an eight bit pixel) to any other pixel value, which is then stored in the frame buffer. This process allows contrast stretching and brightness shift of the incoming image before storage. Also, many boards, instead of simply displaying a pixel as it resides in the frame buffer, use the pixel value to obtain red, green, and blue values in separate look-up tables to display a particular color on a color monitor. This feature, called "pseudocolor," can be used to highlight areas or to indicate, by means of colored graphics, the results of calculations by software.

A television camera using the North American television standard (NTSC) sends a full video frame every one-thirtieth of a second. Many frame grabbers now in common use can digitize frames and display continuously at that same rate. The digitized image in the frame buffer is continuously changing and the image displayed on the monitor reflects this. The effects of changes in lighting, focusing, and camera position on the digitized image, as it resides in the frame buffer, can be instantly seen.

Standard television cameras and monitors are relatively inexpensive because they are manufactured in large numbers for the consumer electronics industry. The more economical frame grabbers use these standard components. It is possible to increase both spatial and gray scale resolution by using higher resolution boards with special cameras and monitors, but at significantly greater cost.

The first step to make in choosing the frame grabber is to decide with which software the frame grabber will be used. This decision on software will usually limit the choice of frame grabbers to relatively few alternatives. There is presently a great deal of activity by hardware manufacturers in frame grabbers that display their image on the microcomputer monitor, rather than on a separate television monitor. This development promises to further reduce the cost of setting up a machine vision workstation. Be aware that not all frame grabbers are designed for the same purposes. Some frame grabbers, especially the ones that capture color images from a color video camera, are more suited for the capture, display, and editing of images for publication or image database applications. They are not well suited for many morphometric applications.

The differing video standards in various parts of the world change some numerical parameters, but are not serious obstacles. NTSC is the color standard in North America, and is a modification of the RS-170 black and white standard. Other video standards used around the world are PAL and SECAM and their variants. PAL is used in most of western Europe and other areas. SECAM is used in France, eastern Europe, the U.S.S.R., and elsewhere. As long as the frame grabber, the video camera, and the monitor all use the same video standard, there is no problem. Board manufacturers distribute appropriate versions of their boards in different geographic areas. With the PAL and SECAM standards, the board receives 25 frames a second (rather than the 30 of NTSC) and typically will have 640 columns of pixels and 512 rows (rather than 480 under NTSC).

The rapid technological progress being made in mass storage devices is especially significant for machine vision. Read-write magneto-optical disks make it possible to archive full gray-scale images of specimens at relatively low cost. Fifty megabyte 3.5 inch diskettes, which can hold 150 8-bit images at 640 by 480 resolution, are now on the market. Further gains are certain in the next decade and will make standard practice the permanent archiving of specimen images along with extracted morphological features and measurements.

2.1 Machine Vision Software for Morphometrics

The goal in designing MorphoSys was to create a program that automates morphometric measurement at the same time that it provides a tool for handling shape data. The user has the capability of intervening in the digitization process, as necessary, to correct errors or make modifications. This is accomplished by allowing direct interaction with shape data in the form of the shapes themselves, not just in the form of numbers describing the shapes. The software is, first of all, a means by which to acquire and manipulate shape information. Secondly, the program performs measurements automatically after the shape information has been extracted. By separating the use of the program into two stages, first, feature extraction, and second, measurement, the program can be used more flexibly. After the feature extraction process is applied to a series of specimens, the extracted features can then be measured in different ways by constructing different measurement schemes.

MorphoSys captures basic shape data in the form of the two-dimensional contours (outlines) of specimens and in the form of selected points that ordinarily correspond to biological landmarks on the specimens. MorphoSys traces contours automatically from a high contrast image at the rate of a few thousand points per second. A typical contour on a 640 by 480 pixel frame grabber consists of 2,000 or fewer points and is traced in less than one second. Highly convoluted contours can be much longer, of course. As MorphoSys traces the contour, the contour is displayed as a colored outline on the monitor. Multiple contours, holes as well as exteriors, can be digitized in a single frame. Contours are identified by number. An editing feature allows manual change of the extracted contour, permitting correction of any inaccuracies that are caused by damage to the specimens, or by shadows, etc. Selected points can be placed manually by moving a cursor controlled by a mouse or trackball. In one mode, the cursor is locked to the contour and follows the contour as the mouse is moved. In this mode, landmarks can be placed precisely on the contour. Also, the cursor can be freed from the contour and any point can be selected by using the mouse. Selected points are displayed on the monitor as bright dots with the number of the selected point centered above. Selected points can be moved by the mouse as desired.

The entire collection of contours and selected points that are extracted from a single specimen are saved together, as a single frame, in a disk file. This disk file, called a "frame file," contains coded information that exactly describes the contours and selected points as they were captured from the image. The frame file contains only text characters (ASCII), allowing these files to be handled relatively easily by other programs. Frame files can be transmitted, by electronic mail for instance, and can be recalled for display, editing, or measurement. Printed copies of the frames can be made on a PostScript®-capable laser printer. Commands written into a text file, by the user, control the action of the software during measurement. The actual measurements are written to a text disk file for further analysis.

The commands that control the measurement functions are of two different types. Measurement commands produce an actual number or set of numbers. For example, the command "distance" calculates the distance between two selected points. Geometric commands calculate new constructed features from existing features. For example, the command "centroid" calculates the centroid of a specified contour and places a specified selected point at that position. There are also measurement commands for the angle formed by three points, the coordinates of a single point, the area enclosed by a contour, and the perimeter of a contour. A more specialized measurement function determines the elliptical Fourier coefficients that describe a contour (Kuhl and Giardina, 1982; Rohlf and Archie, 1984). Besides the centroid command, there are several other geometric commands that create new, constructed features. The command "far" will calculate the farthest point on a contour from an existing selected point. The command "near" will find the nearest point on the contour. The command "project" places a new selected point at a position determined by projecting, at right angles, a selected point onto a line specified by two other selected points. The command "intersection" will find the point that lies at the intersection of two lines, each specified by a pair of selected points. The command "contour intersection" will find the point on a contour at which a line specified by two points crosses. The command "convex hull," given a contour, will calculate a new contour that is the smallest convex polygon that covers the original contour. The command "clip" will remove a section of a contour and designate that section by a new contour number. Other geometric commands are available.

The geometric commands can be used to perform fairly complex operations. As a simple example, say that the angle between a pair of lines that are nearly parallel to each other must be calculated. Three points are needed to calculate an angle, but the point of intersection lies far off the screen. The intersection command can be used to construct a new selected point that lies off the screen, and this constructed point can be used as the vertex in calculating the angle between the original lines. As a second example, consider an elongate leaf whose length must be measured. The centroid command will place a point near the center of the leaf. Using the far command based on this centroid will place a point at one end of the leaf. Then, using the far command based on this end point will place a point at the opposite end of the leaf. Finally, the distance between the selected points, at the two ends of the leaf, is calculated to obtain the length. Although this process seems complicated to describe, once the command file has been created and read by MorphoSys, the length of the leaf can be obtained by just capturing the leaf outline. Thus, the length will be calculated after a click of the mouse button while the cursor is near the leaf rather than by careful positioning of the cursor. Time and tedium are reduced while accuracy is increased.

Geometric commands can be used in combination to produce elaborate measurement schemes. As in the second example above, the contours or selected points that are the results of a geometric command can be used as the operands for other geometric commands and finally, for measurement commands. It is possible to measure ninety radii with a command file that begins with a contour, calculates the centroid, finds the point on the contour that is farthest from the centroid, then using that farthest point as a reference, scans out from the centroid to find ninety points spaced at four degree intervals. Distance commands give the distance from the centroid to each of these ninety points to calculate the ninety radii. These calculations are done in a few seconds rather than the several minutes it would take to perform the measurements by hand and enter them manually in a computer. MorphoSys can yield an increase in rate of data acquisition of more that 100 times over manual measurement in automated measurement schemes like this. This example is, of course, only one way a command file can be constructed to implement a measurement scheme.

MorphoSys shows graphics on the television monitor that indicate the calculations that are performed. The user can check that command functions are working properly; and if not, can intervene to manually alter contours or selected points. Other calculated features or measurements that depend on these edited features will change in response. When a set of commands is active, MorphoSys will calculate as much as possible, given the information it possesses. As selected points are indicated and outlines are captured, the program will perform the calculations that these new data items make possible. If a point or contour is deleted, all points or contours that depend on these will also be deleted. These actions are immediately visible on the television monitor, so errors are easily seen.

The measurements made by MorphoSys are in real units. At the time shape capture begins, the program prompts the user to select a method of standardizing measurements. There are several options. A circular object can be placed in the field and its contour can be captured. MorphoSys will calculate the location of the center of the circle and place spanning diameters across the circle in the x and y directions. The user then enters the actual diameter and the units of measurement. From this information, the program calculates the scaling factors in the x and y directions. A ruler can also be used manually. Scaling factors can be saved in a file and recalled if a compound microscope is used. The scaling information is saved with each frame. When measurements are calculated from the frame, the scaling factors for that frame are used. It is important to note that even for frame grabbers that have "square" pixels, the pixels often deviate by two percent from being perfectly square. The scaling factors in the x and y directions are rarely equal.

Much of the power of this software comes not so much from the image analysis functions, which at present are very simple, as from the interactive access to shape data that is provided to the user and from the control and automation of measurement production by means of command files.

2.2 Imaging

Obtaining a good image is the most critical step in using machine vision successfully. Once the light leaves the specimen, the optics form an image on the sensor of the video camera, and the video camera converts the image to a video signal; further manipulations can only decrease the amount of information in the image. Image enhancement techniques can make an image more pleasing to the eye and make information already present in the image more visible, but even the most sophisticated enhancement techniques do not add information to an image. The information must be present in the image in the first place. Using sophisticated image enhancement is more appropriate if the image was captured by spacecraft, many millions of miles from earth, than if the operator is standing in the same room with the image capture equipment and can improve the image quality by making some adjustments. Efforts invested in designing a machine vision workstation, so that high quality images are obtained to begin with, will prove more successful than efforts invested in improving a lesser quality image by image enhancement techniques. These comments are not meant to disparage the use of image enhancement techniques, but are meant only to point out that the quality of the original image limits the potential for success of all machine vision processes. Try to maximize the information in the image. The requirements for a good image vary greatly, depending on the type of specimen involved and the kind of machine vision techniques that are performed. Optimal imaging for machine vision is clearly an area in which better techniques can be developed by workers concerned with particular kinds of specimens. Three aspects of imaging will be discussed here: illumination, optics, and camera.

2.2.1 Illumination

Getting a good image depends, first of all, on the quality of the lighting. Even the best machine vision equipment will give poor results if insufficient attention is given to illumination. All aspects of this parameter should be controllable. The human eye accommodates a very large range of lighting conditions; a video camera does not. For example, the varying amount of light coming through a window on a partly cloudy day will hardly be noticeable by a person yet will require continual readjustment of the lens aperture. A machine vision workstation should be located in a room that allows control over ambient lighting. A room without windows is best.

To capture the contours of objects like leaves, bones, exoskeletal elements, etc., it is necessary to provide sufficient contrast between the background and the object in order to obtain a clear threshold image. In the case of dark objects, like leaves, it is best to make the background brighter than the object, called "bright field" illumination. For very white or translucent objects, results will be better if the background is made darker than the object, "dark field" illumination.

In bright field illumination, the challenge is to get the background illumination completely uniform across the entire field. Failure to do this will make contour capture difficult. Several things can be done to make the illumination uniform. First, the light source should be large. It should illuminate an area several times larger than the specimen; up to a point, the larger, the better. If, for example, the specimens are about a centimeter in size, try to get even illumination across a few decimeters; reasonable success across this size of field should mean that the central few centimeters are likely to be very evenly illuminated. A light box containing two or more fluorescent tubes is the best. Second, use an extra diffuser of opal lucite or glass above the light box to even out the light intensity further. Third, use a reasonably large distance between this extra diffuser and the light box. By experimenting with different configurations while viewing the results on the monitor, adjustments can be made and success can be judged. With a microscope, either dissecting or compound, the substage illumination is often not uniform enough to meet these requirements; this may require experimentation with a (large) external light source illuminating a diffuser. With dark field illumination, it is the specimen that should be evenly lit. The background should be of a very dark nonreflective material. A shadow box lined with black velvet will work. Shadows on the specimen can be reduced by placing a sheet of paper rolled into a tube on end around the specimen and aiming bright lights at the outside of the tube. The repeated diffuse reflections of light inside the tube will remove most shadows from a white specimen.

Staining of specimens can also increase contrast for contour capture. Fish bones, for example, are often stained with alizarin red dye. By using an interference filter that passes to the camera only light in a very narrow range of green wavelengths, the light reflected by the stained bone, and which is received by the camera, can be greatly decreased. This can make the bone appear distinct from associated unstained tissue even though in bright field illumination the tissue might appear as dark as the bone.

Specimens that are transparent or translucent present difficult problems, but there are possibilities that have not been fully investigated. Some specimens may fluoresce under ultraviolet light. Some transparent specimens may refract light, so that they appear bright under strong glancing illumination. Rotation of the plane of polarization by optically active transparent specimens may be another means of getting a high contrast image.

Landmarks require illumination from above. The requirements are not so strict as for background lighting. Generally, shadows are useful in seeing features of the specimen. Directional lighting in the form of incandescent lights above and to the side of the specimen will produce shadows of the right character. A variable voltage transformer or some other means of controlling the foreground lighting intensity is nearly essential.

Some studies require capture of both contours and landmarks. For this kind of work, a toggle switch that turns off background lights and turns on foreground lights is convenient. The camera lens aperture can be adjusted for the background lights, and then foreground lights can be toggled to adjust the brightness of these lights without changing the aperture. After these adjustments are made initially, a simple flick of the toggle switch will change the lighting from correctly adjusted background to correctly adjusted foreground for an entire series of specimens.

2.2.2 Optics

The purpose of the optics is to form a sharp image with minimal distortion on the sensor of the video camera. One cause of distortion by a lens is the simple geometric distortion that occurs when a spherical surface is flattened. A complete view of all the surroundings, when seen from a single point, is a sphere. A lens, for our purposes, is a device that takes a portion of this spherical view and transfers it as an image to a flat surface. Lenses that cover a large portion of the sphere are called "wide-angle." Wide-angle lenses unavoidably produce distorted images because they flatten a very curved surface and, thus, are completely unsuitable for dimensional measurement unless complicated correction factors are introduced. Short focal length telephoto lenses cover a very small angular portion of the sphere, a portion that is small enough that it is nearly flat already. Other kinds of distortion and aberration may be present in a lens. Dealers specializing in optics can give advice on selection of particular brands for low geometric distortion.

The optical system must be appropriate for the size and nature of the specimen. Generally, the best type of lens for morphometric work on specimens about four centimeters across, or larger, is a short telephoto zoom lens of good quality that is held at a long working distance from the specimen. If a specimen is too large to fit in the field of a short telephoto lens, increase the distance from the lens to the specimen; do not change to a wide-angle lens. A zoom lens can be used to adjust the size of the image, allowing the specimen to fill the frame.

Specimens as small as two millimeters across can be handled by lenses designed for 35 mm single lens reflex cameras. Macro lenses of the better makes are of especially high optical quality. The light sensor of a solid-state video camera is actually smaller than a 35 mm format negative, so the part of the lens that has the greatest distortion, its periphery, is not used. These lenses can be attached to a video camera with a C-mount adapter for the brand of lens used. The use of extension tubes or bellows can increase the effective magnification; therefore, a microscope may not be necessary for specimens that are over two millimeters across. In addition, obtaining a uniform background illumination can be easier with a macro lens configuration than with a microscope. Camera shake is a problem when lens extenders are used. A firm camera mount is important. When using extension tubes or bellows, the distance from the lens to the specimen can be much shorter than the distance from the lens to the camera. Many 35 mm camera lenses are designed for the opposite situation of having a longer object distance and will produce less distortion if reversed. Special mounts called "reversing rings" are available from photographic equipment dealers. This is often not a problem for macro lenses, which are designed for relatively short object distances.

Those who require microscopes have fewer options, but most microscope makers offer configurations specifically designed for attachment to a video camera. Vendors who specialize in microscopes can provide information about video configurations. Again, be aware that for machine vision applications, substage lighting may not be uniform enough and may have to be replaced by a specially designed illumination system.

Check the optics for geometric distortion. Use a ruler to standardize the measurements. Then, place the ruler at various positions and orientations in the field and measure its length. Especially, check measurements across the side of the field versus measurements across the center of the field. This will give a good idea of the repeatability of measurements with the system and should help identify any systematic errors that are due to geometric distortion.

2.2.3 Camera

The third factor in producing a high quality video image is the video camera itself. As in most areas of electronics, the advance of solid-state devices has had great impact on video camera technology. The older technology uses a specialized vacuum tube, the Vidicon, for sensing the image and generating a video signal. The Vidicon depends on the influence of light on a photoconductive surface that is scanned by an electron beam. The basic electronics involved in the operation of the Vidicon make this device decidedly less desirable than more recently developed solid-state devices.

The most serious drawback for morphometrics is that a change in line voltage will influence the circuitry that causes the electron beam to scan the photoconductive surface. The scanning pattern expands with higher voltage, and shrinks with lower voltage. The absolute position of points in the image will, thus, shift with line voltage changes producing errors in measurement. Also, very sharp high contrast images are difficult to obtain with a Vidicon, because they result in regions of high charge juxtaposed to regions of low charge in the photoconductive surface. Large differences in charge over a short distance tend to produce lateral charge movements across the photoconductive surface, reducing the sharpness of the black/white boundary. Also with a Vidicon, the background in a bright field image tends to be nonuniform. Further, very bright images will burn the receptor of a Vidicon, sometimes permanently. An after-image will remain in the video signal produced by the Vidicon after the object has been removed. A Vidicon tube, like most electronic vacuum tubes, has a filament that has a definite life span. When the filament burns out, the tube will have to be replaced. Finally, a glass Vidicon tube is fragile.

In a solid-state sensor, the image is sampled spatially by separate individual cells in a rectangular array instead of by a uniform surface. The relative positions are fixed and are insensitive to fluctuations in line voltage. Very high contrast images do not degrade in sharpness, because the individual cells are insulated from each other. (In extreme cases, the image will "bloom" as electrons move from one cell to another; this can be remedied by decreasing the overall brightness.) It is much easier to obtain a precisely uniform background with a solid-state sensor. The sensor responds instantly to changes in the image. There is no after-image nor is the sensor sensitive to burning by bright images. The sensor does not have a filament and is generally more rugged than a glass vacuum tube. In addition, even the larger solid-state cameras are much lighter and smaller than Vidicon cameras.

Vidicon tubes of the more expensive varieties have somewhat higher resolution than solid-state cameras. However, better solid-state cameras now have sufficient resolution to exceed that of the frame grabber boards that they are used with. Solid-state camera technology continues to gain in resolution and decrease in price. Other attributes listed above make solid-state cameras the choice for morphometric data acquisition. (Merely because a Vidicon tube produces an analog signal does not, in itself, guarantee higher resolution. The rapidity with which the brightness level represented in the analog signal can change limits the resolution of the Vidicon tube. The highest frequency of light and dark that can be transmitted by the tube is its limit of resolution and this may or may not be higher than the highest frequency that can be transmitted by a particular solid-state sensor, even though the signal from the solid-state device is generated from an array of discrete photosensitive sites.)

Many solid-state cameras, particularly the higher resolution kinds that are most desirable for machine vision applications, have automatic functions that should be disabled if they interfere with manual adjustments. Some have an automatic black level or automatic gain that will automatically adjust contrast in the video signal being sent to the frame grabber. In bright field illumination, as the brightness of the image is increased, the camera may compensate for the very bright background by decreasing the contrast or brightness of the image; this compensation defeats the very effect desired. Such cameras usually have some means of disabling automatic functions, often by removing or moving jumpers inside the camera body. Before selecting a camera for the workstation, check whether the camera has a mechanism for disabling automatic functions.

The need of machine vision users for critical control of the video image has prompted manufacturers of video cameras to produce models that allow manual adjustment of the light sensing characteristics of the camera. As opposed to image enhancement techniques performed on a digitized image, which cannot increase the visual information in an image, a change in the characteristics of the light sensor can cause the capture of more information and facilitate subsequent image analysis steps.

2.3 Monitor

The monitor for image display is another important component of the machine vision workstation. Because MorphoSys, as well as many other machine vision programs, uses color to indicate its operation, a studio quality color monitor is the best. Many frame grabber boards produce separate red, green, and blue signals rather than the more common composite color signal that is used by video cassette recorders, for instance. Be sure that the board and monitor are compatible in this regard.

2.4 Specimen Positioning and Orientation

If a microscope is not used, a good solid camera stand will be necessary. Some kinds of copy stands give excellent service. Besides the camera mount that can be racked up and down, some have an opal lucite stage that can also be moved vertically. This opal lucite stage makes both a good diffuser for bright field illumination and a support for the specimen. When looking at the monitor, the specimen is more easily positioned if the camera is mounted backwards (i.e., with the camera base away from the post), because the image shown on the monitor will be in the same orientation as the user's view of the stage. Specimens that are about four centimeters across, or smaller, are easier to position correctly if they are placed on a larger piece of glass that can be slid on the stage. A small lump of clay placed on the sliding glass may aid in holding specimens a few centimeters in size in a standard orientation.

The camera and lens should be adjusted to permit the specimen to fill as much of the frame as possible. The limits of precision are determined by the size of the pixels. When short distances are measured, the size of a single pixel is a significant fraction of the measurement. If the vertical size of the screen is 480 pixels, and a feature one-tenth of the vertical screen dimension is measured, then a single pixel is more than two percent of the distance being measured. With smaller features, the effect becomes more extreme. It is advantageous to sort specimens by size, at least roughly, before digitizing. This allows efficient handling of batches of specimens with approximately equal dimensions. The camera and lens then can be readjusted between batches, so that specimens nearly fill the screen. It is generally not advisable to digitize more than one specimen in a single frame.

Questions may arise about the possibility of differences in positioning of specimens under the camera producing effects that swamp differences in the actual dimensions to be measured. The cosine of eight degrees is 0.99027.... As long as the plane of the dimension to be measured is within plus or minus eight degrees of being perpendicular to the axis of the lens, the error introduced by the difference between the projected measurement and the actual dimension will be less than one percent. The effect of positioning error certainly needs consideration, but in many cases the effect is small. Whether the effect of positioning error is small enough to be ignored depends on the angular error in positioning the specimens, and the accuracy needed for the particular study.

The measurement of living specimens is facilitated by "freezing" the image. With MorphoSys, this can be done with a single keystroke, and it can be done repeatedly if a good image is not obtained at first. Living larval fish, which are rather difficult to measure manually, can be handled in this fashion.

The optics and the sensor should be free of dust. Solid-state sensors pick up dust easily and should be cleaned as necessary following manufacturer's instructions. Lens dust can be distinguished from dust on the sensor by viewing a bright field with the camera and rotating the lens. Spots that rotate are on the lens; those that remain fixed are on the sensor.

3 Using Machine Vision

Users may have widely divergent reasons for taking measurements. A functional morphologist may want to relate aspects of jaw morphology to feeding behavior. A developmental biologist may be interested in allometry during embryogenesis. An ecologist may be interested in characterizing differences in damage done to plants by different insect herbivores. A systematist may want to discover characters that will discriminate similar species. There are appropriate and inappropriate measurements that can be used to achieve these goals. Machine vision is a tool, and like other tools it can be misused. The proper use of machine vision techniques depends on a knowledge, prior to measurement, of how the data will be analyzed. The problem of designing measurement schemes for particular kinds of analyses is far too broad for review here. Other authors in this volume deal with some of these issues. For thorough discussion of the latest methods in depth, examine Rohlf and Bookstein (1990), Bookstein et al. (1985), and Reyment et al. (1984).

The technological breakthrough provided by machine vision presents challenges to users. Machine vision has great power in some circumstances, but at the present time, it has clear limitations. In this regard, machine vision is no different from any other measurement tool. How should this technology be made a part of current practice? Investigators who consider the strengths and weaknesses carefully from the standpoint of their own research questions and materials will benefit the most. Machine vision will be most successful in situations where the investigator has circumscribed problems to be especially amenable to such methods. Machine vision techniques are often less successful in those situations where investigators are constrained to follow preexisting measurement schemes and are forced to measure materials that are more difficult for machine vision techniques. Users who can design automatic measurement schemes on easily imaged objects in ways that serve their research goals will benefit most from the current state of technology. In other words, the tool should fit the problem. The present goal of machine vision research in systematic biology is to extend the techniques to allow automated data acquisition across a broader range of circumstances.

There are many unexplored ways in which machine vision and image analysis techniques can be extended for use in systematic and evolutionary biology. For example, tree bark form and texture is not quantified, because no simple technology exists for measurement. Even basic terminology is too crude to be of much use in tree identification. Yet many tree species can be identified by bark characters alone by students who learn these features by first hand observation (Harlow and Harrar, 1968: 32). There are many such areas in which the possibilities of machine vision can be explored.

4 Conclusions

Many investigators have been successful in applying machine vision in their own work. These gains have been accomplished with simple machine vision techniques. The contribution machine vision can make to progress in systematic and evolutionary biology is clear. The present challenge is to develop new techniques that broaden the range of morphometric problems that can be solved with these methods. The issues of how effectively to use machine vision and how best to develop machine vision methods deserve attention from systematic and evolutionary biologists.

5 Acknowledgments

I thank users of early versions of MorphoSys for their encouragement, suggestions, and advice, especially Howell V. Daly, Cindi Jones, David R. Lindberg, Stuart G. Poss, Bruce Riska, F. James Rohlf, David L. Schultz, and Dennis Slice. I am especially grateful to John Sorensen for his help and patience. I also thank Antonio Valdecasas and Nobuhiro Minaka for comments on early drafts.

6 References

Bookstein, F.L., Chernoff, B., Elder, R.L., Humphries, J.M., Smith, G.R. and Strauss, R.E., 1985. Morphometrics in Evolutionary Biology. Special Publication, 15. Academy of Natural Sciences, Philadelphia, 277 pp.

Harlow, W.M. and Harrar, E.S., 1968. Textbook of Dendrology, (Fifth Edition). McGraw-Hill Book Company, New York, 512 pp.

Kuhl, F.P. and Giardina, C.R., 1982. Elliptic Fourier features of a closed contour. *Computer Graphics and Image Processing*, 18: 236-258.

Lindberg, D.R., 1990. Morphometrics and the systematics of marine plant limpets (Mollusca: Patellogastropoda). In: F.J. Rohlf and F.L. Bookstein (Editors), Proceedings of the Michigan Morphometrics Workshop. Special Publication, 2. Museum of Zoology, The University of Michigan, Ann Arbor, pp. 301-310.

Meacham, C.A. and Duncan, T.O., 1987. MorphoSys, Automated Morphometric System. University Herbarium, University of California, Berkeley.

Reyment, R.A., Blackith, R.E. and Campbell, N.A., 1984. Multivariate Morphometrics, (Second Edition). Academic Press, New York, 233 pp.

Rohlf, F.J. and Archie, J.W., 1984. A comparison of Fourier methods for the description of wing shape in mosquitoes (Diptera: Culicidae). *Systematic Zoology*, 33: 302-317.

Rohlf, F.J. and Bookstein, F.L. (Editors), 1990. Proceedings of the Michigan Morphometrics Workshop. Special Publication, 2. Museum of Zoology, The University of Michigan, Ann Arbor, 380 pp.

A STATISTICAL AND EMPIRICAL EVALUATION OF SOME MORPHOMETRIC VARIABLES OF HONEY BEE CLASSIFICATION

Howell V. Daly

Department of Entomological Sciences, University of California, Berkeley, California 94720, U.S.A.

Abstract

Twenty-five variables used in honey bee classification were examined, particularly for the role of general-size in discrimination. Of these, 11 were size-related. One replicate of these 11 was adjusted to remove size, another was not. Two data sets the 25 variables, one with the 11 size-adjusted values, the other with their raw counterparts, were then tested on two sets of samples using discriminant analysis, and the output values were clustered to assess phenogram credibility. The best classifications and phenograms were produced using larger numbers of combined size-related and unrelated variables, and also by size-related variables that were size-adjusted. Exclusive use of fewer, size-unrelated or size-adjusted variables gave the lowest correct classifications and less credible phenograms in empirical tests.

1 Introduction

In this chapter, the statistical properties and practical use of some classical variables of honey bee classification are scrutinized, especially with regard to the role of general-size in discrimination. This chapter is also a testimony to the "robustness" of multivariate analysis. Honey bees were among the first insects to which morphometric analysis was extensively applied. The 25 variables examined here have been used in identification of Africanized bees (Daly and Balling, 1978; Daly, 1991) and, together with other variables, are the basis for the classification of honey bees of the world (Ruttner, 1988). The variables that we inherited were initially chosen to meet special needs, or because they were conveniently measured. This is in contrast to variables that were selected expressly to describe size and shape for modern morphometric analysis as, for example, described by Bookstein et al. (1985). A search for new variables and methods of analysis is in order, but for the present we have an enormous data base for our current variables; these have proven to be effective for taxonomic purposes with conventional multivariate statistics, and are typical of those widely used in systematic entomology (Daly, 1985).

Morphometric studies on bees began in Russia, where apiculturists searched for bees with long tongues suited to pollinate red clover. According to Alpatov (1929), in his English summary of Russian literature, G. A. Koshevnikov reported the first exact measurements on bees in a publication in 1900. To those measurements were added other distances (distance = the shortest measurement between two points) that could be easily taken from the body, wings, and legs. Some variables were especially useful in identifying certain races. For example, Goetze (1930) found lengths of sections of the cubital vein in the fore wing to differ among European races. This became the cubital index that is still routinely used by bee breeders for distinguishing among European races. Sample statistics and univariate analysis prevailed in bee morphometrics until DuPraw (1965a, 1965b) made two innovations: the use of angles between wing veins as variables to reduce the effect of general body size and the use of discriminant function analysis (DFA). As shown later in this chapter, the exclusive use of angles may not be satisfactory, but the multivariate approach was a major breakthrough. Multivariate analysis of morphometrics now is the foundation for classification of the 24 races of *Apis mellifera*. We are fortunate that F. Ruttner has synthesized our progress to date in his recent monograph (1988).

2 Morphometrics and Classification of Honey Bees

The standard morphometrics for honey bee classification is considered by Ruttner et al. (1978) to be 41 variables: two of pubescence, eight of pigmentation, one distance of proboscis length, four distances of hind leg segments, eight distances on terga and sterna, six distances on the fore wing, 11 angles among veins, and one count of hamuli. Of these, included here are four distances of hind leg segments, four distances on sternum 3 (counting from the petiole), four distances on the fore wing, seven of his angles among veins, and one count of hamuli. Using only fore wings, DuPraw (1965a, 1965b) measured 13 angles among veins and a longitudinal and transverse distance. Of his 15 variables, eight of his angles of venation, some the same as Ruttner's, are included here. In addition, Daly and Balling (1978) added one angle of venation on the fore wing and two distances on the hind wing (Table 1).

The "Western" honey bees, *Apis mellifera*, are native to Africa and Europe. Domesticated stocks of European races have been introduced to most

Table 1.
Morphometric variables of Daly and Balling (1978) used here, and the equivalent variables of other authors. Size-adjusted distances = size-related distances + prefix "*S*-"

Number	Abbrev.	Description	Author and Character Equivalent(s)
(Size-related distances)			
1	*FWLN*	fore wing length	Alpatov (24, 25), Ruttner (21, FL)
2	*FWWD*	fore wing width	Alpatov (2), Ruttner (22, FB)
3	*HWLN*	hind wing length	Daly and Balling (HWLN)
4	*HWWD*	hind wing width	Daly and Balling (HWWD)
5	*TBLN*	hind tibia length	Alpatov (4), Rutter (7, Ti)
6	*FELN*	hind femur length	Alpatov (23), Ruttner (6, Fe)
7	*TRLN*	hind basitarsus length	Alpatov (5), Ruttner (8, Ml)
8	*TRWD*	hind basitarsus width	Ruttner (9, Mt)
9	*STLN*	3rd sternum length	Alpatov (22), Ruttner (15, S3)
10	*WXLN*	3rd sternum, wax mirror length	Alpatov (21), Ruttner (16, Wl)
11	*WXWD*	3rd sternum, wax mirror width	Alpatov (10), Ruttner (17, Wt)
(Size-unrelated distances)			
12	*CUBA*	cubital vein "a" length	Goetze (a), Ruttner (27, a)
13	*CUBB*	cubital "b" vein length	Goetze (b), Ruttner (28, b)
14	*WXDS*	3rd sternum, width between mirrors	Ruttner (18, Wd)
(Angles)			
15	*AN29*	fore wing vein angle	DuPraw (A4), Ruttner (31)
16	*AN30*	fore wing vein angle	DuPraw (B4), Ruttner (32)
17	*AN31*	fore wing vein angle	Ruttner (33)
18	*AN32*	fore wing vein angle	DuPraw (E9), Ruttner (34)
19	*AN33*	fore wing vein angle	DuPraw (G18), Ruttner (35)
20	*AN34*	fore wing vein angle	DuPraw (J10), Ruttner (36)
21	*AN35*	fore wing vein angle	DuPraw (A1)
22	*AN36*	fore wing vein angle	Daly and Balling (36)
23	*AN38*	fore wing vein angle	DuPraw (J16), Ruttner (37)
24	*AN39*	fore wing vein angle	DuPraw (M17)
(Counts)			
25	*HAMU*	hamuli on hind wing	Alpatov (3), Ruttner (42)

agricultural regions of the world. Ruttner (1988) recognizes three major groups of races based on geographic distribution, behavior, and morphometrics: (I) Near East, (II) tropical Africa, and (III) Mediterranean. The last is further subdivided into: (1) west Mediterranean and (2) central Mediterranean and southeast Europe. In this chapter, samples are included from his groups II, III(1), and III(2). These three groups will be referred to here as the "African," "west Mediterranean," and "central Mediterranean" groups, respectively. The last two groups are commonly referred to collectively

Table 2.
OTUs used and their abbreviations. OTU Set = the 11 OTU set or the three-group set; "Genetic" Origin = continents from which the populations were derived ; Total = total number of samples making up the OTU; Origin of Samples = origin and number from subareas.

OTU Set	OTU	"Genetic" Origin	Population Descriptor	Total	Origin of Samples
11	ADA	African	*A. m. adansoni*	12	West Africa (1, Liberia; 1, Mali; 4, Nigeria; 1, Senegal and 5, Upper Volta)
11	AFD, AFT	African	"Africanized bees"	364	Central America (20, Costa Rica; 6, Honduras; 2, Mexico; 2, Nicaragua; 20, Panama) and South America (22, Argentina; 91, Brazil; 15, Colombia; 9, Ecuador; 39, French Guiana; 3, Guyana; 15, Paraguay; 12, Peru; 13, Surinam; 1, Uruguay; 87, Venezuela; 2, West Indies) and 3 swarms intercepted in United States. Divided arbitrarily into a data OTU AFD (53 samples) and a test set AFT (311 samples)
11	CAP	African	*A. m. capensis*	15	Cape region of South Africa
11	SEA	African	*A. m. scutellata*	29	East Africa (25, Kenya; 4, Tanzania). Here divided into eastern and southern geographic OTUs for analysis
11	SSA	African	*A. m. scutellata*	43	southern Africa (4, Zimbabwe; 39, South Africa) (see SEA above)
11	CAD, CAT	European	"European bees"	258	feral colonies in California, U.S.A. Divided arbitrarily into a data OTU CAD (51 samples) and a test set CAT (207 samples)
11	CAR	European	*A. m. carnica*	10	inbred lines at Institut fur Bienenkunde, Oberursel, West Germany
11	IBR	European	Probably *A. m. iberica*	28	Colombia (15) and Venezuela (13)
11	LIG	European	*A. m. ligustica*	17	Kangaroo Island, Australia
11	MEF	European	*A. m. mellifera*	8	Sweden (1) and Tasmania (7)
11	PAN	European	"European bees"	36	man-made hives in vicinity of the Pan-American Highway on the Pacific slope west of the canal in Panama and mostly at lower elevations
3	AFR	--	"African"	total 152	composed of pooled samples from ADA, AFD, CAP, SEA, and SSA
3	MEL	--	"West Mediterranean"	total 36	composed of pooled samples from IBR, MEF
3	EUR	--	"Central Mediterranean"	total 114	composed of pooled samples from CAD, CAR, LIG, and PAN

as "European" bees. Distributed in the three groups are samples from 11 OTUs ("operational taxonomic unit" of Sneath and Sokal, 1973) representing some native races of the Old World and populations derived from them elsewhere in the world (Table 2); refer to Table 2 hereafter for OTU abbreviation codes.

3 Materials and Methods

3.1 Specimens, Measurements, Samples, and OTUs

A fore wing, hind wing, hind leg, and third abdominal sternum of each worker honey bee was placed on a microscope slide. Twenty-five measurements (Table 1) were taken on each bee as described by Daly and Balling (1978) and Daly et al. (1982). These 25 were called the "raw" variables, meaning that they were not transformed and are the same as commonly used in bee morphometrics. A sample was usually 10 worker bees from a managed colony in a man-made hive or from a feral colony, usually in a natural cavity. Some samples were from swarms and a few samples were collected on flowers. In the latter case, the workers may not have been sisters.

The 25 raw variables were represented here by the arithmetic means of the measurements for the sample. Lengths are distances in millimeters on the long axis of a structure or the longitudinal axis of the body; widths are transverse distances; angles are in degrees; hamuli are counts. As explained below, a principal components analysis (PCA) indicated 11 of the raw distance variables were highly correlated with the first component, which is usually considered to be the general-size component. The 11 size-related distances were transformed to natural logarithms and adjusted statistically, by the Burnaby method (Rohlf and Bookstein, 1987), to remove the contribution of the first principal component. These additional 11 variables were used in DFAs in their logarithmic form (equal results were obtained from the antilogs) and were called "size-adjusted" (identified with a prefix *S*-), although one could argue that possibly other components are also related to general-size. The 11 variables were added to the 25 variables for a total of 36 variables in the study, but the original 11 size-related distances, and their corresponding 11 size-adjusted distances, were not used together in the same DFAs.

Of the 820 samples, 302 were used as data samples and the remaining 518 were used as test samples (AFT, CAT). Two data sets were considered here (Table 2): (1) the "11 OTU set," in which ADA, AFD, CAD, CAP, CAR, LIG, MEF, IBR, PAN, SEA, and SSA were entered as separate OTUs in univariate and multivariate analyses; and (2) the "three-group set" of AFR, MEL, and EUR, which were pooled from the 11 OTUs and entered in analyses. A given data sample was a member of the 11 OTU set as well as the three-group set.

The samples in the data OTUs ADA, CAD, CAP, CAR, LIG, PAN, SEA, SSA, and MEF were grouped on geographic origin and the race known to exist in the area at the time of collection, or on information from the collectors about the identity of samples. Samples in data OTU AFD were drawn from a data bank for Africanized bees that initially included samples identified by collectors and to which other samples were added after the morphometric DFA of Daly and Balling (1978). IBR samples were grouped on the basis of their distinctive morphometrics.

3.2 Empirical Evaluations of Discriminant Function Analyses

DFAs were computed using subsets of the 36 variables to examine the discrimination achieved by different selections of variables. After each DFA, the data samples used in the computations were classified according to their discriminant scores. The percentages of the samples correctly assigned to their respective OTUs provided one measure of the discrimination achieved. In this procedure, the percentages of data samples that are correctly classified are expected to be the highest possible because the DFA maximizes the distances among the data groups.

A second evaluation of a DFA is to classify test samples of known origin, but that were excluded from computation of the DFA. This gives a realistic assessment of the potential use of the DFA in classifying new samples of unknown identity, provided that the test samples are representative of the population to be sampled in the future. The large groups of samples, AFT and CAT, were used here as tests for this purpose and given the suffix "T" for tests. The samples were drawn arbitrarily from the same data banks as the data OTUs AFD and CAD (suffix "D" for data). The samples in the data banks were in a sequence determined by the order in which they were measured. Every seventh sample in the data bank for Africanized bees was assigned to AFD, and the remaining samples were assigned to the test group AFT. Similarly, every fifth sample of the data bank for feral European bees from California was assigned to CAD, and the remaining samples were assigned to the test group CAT.

Statistical comparisons of CAD:CAT and AFD:AFT indicated that the method of drawing data and test samples was largely, but not completely satisfactory (see also Fig. 5C, below). Univariate ANOVA of each of 36 variables for CAD:CAT gave no significant differences (*P*->-0.05) except for *S-HWLN* ($P = 0.04$). Likewise, for AFD:AFT only *AN35* ($P = 0.01$) and *AN36* ($P = 0.04$) were significantly different. The covariance matrices for CAD:CAT were equal ($P = 0.84$), but not so for AFD:AFT ($P = 0.005$). Oddly, the samples grouped by variables in OTU AFD had normal distributions, but AFT had the highest numbers of variables with nonnormal distributions (Table 3).

In the 11 OTU set, a "correct" classification of a sample of AFT or CAT was restricted to the data OTUs AFD and CAD, respectively. For this reason, percentages of correct classification were lower than in the three-group set, where a correct classification of a sample involved assignment to one of the larger, more diverse, pooled AFR or EUR groups, respectively.

A third empirical evaluation was based on how accurately morphometric relationships, based on a subset of variables, reflected known relationships among OTUs. Phenograms were prepared for the 11 OTUs in each of six DFAs. Certain clusters of OTUs were expected if the phenogram depicted relationships known to exist among the OTUs; a phenogram that exhibited more of the expected clusters of OTUs was considered superior to a phenogram with fewer such clusters. Two kinds of comparisons were involved, those dependent on other morphometric studies, and those largely independent of previous morphometric studies. For the first kind of comparison, three major "stems" in the phenograms were expected because Ruttner's three major taxonomic groups were based on morphometrics of some of the same variables used here. The expected major groups and their included OTUs are: African (ADA, CAP, SEA, SSA, and AFD), west Mediterranean (MEF and IBR), and

Table 3.
Kolmogorov-Smirnov goodness of fit tests for normal distributions with intrinsic hypotheses. Distributions of variables tested individually by OTU. Asterisks (*) = absolute maximum difference exceeded critical value ($P = 0.05$) for Lilliefors procedure (Sokal and Rohlf, 1981); hyphens (-) = nonsignificant. Subtotal columns (ST) are for the number of significant differences per variable across the 11 OTU (European + African) and three-group sets.

Variable	European OTUs						African OTUs					ST	Three Group Set			ST	Test Set	
	CAD	CAR	PAN	LIG	MEF	IBR	ADA	AFD	CAP	SEA	SSA		AFR	MEL	EUR		CAT	AFT
FWLN	-	-	-	-	-	-	-	-	-	*	-	1	*	-	-	1	*	*
FWWD	-	-	-	-	-	-	-	-	-	*	-	1	-	-	-	0	-	-
HWLN	-	*	-	*	-	-	-	-	-	-	-	2	*	-	-	1	-	-
HWWD	-	-	-	-	-	-	-	-	-	-	-	0	-	*	-	1	-	-
TBLN	-	*	-	-	*	-	-	-	-	-	-	2	-	-	-	0	-	-
FELN	*	*	-	-	*	-	-	-	-	-	-	3	-	*	*	2	-	*
TRLN	-	-	*	-	-	-	-	-	-	-	-	1	-	-	*	1	-	*
TRWD	-	*	-	-	-	-	-	-	-	-	-	1	-	-	-	0	-	-
STLN	-	*	-	-	*	-	-	-	-	-	-	2	-	-	-	0	-	*
WXLN	-	-	-	-	*	-	-	-	-	-	-	1	-	-	-	0	-	*
WXWD	-	-	-	-	-	-	-	-	-	-	*	1	-	-	*	1	-	*
CUBA	-	-	-	-	-	*	-	-	-	-	-	1	-	-	-	0	-	*
CUBB	-	-	-	-	-	-	-	-	-	-	-	0	-	-	-	0	-	-
WXDS	-	-	-	-	-	-	-	-	-	-	-	0	*	-	-	1	*	*
AN29	-	-	-	*	-	-	-	-	-	-	-	1	*	-	-	1	-	-
AN30	-	-	-	-	-	-	*	-	-	-	-	1	-	-	-	0	-	-
AN31	-	-	-	-	-	-	-	-	-	-	-	0	*	-	-	1	-	-
AN32	-	-	*	-	-	-	-	-	-	-	-	1	-	-	-	0	-	*
AN33	-	-	-	-	-	-	-	-	-	-	-	0	-	-	-	0	-	-
AN34	*	-	-	-	-	-	-	-	-	-	-	1	-	-	-	0	-	*
AN35	-	-	-	*	-	*	-	-	-	-	-	2	-	*	-	1	-	-
AN36	-	-	*	-	-	-	-	-	-	-	-	1	-	-	*	1	-	-
AN38	-	-	-	-	*	-	-	-	-	-	-	1	-	-	-	0	-	-
AN39	-	-	-	-	-	-	-	-	-	-	-	0	*	-	-	1	-	*
HAMU	-	-	-	-	-	*	-	-	-	-	-	1	*	*	-	2	-	*
S-FWLN	-	-	-	-	-	-	*	-	-	-	-	1	*	-	-	1	*	-
S-FWWD	-	-	-	*	-	-	-	-	-	-	*	2	-	-	-	0	-	-
S-HWLN	-	-	-	-	-	-	-	-	-	-	-	0	-	-	-	0	-	-
S-HWWD	-	-	-	-	-	*	-	-	-	-	-	1	-	*	-	1	-	*
S-TBLN	-	-	-	-	-	-	*	-	*	-	-	2	-	-	-	0	-	-
S-FELN	*	-	-	-	-	-	-	-	-	-	-	1	-	-	*	1	-	-
S-TRLN	-	-	-	-	-	-	-	-	*	-	-	1	-	-	-	0	-	*
S-TRWD	-	-	-	-	-	-	-	-	-	-	-	0	-	-	-	0	-	-
S-STLN	-	-	-	-	-	-	-	-	-	-	-	0	-	-	-	0	*	*
S-WXLN	-	*	-	-	-	-	-	-	-	-	-	1	-	-	-	0	-	-
S-WXWD	-	*	-	-	-	-	-	-	*	-	-	2	*	-	-	1	-	*
Totals	3	7	2	5	5	4	3	0	3	2	2	36	9	5	5	19	4	16

central Mediterranean (CAR, LIG, CAD, and PAN). The race *Apis m. iberica* occupies the Iberian Peninsula, separated by the Pyrenees from the closely similar race, *A. m. mellifera*, which is distributed over northwestern Europe (Cornuet and Fresnaye, 1989). The latter are known in Europe as the "common black bees." Certain bees from Colombia and Venezuela are known locally as "black bees" by beekeepers. Samples measured by Daly (1975) matched the distinctive morphometrics of Ruttner's group III(1). The black bees were probably introduced to the New World by Spanish colonists and are likely to be *Apis m. iberica*. The expected cluster for the west Mediterranean group would be IBR+MEF, based on previous studies of morphometrics.

In the second kind of comparison, certain clusters could be expected on the basis of the geographic proximity and evolutionary history of OTUs in the Old World or the history of introduced populations.

In Ruttner's group II, the races of subsaharan Africa are distributed as follows: west Africa (*Apis m. adansoni*); eastern Africa from Ethiopia to South Africa (*Apis m. scutellata*), except near Capetown (*A. m. capensis*); higher elevations in rain forests of east Africa (*A. m. monticola*); and the coast of east Africa (*A. m. litorea*). Two possible clusters were expected for the native African OTUs used here: ADA+(SEA+[SSA+CAP]) when based on geographic proximity, and a north-south distribution or ADA+(CAP+[SSA+SEA]), when based on formal taxonomy, assuming that *A. m. capensis* evolved from *Apis m. scutellata*. In all such clusters, ADA is expected to be less closely related to the other three African OTUs because it is separated from the others by a greater geographic distance.

"Africanized" bees of South and Central America were derived from queens introduced to southern Brazil in 1956 from South Africa (46 colonies) and Tanzania (one colony). Bees in both source areas of Africa are *A. m. scutellata* according to Ruttner (1988). Controversy continues as to whether Africanized bees have remained essentially African in genetics (Hall and Muralidharan, 1989; Smith et al., 1989) or have hybridized with European bees to an unknown extent (Rinderer, 1986). If Africanized bees still resemble *Apis m. scutellata* in morphometrics, the expected clusters would be AFD+SSA or AFD+(SSA+SEA).

European bees that are domesticated in North America were derived from colonies originally introduced from Europe. According to the review by Sheppard (1989a, 1989b), the first race introduced was *Apis m. mellifera* from England to Virginia in 1622 and was presumably the first race introduced to California in 1853. *Apis m. mellifera* is a member of Ruttner's group III(1) and differs morphometrically, especially in certain characters of the fore wing venation, from *Apis m. carnica* and *Apis m. ligustica* of group III(2). The latter race, known as Italian, was introduced to the east coast beginning in 1859 and subsequently became the favored stock of beekeepers throughout North America. Domesticated bees in South and Central America commonly have been imported from breeders in North America and often resemble the Italian stocks. The samples from Panama used here were probably from such sources in North America and were collected in 1979, before the invasion by Africanized bees. The expected cluster within the central Mediterranean group would be CAR+(LIG+[PAN+CAD]).

3.3 Computation

The statistical analyses were performed on IBM 3090 and IBM PC/AT computers with several program packages: SAS 5.16 (SAS Institute Inc., 1979) and SPSSX 3.1 (Norusis, 1985) for mainframe computers; and BIOM (Rohlf, 1989), BIOΣTAT II (Pimentel and Smith, 1986), NTSYS 1.5 (Rohlf, 1988) and STATGRAPHICS 3.0 (Statistical Graphics Corporation, Rockville, Maryland) for IBM PC. Statistical procedures followed Sokal and Rohlf (1981).

4 Results

4.1 Univariate Analysis

The purposes of the univariate analyses were to determine whether : (1) the distributions of the samples as grouped were normal; (2) differences in the distributions of a variable existed among OTUs, as determined by ANOVA; and (3) whether the variances in the ANOVAs were homogeneous. Thus, for each data group of the 11 OTU set, the three-group sets and the two test sets (CAT, AFT), the distribution of each variable was compared to the expected normal distribution, for a total of 576 tests. The Kolmogorov-Smirnov goodness of fit test (K-S Z) provided by SPSSX indicated that only 0.7% of these had two-tailed probabilities exceeding 0.05: *FWLN* and *AN34* in OTU AFT and *AN35* in OTUs IBR and MEL. However, the procedure of Lilliefors, recommended by Sokal and Rohlf (1981: 718), indicated that a much larger number of test statistics were different from those expected for normal distributions.

The Lilliefors tests are shown in Table 3. In summary, only the variables *CUBB*, *AN33*, *S-HWLN*, *S-TRWD*, and *S-STLN* had normal distributions among data OTUs. Out of 396 tests in the 11 OTU set, 36 (9.1%) had nonnormal distributions compared to 19 (17.6%) out of 108 tests for the three-group set. For comparisons within the 11 OTU set, the following numbers of tests exceeded the 5% critical values: for 11 size-related distances, 15 (12.4%) of 121 tests were significant; for 11 size-adjusted distances, 11 (9.1%) of 121 tests; for the three distances *CUBB*, *CUBA*, and *WXDS*, one (3.0%) of 33 tests; for the 10 angles, eight (7.3%) of 110 tests; and for *HAMU*, one (9.1%) of 11 tests. The same comparisons within the three-group set are: for 11 size-related distances, seven (21.2%) of 33 tests were significant; for 11 size-adjusted distances, four (12.1%) of 33 tests; for the three distances *CUBB*, *CUBA*, and *WXDS*, one (11.1%) of nine tests; for the 10 angles, five (16.7%) of 30 tests; and for *HAMU*, two of three tests. Common logarithmic transformation of the 25 raw variables gave virtually the same nonnormal distributions. The effect of adjusting the 11 size-related distances for general-size indicated that eight variables that were nonnormal before adjustment were also nonnormal afterwards, 12 that were normal became nonnormal, and 21 that were nonnormal became normal after size adjustment.

Univariate ANOVA for each of the 36 variables in each of the 11 OTU and three-group sets uniformly gave highly significant differences among the OTUs within the sets ($P < 0.0000$). Three tests for homogeneity of variances provided by SPSSX were performed: Cochran's C, Bartlett-Box F, and maximum variance/minimum variance. Each analysis resulted in no, one, two or three critical values indicating $P < 0.05$. I have chosen to consider

Table 4.
Tests for homogeneity of variances, where $P < 0.05$ for Cochrans C test (= C), Bartlett-Box F test (= B), and maximum variance/ minimum variance test (= M); n.s. = not significant.

Var.	OTU Sets		Var.	OTU Sets		Var.	OTU Sets	
	11	3		11	3		11	3
FWLN	C, B	n.s.	*CUBB*	B	B	*HAMU*	C, B	n.s.
FWWD	C, B	n.s.	*WXDS*	n.s.	n.s.			
HWLN	C, B	C, B				*S-FWLN*	C, B	n.s
HWWD	C, B	n.s.	*AN29*	n.s.	n.s.	*S-FWWD*	C, B	n.s.
TBLN	C	C, B	*AN30*	B	n.s.	*S-HWLN*	C, B	n.s.
FELN	C, B	C, B	*AN31*	C, B	B	*S-HWWD*	C, B	C
TRLN	C	n.s.	*AN32*	C, B	C, B, M	*S-TBLN*	B	n.s.
TRWD	C, B	n.s.	*AN33*	C, B	B	*S-FELN*	B	n.s.
STLN	n.s.	C, B	*AN34*	n.s.	n.s.	*S-TRLN*	C	C, B
WXLN	n.s.	B	*AN35*	B, M	B	*S-TRWD*	B	n.s.
WXWD	C, B	n.s.	*AN36*	n.s.	n.s.	*S-STLN*	C	C, B
			AN38	n.s.	n.s.	*S-WXLN*	n.s.	n.s.
CUBA	C	C, B	*AN39*	C, B	n.s.	*S-WXWD*	B	n.s.

significant those analyses with two or three such values (Table 4). In summary, of the 36 variables, the following 14 had homogeneous variances in the ANOVAs: *AN29*, *AN30*, *AN34*, *AN36*, *AN38*, *CUBB*, *TRLN*, *WXLN*, *WXDS*, *S-TBLN*, *S-FELN*, *S-TRWD*, *S-WXLN*, and *S-WXWD*. For the 11 OTU set, 17 variables were heterogeneous and for the three-group set, eight variables were heterogeneous. Common logarithmic transformation of the 25 variables, not adjusted for size, gave the same result or heterogeneous variances, except for *CUBA*, which, after transformation, had homogeneous variances in the three-group set. Here again, adjustment of the 11 distances for general-size had an effect on the variances: four analyses that had heterogeneous variances before size adjustment remained so after adjustment, one that was homogeneous became heterogeneous, and six that were heterogeneous became homogeneous after adjustment.

4.2 Multivariate Analysis

4.2.1 Principal Components Analysis

The PCAs were performed with the programs FACTOR of SPSSX and MPCA of BIOΣTAT. The purposes of PCAs were: (1) to identify size-related variables among the 25 raw variables (based on 820 samples), so that they could be adjusted for size; (2) to determine the number of nonzero eigenvalues in each of the six data matrices used in DFAs of the 11 OTUs described below; and (3) to visually search for groups among the scatter of sample points, when their scores were plotted on components 1-3. For the last, the PCA involved 820 samples and 25 raw variables, of which the 302 samples that were plotted corresponded to the data for the 11 OTU and three-group sets. The dispersion

Table 5.
Correlations of 25 raw variables with principal components 1-3 for 820 samples; the dispersion matrix was a correlation matrix; the percentage of variance accounted for by each component is in parenthesis (see also Fig. 1).

Variables		Components		
No.	Abberviation	PC-1 (42.5%)	PC-2 (15.1%)	PC-3 (7.1%)
1	*FWLN*	0.963	- 0.072	- 0.050
2	*FWWD*	0.892	0.133	- 0.109
3	*HWLN*	0.902	- 0.152	- 0.021
4	*HWWD*	0.848	0.239	- 0.133
5	*TBLN*	0.891	- 0.249	- 0.001
6	*FELN*	0.942	- 0.168	- 0.025
7	*TRLN*	0.911	- 0.132	- 0.005
8	*TRWD*	0.861	0.050	- 0.085
9	*STLN*	0.925	- 0.028	- 0.058
10	*WXLN*	0.894	- 0.165	- 0.075
11	*WXWD*	0.945	- 0.072	- 0.085
12	*CUBA*	0.380	0.636	- 0.128
13	*CUBB*	0.351	- 0.682	0.131
14	*WXDS*	- 0.572	0.267	0.109
15	*AN29*	- 0.461	- 0.429	- 0.260
16	*AN30*	0.523	0.238	0.233
17	*AN31*	- 0.034	- 0.683	0.029
18	*AN32*	0.561	0.597	- 0.080
19	*AN33*	- 0.191	- 0.564	- 0.628
20	*AN34*	- 0.031	0.453	- 0.345
21	*AN35*	- 0.050	0.718	- 0.332
22	*AN36*	0.161	0.506	0.723
23	*AN38*	0.069	- 0.451	0.410
24	*AN39*	0.320	- 0.147	0.445
25	*HAMU*	0.088	- 0.085	0.168

matrices were correlation matrices because the variables included distances, angles, and a count.

The correlation matrix of the 25 raw variables (820 samples) gave 300 correlations (half-matrix) of which 10 were greater than 0.90 and with the highest correlation of 0.928 between *FWLN* and *HWLN*. Although this evidence of linear dependence among some variables was present, the matrix had 25 nonzero eigenvalues, of which the 25th was 0.024. Components 1-3 accounted for a total of 64.7% of the variance (Table 5).

For 11 of the distances, the correlations with component 1 were high and positive, but low with all subsequent components (Table 5, Fig. 1). These variables were clearly associated with general-size and were designated "size-related" (Table 1). Of the remaining distances, *CUBA* and *CUBB* also had positive correlations with component 1, but these were lower in value and some of the correlations with subsequent components were absolutely higher.

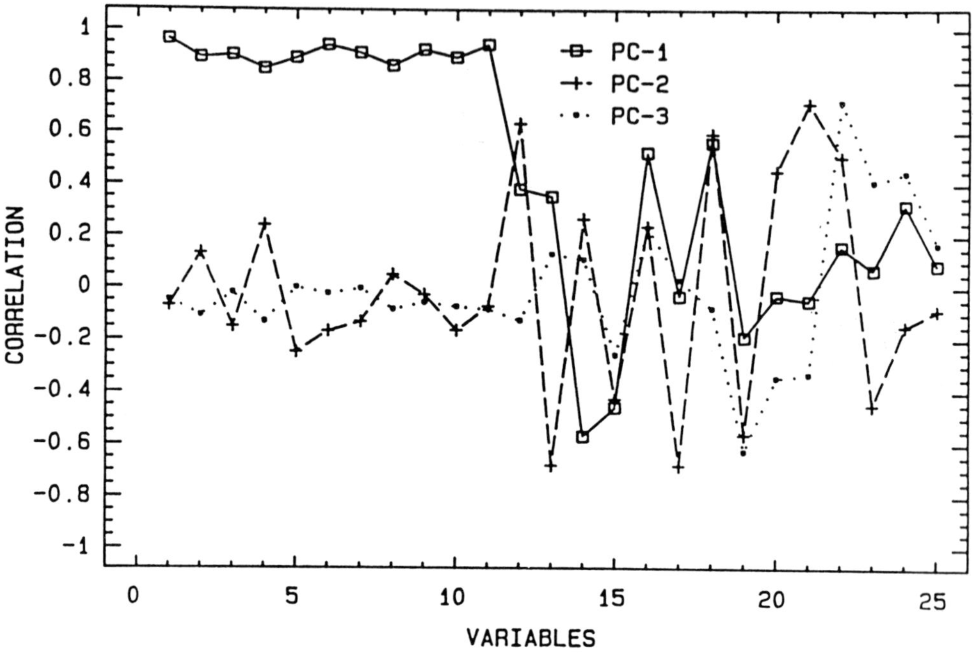

Figure 1. Correlations of 25 raw variables with principal components 1-3 (Table 5). Variables arranged from left to right in the order of Table 1.

WXDS had its highest correlation with component 1, but it was negative. The 10 angles had low to medium, positive and negative correlations with component 1 and some had absolutely higher correlations with subsequent components. *HAMU* had low correlations with all components except the sixth, on which it was uniquely high. For these reasons, *CUBA*, *CUBB*, *WXDS*, the 10 angles, and *HAMU* were not considered to be among the variables directly related to general-size and were designated "size-unrelated."

Visual inspection of the scatter plots of 302 samples on components 1-3 indicated that at most three discrete clusters could be recognized (Fig. 2). These corresponded to the African, west Mediterranean, and central Mediterranean groups.

4.2.2 Discriminant Function Analysis of the Three-Group Set

The purpose of the DFAs of the three-group set was to determine how reduced numbers of variables, with and without other desirable statistical properties, would affect the percentages of correctly classified data and test samples (Fig. 5C). Comparisons were made among subsets of variables, including two series of two to eight raw variables (each series with different univariate properties), 17 variables selected for reduced linear dependence,

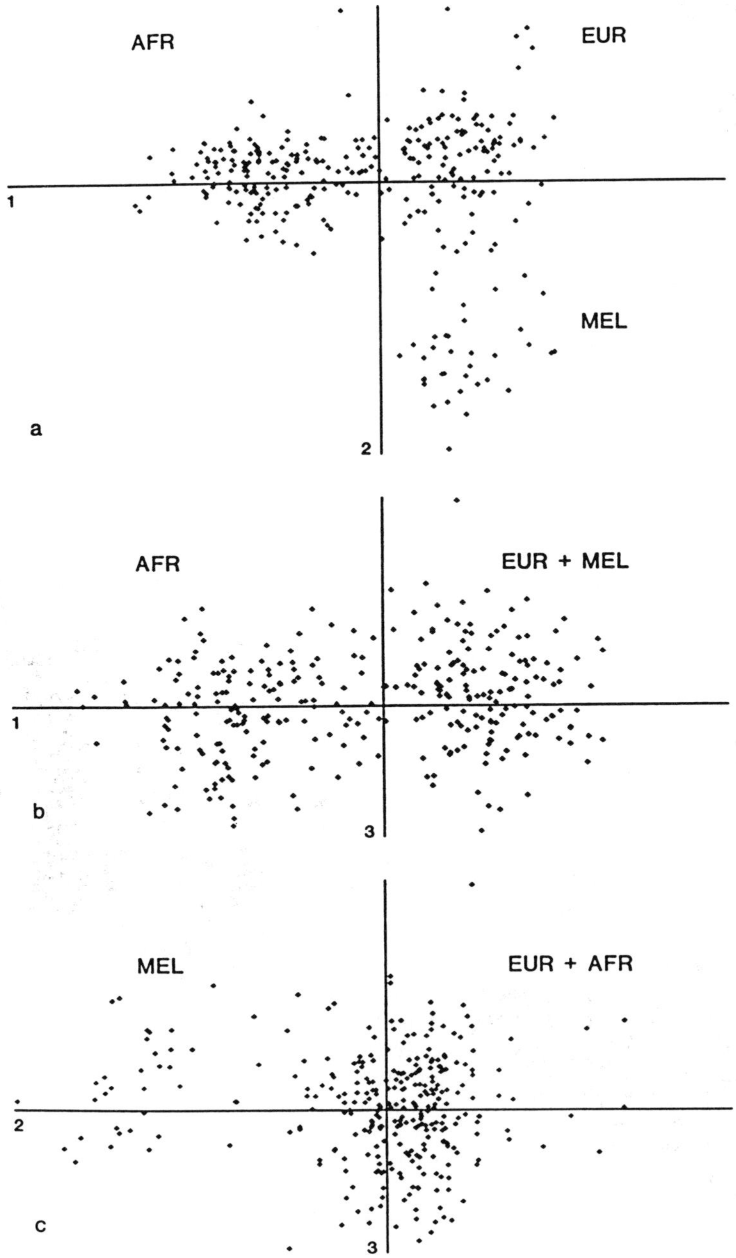

Figure 2. Plots of 302 samples on principal components 1-3, using 25 raw variables. (A) Axes 1 versus 2. (B) Axes 1 versus 3. (C) Axes 2 versus 3. Clusters representing the three major groups are labeled.

and the 25 raw variables. DFA of the 25 raw variables by the direct selection method gave a very slightly higher overall percentage of correct classifications (AFR, 97.4%; EUR, 100%; MEL, 100%; CAT, 97.1%; and AFT, 92.0%) in comparison to DFA by the stepwise method with Wilks' lambda (AFR, 96.7%; EUR, 100%; MEL, 100%; CAT, 97.5%; and AFT, 91.3%). The latter DFA eliminated the variables *FWLN*, *HWLN*, and *AN33*. According to Box's *M* test, all the covariance matrices that are reported here for DFAs of the three-group set were unequal ($P < 0.01$), except for the DFA in series II with *WXLN* and *FWWD* ($P = 0.13$).

The two series of variables were selected from the 25 raw variables as follows. All 25 were used in an initial DFA. Each variable was then ranked according to the pooled within-group correlation between the variable and the canonical discriminant functions. Of the 25 variables, 14 had their highest correlation with function 1, and 11 had their highest correlation with function 2 (Table 6). Series "I" of the DFAs was based on the highest ranked variables from each function, beginning with one variable from each function, then the highest two variables from each function, and so on, giving subsets of two, four, six, and eight variables. Series "II" of the DFAs was based on the highest ranked variables from each function, but only if the variable also exhibited the desired normal distribution and homogeneous variance in a univariate, three group ANOVA (Tables 3, 4). These univariate properties, however, do not guarantee that the variables in multivariate analysis will be multivariate normal.

The two variables with the highest correlations with the functions (Table 6), *FELN* (0.690 on function 1) and *AN32* (0.657 on function 2), gave the following counts and percentages of correctly classified data samples: AFR (141 or 92.8%), EUR (108 or 94.7%), and MEL (35 or 97.2%); and for test samples: AFT (251 or 80.7%) and CAT (188 or 90.8%) (Fig. 3). The univariate analyses of *FELN* indicated nonnormal distributions in AFT, EUR, and MEL (Table 3); both *FELN* and *AN32* had heterogeneous variances in the three-group sets (Table 4); and the covariance matrices in the DFA were unequal ($P < 0.0000$).

For comparison, multivariate normal data with the same sample sizes and nearly the same means, standard deviations and correlation as the actual data for *FELN* and *AN32* were generated by the program GEN in BIOΣTAT, using the actual square root of within-group mean square for each variable for the three groups as the standard deviations. This artificial data was normally distributed with homogeneous variances, and the covariance matrices were equal ($P = 0.54$). DFA, based on the artificial data, gave the following counts and percentages of correctly classified actual data: AFR (141 or 92.8%), MEL (35 or 97.2%), and EUR (108 or 94.7%); and for test samples of actual data: AFT (253 or 81.4%) and CAT (187 or 91.3%). The observed counts of correct classifications of actual data for the three groups and two test groups by the DFA based on artificial data was nearly identical to the counts of correctly classified samples by the DFA based on actual data (see counts in paragraph above; $G = 0.0099$, df = 4).

In series II, the variables *WXLN* (0.542 on function 1) and *FWWD* (0.313 on function 2) had normal distributions (Table 3), homogeneous variances in univariate ANOVA (Table 4), and equal covariance matrices in the DFA. The percent correct classification using these two variables was as follows for data samples: AFR (87.5%), EUR (70.2%), MEL (77.8%); and for test samples: AFT (60.4%) and CAT (86.0%) (Fig. 4).

Table 6.
Pooled within-group correlations between discriminating variables and canonical discriminant functions for analysis of three-group set with 25 raw variables; ranked by absolute magnitude of the correlation on function 1 (upper group) and function 2 (lower group); variables marked with an asterisk (*) have, in univariate analysis, either nonnormal distributions or heterogeneous variances or both. Reduced variable sets are designated as "I" = variable selected on correlation rank only and "II" = variable selected on correlation rank, normal distribution, and homogeneous variance in three-group set.

Variable	Function 1	Function 2	Members of Reduced Variable Sets
FELN *	0.690	0.120	I-2 to I-8
WXWD *	0.628	0.288	I-4 to I-8
TBLN *	0.572	0.069	I-6 to I-8
FWLN *	0.562	0.307	I-8
WXLN	0.542	0.180	II-2 to II-8
TRLN *	0.519	0.151	
HWLN *	0.516	0.185	
STLN *	0.459	0.280	
WXDS *	- 0.402	0.000	
TRWD	0.325	0.261	II-4 to II-8
CUBB	0.292	- 0.275	II-6, II-8
AN38	0.234	- 0.160	II-8
AN39 *	0.198	- 0.065	
HAMU *	0.122	- 0.067	
AN32 *	0.012	0.657	I-2 to I-8
CUBA *	0.002	0.540	I-4 to I-8
HWWD *	0.245	0.407	I-6 to I-8
AN31 *	0.175	- 0.401	I-8
AN29 *	- 0.091	- 0.314	
FWWD	0.304	0.313	II-2 to II-8
AN35 *	- 0.190	0.266	
AN33	0.080	- 0.255	II-4 to II-8
AN36 *	- 0.036	0.224	
AN34	- 0.132	0.200	II-6, II-8
AN30	0.158	0.173	II-8

Following the two variables in each series, the effects of four to eight variables were examined (Figs. 3, 4). To summarize, small subsets of the variables most highly correlated with the discriminating functions gave the highest percentages of correctly classified samples. Despite undesirable univariate properties, just two such variables gave percentages greater than 90 for correctly classified data samples, and the counts of both data and test samples did not differ from those of artificial multivariate normal variables. Additional variables in series I increased by 4.4-5.3 percentage points the percentages of samples correctly classified as OTUs EUR, CAT, and AFT;

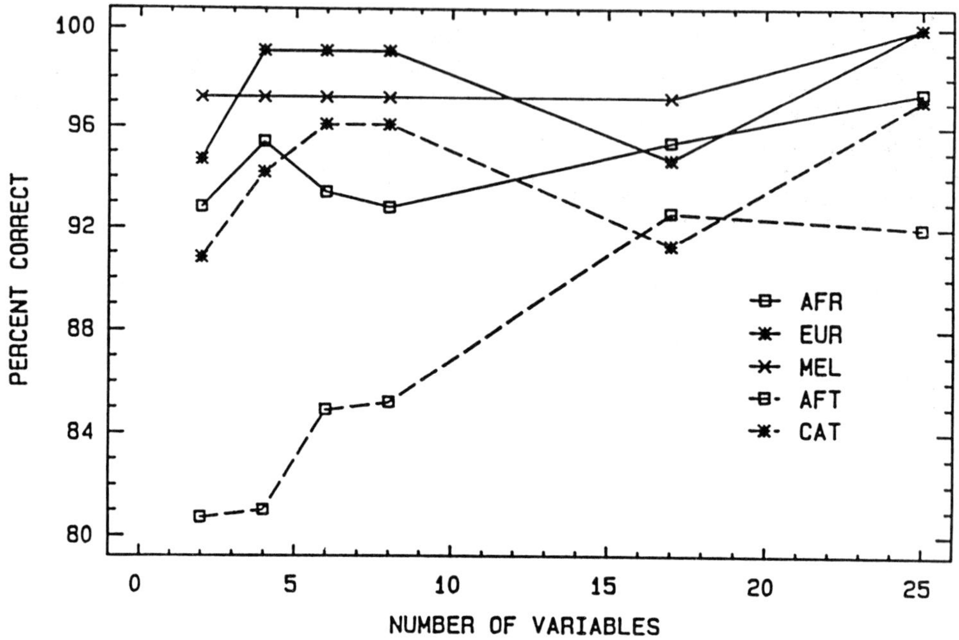

Figure 3. Percentages of correctly classified data [= solid lines] and test [= broken lines] samples of the three-group set for DFAs of series I (two to eight variables), 17 variables with reduced linear dependence, and 25 raw variables.

percentages remained high and unchanged for MEL and varied within 2.6 percentage points for AFR (Fig. 3). Additional variables in series II increased by 3.3-21 points the percentages of correctly classified samples in all OTUs (Fig. 4). The improvement in the percentage of correct classification increased with the number of variables used, especially for AFT in series I, where: two variables yielded 80.7%, eight variables gave 85.2%, and all 25 raw variables showed 92.0% (Fig. 3).

The final comparison was of the 25 raw variables and a subset of variables selected for reduced linear dependence. To accomplish the latter, the 11 size-adjusted distances were substituted for the size-related distances and the new correlation matrix, based on 302 samples, was searched for high correlations. The variables *AN33*, *AN35*, *S-FELN*, *AN31*, *AN29*, and *S-HWWD* had correlations greater than an absolute value of 0.71, and had less desirable univariate properties than the other variables with which they were correlated. The six variables, ranked in descending order according to their highest correlation, were deleted in succession and cumulatively. At each deletion, the eigenvalues of the matrix were inspected. With the addition of *S-HWWD* to those deleted, the least eigenvalue changed from 0.008 to 0.09. At this point, the correlation matrices for each of the three groups were inspected and *S-FWLN* deleted, because it had a value of 0.78 in group EUR and a nonnormal distribution in group AFR.

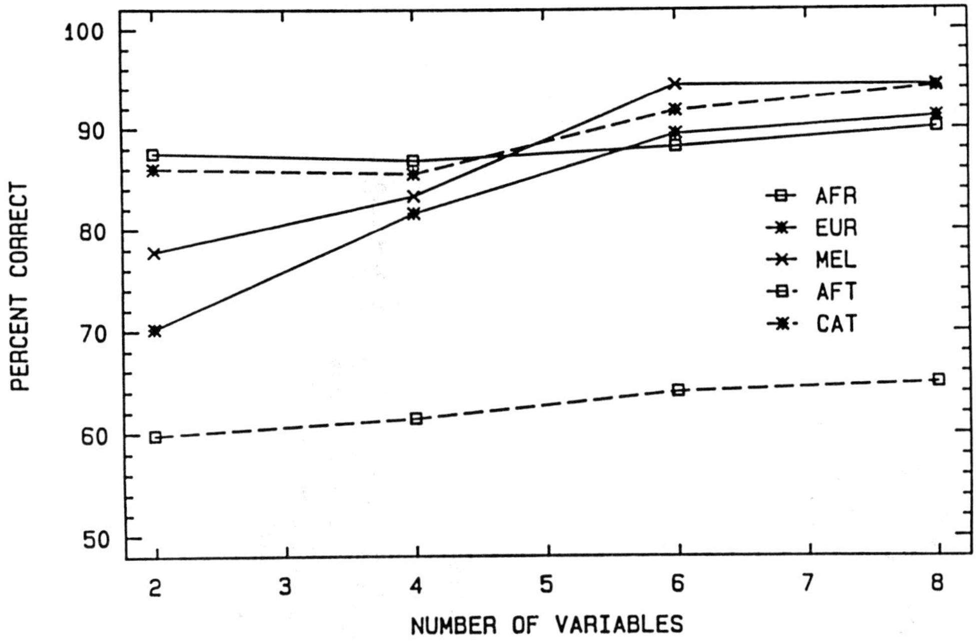

Figure 4. Percentages of correctly classified data [= solid lines] and test [= broken lines] samples of the three-group set for DFAs of series II.

The final selection of 18 variables was *HAMU*, *AN30*, *AN32*, *AN34*, *AN36-AN39*, *CUBB*, *CUBA*, *WXDS*, *S-FWWD*, *S-HWLN*, *S-HWWD*, and *S-WXWD*. All correlations in the total correlation matrix and in each of the three group matrices were now less than the absolute value of 0.66, and the determinant for each matrix was nonzero. This set of 18 variables also was used below as DFA-3 (defined later) for the 11 OTU set.

DFA, with stepwise selection of the 18 variables, resulted in the further removal of *S-HWLN* in the three-group set. Comparison of the DFAs of 25 raw variables versus the 17 variables with reduced linear dependence, indicated that the 25 raw variables gave the highest percentages of correct classification by 2-5.8 percentage points except for OTU AFT, where the latter data matrix gave 0.6 points higher (Fig. 3).

4.2.3 Discriminant Function Analysis of the 11 OTU Set

DFAs were performed with the program DISCRIMINANT in SPSSX 3.1. The direct selection method with Wilks' lambda as the test criterion was used, except for DFA-3 below. DFA of the 25 raw variables by the stepwise method with Wilks' lambda eliminated the variable *AN31*, but produced almost identical percentages of correctly classified samples as the direct method with all 25 variables. Box's *M* test indicated the covariance matrices for all DFAs reported below had unequal covariances ($P < 0.0000$). The final number of variables in each of the six subsets below equals the number eigenvalues greater than or equal to 0.02 in each total correlation matrix, as determined by PCA.

For the DFA and phenograms of the 11 OTU set, subsets of the 36 variables were selected as follows: (1) 25 raw variables including the 11 size-related distances and 14 size-unrelated variables (Table 1); (2) 25 variables, substituting 11 size-adjusted distances for the 11 size-related distances, of which the variable *S-WXWD* failed the tolerance test in DFA, leaving 24 variables; (3) 18 variables selected for reduced linear dependence, of which stepwise selection retained all 18 in DFA of 11 OTUs (see description of this matrix above under DFA of three-group set; no further effort was made to remove high correlations from each of the 11 group matrices); (4) 11 size-related distances with high linear dependence; (5) 11 size-adjusted distances, of which the variable *S-WXWD* failed the tolerance test in DFA, leaving 10 variables; and (6) 10 angles of venation. These analyzed subsets are labeled DFAs-1 through -6, respectively.

Plots of the 302 samples according to their discriminant scores on functions 1-2 for DFA-1 (25 raw variables) showed correct assignment of the 11 OTUs to the three major clusters seen in the PCA (Fig. 2). The overlap of related OTUs seen on functions 1-2 (Figs. 5A, 5B), and functions 1-3 (not illustrated) as well, prevented a visual determination of possible separations among OTUs within major groups.

The degree of separation of the OTUs was examined in detail by classifying each data and test sample in each of the six DFAs. With 11 OTUs, the data samples could be classified correctly, by chance alone, at a probability of 9.1%. The highest percentages of correctly classified samples in OTUs IBR (100%), LIG (100%), ADA (100%), PAN (97.2%), CAD (96.1%), and AFD (79.2%) were in DFA-1; SEA (100%) in DFAs-1 and -2; SSA (90.7%) in DFAs-1 and -3; CAP (93.3%) in DFA-2; CAR (90%) in DFAs-1, -2, and -6; and all samples of MEF were correctly classified in all DFAs (Fig. 6). OTU ADA had the lowest percentage correctly classified at 33.3% in DFA-5, becoming largely indistinguishable from other African OTUs when the 10 size-adjusted distances were employed.

In summary, the overall percentages of data OTUs that were correctly classified in DFAs-1 through -6 were 93.05%, 90.7%, 84.1%, 78.8%, 66.9%, and 65.6%, respectively (Fig. 7, TOT). The percentages of correct classification of the test samples AFT and CAT paralleled the classification of the data OTUs AFD and CAD, but averaged 6.3 and 7.8 percentage points lower, respectively (Fig. 7).

4.2.4 **Phenograms of the 11 OTU Set**

Phenograms were produced by NTSYS-PC 1.5, using as data the values from SPSSX for the canonical discriminant functions at the group centroids. Euclidean distances among the centroids were computed, followed by clustering with UPGMA. Six phenograms were prepared, based on the six DFAs described above. The correlations for the six phenograms with the original distance matrices were 0.89, 0.89, 0.89, 0.80, 0.86, and 0.89, respectively.

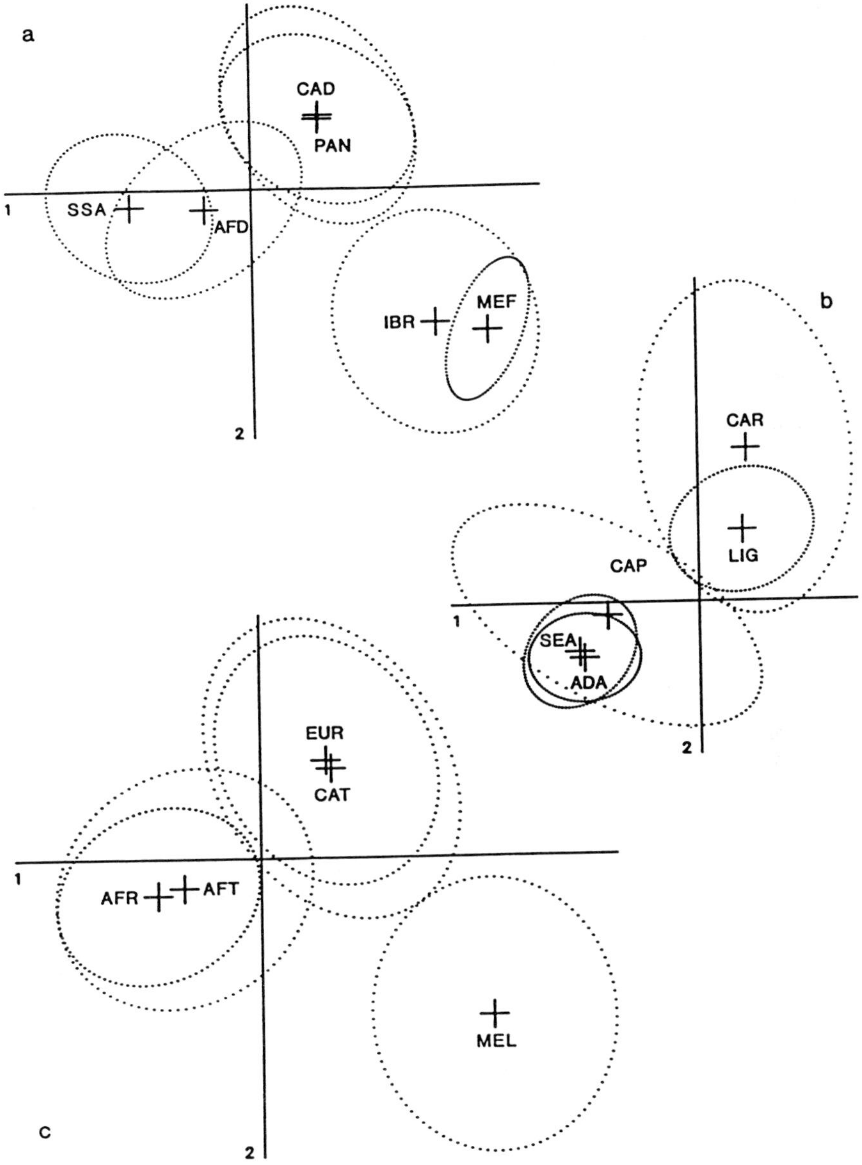

Figure 5. Plots of 95% confidence ellipses and centroids [= crosses] for OTUs on discriminant functions 1 and 2, using the algorithm of Cornuet (1982). (A) The 11 OTU set of DFA-1, showing three pairs of OTUs that have presumed descendent relationships; (B) The 11 OTU set of DFA-1, showing five OTUs of Old World origin; the variance accounted for by functions 1 and 2 is 49.5% and 24.2%, respectively. (C) The DFA of the three-group set with 25 raw variables, showing also the superimposed test samples AFT and CAT; variance accounted for by functions 1 and 2 is 64.6% and 35.4%, respectively.

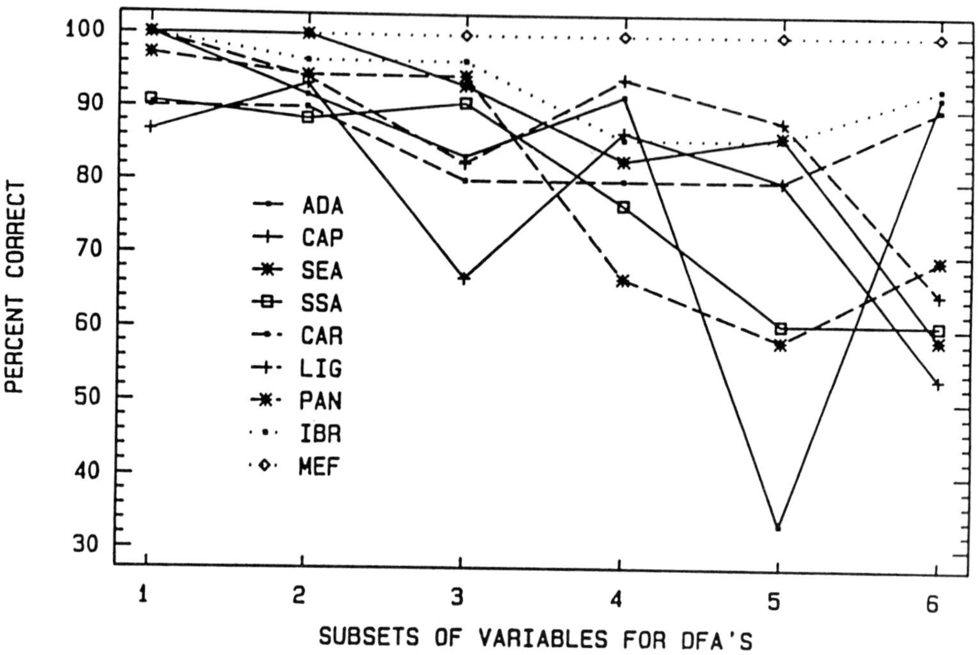

Figure 6. Percentages of correctly classified data samples in six DFAs for four African OTUs [= solid lines], three central Mediterranean OTUs [= broken lines], and two west Mediterranean OTUs [= dotted lines]. Numbered subsets of variables for DFAs-1 through -6 correspond to those described in text.

In phenogram-1, based on DFA of 25 raw variables, the stems of the three major groups, African, west Mediterranean (= IBR+MEF), and central Mediterranean, were distinct at about the 6.5 phenon and the expected cluster CAR+(LIG+[CAD+PAN]) was present. The native African OTUs were clustered, as would be expected, on the basis of their formal taxonomy, ADA+(CAP+[SSA+SEA]), but AFD did not directly join SSA+SEA in the cluster AFD+(CAP+[SSA+SEA]) (Fig. 8).

In phenogram-2, based on DFA of 10 size-adjusted distances and 14 size-unrelated variables, the stems of the three major groups were distinct at about the 6.4 phenon. The native African OTUs appear here as a series of geographically more distant OTUs in the cluster ADA+(SEA+[SSA+CAP]). The expected clusters AFD+SSA and PAN+CAD were present. The close relation of LIG to CAD+PAN was not indicated, because CAR was placed as the closest central Mediterranean race. Phenogram-3, based on DFA of 18 variables with reduced linear dependence, was similar to phenogram-2 in these details, except that IBR and MEF join relatively lower, at the phenon joining the African and central Mediterranean groups, so that the west Mediterranean stem does not appear equivalent to other major groups (Fig. 8).

In phenogram-4, based on DFA of 11 size-related distances, clusters of the African and central Mediterranean groups were present, but the latter is

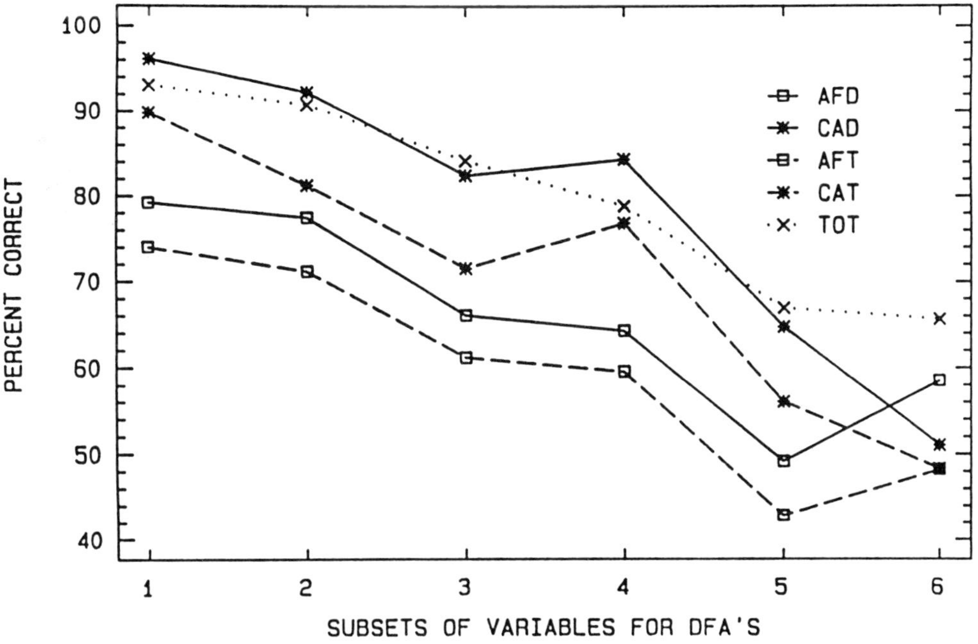

Figure 7. Percentages of correctly classified data samples for OTUs CAD and AFD [= solid lines], the corresponding test samples CAT and AFT [= broken lines], and the overall percentages for data OTUs [= TOT, dotted line]. Numbered subsets of variables for DFAs-1 through -6 correspond to those described in text.

nested within a disrupted west Mediterranean group. The expected clusters CAR+(LIG+[PAN+CAD]) and AFD+SSA were present. The remote position of ADA, relative to other native African OTUs, was present, but among the latter the cluster CAP+SEA was not expected.

In phenogram-5, based on DFA of 10 size-adjusted distances, the stem of the west Mediterranean group (= IBR+MEF) was distinct, but not the central Mediterranean and African groups, which were mixed together. The expected clusters, PAN+CAD and AFD+SSA, were present, but not others.

In phenogram-6, based on DFA of 10 angles of venation, the stems of west Mediterranean (= IBR+MEF) and African groups were distinct at phenon 3.20, but the central Mediterranean group had the African group included within it. Also present was the geographic series ADA+(SEA+[CAP+SSA]), but no other expected clusters occurred (Fig. 8).

In summary, the phenograms were ranked subjectively in descending order of expected clusters as: 1 = 2 > 3 > 4 > 5 = 6. The three major groups appeared with the correct OTUs, and with stems at the same phenon (within each phenogram), in phenograms-1 and -2, and almost so in phenogram-3; however, the integrity of one or more of the major groups was lost in the other phenograms. The expected cluster PAN+CAD appeared in all phenograms except the sixth, but its anticipated relation in the cluster

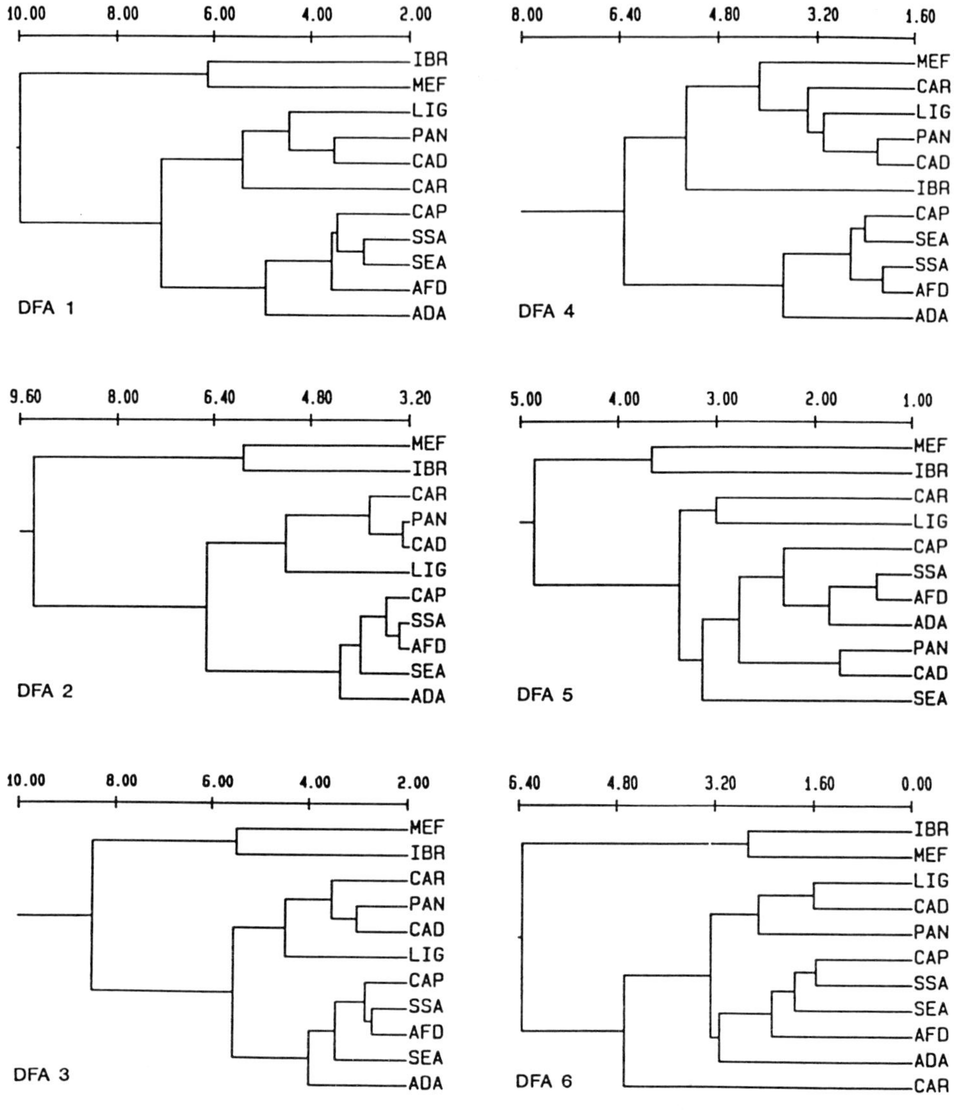

Figure 8. Phenograms of the morphometric relationships of 11 OTUs based on six DFAs, as described in the text.

CAR+(LIG+[PAN+CAD]) was indicated only in phenograms-1 and -4. The expected cluster AFD+SSA appeared in phenograms-2 through -5. A remote position for ADA appeared all phenograms except the fifth. The relations among the native African OTUs, although plausible in phenograms-1 through -3, and phenogram-6, differed in relative position and a posteriori explanation.

5 Discussion

5.1 General-Size and Infraspecific Classification

The infraspecific classification of honey bees has been difficult because the bees lack a variety of useful taxonomic characters. They exhibit close similarity in anatomical details of the adults and larvae, variable behavior and color patterns, a low level of isozyme polymorphism, and tiny, similar chromosomes. Although molecular genetics offers great potential for the future (Hall, 1986, 1988; Smith, 1988), morphometric characters, until now, have proven to be the most effective to distinguish races. Morphometric analysis is relatively quick, inexpensive, and can be applied to specimens preserved and stored indefinitely by conventional museum methods. The quantitative framework has allowed comparison of "gaps" among populations to justify recognition of formal races and has correlated well with biogeography and biology (Ruttner, 1988).

The confounding effect of general-size on morphometric relationships, however, is of special concern to morphometricians (see Bookstein et al., 1985). The problem is acute in studies of organisms with indeterminate growth or those whose body sizes are regularly influenced by environment. In the context of modern morphometrics, the use of raw, size-related variables in the classification of honey bees could be considered a mistake. Before drawing such a conclusion, the following facts about the biology of honey bees and empirical tests of DFAs, including size-related variables, should be considered.

Honey bees, as social insects, rear progeny to maturity with a common diet, in comb cells of regular dimensions, and with temperature closely controlled by adult bees. Once the adult instar is reached and the exoskeleton sclerotized, size and shape are fixed for the life of the bee. As a consequence of this and their shared maternal genetics, healthy workers from a colony are relatively uniform in size and shape.

Size-related, racial characteristics of native bees are only slightly changed in descendent populations that were introduced abroad to new environments many decades ago. The best documented instances are some early introductions from Europe to Australia that were subsequently isolated from contact with other bee races. *Apis m. ligustica* was introduced from Italy to Kangaroo Island in 1884, and *Apis m. mellifera* from England to Tasmania, probably as early as 1814. Both populations were shown by Ruttner (1976) to retain most of the morphometric features of their native races after 90 and 160 years, respectively. In this chapter, a close morphometric relationship is still evident among colonies from Panama, California, and Kangaroo Island, Australia; all of these colonies allegedly trace their origin mainly, if not wholly, to *Apis m. ligustica* in Italy (Figs. 5A, 5B). Africanized bees in Latin America still resemble *Apis m. scutellata* from southern Africa (Fig. 5A). Finally, "black bees" in Latin America and Tasmania, Australia, still have the distinctive morphometrics of *Apis m. iberica* and *Apis m. mellifera* of western Europe (Fig. 5A).

Size-related variables, such as wing width and tongue length, have been shown to have heritabilities of up to 0.85 (Roberts, 1961). Within colonies, body size has been found to play a role in the recruitment system of individual foraging workers (Waddington, 1989). Among populations, body size and the relative sizes of appendages of honey bees exhibit a north-south cline from 60°N to 34°S in latitude, that corresponds to Bergmann's and Allen's rules (Ruttner, 1988). Thus, body size in honey bees is evidently heritable, adaptive, and relatively stable in new environments.

For ordinary taxonomic purposes, size should be considered along with shape and other taxonomic characters in the infraspecific classification of honey bees. Some percentage of samples is usually expected to be unidentified in identifying subspecies of animals, because the clusters of samples of OTUs, even in multivariate space, often overlap; hence the "75 percent rule" of Mayr (1969). Size becomes a special consideration if the cost of misidentification is high as, for example, in regulatory entomology. In this case, the use of several independent sources of information for honey bees, as demonstrated by Spivak et al. (1987), should reduce the risk of misidentification.

Environmental effects on the body sizes of bees were demonstrated long ago in experiments (Alpatov, 1929; Grout, 1937). The usual risk in using size-related morphometrics in identification of honey bees is that abnormally small individuals may be misidentified because they are in the size range of a smaller race. Size-unrelated measurements included in the DFA may or may not outweigh the contribution of abnormally small, size-related measurements to the discriminant score. For example, Daly and Morse (1991) report that dwarf European workers experimentally reared from drone cells were 9.1% smaller than normal and were incorrectly identified as the smaller Africanized bees by morphometric methods.

Other authors report better success in identification of abnormal bees. Herbert et al. (1988) induced nutritional stress with various diets fed to bee larvae, such that small adult bees were produced. Except for the smallest progeny, the bees were correctly identified as European, as opposed to Africanized, by morphometric methods. Rinderer et al. (1986) reared workers of European and Africanized bees in cross-fostering experiments. Although the larger cell size of European comb resulted in larger bees of both types and Africanized comb resulted in smaller bees, the progeny were correctly identified by morphometric methods.

5.2 Univariate Analysis

Turning now to the present study, 36 morphometric variables were examined in univariate and multivariate analyses. Each of the 36 variables was grouped into 11 OTU and three-group sets and tested for a normal distribution. With the preferred goodness of fit test, a much higher number of the distributions were nonnormal than would be expected by chance alone (Table 3). Although in univariate ANOVA each variable exhibited highly significant differences among the OTUs, the number of ANOVAs with heterogeneous variances also were much higher than expected (Table 4). In this study, a larger number of groups (the 11 OTU set), each with smaller numbers of samples, gave a lower percentage of nonnormal distributions but a higher number of variables with heterogeneous variances; fewer groups (the three-group set), each with larger numbers of samples, gave the reverse.

In summary, the morphometric variables routinely used in honey bee classification did not consistently exhibit the features recommended for univariate statistical analysis. The samples were grouped taxonomically and without regard to such potential problems. Logarithmic transformations failed to make the distributions and variances acceptable, although the statistical adjustment of size-related variables did tend to improve some, but not all, of the variables in this regard. This frequency of nonnormal distributions and heterogeneous variances is a warning not to trust significance tests when based on marginal values without also verifying the assumed univariate properties.

5.3 Multivariate Analysis

A visual search for clusters in scatterplots revealed, at most, three clusters when 302 data samples were examined with reference to principal components 1-3 (based on PCA of 820 samples and 25 raw variables). These clusters corresponded to three of the four major taxonomic groups of Ruttner's (1988) classification of honey bee races (Fig. 2). The same was true for a DFA of the 11 OTU set, using all 25 raw variables, in which the plots of grouped samples on discriminant functions 1 and 2 were largely superimposed in three groups (Figs. 5A, 5B). The classification of this DFA, however, indicated that 93% of 302 data samples were assigned correctly to 11 OTUs. Therefore, it was not apparent by visual examination of the PCA and DFA plots on axes 1-3, that many more clusters existed, which were biologically meaningful and could be distinguished in multivariate space. This is a warning not to trust a few scatterplots to reveal all the interesting clusters.

5.3.1 Discriminant Function Analysis of the Three-Group Set

Using the DFA of 25 raw variables for the three-group set as a starting point, the effects of a reduced number of variables on percentages of correct classifications were examined. Two series of DFAs were made, one selecting variables that were highly correlated with the two canonical discriminant functions, and a second series of variables that were ranked less highly, but also had normal distributions and homogeneous variances in univariate ANOVA (Table 6). The two to eight variables of series I clearly gave the highest percentages of correct classifications of both data and test samples, irrespective of the other statistical properties of the variables (Fig. 3). The addition of variables within series I, and the use of 17 variables with reduced linear dependence, gave some erratic changes in the already high percentages of correctly classified samples. The highest percentages of correct classification for OTUs AFR, EUR, MEL, and the test group CAT, were with the 25 raw variables. The test group AFT had large increases in correctly classified samples with an increase in the number of variables, but the highest percentage was with 17 variables. The differences between the percentages of correct classifications of the data samples and the test samples in group AFT also tended to decrease with the addition of variables (Fig. 3).

Lower percentages of correct classification were obtained in series II using variables with lower correlations with the discriminating functions, and with normal distributions and homogeneous variances in univariate ANOVA (Fig. 4). In this series of DFAs, the percentages of correctly classified test samples of group CAT were actually higher than for the data samples. The addition of variables here steadily increased the percentages of correctly classified data and test samples.

Evidently variables that could have been rejected as unsuitable after univariate analysis (nonnormal distributions, heterogeneous variances in group ANOVAs) and multivariate analysis (linear dependence, unequal covariance matrices) were highly effective in DFAs. Furthermore, a larger number of variables, a maximum of 25 in this study, apparently can compensate for the lack of recommended statistical properties at both the univariate and multivariate levels and yield high percentages of correct classifications in empirical tests.

5.3.2 **Discriminant Function Analysis of the 11 OTU Set**

In a search for subsets of variables that were effective discriminators and correlated with general-size or not, six DFAs were made to compare classifications of 11 OTUs using: (1) 25 raw variables, including 11 size-related distances and 14 size-unrelated variables (Table 1); (2) 24 variables, including 10 size-adjusted distances and 14 size-unrelated variables; (3) 18 variables selected for reduced linear dependence; (4) 11 size-related distances with high linear dependence; (5) 10 size-adjusted distances; and (6) 10 angles of fore wing venation unrelated to general-size. Empirical evaluations of the classifications were made in terms of: (1) the percentages of correct classifications for data samples (Fig. 6); (2) the same for test samples (Fig. 7); and (3) the credibility of relationships among OTUs indicated by phenograms (Fig. 8).

DFAs-1 and -2 were superior and about equal in all three evaluations. These DFAs differed in whether some distance variables were size-related or size-adjusted, respectively. This did not have an appreciable effect on the empirical evaluations. With the current programs, considerable extra effort was required to compute the size-adjusted data matrices and to combine them with other data. The results of these manipulations follow, with the percentages of correct classifications for each DFA given in parentheses for the overall set, and for test samples CAT and AFT: DFA-1, with 302 data samples correctly classified in 11 OTUs, overall (93.0%), CAT (89.8%) and AFT (74.0%); DFA-2, overall (90.7%), CAT (81.2%) and AFT (71.1%); DFA-3, with 18 variables with reduced linear dependence, overall (84.1%), CAT (71.5%) and AFT (61.1%); DFA-4, with 11 size-related distances, overall (78.8%) CAT (76.8) and AFT (59.5%); DFA-5, with 11 size-adjusted distances, overall (66.9%), CAT (56.0%) and AFT (42.9%); and DFA-6, with 10 angles of venation, overall (65.6%), CAT(48.3%) and AFT (48.2%).

The phenograms, based on DFAs-1 and -2, differed in details, but were comparable and depicted the expected three major groups and the highest number of expected clusters of OTUs within the major groups (Fig. 8). Phenograms for DFAs-3 and -4 ranked next in order, followed by those for DFAs-5 and -6, because they failed to indicate all three major groups and were missing some of the expected clusters of OTUs.

In conclusion, the highest percentages of correct classification and most credible phenograms of the 11 OTUs were produced by larger numbers of variables that combined size-unrelated and size-related variables, or by a combination with the latter variables adjusted for size. The exclusive use of fewer, size-unrelated or size-adjusted variables gave the lowest percentages of correct classification and the least credible phenograms in the empirical tests.

6 Acknowledgments

Special thanks go to Christopher Meacham and F. James Rohlf, who were always ready to answer questions and to assist me with computer programs. Christina Jordan kindly helped prepare the illustrations.

7 References

Alpatov, W.W., 1929. Biometrical studies on variation and races of the honey bee (*Apis mellifera* L.). *Quarterly Review of Biology*, 4: 1-58.

Bookstein, F.L., Chernoff, B., Elder, R.L., Humphries, J.M., Smith, G.R. and Strauss, R.E., 1985. Morphometrics in Evolutionary Biology. Special Publication, 15. Academy of Natural Sciences, Philadelphia, 277 pp.

Cornuet, J.M, 1982. Representation graphique de populations multinormales par des ellipses de confiance. *Apidologie*, 13: 15-20.

Cornuet, J.M. and Fresnaye, J., 1989. Etude biometrique de colonies d'abeilles d'Espagne et du Portugal. *Apidologie*, 20: 93-101.

Daly, H.V., 1975. Identification of Africanized bees by multivariate morphometrics. In: Proceedings of the 25th International Apiculuture Congress, (Grenoble, France). Apimondia, Bucharest, pp. 356-358.

Daly, H.V., 1985. Insect morphometrics. *Annual Review of Entomology*, 30: 415-438.

Daly, H.V., 1991. Systematics and identification of Africanized honey bees. In: M. Spivak and M. Breed (Editors), The African Honey Bee. Westview Press, Boulder, Colorado, pp. 13-44.

Daly, H.V. and Balling, S.S., 1978. Identification of Africanized honey bees in the Western Hemisphere by discriminant analysis. *Journal of the Kansas Entomological Society*, 51: 857-869.

Daly, H.V., Hoelmer, K., Norman, P. and Allen, T., 1982. Computer-assisted measurement and identification of honey bees (Hymenoptera: Apidae). *Annals of the Entomological Society of America*, 75: 591-594.

Daly, H.V. and Morse, R.A., 1991. Abnormal sizes of worker honey bees (*Apis mellifera* L.) reared from drone comb (Hymenoptera, Apoidea). *Journal of the Kansas Entomological Society*, 64: 193-196.

DuPraw, E.J., 1965a. The recognition and handling of honeybee specimens in non-Linnean taxonomy. *Journal of Apicultural Research*, 4: 71-84.

DuPraw, E.J., 1965b. Non-Linnean taxonomy and the systematics of honeybees. *Systematic Zoology*, 14: 1-24.

Goetze, G.K.L., 1930. Variabilitats und Zuchtungsstudien an der Honigbiene mit Besonderes Berucksichtung der Langrusseligkeil. *Archiv fur Bienenkunde*, 11: 185-236.

Grout, R.A., 1937. The influence of size of brood cell upon the size and variability of the honeybee (*Apis mellifera* L.). *Research Bulletin, Agricultural Experiment Station, Iowa State College*, 218: 260-280.

Hall, H.G., 1986. DNA differences found between Africanized and European honeybees. *Proceedings of the National Academy of Sciences of the United*

States of America, 83: 4874-4877.

Hall, H.G., 1988. Characterization of the African honey-bee genotype by DNA restriction fragments. In: , G. Needham, R. Page, M. Delfinado-Baker and C. Bowman (Editors), Africanized Honey Bees and Bee Mites. Ellis Horwood Publishing Company, Chichester, U.K., pp. 287-293.

Hall, H.G. and Muralidharan, K., 1989. Evidence from mitochondrial DNA that African honey bees spread as continuous maternal lineages. *Nature*, 339: 211-213.

Herbert, E.W., Sylvester, H.A., Vandenberg, J.D. and Shimanuki, H., 1988. Influence of nutritional stress and the age of adults on the morphometrics of honey bees (*Apis mellifera* L.). *Apidologie*, 19: 221-220.

Mayr, E, 1969. Principles of Systematic Zoology. McGraw-Hill Book Company, New York, 428 pp.

Norusis, M.J., 1985. SPSSX: Advanced Statistics Guide. McGraw-Hill Book Company, New York, 505 pp.

Pimentel, R.A. and Smith, J.D., 1986. BIOΣTAT II, a Multivariate Statistical Toolbox. Sigma Soft, Placenta, California.

Rinderer, T.E., 1986. Africanized bees: the Africanization process and potential range in the United States. *Bulletin of the Entomological Society of America*, 32: 222-227.

Rinderer, T.E., Sylvester, H.A., Collins, A.M. and Pesante, D., 1986. Identification of Africanized and European honey bees: effects of nurse-bee genotype and comb size. *Bulletin of the Entomological Society of America*, 32: 150-152.

Roberts, W.C., 1961. Heterosis in the honey bee as shown by morphological characters in inbred and hybrid bees. *Annals of the Entomological Society of America*, 54: 878-882.

Rohlf, F.J., 1988. NTSYS-PC, Numerical Taxonomy and Multivariate Analysis System, (Version 1.50). Exeter Publishing Ltd., Setauket, New York.

Rohlf, F.J., 1989. BIOM, a Package of Statistical Programs to Accompany the Text Biometry. Exeter Publishing Ltd., Setauket, New York.

Rohlf, F.J. and Bookstein, F.L., 1987. A comment on shearing as a method for "size correction." *Systematic Zoology*, 36: 356-367.

Ruttner, F., 1976. Isolated populations of honeybees in Australia. *Journal of Apicultural Research*, 15: 97-104.

Ruttner, F., 1988. Biogeography and Taxonomy of Honeybees. Springer-Verlag, New York, 284 pp.

Ruttner, F., Tassencourt, L. and Louveaux, J., 1978. Biometrical-statistical analysis of the geographic variability of *Apis mellifera* L. I. Material and methods. *Apidologie*, 9: 363-381.

SAS Institute Inc., 1979. SAS User's Guide: Statistics, (1979 Edition). SAS Institute Inc., Cary, North Carolina, 494 pp.

Sheppard, W.S., 1989a. A history of the introduction of honey bee races into the United States. Part I. *American Bee Journal*, 129: 617-619.

Sheppard, W.S., 1989b. A history of the introduction of honey bee races into the United States. Part II. *American Bee Journal*, 129: 644-667.

Smith, D.R., 1988. Mitochondrial DNA polymorphisms in five Old World subspecies of honey bees and in New World hybrids. In: G. Needham, R. Page, M. Delfinado-Baker and C. Bowman (Editors), Africanized Honey Bees and Bee Mites. Ellis Horwood Publishing Company, Chichester,

U.K., pp. 303-312.

Smith, D.R., Taylor, O.R. and Brown, W.M., 1989. Neotropical Africanized honey bees have African mitochondrial DNA. *Nature*, 339: 213-215.

Sneath, P.H.A. and Sokal, R.R., 1973. Numerical Taxonomy, the Principles and Practice of Numerical Classification. W.H. Freeman and Company, New York, 573 pp.

Sokal, R.R., and Rohlf, F.J., 1981. Biometry, (Second Edition). W.H. Freeman and Company, New York, 859 pp.

Spivak, M., Ranker, T., Taylor, O.R. and Taylor, W., 1987. Discrimination of Africanized honey bees using behavior, cell size, morphometrics, and a newly discovered isozyme polymorphism. In: G. Needham, R. Page, M. Delfinado-Baker and C. Bowman (Editors), Africanized Honey Bees and Bee Mites. Ellis Horwood Publishing Company, Chichester, U.K., pp. 311-324.

Waddington, K.D., 1989. Implications of variation in worker body size for the honey bee recruitment system. *Journal of Animal Behavior*, 2: 91-103.

LEPIDOPTERAN WING MORPHOLOGY: THE MULTIVARIATE ANALYSIS OF SIZE, SHAPE, AND ALLOMETRIC SCALING

Richard E. Strauss

Department of Biological Sciences, Texas Tech University, Lubbock, Texas 79409, U.S.A.

Abstract

Wing morphology has historically been a major focus in taxonomic and evolutionary studies of lepidopterans. However, general patterns of quantitative variation and diversification in wing sizes and shapes, and the factors underlying them, have been largely unexplored. A morphometric study of wing and body forms among six taxa of nymphalid butterflies (the Nymphalini, Heliconiini, Amathusiinae, Morphinae, Danainae, and Ithomiinae) revealed that allometric size-scaling variation accounts for more than 90% of total morphological variation and more than 96% of wing variation; thus, overall body size seems to be the major determinant of wing shape. Fore wings and hind wings are isometric in size (area) with respect to one another; however, wing shape within and among groups is significantly allometric, resulting in considerable shape differences between small and large species. Intertaxon differences in allometric trends are evident, but differences in wing form among groups are subtle and complex. Wing and body characters can be grouped into four integrated character suites. A mapping of the hypothesized phylogenetic relationships among the taxa onto the discriminant function space suggests that the amathusiine-morphine lineage has evolved by simple divergence from the presumed ancestral wing form, but that morphological form changes within other lineages have involved significant amounts of homoplasy.

1 Introduction

Wing morphology has played a major role in taxonomic and evolutionary studies of the Lepidoptera and other insects. Such studies have typically concentrated on pigmentation and venation patterns, which are the most obvious qualitative aspects of wing morphology. Morphometric analyses of wing morphology have been much rarer. For example, although quantitative measures have sometimes been considered in studies of spot patterns (Mason et al., 1968; Brakefield, 1979; Kingsolver and Wiernasz, 1987; Wiernasz, 1989), of sexual dimorphism (Pearse and Murray, 1982), and of flight behavior (Nachtigall, 1967; Alexander, 1968; Corkum, 1987), only qualitative or crude numeric descriptors, such as length/width ratios or wing margin angles, have typically been applied in analyses of evolutionary or ecological diversification (Chang, 1963; Ricklefs and O'Rourke, 1975; Johnson and Walter, 1978; Curtsinger, 1986).

There has been much interest in the aerodynamic properties of various wing forms, but the adaptive significance of particular interspecific patterns of wing variation is not well understood. Measures of aspect ratio, for example, which are widely used in studies of bird and bat wings (Greenewalt, 1962; Norberg, 1981; Norberg and Rayner, 1987), are far too simplistic to explain even the basic elements of lepidopteran flight patterns (Betts and Wootton, 1988), let alone the evolution of butterfly wings. Until very recently, the study of functional and architectural constraints as limitations on the expression of interspecific variability has been largely neglected. Such factors may be important in comparative studies, however, because if present they should produce stable and conservative integrative patterns of covariation among measures of wing size and shape.

The tremendous diversity of lepidopteran wing and body form is a valuable source of information about morphological variation in a monophyletic group of organisms in the context of functional and phylogenetic limitations. Analysis of such variation must initially be conducted at a level of generality that allows us to discover the commonalities of form that underlie variations on basic themes, and such analyses in turn should be founded upon basic biological principles or constructs such as homology and allometry. Such principles are important because they prevent the methodology of morphometrics from being a biologically empty statistical exercise.

There are a number of important questions that are difficult to answer by traditional means but that can easily be approached using morphometric methods (Strauss, 1990). These include the following: (1) Which aspects of wing form are conservative, and which are "plastic" or labile within and among species and higher taxa? To what extent are such patterns concordant with prior hypotheses of aerodynamic efficiency or functional limitation? (2) Do allometric scaling effects account for a significant portion of shape variation in lepidopterans, as they do almost universally in other organisms? If allometric scaling is important, do the magnitudes of the scaling coefficients depend upon the taxonomic level within which they are estimated? That is, is there a "taxon-level problem" in lepidopteran wing form analogous to that observed in the scaling of brain weight to body weight among mammals (Pagel and Harvey, 1989)? (3) To what degree are fore wing, hind wing, and body form characteristics correlated among species? Is the ostensibly greater variation in hind wing form in many groups of butterflies significant with respect to relative size? (4) Are taxonomic groups (at whatever categorical level) distinguishable on the basis of wing morphology? That is, have there been historical limitations on variation, or are wing forms sufficiently plastic (genetically or epigenetically) that they conceal phylogenetic differences? (5) Whether or not there are significant allometric or historical components to variation among species, can patterns of residual individual variation be correlated with ecological factors?

The purpose of this paper is to summarize the results of a comparative study of variation in size and shape among six higher taxa (tribes and subfamilies) of nymphalid butterflies, with particular emphasis on allometric scaling and on nonallometric patterns of diversification in wing form. This study will be discussed within the more general context of quantitative analyses of biological form, emphasizing the use of landmark-based models of form and the fundamental decomposition of variation into size and shape components.

2 Methods

2.1 Taxa Studied

Body and wing forms were quantified from high quality photographs (Smart, 1984) of 368 exemplar specimens of nymphalid butterflies representing six higher taxa (Fig. 1): the subfamilies Amathusiinae (n = 35), Morphinae (n = 37), Danainae (n = 53), and Ithomiinae (n = 53), and the tribes Heliconiini (n = 110) and Nymphalini (n = 76) of the subfamily Nymphalinae. The ithomine and heliconine data used in this study were those used by Strauss (1990) to study convergent evolution in these two groups.

The genera represented in this study (followed respectively by numbers of species and specimens in parentheses) are the following: Amathusiinae: *Aemona* (1, 1), *Amathusia* (2, 2), *Amathuxidia* (1, 1), *Discophora* (1, 1), *Enispe* (2, 2), *Faunis* (3, 3), *Hyantis* (1, 1), *Morphopsis* (1, 1), *Morphotenaris* (1, 1), *Stichophthalma* (4, 4), *Taenaris* (9, 9), *Thaumantis* (4, 4), *Thauria* (1, 1), *Xanthotaenia* (1, 1), *Zeuxidia* (3, 3). Danainae: *Amauris* (4, 4), *Danaus* (21, 24), *Euploea* (18, 20), *Idea* (4, 4), *Ituna* (1, 1), *Lycorea* (1, 1). Heliconiini: *Agraulis* (1, 1), *Dione* (3, 3), *Dryadula* (1, 1), *Dryas* (1, 3), *Eueides* (6, 12), *Heliconius* (26, 86), *Philaethria* (1, 1), *Podotricha* (2, 2). Ithomiinae: *Aeria* (2, 2), *Athesis* (1, 1), *Callithomia* (2, 2), *Ceratinia* (1, 1), *Corbulis* (1, 1), *Dircenna* (1, 1), *Ekpityches* (1, 1), *Elzunia* (2, 2), *Episcada* (2, 2), *Eutresis* (1, 1), *Godyris* (3, 3), *Hyalyris* (5, 5), *Hypoleria* (2, 2), *Hypomenitis* (1, 1), *Hypothris* (5, 5), *Ithomia* (4, 4), *Mechanitis* (4, 4), *Melinaea* (4, 4), *Napeogenes* (3, 3), *Oleria* (3, 3), *Pteronumia* (2, 2), *Sais* (2, 2), *Thyridia* (1, 1), *Tithorea* (2, 3), *Xanthocleis* (1, 1). Morphinae: *Morpho* (31, 41). Nymphalini: *Argyronome* (1, 1), *Argynnis* (1, 1), *Boloria* (1, 2), *Brenthis* (2, 3), *Childrena* (1, 1), *Chlosyne* (7, 7), *Clossiana* (4, 4), *Damora* (1, 1), *Dynamine* (7, 7), *Euphydryas* (6, 6), *Euptoieta* (1, 1), *Fabriciana* (4, 4), *Gnathotriche* (1, 1), *Issoria* (4, 4), *Melitaea* (5, 5), *Mellicta* (1, 1), *Mesoacidalia* (1, 1), *Microtia* (1, 1), *Phalanta* (1, 1), *Phyciodes* (14, 14), *Pandoriana* (1, 1), *Proclossiana* (1, 1), *Speyeria* (7, 8), *Timelaea* (1, 1).

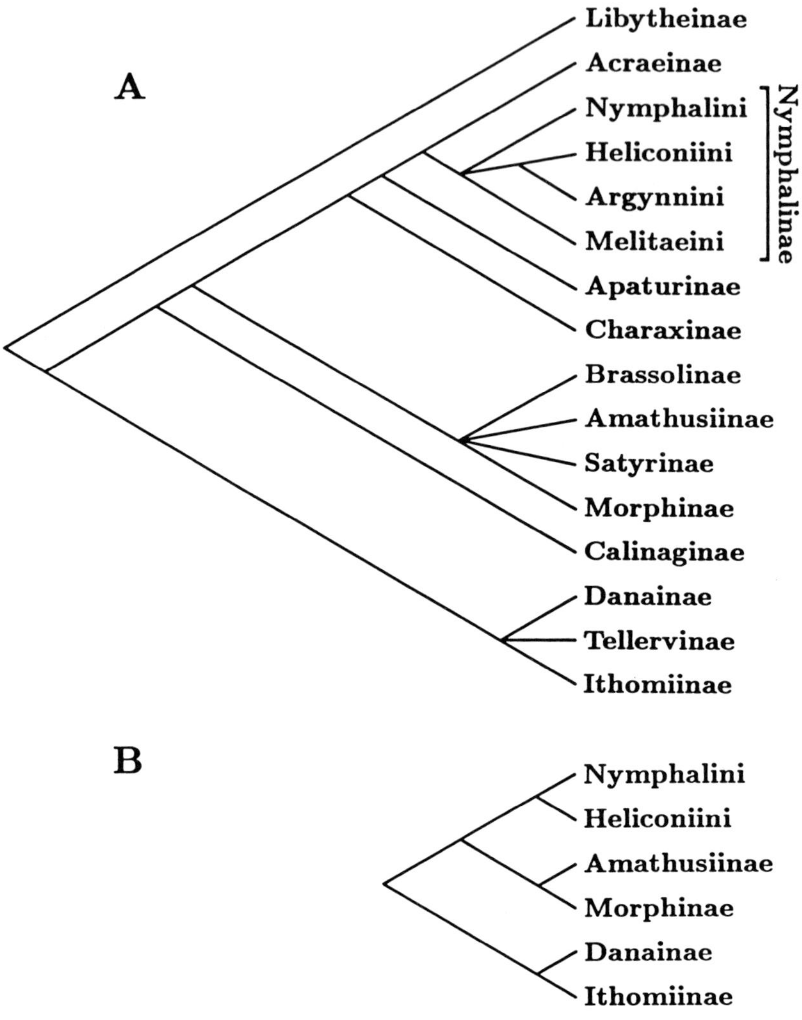

Figure 1. (A) Hypothetical phylogenetic relationships among the subfamilies of nymphalid butterflies (after Ackery [1984] and Scott [1984]). (B) Phylogenetic tree, reduced from Fig. 1A, of the six higher taxa used in the present study.

2.2 Morphological Measurements

Morphological landmarks and outlines of the wings (Fig. 2), body, and antennae of each specimen were digitized, along with a metric scale bar for each specimen, using a Houston Instruments Hipad DT-114 digitizer linked to an IBM PC/AT microcomputer. The acquisition program DIGITIZE (Strauss, unpublished) was used to capture points and organize them into data files.

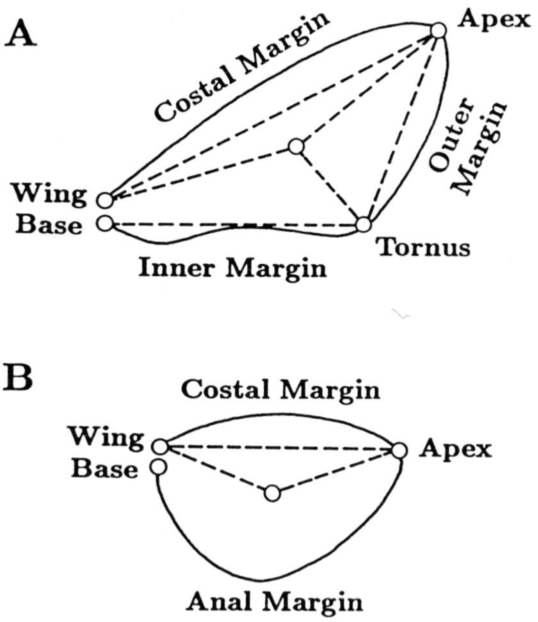

Figure 2. Morphological points (circles), boundary margin-lengths (solid lines), and linear interpoint distances (dotted lines) used as morphometric characters. Open circles on boundary represent landmarks; interior circles represent wing centroids (centers of area) estimated from boundaries.

Digitized points were of two kinds (Bookstein et al., 1985): (1) anatomically comparable "landmarks," hypothesized to be topologically homologous among individuals based on intersections of wing veins with the wing margin; and (2) "helping points," used only to indicate the position of the wing margin between landmarks.

Points were then converted to sets of linear measures (in mm) between selected pairs of points, and to centroids and areas of point-polygons, using the program DISTANCE (Strauss, unpublished). Centroids and areas were estimated by approximating wing outlines as polygons (Harvey, 1981), derived by connecting all consecutive wing margin landmarks and helping points by straight line segments. The centroid is the center of wing area, used as an approximation to the center of wing mass. Wing areas were converted to square roots to be consistent in dimension with linear distances.

In addition to the areas of the fore wings and hind wings (called *FSqArea* and *HSqArea*, respectively), the following 19 interpoint distances (Fig. 2) were measured for each individual (Strauss, 1990): Fore wing characters: *FCostD* - linear distance between wing base and apex; *FCostM* - distance along the wing margin between wing base and apex; *FOuterD* - linear distance between apex and tornus; *FOuterM* - wing-margin distance between apex and tornus; *FInnerD* - linear distance between wing base and tornus; *FInnerM* - wing-margin distance between wing base and tornus; *FCentB* - distance from centroid to wing base; *FCentA* - distance from centroid to apex; *FCentT* - distance from centroid to tornus. Hind wing characters: *HCostD* - linear distance between wing base and apex; *HCostM* - costal wing-margin distance between wing base and apex; *HAnalM* - anal wing-margin distance between wing base and apex; *HCentB* - distance from centroid to wing base; *HCentA* - distance from centroid to apex. Body characters: *BodLen* - body length along the midline; *HeadWd* - head width; *ThrxLn* - thorax length; *ThrxWd* - thorax width; *AntLen* - antenna length.

All wing characters were measured separately for left and right wings and then averaged for each individual. For antenna lengths, the longer of the two measurements was used rather than the average because of possible breakage or photographic foreshortening.

2.3 Statistical Methods

For multivariate analyses, all distance measurements were converted to natural logarithms. The logarithmic transformation has several beneficial effects on data structure: for individual characters, the variances become independent of their mean values (thus comparable to one another in terms of intrinsic variability) to the extent that the individual coefficients of variation are constant (Lewontin, 1966; Lande, 1977); among characters, exponential bivariate and multivariate allometric relationships become collinear, with slopes that characterize relative growth rates with respect to body size (Strauss, 1987). A general-size factor was estimated separately for each group as the first principal component (PC-1) of the within-group covariance matrix (Bookstein, 1989).

Multiple group principal components analysis (PCA) and discriminant function analysis (DFA) were used for interspecific comparisons. Because the six nymphalid taxa differ significantly in mean body size, the multiple group principal components were "sheared" (Bookstein et al., 1985), using the reformulated procedure of Rohlf and Bookstein (1987), to partition the effects of within-group size variation from the second and subsequent components. For the same reason, discriminant functions were calculated using residuals from regressions of each character separately on the pooled within-group PC-1. Pairwise Pearson product-moment correlations among these residuals were subsequently clustered using the unweighted pair group algorithm (UPGMA) in order to examine size-independent correlation patterns among characters. Absolute values of the correlation coefficients were used for clustering so as to represent the strengths of the relationships among characters rather than their directions (i.e., positive or negative).

To evaluate the contributions of individual characters to principal components and discriminant functions, their loadings are portrayed as vector correlations (directional cosines: Bryant, 1984; Strauss, 1985) of characters with the multivariate axes. Discriminant function loadings were derived from the total canonical structure (Klecka, 1980).

The principal components and discriminant function analyses depict patterns of variation and differentiation among taxa. In order to evaluate historical patterns of diversification, a cladogram summarizing the hypothetical evolutionary relationships among the nymphalid taxa was mapped onto multivariate spaces using the multidimensional optimization

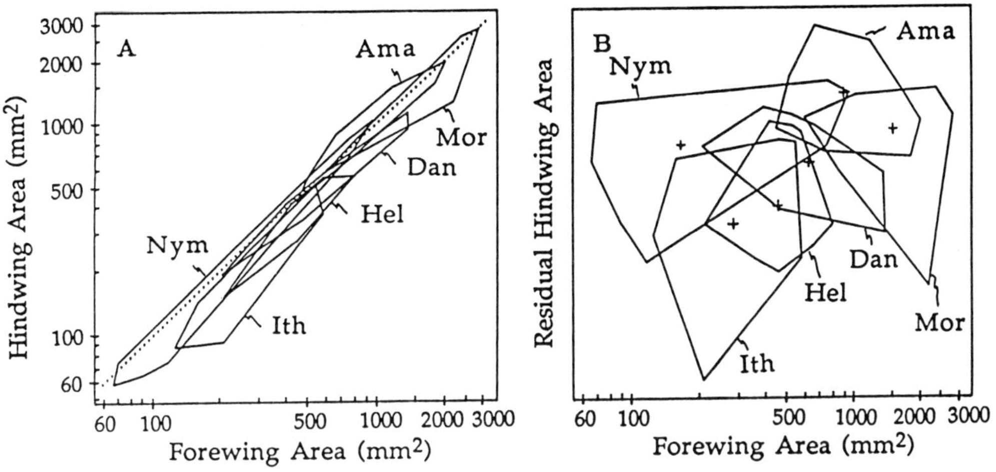

Figure 3. (A) Log-linear relationship between fore wing and hind wing sizes. Polygons enclose points for each taxon. Dotted line indicates equal fore wing and hind wing areas (isometry; slope = 1). Within-taxon regression lines are not significantly different from each other or from isometry. (B) Residual hind wing areas (arbitrary units) from a sheared regression on pooled within-taxon fore wing areas, plotted as a function of fore wing area. Crosses represent taxon centroids.

procedure (program CONTMAP) of Strauss (in press). The program implements Powell's optimization procedure (Polak, 1971; Brent, 1973) to find the node values that minimize the sum of the squared branch lengths of the tree, producing a least-squares solution that has been shown to be a reasonable model for the evolution of continuous characters.

3 Results

3.1 Bivariate Comparisons

Most morphometric variation among species and among nymphalid taxa is significantly correlated with size, and many ostensibly significant differences in form are not significant when size effects are considered. For example, the relationship between fore wing and hind wing sizes, which on casual observation seems to be relatively uncoupled in many groups of lepidopterans, is highly predictable among these nymphalids, which display little interspecific variation (Fig. 3A). Furthermore, this relationship is isometric (bivariate log-linear slope = 0.999, not significantly different from 1.0), indicating that the relative sizes of fore wing and hind wing do not vary systematically with size across the approximately 60-fold range of areas (eight-fold linear range) for these species. None of the within-group regression slopes, estimated separately by taxon, are significantly different from one another or from isometry.

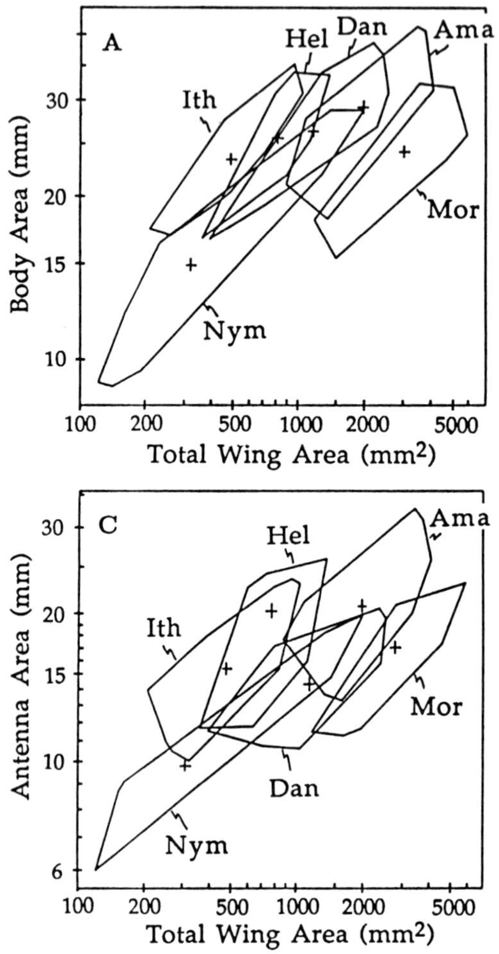

Figure 4. (A) Log-linear relationship between body length and total wing size (fore wing + hind wing). Wing size is expressed as the square root of wing area to be consistent in dimension with body length. Polygons enclose points for each taxon. (B) Residual body lengths (arbitrary units) from sheared regression on pooled within-taxon wing size, plotted as a function of wing size. Crosses represent taxon centroids. (C) Log-linear relationship between antennal length and total wing size. (D) Residual antennal lengths (arbitrary units) from the sheared regression on pooled within-taxon wing size, plotted as a function of wing size.

The small amount (6.2%) of size-independent variation in fore wing and hind wing sizes among groups can be examined by "shearing" fore wing area from hind wing area (using the pooled within-group size axis) and examining the patterns of residual variation (Fig. 3B). Despite the significant size-dependency of wing and body form, an analysis of variance of the resulting residuals (equivalent to an analysis of covariance of the original data, with fore wing area as the covariate) indicated that: (1) there is significant residual variation among group means ($F_{5,32} = 113.8$, $P < 0.001$), and (2) based on a Scheffe multiple-comparisons test, the amathusiines have significantly larger hind wings (relative to fore wing size) than the other groups, and the heliconines and ithomines have significantly smaller hind wings than the others. The relative hind wing sizes of the latter two groups are not significantly different from one another.

Similar patterns of size-dependent scaling of body form are observed with body length (Fig. 4A) and antenna length (Fig. 4C), although in these cases the degree of size-dependency is much less. Using total wing area as a size reference, these analyses reveal that there is significant size-independent variation in both body length ($F_{5,32} = 347.9$, $P < 0.001$) and antennal length ($F_{5,32} = 255.2$, $P < 0.001$). All pairs of taxa are significantly different in relative body length, except danaines and nymphalines. All are different in relative antennal length except for two pairwise combinations: heliconines-ithomines, and amathusiines-nymphalines. As noted below, these size-independent, univariate differences in form are important factors in the multivariate analyses.

3.2 **Multivariate Discrimination**

The importance of size-scaling as a major determinant of morphological form in nymphalids is reinforced by analyses of multivariate variation within and among taxa. For example, an among-group PCA of wing and body characters revealed that most variation (91%) in form can be accounted for by allometrically consistent differences in body proportions (Figs. 5A, 5C). In particular, nonallometric variation in wing form seems to be negligible among taxa. Most of the residual variation is due to relative body and antennal lengths (Figs. 5B, 5D), which are longer than average in heliconines and ithomines, shorter than average in morphines and nymphalines, and intermediate in amathusiines and danaines. There is some secondary differentiation on PC-3 between these taxon-pairs in head width and thorax dimensions, but it accounts for less than 2% of total variation and is not significant based on ANOVAs.

The same patterns are obtained in a parallel analysis of body and antennal characters only, although the proportion of total variance that is due to size is much less in this case (77%, down from 91%). Sheared principal components and size-independent discriminant analyses of body and antennal characters differentiate taxa no better than the among-group PCAs of all characters (Fig. 5), and, therefore, are not described here. Thus, these multivariate analyses of wing and body characters mirror the bivariate results of Figs. 3 and 4, indicating that body form (primarily relative head and antennal lengths) partially differentiates taxa but cannot discriminate them. Note that this pattern of character covariance is substantially different than that described by Strauss (1990) in a study of only heliconines and ithomines, in which wings and bodies tended to be uncoupled in their patterns of variation and to behave as separate "character suites" in the classical sense (Olsen and Miller, 1958; Cheverud, 1982b), with antennal length being more highly correlated with

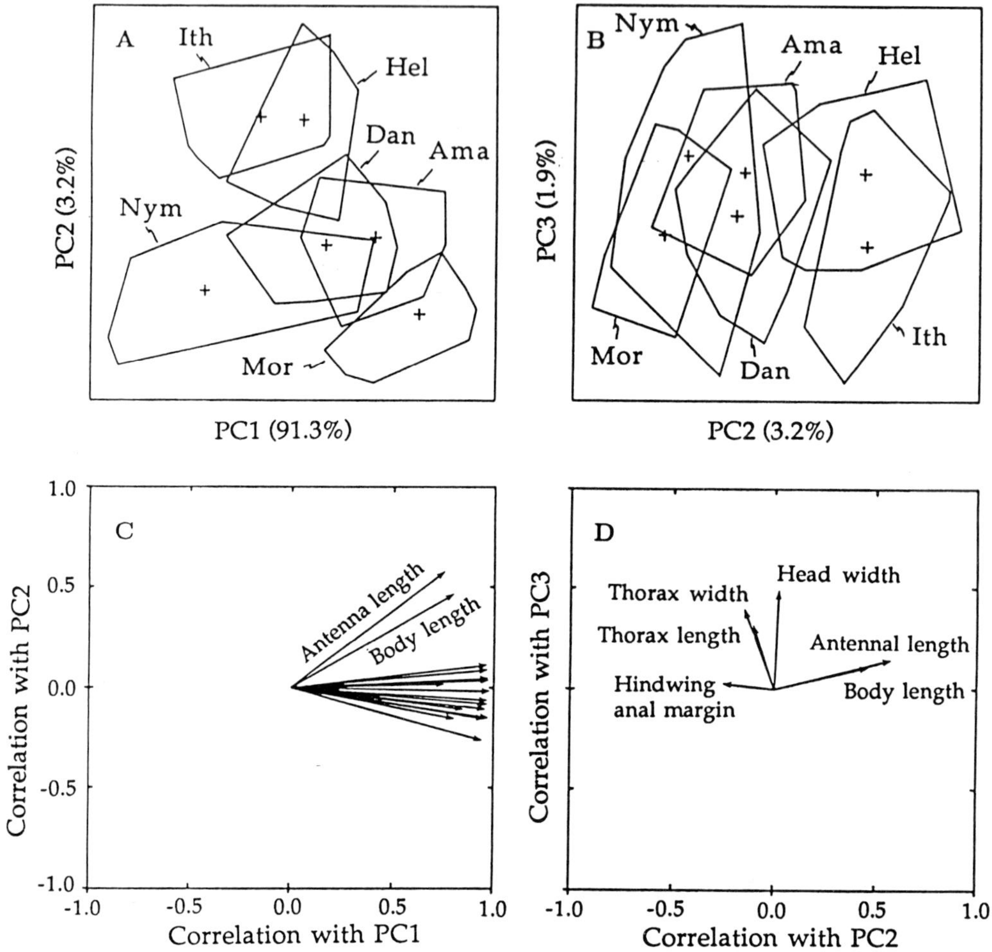

Figure 5. Principal components analysis of all wing and body characters for the six pooled nymphalid taxa. (A) Scatter plot of specimen scores on the first two components (arbitrary units). Polygons enclose points for each taxon; crosses represent taxon centroids. (B) Scatter plot of specimen scores on the second and third components. (C, D) Corresponding correlations of characters with principal components. Vectors indicate the directions of maximum variation for each character within the principal-component planes.

wing characters than with body characters.

Given these results, it still might be possible to detect subtle wing form differences among the nymphalid groups by concentrating on wing characters alone. The first principal component from an among-group PCA accounts for 96.4% of total variation, which is almost completely size-dependent scaling. Although the taxa cannot be distinguished within the space of the first few components (which are, therefore, not shown here), they can be at least

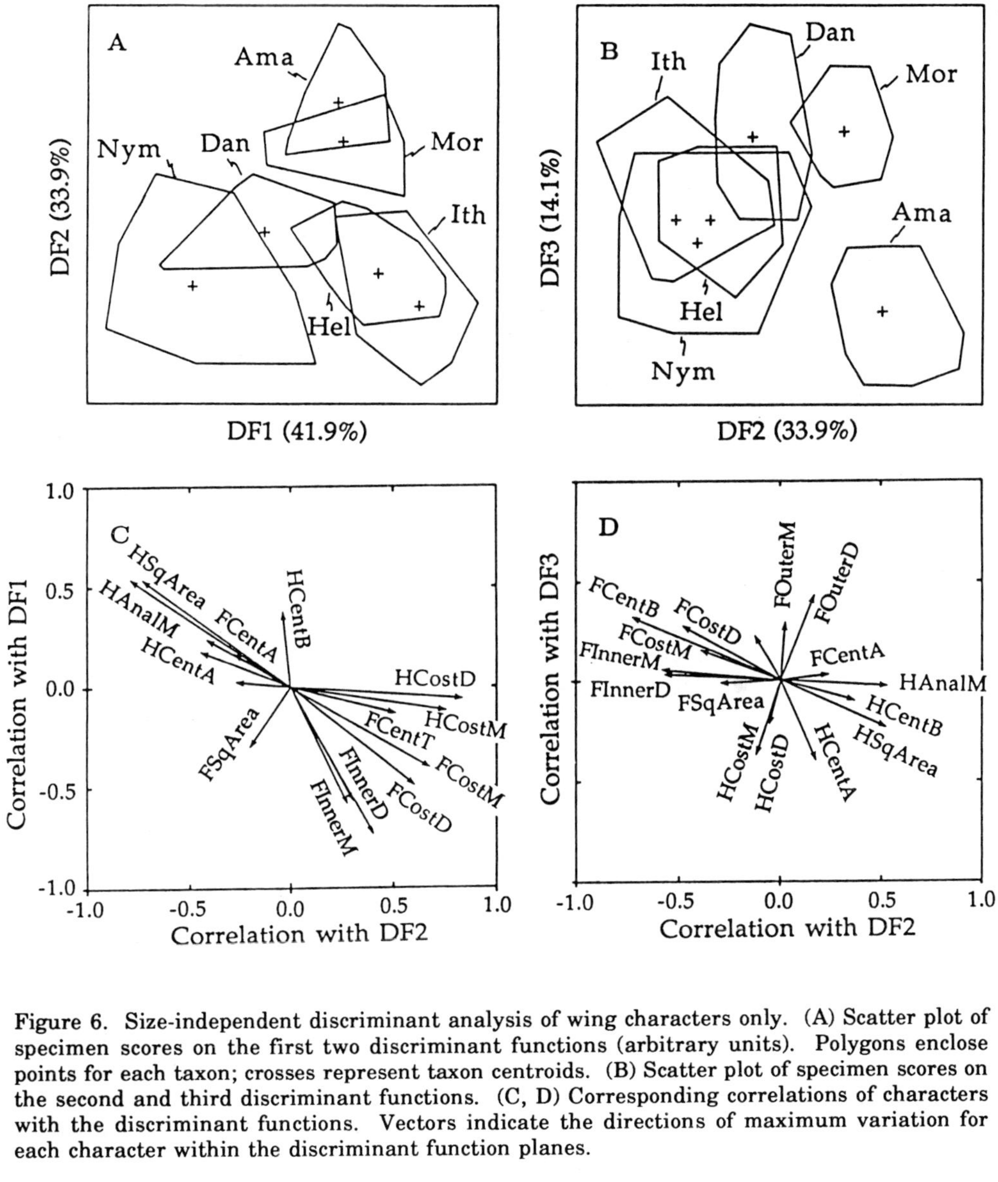

Figure 6. Size-independent discriminant analysis of wing characters only. (A) Scatter plot of specimen scores on the first two discriminant functions (arbitrary units). Polygons enclose points for each taxon; crosses represent taxon centroids. (B) Scatter plot of specimen scores on the second and third discriminant functions. (C, D) Corresponding correlations of characters with the discriminant functions. Vectors indicate the directions of maximum variation for each character within the discriminant function planes.

partially discriminated by a size-independent discriminant analysis of wing characters (Fig. 6) based on residuals from regressions of each character separately on the pooled within-group PC-1. The first two discriminant functions (Fig. 6A) distinguish the taxa in three pairs: nymphalines-danaines, heliconines-ithomines, and amathusiines-morphines; the third function (Fig. 6B) serves to discriminate the latter pair. It is notable that the taxa are distinguished not by any particular wing characters, but rather on the basis of subtle, coordinated variation in many characters.

3.3 Size-Independent Character Variation

Vector plots of characters (Figs. 5C, 5D, 6C, 6D) portray patterns of variation only within the corresponding principal component and discriminant function planes, and it is difficult to comprehend total patterns of variation from such limited perspectives. A UPGMA cluster analysis of size-independent correlations among all characters, both wing and body (Fig. 7), suggests the existence of four character suites: (1) a suite consisting of thorax dimensions and head width associated with the lengths of the trailing fore wing and leading hind wing edges, which are tightly coupled during flight; (2) the length of the outer edge of the fore wing, which is most closely associated with fore wing area; (3) the length of the leading fore wing margin, which seems to primarily determine the position of the fore wing center; and (4) the length of the trailing hind wing margin, correlated with hind wing area and position of the hind wing center, and also associated with antennal and body lengths. Body length in the latter case is primarily a function of abdominal length because thorax and head dimensions vary as components of the first suite.

Thus, the three edges of the fore wing seem to assort more or less independently among species, as do the leading and trailing edges of the hind wing. Variation in the two wings is constrained at their edges of contact and ostensibly linked to variation in thorax dimensions, which are possibly mediated by flight muscle volumes and areas of attachment. Abdomen length is linked to hind wing size, and the correlation of antennal length with the latter may be a function of balance constraints.

3.4 Differences in Allometric Scaling Among Taxa

The large amounts of size-dependent variation among these species includes the significant differences in body and wing form between small and large species that are correlated with variation in size. Such differences arise from allometric scaling effects, as opposed to isometric patterns, and can be estimated from rates of change of individual characters as a function of change in overall size ("multivariate allometries": Jolicoeur, 1963; Strauss, 1985, 1987; Bookstein, 1989). Because "change in size" in this case represents size variation among species, scaling effects are measured as "static" allometric coefficients (Cheverud, 1982a).

Static allometries were estimated separately by taxon with respect to their respective within-group PC-1 axes, which in turn are robust estimates of size vectors in this study due to the wide ranges in size. Despite the fact that the size vectors are almost perfectly correlated among taxa (i.e., parallel in the multivariate character space; Table 1), there are a number of significant intertaxon differences in allometric coefficients (Fig. 8). Differences in character allometries reflect ways in which large species differ from small species within each group. For example, the outer fore wing margin is

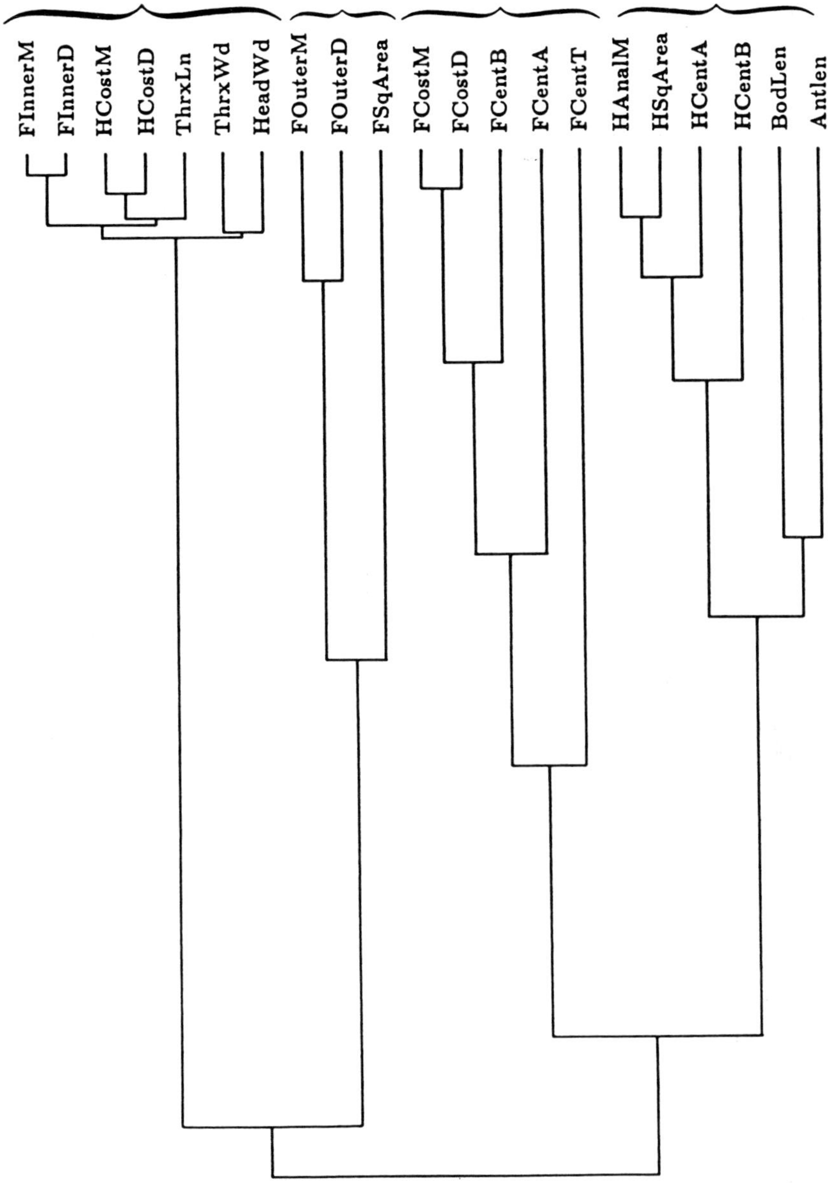

Figure 7. UPGMA cluster analysis of size-independent character correlations across all individuals and taxa.

positively allometric in danaines and morphines (reflecting a proportionate increase in outer margin length with increasing size), negatively allometric in nymphalines and amathusiines (reflecting a proportionate decrease with

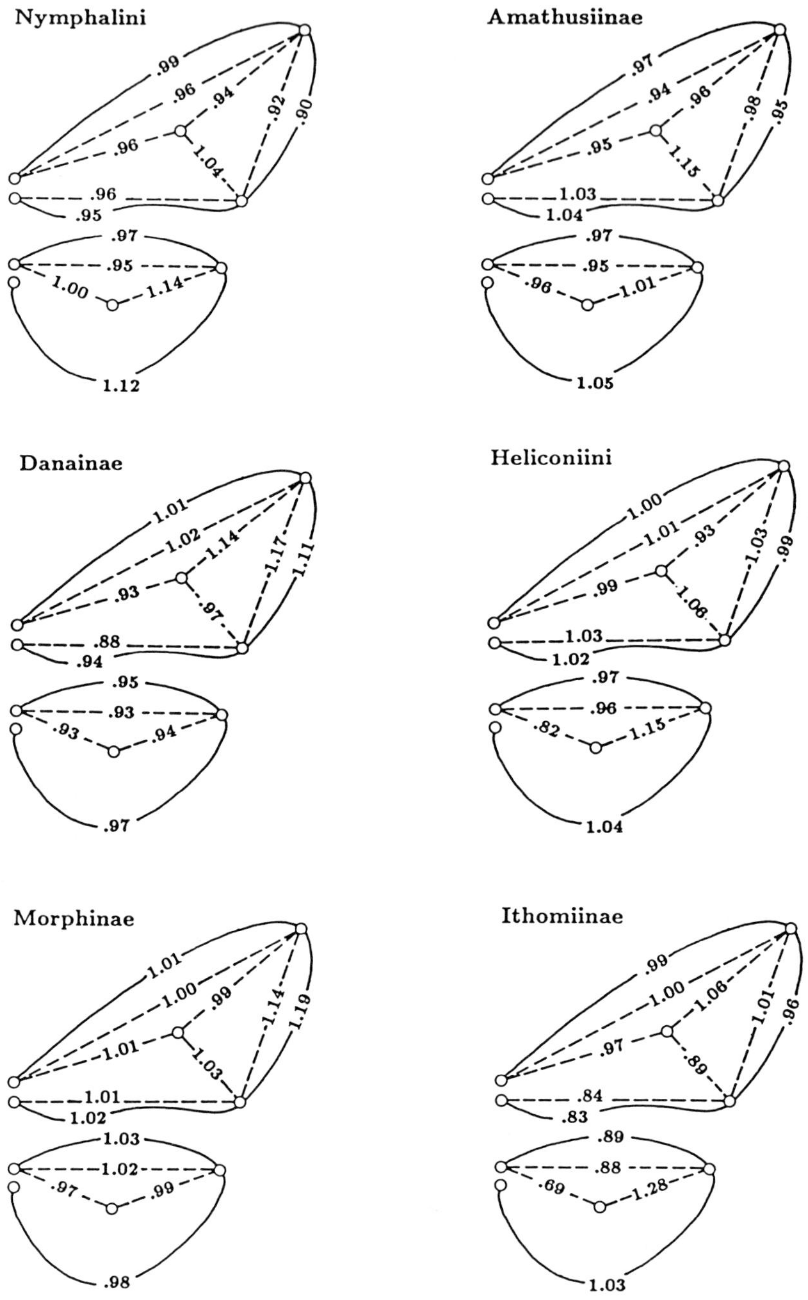

Figure 8. Multivariate static allometries with respect to wing size, by taxon.

Table 1.
Vector correlations among within-group principal components: PC-1 above diagonal, PC-2 below diagonal.

	Nym	Hel	Ama	Mor	Dan	Ith
Nymphalini	--	0.999	1.000	1.000	0.999	0.999
Heliconiini	0.334	--	0.998	0.999	0.998	0.999
Amathusiinae	0.362	0.098	--	1.000	1.000	1.000
Morphinae	0.762	0.045	0.047	--	0.999	1.000
Danainae	0.781	0.395	0.632	0.549	--	1.000
Ithomiinae	0.916	0.150	0.220	0.918	0.688	--

increasing size), and approximately isometric in heliconines and ithomines. More subtle aspects of wing shape change can be deduced from the relative allometric values of geometrically related interpoint distances. For example, the allometries of the distances from the fore wing centroid to apex and tornus indicate that the outer fore wing margins of nymphalines, amathusiines, heliconines, and morphines tend to be more parallel with the body axis in large species than in small ones, although the opposite is true in danaines and ithomines. The nearly isometric costal margin of the fore wing probably reflects an aerodynamic constraint on interspecific variability (Strauss, 1990).

3.5 **Mapping Character Evolution**

The evolution of continuous size and shape characters is often viewed as movement in evolutionary time through a "morphospace" (morphological space), the dimensions of which are the character vectors (Bookstein et al., 1985). Linear functions of the original characters, such as principal components and discriminant functions, represent particular projections of the complete character space that characterize patterns of maximum variation or discrimination. In order to evaluate historical patterns of diversification, a hypothesis of phylogenetic relationships must be mapped onto such projections (Strauss, in press).

For example, Fig. 6A portrays the patterns of maximum discrimination among the six nymphalid taxa, and the cladogram of Fig. 1B diagrams their hypothetical relationships. In Fig. 9, the cladogram has been mapped onto the discriminant space by computing the shortest tree connecting the group centroids that is identical in topology with the cladogram, providing a parsimonious model of evolutionary diversification that is consistent with all of the available data. Because the discriminant space was estimated from size-independent wing characters, Fig. 9 is a model of the evolution of nymphalid wing shape from the most recent common ancestor of the six taxa. It implies that, although the amathusiine-morphine lineage has evolved by simple divergence into a new region of the morphospace, changes in wing form of the other lineages have involved significant amounts of homoplasy (convergence of danaines with nymphalines, and of heliconines with ithomines).

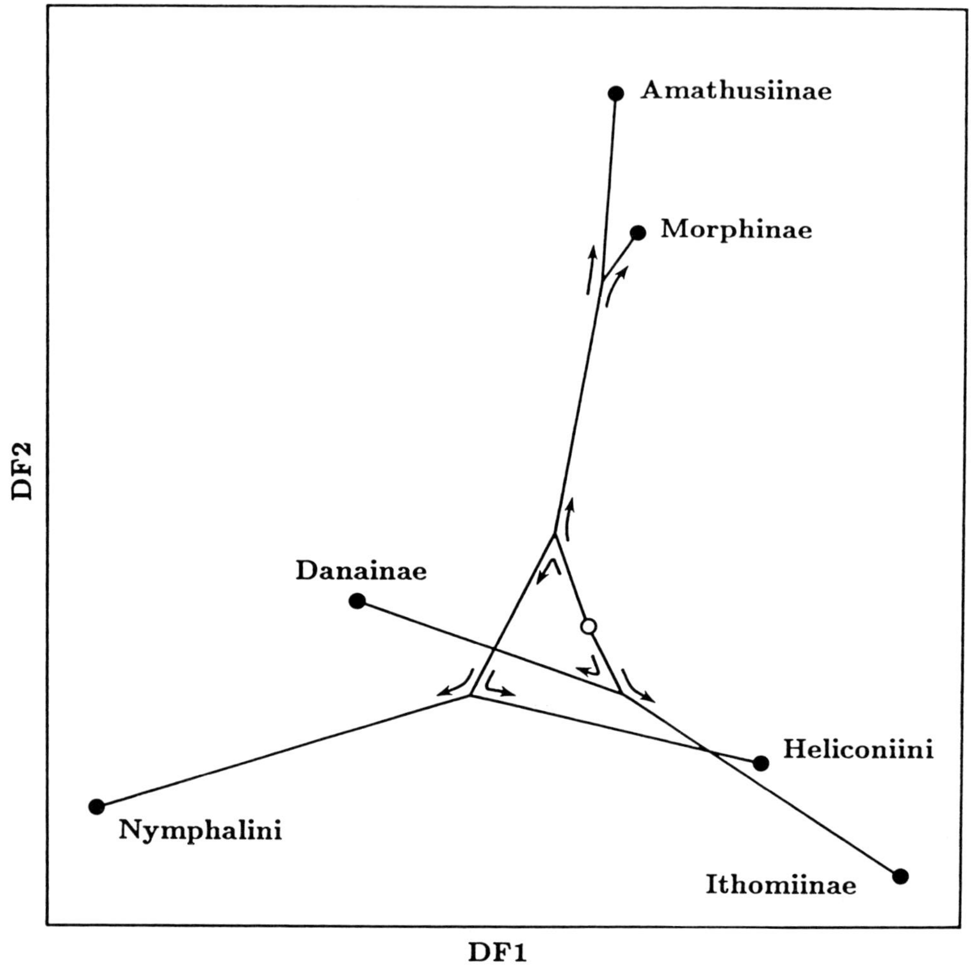

Figure 9. A mapping of the phylogenetic tree for the six taxa [from Fig. 1B] onto the size-independent discriminant space of wing characters [from Fig. 6A]. Taxa are represented by their group centroids. Open circle represents the root of the tree. Arrows represent inferred directions of morphological change.

4 Discussion

4.1 Advantages of Landmark-Based Morphometry

The discipline of biological morphometry was originally an outgrowth of numerical taxonomy, which sought objective means of classifying organisms based on differences in body form. Numerical taxonomy in turn derived its methodology from several other quantitative disciplines, notably psychometrics and econometrics. Many of the descriptive and statistical techniques still in common use were borrowed wholesale from these other disciplines and applied directly to mensural measurements of organisms with little regard for the quality or importance of the measurements themselves (Strauss and Bookstein, 1982). For example, neither the classic text by Sneath and Sokal (1973), nor any of the more recent texts on morphometry (Pimentel, 1979; Neff and Marcus, 1980; Reyment et al., 1984), provided guidelines for selecting characters to be studied other than that they be numerous and sample the whole form.

Biological morphometry has progressed significantly in the past ten years, however, primarily by the infusion of a geometric perspective of form and a renewed emphasis on explicit assumptions about anatomical homologies and on allometric models of growth and size-variation. Such models and assumptions are necessary for scientific comparisons (as opposed to mere statistical contrasts) of organismic forms because they furnish us with a firm biological context in which to interpret the results. Hypotheses about homologies, which are anatomical correspondences derived from common ancestry, provide a basis for inferring that observed patterns of variation and differentiation are directly comparable among organisms and are not the result of methodological artifact. Allometric models provide a basis for partitioning size-correlated sources of shape variation from size-independent sources, once we have agreed on a suitable measure of "size" (which itself is no trivial problem: Bookstein, 1989).

Not all morphometric methods can directly utilize information about allometry and homology. For example, insect wings have the advantages for biometric purposes of being sufficiently flat and of having smoothly contoured boundaries. Thus, wing outlines can be efficiently described and filtered (i.e., smoothed) by means of radial or tangential functions. Procedures for analyzing such functions include Fourier, elliptical Fourier, tangent-angle, and "eigenshape" methods (Benson et al., 1982), all of which have been used and evaluated for this purpose by Rohlf and Archie (1984) (see also Ferson et al. [1985] and Rohlf [1986 and this volume]). Although such functions may capture subtle variations in wing shape and allow forms to be smoothed, reconstructed and discriminated, they do not permit the utilization of anatomical landmarks when they are present, such as those provided by wing veins, for the purpose of detecting localized regions of variation (Bookstein et al., 1982). Nor do they allow the decomposition of form differences into allometric (size-dependent) and size-independent components. Scaling effects have been shown to account for a significant portion of shape variation in lepidopterans (Petersen, 1952; Brown and Davies, 1972; Sylven and Carlbäcker, 1981; Strauss, 1990), as they do almost universally in other organisms (Gould, 1966, 1971; Sweet, 1980; Houck et al., 1990). The presence of allometric scaling may be used as an objective criterion by which to determine whether observed variation in shape is a developmentally mandated correlate of trends in size variation, or instead is discrete, representing sets of ostensible evolutionary adaptations (Strauss, 1984). This is particularly important when shape differences are used diagnostically in taxonomic descriptions (Strauss, 1985; Janzon, 1986).

4.2 Morphological Variation and Differentiation

The primary objective of this study was to examine patterns of variation within and among six groups of nymphalids, a very diverse group of butterflies. The major conclusion has been that allometric scaling is by far the most important source of variation in wing and body form, both among individual species and among higher taxa (tribes and subfamilies). More than 90% of variation in wing and body characters, and more than 96% in wing characters alone, can be accounted for by allometric effects. The residual variation can be used to detect mean differences among taxa, but with large amounts of overlap in almost all cases. Even a discriminant analysis will separate taxa only into pairs of species. However, despite the fact that the within-taxon size vectors are highly correlated (i.e., parallel within the character space), many systematic differences in allometric patterns are evident among taxa. Observable differences in form among nymphalid butterflies are likely to be as much a result of such differential allometric patterns within taxa as they are of size-independent differences in form between taxa.

The size-independent differences in form that can be detected are the product of minor, but coordinated, differences among almost all of the metric characters examined. The studies of character variation reported here suggest two major conclusions. First, body characters are relatively more variable than wing characters; the primary patterns of size-independent variation are almost solely a product of body characters. Second, independent of their relative variabilities, the characters examined fall into four covariance sets (Gould, 1984), which can be interpreted as "character suites" in the sense of Olsen and Miller (1958). Together these represent four more-or-less independent sources of character variation that combine (with the addition of unique variance) to give rise to individual morphologies. Although the mechanisms or processes underlying such evolutionary uncouplings are not obvious, empirical discoveries of covariance sets can lead to testable hypotheses about genetic and functional correlations among body parts (Zelditch, 1987).

4.3 Morphological Evolution

An important problem in all comparative morphological studies is that of distinguishing between historical and adaptive effects on organismic design (e.g., Clutton-Brock and Harvey, 1979; Harvey and Mace, 1982). Although patterns of morphological variability and discrimination can be assessed solely on the basis of morphometric analysis, studies of morphological evolution must be firmly grounded within a tight phylogenetic framework, wherein the phylogeny is conclusively supported by characters other than those under study (Felsenstein, 1985; Donoghue, 1989). The hypothetical evolutionary relationships used in this study (Fig. 1) are based on phylogenetic studies that were explicitly cladistic in approach (Ackery, 1984; Scott, 1984), utilizing a variety of discrete adult, pupal, and larval characteristics. Thus, although one set of morphological data can never be said to be truly independent of another set, the morphometric characters used in this study should be effectively uncorrelated with the traits upon which the phylogeny was based.

The mapping of the hypothetical phylogenetic relationships onto the discriminant space (Fig. 9) represents a model of wing shape evolution among the six nymphalid taxa that were surveyed. Although both the particular morphospace projection portrayed and the positions of the tree nodes (i.e., the putative ancestors) within the space would change, perhaps considerably, if more taxa were included, this simplified model implies a complex pattern of evolution with simple divergence in one lineage (the amathusiines-morphines) and both divergence and convergence in overall form in the other lineages. To the extent that the hypothesized phylogeny is true, this example illustrates the danger in relying on a purely phenetic approach to assessing evolutionary relationships. A simple spanning tree (such as a Prim network: Prim, 1957) applied to the discriminant space would yield a model of evolution in which all morphological change was divergent. Only with an independently derived phylogenetic hypothesis can the extent and direction of homoplasious evolution be judged.

The results of this study suggest that significant ontogenetic and evolutionary constraining forces exist that govern major sources of variation, but also indicate that additional comparative quantitative studies are needed. An adequate assessment of allometric and functional morphology will substantially improve our understanding of lepidopteran morphology in particular, and of processes of morphological divergence in general.

5 References

Ackery, P.R., 1984. Systematic and faunistic studies on butterflies. In: R.I. Vane-Wright and P.R. Ackery (Editors), The Biology of Butterflies. Symposium of the Royal Entomological Society, London, Volume 11. Academic Press, New York, pp. 9-21.

Alexander, R.McN., 1968. Animal Mechanics. University of Washington Press, Seattle, 346 pp.

Benson, R.H., Chapman, R.E. and Siegel, F., 1982. On the measurement of morphology and its change. *Paleobiology*, 8: 328-339.

Betts, C.R. and Wootton, R.J., 1988. Wing shape and flight behavior in butterflies (Papilionoidea and Hesperioidea). *Journal of Experimental Biology*, 138: 271-288.

Bookstein, F.L., 1989. "Size and shape": a comment on semantics. *Systematic Zoology*, 38: 173-180.

Bookstein, F.L., Chernoff, B., Elder, R.L., Humphries, J.M., Smith, G.R. and Strauss, R.E., 1985. Morphometrics in Evolutionary Biology. Special Publication, 15. Academy of Natural Sciences, Philadelphia, 277 pp.

Bookstein, F.L., Strauss, R.E., Humphries, J.M., Chernoff, B., Elder, R.L. and Smith, G.R., 1982. A comment upon the uses of Fourier methods in systematics. *Systematic Zoology*, 31: 85-92.

Brakefield, P.M., 1979. Spot number in *Maniola jurtina* -- variation between generations and selection in marginal populations. *Heredity*, 42: 259-266.

Brent, R.P., 1973. Algorithms for Minimization without Derivatives. Prentice-Hall Publishers, Englewood Cliffs, New Jersey, 326 pp.

Brown, V. and Davies, R.G., 1972. Allometric growth in two species of *Ectobius* (Dictyoptera: Blattidae). *Journal of Zoology*, 166: 97-132.

Bryant, D., 1984. Geometry, statistics, probability: variations on a common theme. *American Statistician*, 38: 38-48.
Chang, V.C., 1963. Quantitative analysis of certain wing and genitalia characters of *Pieris* in western North America. *Journal of Research on the Lepidoptera*, 2: 97-125.
Cheverud, J.M., 1982a. Relationships among ontogenetic, static, and evolutionary allometry. *American Journal of Physical Anthropology*, 59: 139-149.
Cheverud, J.M., 1982b. Phenotypic, genetic, and environmental morphological integration in the cranium. *Evolution*, 36: 499-516.
Clutton-Brock, T.H. and Harvey, P.H., 1979. Comparison and adaptation. *Proceedings of the Royal Society of London*, 205: 547-565.
Corkum, L.D., 1987. Patterns in mayfly (Ephemeroptera) wing length: adaptation to dispersal? *Canadian Entomologist*, 119: 783-790.
Curtsinger, J.W., 1986. Quantitative wing variation in inbred and outbred lines of *Drosophila melanogaster*. *Journal of Heredity*, 77: 267-271.
Donoghue, M.J., 1989. Phylogenies and the analysis of evolutionary sequences, with examples from seed plants. *Evolution*, 43: 1137-1156.
Felsenstein, J., 1985. Phylogenies and the comparative method. *The American Naturalist*, 125: 1-15.
Ferson, S., Rohlf, F.J. and Koehn, R.K., 1985. Measuring shape variation among two-dimensional outlines. *Systematic Zoology*, 34: 59-68.
Gould, S.J., 1966. Allometry and size in ontogeny and phylogeny. *Biological Review of the Cambridge Philosophical Society*, 41: 587-640.
Gould, S.J., 1971. Geometric similarity in allometric growth: a contribution to the problem of scaling in the evolution of size. *The American Naturalist*, 105: 113-136.
Gould, S.J., 1984. Covariance sets and ordered geographic variation in *Cerion* from Aruba, Bonaire and Curaçao: a way of studying nonadaptation. *Systematic Zoology*, 33: 217-237.
Greenewalt, C.H., 1962. Dimensional relationships for flying animals. *Smithsonian Miscellaneous Collections*, 14: 41-46.
Harvey, P., 1981. A simple algorithm for the unique characterization of convex polygons. *Computers & Geosciences*, 7: 387-392.
Harvey, P. and Mace, G., 1982. Comparisons between taxa and adaptive trends: problems of methodology. In: King's College Sociobiology Group (Editors), Current Problems in Sociobiology. Cambridge University Press, Cambridge, U.K., pp. 343-361.
Houck, M.A., Gauthier, J.A. and Strauss, R.E., 1990. Allometric scaling in the earliest fossil bird, *Archaeopterx lithographica*. *Science*, 247: 195-198.
Janzon, L.A., 1986. Morphometric studies of some *Pteromalus* Swederus species (Hymenoptera: Chalcidoidea) with emphasis on allometric relationships, or: are ratios reliable in chalcid taxonomy? *Systematic Entomology*, 11: 75-82.
Johnson, J.W. and Walter, E., 1978. Similarities and differences in forewing shape of six California *Catocala* species (Noctuidae). *Journal of Research on the Lepidoptera*, 17: 231-239.
Jolicoeur, P., 1963. The multivariate generalization of the allometry equation. *Biometrics*, 19: 497-499.
Kingsolver, J.G. and Wiernasz, D.C., 1987. Dissecting correlated characters: adaptive aspects of phenotypic covariation in melanization pattern of

Pieris butterflies. *Evolution*, 41: 491-503.
Klecka, W.R., 1980. Discriminant Analysis. Quantitative Applications in the Social Sciences, Volume 19. Sage Publications, Beverly Hills, California, 71 pp.
Lande, R., 1977. On comparing coefficients of variation. *Systematic Zoology*, 26: 214-217.
Lewontin, R.C., 1966. On the measurement of relative variability. *Systematic Zoology*, 15: 141-142.
Mason, L.G., Ehrlich, P.R. and Emmel, T.C., 1968. The population biology of the butterfly *Euphydryas editha*. VI. Phenetics of the Jasper Ridge colony 1965-1966. *Evolution*, 22: 46-54.
Nachtigall, W., 1967. Aerodynamische Messungen am Tragflugelsystem segelner Schmetterlinge. *Zeitschrift fur Vergleichende Physiologie*, 54: 210-231.
Neff, N.A. and Marcus, L.F., 1980. A Survey of Multivariate Methods for Systematics. American Museum of Natural History, New York, 221 pp.
Norberg, U.M., 1981. Allometry of bat wings and legs and comparison with bird wings. *Philosophical Transactions of the Royal Society, Series B*, 292: 359-398.
Norberg, U.M. and Rayner, J.M.V., 1987. Ecological morphology and flight in bats (Mammalia; Chiroptera): wing adaptations, flight performance, foraging strategy and echolocation. *Philosophical Transactions of the Royal Society, Series B*, 316: 335-427.
Olsen, E.C. and Miller, R.L., 1958. Morphological Integration. University of Chicago Press, Chicago, 242 pp.
Pagel, M.D. and Harvey, P.H., 1989. Taxonomic differences in the scaling of brain on body weight among mammals. *Science*, 244: 1589-1593.
Pearse, F. Kay and Murray, N.D., 1982. Sex and variability in the common brown butterfly *Heteronympha merope merope* (Lepidoptera: Satyrinae). *Evolution*, 36: 1251-1264.
Petersen, B., 1952. Studies on geographic variation of allometry in some European Lepidoptera. *Zoologiska Bidrag från Uppsala*, 29: 1-38.
Pimentel, R.A., 1979. Morphometrics, the Multivariate Analysis of Biological Data. Kendall/Hunt Publishing Company, Dubuque, Iowa, 276 pp.
Polak, E., 1971. Computational Methods in Optimization. Academic Press, New York, 327 pp.
Prim, R.C., 1957. Shortest connection networks and some generalizations. *Bell System Technological Journal*, 36: 1389-1401.
Reyment, R.A., Blackith, R.E. and Campbell, N.A., 1984. Multivariate Morphometrics, (Second Edition). Academic Press, New York, 233 pp.
Ricklefs, R.E. and O'Rourke, K., 1975. Aspect diversity in moths: a temperate-tropical comparison. *Evolution*, 29: 313-324.
Rohlf, F.J., 1986. Relationships among eigenshape analysis, Fourier analysis, and the analysis of coordinates. *Mathematical Geology*, 18: 845-854.
Rohlf, F.J., (this volume). The analysis of shape variation using ordinations of fitted functions.
Rohlf, F.J. and Archie, J.W., 1984. A comparison of Fourier methods for the description of wing shape in mosquitoes (Diptera: Culicidae). *Systematic Zoology*, 33: 302-317.
Rohlf, F.J. and Bookstein, F.L., 1987. A comment on shearing as a method for "size correction." *Systematic Zoology*, 36: 356-367.

Scott, J.A., 1984. The phylogeny of butterflies (Papilionoidea and Hesperioidea). *Journal of Research on the Lepidoptera*, 23: 241-281.
Smart, P., 1984. The Illustrated Encyclopedia of the Butterfly World. Salamander Books Limited, London, 274 pp.
Sneath, P. and Sokal, R., 1973. Numerical Taxonomy, the Principles and Practice of Numerical Classification. W.H. Freeman and Company, New York, 573 pp.
Strauss, R.E., 1984. Static allometry and functional feeding morphology in haplochromine cichlids. In: A.A. Echelle and I.L. Kornfield (Editors), Evolution of Fish Species Flocks. University of Maine at Orono Press, Orono, Maine, pp. 217-229.
Strauss, R.E., 1985. Evolutionary allometry and variation in body form in the South American catfish genus *Corydoras* (Callichthyidae). *Systematic Zoology*, 34: 381-396.
Strauss, R.E., 1987. Allometry and relative growth in evolutionary studies. *Systematic Zoology*, 36: 72-75.
Strauss, R.E., 1990. Patterns of quantitative variation in lepidopteran wing morphology: the convergent groups Heliconiinae and Ithomiinae (Papilionoidea: Nymphalidae). *Evolution*, 44: 86-103.
Strauss, R.E., (in press). Parsimonious mapping of continuous characters onto evolutionary trees for studies of character evolution. *Paleobiology*.
Strauss, R.E. and Bookstein, F.L., 1982. The truss: body form reconstructions in morphometrics. *Systematic Zoology*, 31: 113-135.
Sweet, S.S., 1980. Allometric inference in morphology. *American Zoologist*, 20: 643-652.
Sylven, E. and Carlbäcker, U., 1981. Morphometric studies on Oligotrophini adults (Diptera: Cecidomyiidae) including an attempt to correct for allometric deviations. *Entomologica Scandinavica, Supplement*, 5: 185-210.
Wiernasz, D.C., 1989. Female choice and sexual selection of male wing melanin pattern in *Pieris occidentalis* (Lepidoptera). *Evolution*, 43: 1672-1682.
Zelditch, M.L., 1987. Evaluating developmental models of integration in the laboratory rat using confirmatory factor analysis. *Systematic Zoology*, 36: 368-380.

MULTIVARIATE STUDY OF TEMPORAL VARIATION IN MORPHOLOGY OF A GALL-FORMING APHID

David Wool and Ora Manheim

Department of Zoology, George S. Wise Faculty of Life Sciences, Tel Aviv University, Tel Aviv 69978, Israel

Abstract

Gall-forming aphids are attractive for morphometric studies of geographical and temporal variation. The clonal (parthenogenetic) reproduction within galls enables the estimation of the relative magnitude of genetic and environmental effects on variation in each morphological character, and the separate investigation of processes that take place within and among clones. Observations on a gall-forming aphid, *Aploneura lentisci*, at a single locality revealed a seasonal trend of increasing average alate size. The present morphometric analysis of measurements of these aphids, taken during five years of sampling, leads to the conclusion that this trend is not due to an increase in size of individual aphids maturing on different dates within each gall (a nongenetic trend). The trend is found when different galls are sampled at each date, implying that galls with smaller aphids mature earlier in the season than galls with larger aphids. This tendency may be due to a genetic difference among clones, or to differences in environmental conditions (e.g., nutritional value of the plant sap) affecting aphid growth in different galls.

1 Introduction

Most studies of morphological variation in insects seek answers to questions about the spatial differences among populations, and the possible differentiation into host races, local races and subspecies, and finally, species. Another aspect, less frequently reported, is temporal variation within the same locality. Although both geographic and temporal variation in insect morphology may have a genetic component, both may be strongly affected directly by changes in the environment.

These two components of phenotypic variance cannot usually be measured separately, except by elaborate breeding designs (see Falconer, 1981). One of the rare cases where the biology of the organisms makes this separation possible is morphological variation in gall-forming aphids.

Gall-forming aphids of the genus *Pemphigus* (Pemphigidae) make galls on *Populus* trees. Investigations of morphological variation of these aphids since the 1950s, were instrumental in developing the methodology for large scale geographical variation studies (Sokal 1952, 1962; Sokal et al., 1980; Sokal and Riska 1981; and other publications by Sokal and associates listed in the two latter papers). Similar studies on a smaller scale were carried out on three species of related gall aphids (subfamily Fordinae), which infest *Pistacia* trees in Israel (Wool and Koach, 1976; Wool, 1977). The advantage of these insects as research organisms stems from their uncommon mode of reproduction: each gall is formed by a single female and contains a clone of female aphids, produced parthenogenetically (see review by Wool, 1984). Variation among aphids within galls thus cannot be genetic. As indicated by Sokal and his associates, the variance of each morphological character can be partitioned, by appropriate statistical techniques, into components: among aphids within galls, among gall means within trees or localities, and among locality means. The variance components give an indication of the relative roles of genetic and environmental factors in aphid morphology.

Once differences in morphology are detected, research may proceed in several directions. One is to search for the geographical pattern of these differences. This line of investigation was developed by Sokal and Oden (1978a, 1978b) and Oden and Sokal (1986), and applied to human data (e.g., Sokal and Winkler, 1987). Another direction is to look for correlations of the morphological characters, or of factors representing groups of characters (extracted by ordination techniques), with environmental variables and try to understand the underlying causative factors (e.g., Wool and Koach, 1976; Wool, 1977).

This paper presents a quantitative analysis of morphological characters in a gall-forming aphid, *Aploneura lentisci* (Pass.) (Fordinae), with an emphasis on temporal (seasonal), rather than spatial, variation. Examples of seasonal changes in morphology are reported by Bodenheimer and Swirski (1957) in some free-living aphids and by Heryford and Sokal (1971) in the gall aphid, *Pemphigus populitransversus*. To the extent that the measured characters are size-related, the seasonal patterns may reflect changes in plant sap quantity or quality; for a recent example of tree effects on stem-mother and alate morphology in a gall-forming aphid see Bingham and Sokal (1986). However, the possibility of seasonal changes in the genetic composition of the populations cannot be ruled out.

Aploneura lentisci makes its galls exclusively on *Pistacia lentiscus*, a common bush in the Mediterranean forest of Israel. The aphid is widespread in climatically similar habitats in Spain, France, Italy, Turkey, Syria and Lebanon (Bodenheimer and Swirski, 1957; Wool, 1984). Investigations on the ecology and population dynamics of this species have been carried out since 1980 (Wool and Manheim, 1986, 1988).

2 Materials and Methods

"Kidney-shaped" galls of *A. lentisci* (Fig. 1) were collected in the Botanical Gardens of Tel Aviv University, which served as the site of our long-term study of population dynamics in this species. Five collections were made at two week

Figure 1. Mature galls of *A. lentisci*.

intervals, beginning on October 15, 1981. In addition, galls were collected at seven other localities, five in October and two in December of the same year. Additional small samples were obtained in the Botanical Gardens in 1982 and 1983. In 1984 and 1985, a number of galls were covered by nylon screen bags and inspected at intervals; we collected and preserved the alates of these galls as they emerged.

2.1 Aphids

Alates from each gall were preserved in 70% ethanol. Eight alates per gall or collection date (if available) were mounted on microscope slides. The procedure involved clearing for 10 hours in 8% potassium hydroxide, washing with 1% hydrochloric acid in 96% ethanol to neutralize the potassium hydroxide, and storage for 48 hours in 96% ethanol before mounting in Euparal (Chroma, Stuttgart). The same procedure was used in our previous studies (Wool and Koach, 1976; Wool, 1977; Wool and Manheim, 1983).

2.2 Measurements

Twenty-one morphological characters were measured on each alate, using a Visopan projection microscope (Reichert, Austria). Character codes, descriptions, means and standard deviations are listed in Table 1. Measurements in the table are given in microns for comparison with data on other species.

Table 1.
Character codes, description, and overall means (± SE). Means were calculated using gall medians as data and transformed to microns (see text for explanation).

Code	Description	Dimensions in microns mean	SD	*n*
WTL	Fore wing total length (base to tip)	2306	267	228
WL	Fore wing length (wing base to *C-Sc* intersection)	1696	208	228
W W	Fore wing width (at widest point)	882	109	197
HWL	Hind wing length (base to tip)	1694	197	197
H W W	Hind wing width (at widest point)	519	65	193
H W	Head width (at widest point)	287	18	200
T W	Width of the large thoracic sternite	403	39	230
MPTS	Length of clypeus + rostrum	281	23	181
A1	Length of third antennal segment	79	10	200
A2	Length of fourth antennal segment	67	8	200
A3	Length of fifth antennal segment	71	9	200
A4	Length of sixth (terminal) segment	134	15	229
F1	Fore leg femur length	306	29	200
Ti1	Fore leg tibia length	351	36	223
Tar1	Fore leg tarsus length	114	10	198
F2	Middle leg femur length	226	19	230
Ti2	Middle leg tibia length	324	34	200
Tar2	Middle leg tarsus length	112	9	200
F3	Hind leg femur length	306	29	201
Ti3	Hind leg tibia length	421	45	230
Tar3	Hind leg tarsus length	130	11	200

2.3 Statistical Analysis

The median of the eight alates per gall was used for calculations. Because reproduction within galls is parthenogenetic, variation among aphids within galls is assumed here to have no genetic component. The choice of the gall median (rather than gall mean) to represent a clone avoids the effect of single aberrantly small or large individual aphids and is probably a good estimate of the genetic properties of the clone.

The data base was a 231 x 21 matrix of medians of the 21 characters. Data were analyzed using the NTSYS program package (Rohlf, 1985). The analysis was similar to our former procedure (Wool and Koach, 1976) and proceeded in the following steps: (1) Cluster (UPGMA) and factor analyses (principal components analysis) were carried out on the 21 x 21 character correlation matrix to detect covarying groups of characters. (2) The data matrix was standardized by subtracting the overall mean for each character and dividing by its standard deviation. The resulting matrix had a mean of zero and a standard deviation of one, regardless of the units of measurements of the individual characters. To avoid negative numbers, the standard deviation

Table 2.
Variance components of the morphological characters of *A. lentisci*. Averages over eight localities (analyzed separately). Numbers will not add up to 100% because of averaging and rounding errors.

Character	Mean variance components (%)		
	Within-galls	Among-galls, Within-trees	Among-trees
WTL	9.5	68.0	22.4
WL	10.4	70.4	19.4
WW	15.4	65.3	18.2
HWL	17.9	58.7	23.5
HWW	26.5	47.7	25.8
HW	34.2	40.0	25.8
TW	17.3	64.3	18.4
MPTS	52.2	41.2	6.6
A1	27.1	63.6	9.3
A2	30.3	50.0	19.6
A3	29.1	54.6	16.3
A4	23.8	61.0	13.1
F1	19.8	56.1	24.1
Ti1	20.6	53.5	25.9
Tar1	35.5	48.1	16.4
F2	17.8	60.0	22.2
Ti2	19.3	60.7	20.0
Tar2	35.4	48.5	16.1
F3	15.4	58.6	26.0
Ti3	16.8	61.9	21.3
Tar3	29.1	49.7	21.2

units were transformed to the probit scale by adding five. (3) The means of probits were calculated for each character group and sampling date, combining all characters in the cluster. Means of character groups were plotted along the time axis to illustrate the seasonal trends.

Although the emphasis in this study was on temporal variation, we followed our previous procedures and partitioned the variation in each character into components using a "nested" analysis of variance (Sokal and Rohlf, 1981). The variance components obtained in *A. lentisci* were compared with the three species which we studied in the past (see Wool, 1977).

3 Results

The list of characters, their means and standard deviations is given in Table 1. For comparison with our previous studies, we list (Table 2) the variance components due to differences within and among galls and trees within localities; the analysis was done separately for each locality and the table presents the averages over eight localities for which sample sizes were adequate. Among-locality differences were not analyzed in this paper because we were only interested in the processes taking place within localities.

In all characters, except *MPTS* (see Table 1 for the definitions of character codes), the largest variance component was among galls on the same tree (ranging from 40 to 70% of the total, Table 2). We have suggested in the past (Wool and Koach, 1976; Wool, 1977), that this component measures primarily the genetic differences among fundatrices, but this suggestion may be disputed (see Discussion).

Variation among tree means contributed about 20% of the total. This component is probably a measure of differences in growth conditions between trees, which in turn affect the nutritional conditions for the aphids in the galls.

3.1 Correlations Among Characters

Principal components analysis was performed on the 21 x 21 character correlation matrix. The characters are plotted in the field of the first two principal components in Fig. 2. The first principal component explained 86.9% of total variation, and the first two together accounted for 93% of the overall variance. All characters were strongly associated with factor 1, which represents size. Only character *A2* was strongly associated with factor 2.

The finer details of the character distribution do indicate some order, but this is much better illustrated by the results of UPGMA cluster analysis on the same correlation matrix (Fig. 3). The character clusters are readily interpretable biologically: the wing characters (fore and hind wings) form one cluster; femora and tibiae of the three pairs of legs form two adjacent clusters, as do the tarsi and antennal segments (the latter clusters are less tight). The cophenetic correlation coefficient for the phenogram is 0.76, indicating reasonably strong distortion in the dendrogram.

3.2 Temporal Variation in Size-Related Characters

From mid-October to mid-December, 1981, there was a steady increase in size-related alate characters. Fig. 4 illustrates the frequency distributions and the means (arrows) of three characters (*F2*, *Ti3*, *WTL*), as an example. The change in the mean is somewhat obscured by wide variation among galls at each collection date. We were unaware of the seasonal trend until a preliminary analysis of the 1981 data was completed in 1983. We then wished to find out whether the seasonal increase in size was a result of sampling different galls, or the change was occurring within every gall, with the earlier alate aphids being smaller than later maturing ones. In 1984, and again in 1985, we covered a number of galls with nylon-net bags and collected the aphids emerging from each gall separately, from September through December. The medians for these collection dates were combined with the previous data to create the final matrix for five years (1981-1985) and analyzed.

Based upon the cluster analysis (Fig. 3), we decided to use four character groups: wings, femora, tibiae and antennal segments. The mean probit score (over all characters) was calculated for each of the four character groups at each collection date, and these mean scores were plotted against time. Fig. 5

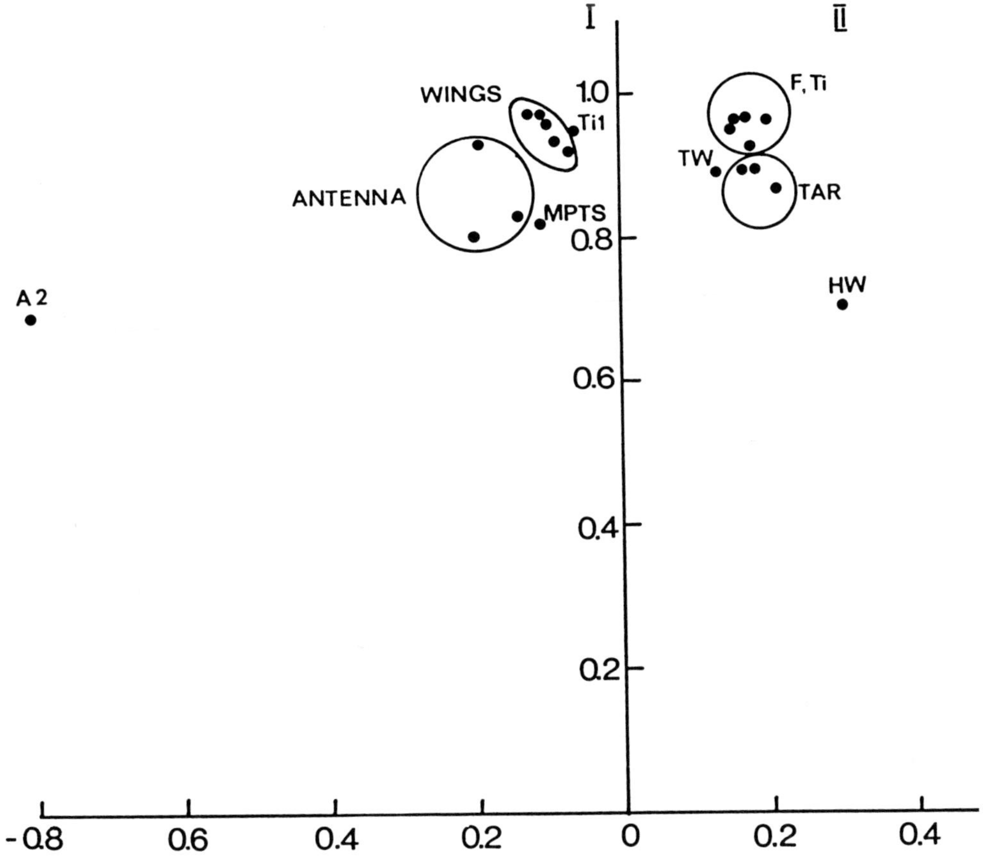

Figure 2. Principal components analysis of the character correlation matrix (1981 data). The 21 characters are plotted in the field of the first two principal components. All characters are strongly associated with factor 1 (size).

illustrates the trend in the femur group as an example. The trends for the other three groups of characters were almost identical. The figure confirms that in 1981, aphid dimensions increased linearly during the season.

Seasonal data for 1982 and 1983 were not collected. In 1984 and 1985, no trend is discernible; however, note that the method of sampling in 1984 and 1985 was not the same as in 1981. In 1981, *different* galls were sampled and removed at each date; whereas, in 1984, and again in 1985, the *same* galls were sampled at different dates. The absence of pattern suggests that the size increase trend is not a within-gall phenomenon, but resulted from sampling different galls, a phenomenon which may have a genetic component.

Another result that is apparent from the figure is the size differences of aphids in different localities. Femur data for all 1981 localities, other than the Botanical Gardens, are plotted in Fig. 5. The aphids in the Botanical Gardens

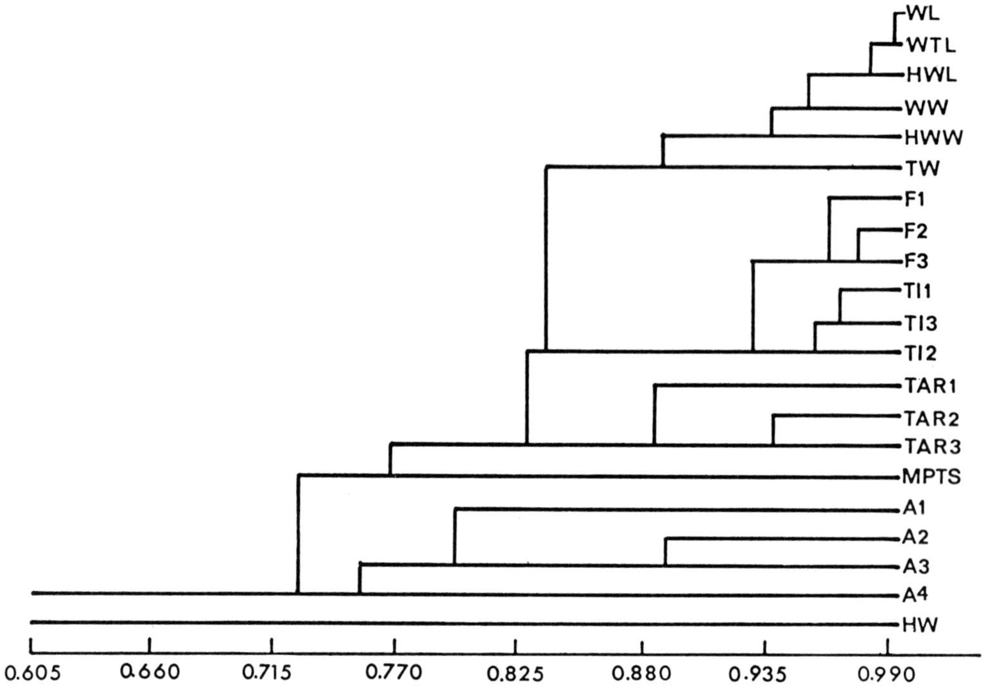

Figure 3. Cluster analysis (UPGMA) of the characters. Five clusters are indicated, representing groups of related limbs (wings, legs and antennae).

were smaller than in all other localities collected from at the same dates. This may be due to the environmental differences among sites, although it should be emphasized that growth conditions for the host plants in the Gardens are probably better than in the natural localities due to irrigation during the dry summer.

4 Discussion

Morphological studies of organisms for taxonomic, ecological and evolutionary research involve the comparison of sizes and shapes. "Shape" in particular is a problematic term. When organisms differ in shape, it is not always possible to quantify the difference, and it may well be a simultaneous difference in more than one character which catches the eye and causes them to appear different.

Quantitative comparisons of shapes were perhaps first attempted by D'Arcy Thompson early in this century (Thompson, 1917). He showed that very different final shapes of organisms result from changes in growth patterns, which can be expressed by relatively simple mathematical functions of a small number of variables. Ordination techniques are used today to perform this

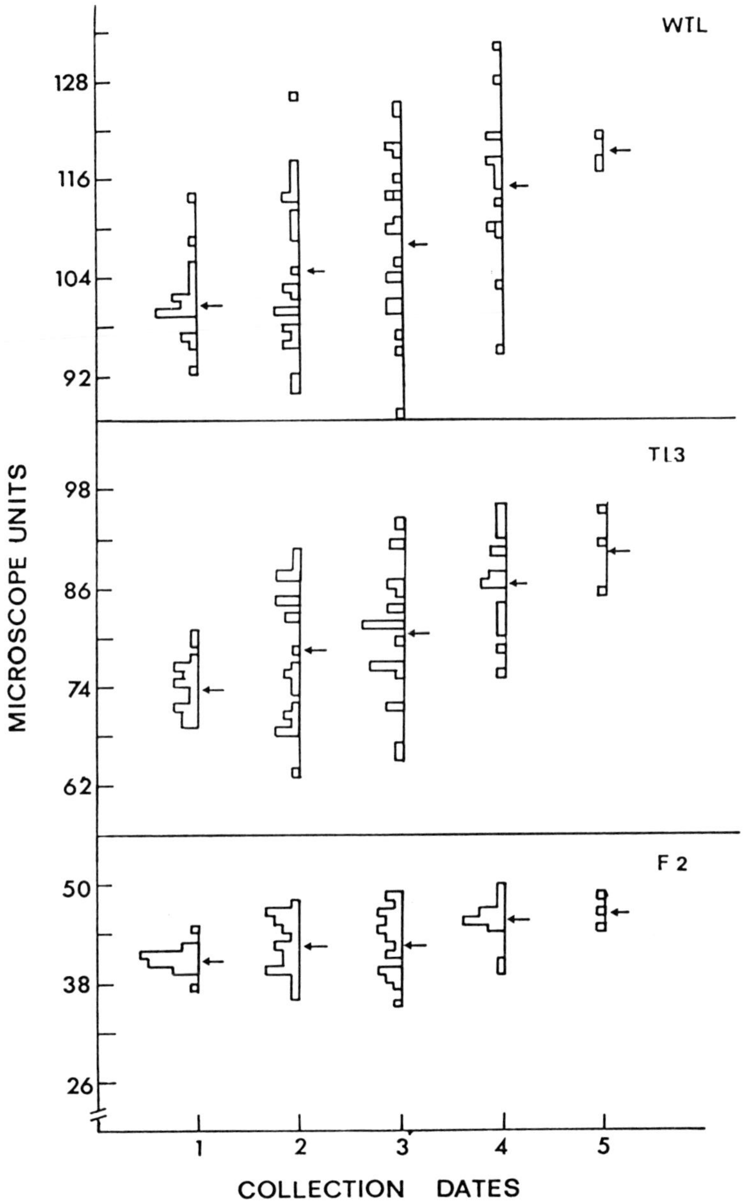

Figure 4. Frequency distributions of three alate characters (*F2*, *Ti3*, *WTL*) in five sampling dates (1981, Botanical Gardens). Size scale is in microscope units. Arrows represent means for each distribution. Each point represents one gall. The means are seen to increase with time.

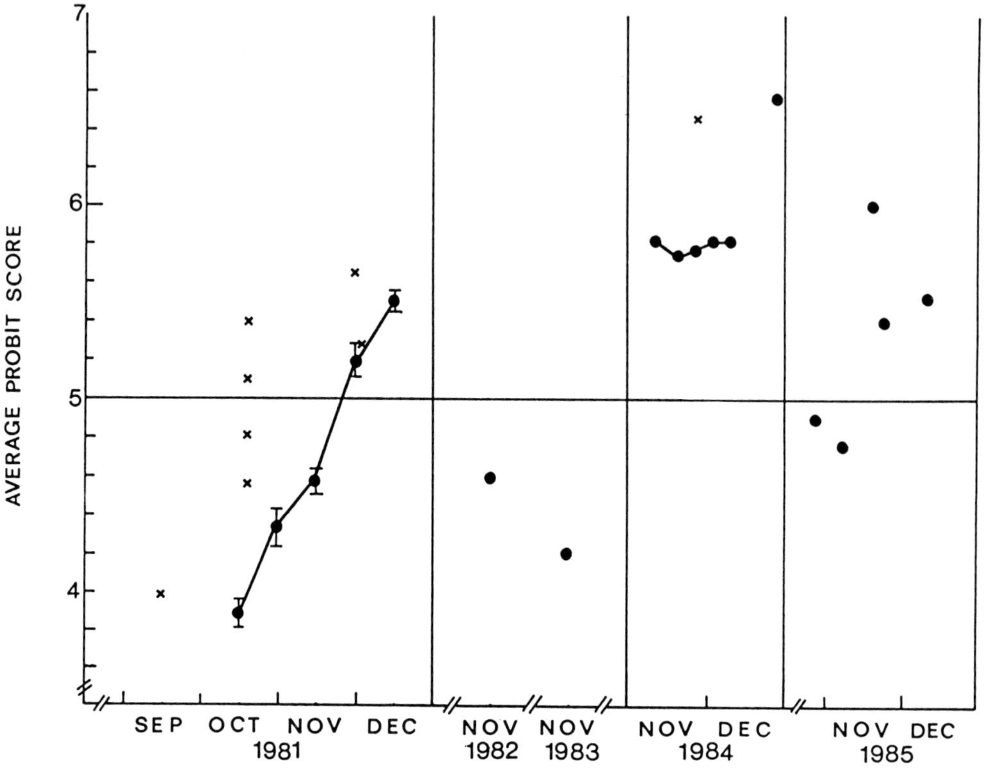

Figure 5. Average probit scores for the femur character group during five sampling years (vertical bars indicate ± 1 standard errors around the means, using the three characters within the cluster - *F1, F2, F3* - as separate data). In 1981, aphid size-related characters in the Botanical Gardens (dots) increased during the alate emergence season. No such trend occurred in 1984-85. For interpretation see text. Average scores for localities other than the Botanical Gardens are marked by X.

reduction of space to a few dimensions, chosen so that the aspects in which the organisms are not the same may be brought out more clearly.

This paper describes biologically meaningful phenomena which were discovered by the application of ordination techniques to a large data matrix. The combined use of factor analysis and UPGMA cluster analysis indicated four groups of correlated characters in the 21 linear dimensions of *A. lentisci* alates. These groups are readily interpretable biologically; although all of them seem to be associated with aphid size (factor analysis, Fig. 2), still they assemble into distinct clusters of legs, wings, and antennal segments, perhaps due to the developmental constraints during ontogeny or to (partial) control by separate groups of genes.

Correlated characters present a dilemma when searching for temporal trends in the data. One can, of course, choose a single character from each cluster to illustrate the pattern, but in doing so, much of the information from the characters, as a group, is lost. The solution taken here (and by Wool and Koach, 1976) was to standardize the data matrix so that all characters had a mean of zero and a standard deviation of one, and then use the mean score of each group of characters as the representative of the cluster.

As a result, the increase in aphid dimensions indicated in the raw data (Fig. 4) is much more clearly expressed, and can also be tested for significance. As an example, for the femur character cluster, illustrated in Fig. 5, the probit scores of the three femora in the cluster were used as data for each collection date, and subjected to single-classification ANOVA. The differences among dates were very highly significant ($F_{4,10}$ = 188.8; $P << 0.001$). (We could decrease the error mean square even further by using the three probits for each gall as separate data.) The analysis of the wings, antennae, tibia, and tarsus character clusters all yielded highly significant differences among collection dates.

Bodenheimer and Swirski (1957) list several biometrical investigations of different aphids in Israel. They noted seasonal changes in aphid size, with the biggest aphids in the spring (March) and the smallest at the end of summer (August to October). They suggest that aphid size may be a measure of aphid growth conditions (late winter and spring conditions in Israel are more favorable for aphid growth).

In North America, *Pemphigus populitransversus* aphids leave the galls from mid-July through September. Heryford and Sokal (1971) detected a seasonal size increase of nine characters in that species. In our Fordinae, alates appear only as late as September or October. The analysis indicates that within the two months during which alate aphids leave the galls, in 1981, there was a trend of size increase. The absence of such a trend in 1984 and 1985, when the sampling scheme was altered, suggests that within each clone, early alates are not necessarily smaller than late-emerging ones. This size increase may have a partly genetic basis (i.e., clones which mature earlier may contain genetically smaller aphids).

Unfortunately, no proper control is available to test this hypothesis, because the 1981 sampling method was discontinued in 1984. Contrary to the present argument, Heryford and Sokal (1971) reported that the trend of increasing size was found within galls in *Pemphigus populitransversus*. Clearly, more data are needed to resolve these discrepancies.

Size in aphids, as in other organisms, results from a suite of quantitative characters that are affected by both genetic and nongenetic factors; as an attribute, size is determined by the relative magnitude of these components. Size-related linear measurements of organisms, therefore, may be highly correlated. For example, in another gall aphid, *Geoica utricularia*, the characters could be divided into four groups by factor analysis (Wool and Koach, 1976; Wool, 1977): one group included strongly size-dependent characters, another contained size-independent characters (i.e., numbers of sensoria on antennal segments, which cannot be measured in *Aploneura*), and the two remaining character groups were intermediate. In contrast, for two other Fordinae, *Baizongia pistaciae* and particularly *Slavum wertheimae*, all characters were strongly size-dependent (Wool, 1977; Wool and Manheim, 1983).

Superficially, *A. lentisci* alates resemble the alates of the other Fordinae, but there are some differences. *Aploneura lentisci* is smaller and has shorter legs and somewhat narrower wings than three other species of Fordinae studied by Wool (1977). In *A. lentisci* the fore wings at rest are placed flat, parallel to the body, but in the other three species they meet at an angle, forming a "roof" above the body.

Another difference is that *A. lentisci* produces galls that are not sealed by the plant tissues, in contrast to the other three Fordinae studied (Wool, 1977). In principle, this could allow small nymphs to migrate between galls through the opening slit. Evidence for intergall migration has been detected by electrophoresis in the related aphid *Pemphigus populitransversus* (Setzer, 1980), and such migration should increase the variance among aphids within galls. Our data, however, show that the magnitude of the within-gall variance component in *A. lentisci* is not different from that of the other species in any consistent way (compare Table 2 with Wool, 1977: table 3). Based upon our morphological evidence, we have no reason to suspect that intergall migration occurs in *A. lentisci*.

Variation among aphids within galls contributed 10-35% of total variation, with the exception of character *MPTS*, where it gave 52%. In the parthenogenetic clones, within-gall variation is probably due to developmental, ontogenetic differences among individuals. Microenvironmental variation within the gall could also affect this component of morphological variation, but such effects are less likely in view of evidence from Wool and Manheim (1983). This component was removed from all further analysis by using the gall median of each character to represent the gall.

In our former studies (Wool and Koach, 1976; Wool, 1977; Wool and Manheim, 1983), variation among gall means on the same tree was considered to be an estimate of mainly genetic variation among fundatrices. This interpretation, however, may not be entirely justified. Whitham and Slobodchikoff (1981) and Whitham et al. (1984) suggest that each tree can itself be considered a colony of "individuals," an idea first expressed by Lamarck (1809 [1984: 275-276]). In genetic terms, a tree may be a mosaic of differing genotypes due to somatic mutations in the meristematic tissues of different buds and, therefore, may present nutritionally different environments for insect herbivores. If this is true, the among-gall variance component here also may include varying environmental effects of nutrition in different galls. Moreover, physiological differences among branches need not be of genetic origin to affect the aphids. Shaded branches may contain less nutrients than branches exposed to the sun, upper branches may differ from lower ones and differences may also exist between leaf positions on the same branch. Therefore, the tree may control, in part, the differences in size and shape among the alates emerging from different galls. We do not yet have the means for measuring this effect quantitatively.

5 Acknowledgments

We thank A. Shouv for the photograph (Fig. 1) and Mrs. R. Suzin for the drawings; G. Cohen collected and partly analyzed the data for 1982-83.

6 References

Bodenheimer, F.S. and Swirski, E., 1957. The Aphidoidea of the Middle East. The Weizmann Science Press, Jerusalem, 378 pp.

Bingham, B. and Sokal, R.R., 1986. Host tree effects on the morphology of *Pemphigus populitransversus* (Homoptera: Aphididae). *Journal of the Kansas Entomological Society*, 59: 287-295.

Falconer, D.S., 1981. Introduction to Quantitative Genetics, (Second Edition). Longman, London, 340 pp.

Heryford, N.N. and Sokal, R.R., 1971. Seasonal morphometric variation in *Pemphigus populitransversus*. *Journal of the Kansas Entomological Society*, 44: 384-390.

Lamarck, J.B., 1809, (English Edition, 1984). Zoological Philosophy. University of Chicago Press, Chicago, 458 pp.

Oden, N.L. and Sokal, R.R., 1986. Directional autocorrections: an extension of spatial correlograms to two dimensions. *Systematic Zoology*, 35: 608-617.

Rohlf, F.J., 1985. Numerical Taxonomy System of Multivariate Statistical Programs. State University of New York at Stony Brook, Stony Brook, New York, 35 pp.

Setzer, R.W., 1980. Intergall migration in the aphid genus *Pemphigus*. *Annals of the Entomological Society of America*, 73: 327-331.

Sokal, R.R., 1952. Variations in a local population of *Pemphigus*. *Evolution*, 6: 296-315.

Sokal, R.R., 1962. Variation and co-variation of characters of alate *Pemphigus populitransversus* in eastern North America. *Evolution*, 16: 227-245.

Sokal, R.R., Bird, J. and Riska, B., 1980. Geographic variation in *Pemphigus populicaulis* (Insecta, Aphididae) in eastern North America. *Biological Journal of the Linnean Society*, 14: 163-200.

Sokal, R.R. and Riska, B., 1981. Geographic variation in *Pemphigus populitransversus* (Insecta, Aphididae). *Biological Journal of the Linnean Society*, 15: 201-234.

Sokal, R.R. and Rohlf, F.J., 1981. Biometry, (Second Edition). W.H. Freeman and Company, New York, 859 pp.

Sokal, R.R. and Oden, N.L., 1978a. Spatial autocorrelations in biology. I. Methodology. *Biological Journal of the Linnean Society*, 10: 199-228.

Sokal, R.R. and Oden, N.L., 1978b. Spatial autocorrelations in biology. 2. Some biological implications and four applications of evolutionary and ecological interest. *Biological Journal of the Linnean Society*, 10: 229-249.

Sokal, R.R. and Winkler, E.M., 1987. Spatial variation among Kenyan tribes and subtribes. *Human Biology*, 59: 147-164.

Thompson, D.W., 1917. On Growth and Form, (Abridged Edition,1984: J.T. Bonner, Editor), Cambridge University Press, Cambridge, U.K., 346 pp.

Wool, D., 1977. Genetic and environmental components of morphological variation in gall-forming aphids (Homoptera, Aphididae, Fordinae) in relation to climate. *Journal of Animal Ecology*, 46: 875-889.

Wool, D., 1984. Gall-forming aphids. In: T.N. Anathakrishnan (Editor), Biology of Gall Insects. Oxford and IBH Publishing Company, New Delhi, pp. 11-58.

Wool, D. and Koach, J., 1976. Morphological variation of the gall-forming aphid, *Geoica utricularia* (Homoptera) in relation to environmental variation. In: S. Karlin and E. Nevo (Editors), Population Genetics and

Ecology. Academic Press, New York, pp. 239-272
Wool, D. and Manheim, O., 1983. The effects of environment subdivision on morphological variation in the "cauliflower" galls of the aphid *Slavum wertheimae* (Homoptera, Aphididae, Fordinae). *Israel Journal of Entomology*, 17: 95-104.
Wool, D. and Manheim, O., 1986. Population ecology of the gall-forming aphid *Aploneura lentisci* (Pass.) in Israel. *Researches on Population Ecology*, 28: 151-162.
Wool, D. and Manheim, O., 1988. The effects of host-plant properties on gall density, gall weight and clone size in the aphid, *Aploneura lentisci* (Pass.) (Aphididae, Fordinae) in Israel. *Researches on Population Ecology*, 30: 227-234.
Whitham, T.G. and Slobodchikoff, C.N., 1981. Evolution by individual, plant-herbivore interactions, and mosaics of genetic variability: the adaptive significance of somatic mutations in plants. *Oecologia*, 49: 287-292.
Whitham, T.G., Williams, A.C. and Robinson, A.M., 1984. The variation principle: individual plants as temporal and spatial mosaics of resistance to rapidly evolving pests. In: P.W. Price, C.N. Slobodchikoff and W.S. Gaud (Editors), A New Ecology: Novel Approaches to Interactive Systems. John Wiley and Sons, New York, pp. 15-51.

THE USE OF ORDINATION METHODS TO RESOLVE PROBLEMS OF SPECIES DISCRIMINATION IN THE GENUS *CINARA* CURTIS (HOMOPTERA: APHIDOIDEA: LACHNIDAE)

Robert G. Foottit

Biological Resources Division, Centre for Land and Biological Resources Research, Agriculture Canada, Research Branch, C.E.F., Ottawa, Ontario K1A 0C6, Canada

Abstract

Morphological variation among species of aphids of the genus *Cinara* Curtis, on western North American pines, was characterized using univariate and multivariate analyses. Patterns of morphometric variation were shown to be complex and unique for many of the species. Functional groups of characters exhibited different internal relationships and different degrees of association with the main components of variation. Information on the covariation among the characters allowed for selection of those characters that best discriminated among the species. Knowledge of the variability of characters within and among taxa, at all levels within the hierarchy, would facilitate the development of a sound phylogenetic scheme for the tribe Cinarini.

1 Introduction

Aphids of the genus *Cinara* Curtis feed on a wide range of conifer species. In temperate North America, they are a common component of forest ecosystems. As is the case with many aphid genera, the taxonomy of *Cinara* is generally considered to be difficult. There are a large number of species that are morphologically similar although their biology is known to differ (Fedde, 1965). The majority of species of *Cinara* are usually dark brown to black, often extremely hirsute, and show a large amount of size variation. In addition, their distributions and host plant relations are poorly understood. Many previously used taxonomic characters are of limited use for discriminating between species, due in large part to reliance on characters that are likely to be influenced by size variation. Over 200 species of *Cinara* have been described (Eastop and Hille Ris Lambers, 1976), but a number are known only from their original descriptions (Voegtlin, 1976).

Because different *Cinara* taxonomists have favored the use of particular characters, their results often have been contradictory. For example, Oestlund (1942) devised a phylogenetic scheme based partly on the frequency of occurrence of the sensoria on the antennal segments, but Fedde (1967) found sensorial counts in some species to be unstable. Many workers have used ratios of morphological features, particularly those of the antennal segments, to designate new species (Eastop, 1972) although others have found ratios to be unsatisfactory for the separation of species (Fedde, 1967). Bradley (1959, 1961) attempted to separate species based on the overall length of the rostrum, noting that there was a significant positive relationship between the length of the rostrum and the feeding site, and hence the bark thickness. However, in some species the feeding sites are known to change with season. A number of taxa were erected by Hottes (e.g., 1955) on the basis of differences in the length and number of setae, without a comparison on a geographic basis of the range of variation of these structures.

It is clear that the taxonomic difficulties that have arisen in the study of North American *Cinara* aphids are due in large measure to the lack of critical evaluation of the taxonomic characters previously used and to the lack of any quantitative evaluation of classification schemes of *Cinara* species (Bradley, 1961; Fedde, 1965, 1967; Pepper and Tissot, 1973). One of the greatest problems in evaluating the taxon designations of other workers, particularly in a morphologically variable group such as aphids, is the frequent absence of statements of sample size and of measures of the variation of characters. Quantitative assessment of individual and geographic variation in morphological characters, and observations of biological phenomena of potential systematic use, such as differences in feeding site preferences, are needed for the separation of species. The grouping of *Cinara* species into more tractable generic and/or subgeneric categories would help to bring order into the morphological and biological diversity of this group of aphids.

The purpose of this review is to outline the use of ordination techniques in the taxonomic analysis of the genus *Cinara*. These approaches have been used to describe components of morphological character variation in selected species, as an aid in the definition of species boundaries within taxonomically difficult complexes and host plant groups, and to evaluate quantitatively the ability of characters to discriminate between species. The use of ordination techniques in the resolution of the phylogenetic relationships among species of *Cinara* is also discussed.

Although there are many species of *Cinara* in North America, I studied a more limited group on lodgepole pine, *Pinus contorta* Douglas ex Loudon, and other ecologically associated species, in order to examine morphological character variation in detail. The first part of this work consisted of an examination of variation within, and an analysis of geographic variation among, samples of a single species. Based on the results of this work, morphometric trends in a number of species were examined; in addition, attempts were made to establish boundaries between the species.

2 Methods and Materials

Lodgepole pine, which is widespread in western North America (Fig. 1) (Mirov, 1967) and exhibits considerable morphological variation and ecological amplitude (Wheeler and Guries, 1982a, 1982b), was the main host plant that was studied. However, *Cinara* from two other pine species, (*P. monticola* Douglas ex D. Don and *P. ponderosa* Douglas ex P. and C. Lawson) were also studied, as these host plants were found in association with *P. contorta*.

The life cycles of species of *Cinara* are autoecious, that is, the aphids do not migrate between primary and secondary hosts. *Cinara* are also holocyclic, with sexual forms produced every year; however, anholocyclic life cycles are known for some *Cinara* species. The summer apterous, viviparous morph (the virginoparae) (Fig. 2) are often the most taxonomically distinct morph due in part to the development of sclerotized areas on the dorsum of the abdomen and other structures.

Aphids of the genus *Cinara* may feed on the new growth shoots, needle fascicles, small branches, roots, and on the main stem of the tree. They are usually restricted to a particular feeding site at the start of the summer. However, some species show a tendency to gradually move as a colony to other feeding sites, on the same or different trees, as the season progresses, in order to exploit new food sources (Kidd et al., 1990) or acquire protection from such factors as weather (Bradley, 1961). The procedures used to collect and process the samples, prepare slide-mounted material, and define characters and their operational dimensions are given in Foottit and Mackauer (1990). Within-sample and between-sample variability were characterized using techniques of multivariate morphometrics (Pimentel, 1979; Neff and Marcus, 1980; Dunn and Everitt, 1982; Reyment et al., 1984). These methods are now widely used in biosystematic studies to describe patterns of morphological variation and to quantify the discriminatory ability of taxonomic characters (Foottit and Sorensen, this volume).

3 Character Selection

The first phase of this study involved the analysis and description of variation and covariation of morphological characters within and between samples of a single *Cinara* species. By examining the morphometric characteristics of a large set of variables from different body areas of a single species, it was hoped that a tractable set of characters that included a maximum amount of information content, and hence discriminatory ability, for the subsequent analysis of species differences could be identified.

A number of morphometric studies have been carried out on geographic samples of aphids. However, most of these works (Wool, 1977; Sokal et al., 1980; Sokal and Riska, 1981; Wool and Manheim, 1983) have centered on the analysis of variation in aphid systems where galls are produced. As a result, some partitioning of the genetic and environmental components of morphological variation is obtained. However, many aphid taxa do not fall into this category and, in fact, some taxa are totally parthenogenetic. My interest was to use morphometrics as an exploratory tool in aphid taxonomy to describe and quantify the morphological variation in field collections of aphids, to search for taxonomic characters, and to quantify the degree to which these

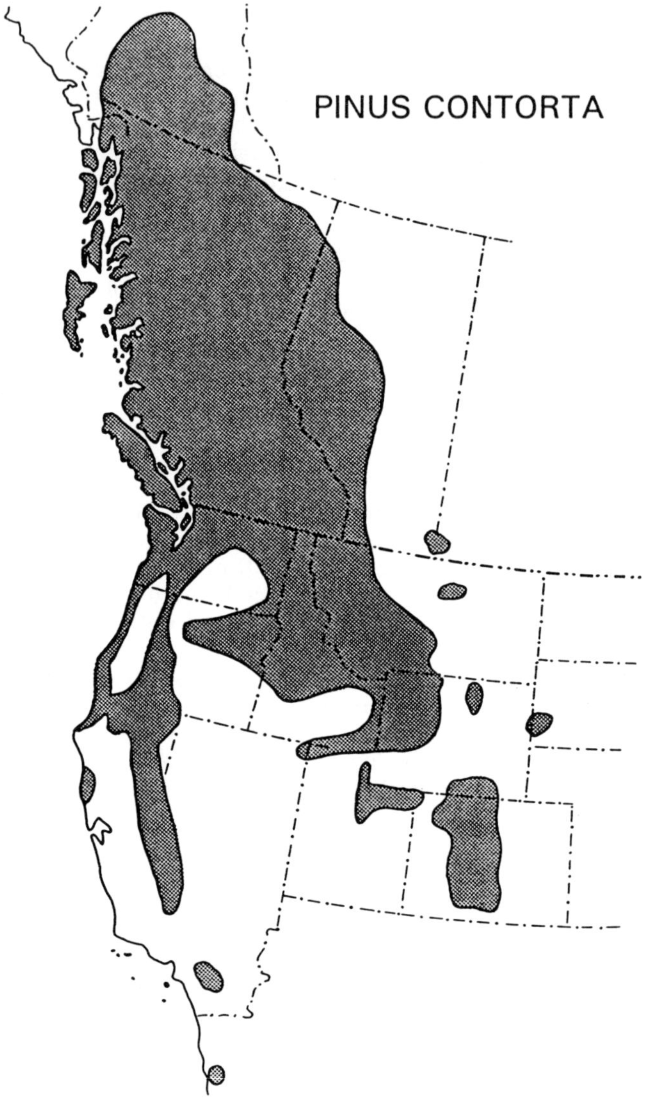

Figure 1. Geographical distribution of *Pinus contorta* Douglas ex Loudon. (Redrawn from Wheeler and Guries, 1982a, 1982b)

characters could discriminate between species in the face of the many factors that influence the morphology of aphids.

A morphologically distinct species of *Cinara*, the identity of which was not in doubt, was selected for the initial characterization of infraspecific variation in order that samples of other, subsequently determined species, were not

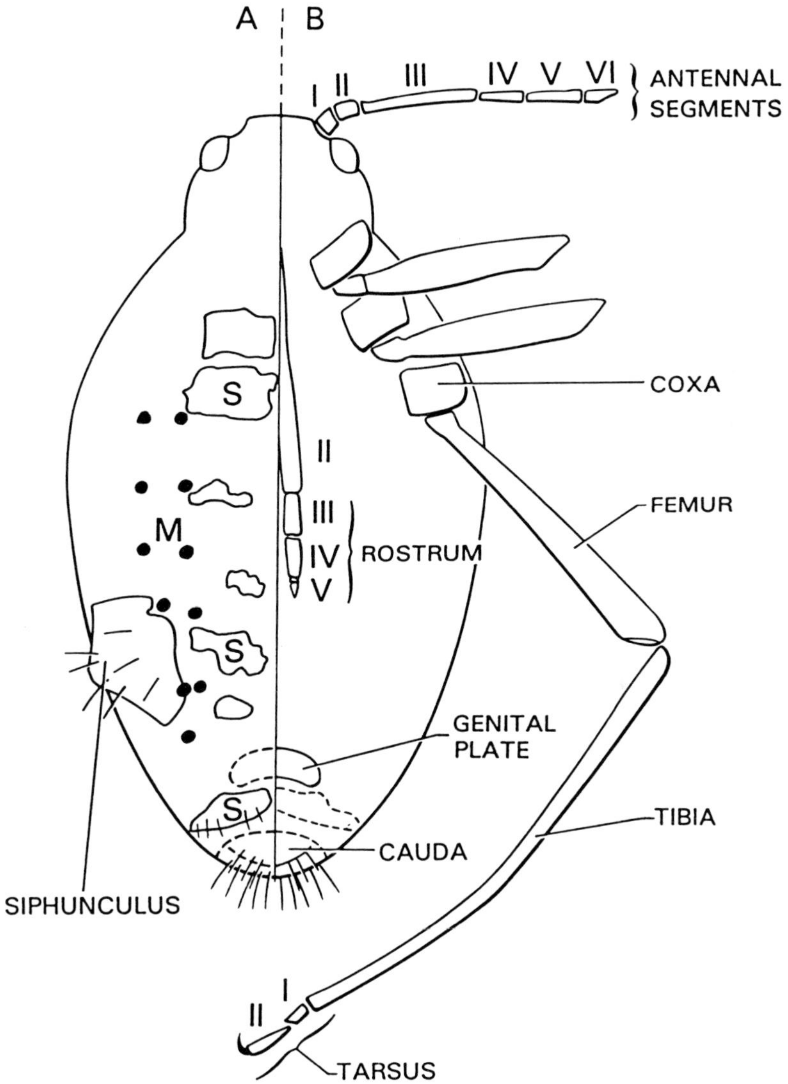

Figure 2. General morphological features of the adult virginoparous morph of *Cinara* (Homoptera: Aphididae). (A) Dorsal view. (B) Ventral view. M = muscle attachment plates; S = sclerotized areas on the dorsum of the abdomen.

included in the initial analysis of variation. *Cinara nigra* (Wilson) as chosen for this analysis because it has a distinctive, large sclerotized patch on the dorsal surface of the abdomen. A single sample of 20 adult virginoparae was used to gather measurement data. A series of continuous measurements was taken in order to estimate differences in size and shape, and counts of setae on

Table 1.
Continuous variables and setal counts used in the morphometric analysis of *Cinara* populations and species. See Foottit and Mackauer (1990) for operational dimensions of variables and locations of setal counts. (* = variable included in restricted data set.)

Abbrev.	Variable	Abbrev.	Variable
BL*	Body length	TS1VL*	Hind leg, tarsus seg. 1, ventral length
DE	Distance between eyes	TS1DL	Hind leg, tarsus seg. 1, dorsal length
DHC	Distance between hind coxae	TS2L*	Hind leg, tarsus seg. 2, length
FRW*	Frons width	TS2W	Hind leg, tarsus seg. 2, width
HL	Head length	SLH	Setal length, head
A1L	Antennal seg. 1, length	SLA3*	Setal length, antennal seg. 3
A1W	Antennal seg. 1, width	SLT*	Setal length, hind leg, tibia
A2L*	Antennal seg. 2, length	SLTS2	Setal length, hind leg, tarsus seg. 2
A2W	Antennal seg. 2, width	SLS	Setal length, siphunculus
A3L*	Antennal seg. 3, length	SLCA	Setal length, cauda
A3W	Antennal seg. 3, width	SLGP*	Setal length, genital plate
A4L*	Antennal seg. 4, length	SLAT8	Setal length, abdominal tergite 8
A4W	Antennal seg. 4, width	SLAT5*	Setal length, abdominal tergite 5
A5L*	Antennal seg. 5, length	SNA6SA*	Setal number, antennal seg. 6, subapical setae
A5W	Antennal seg. 5, width	SNA6B*	Setal number, antennal seg. 6, base
A6BL*	Antennal seg. 6, base, length	SNA5*	Setal number, antennal seg. 5
A6BW*	Antennal seg. 6, base, width	SNA2*	Setal number, antennal seg. 2
A6PTL*	Antennal seg. 6, processus terminalis, length	SNR4*	Setal number, rostrum seg. 4, accessory setae
R5L*	Rostrum seg. 5, length	SNGP*	Setal number, genital plate
R4L*	Rostrum seg. 4, length	SNAT5*	Setal number, abdominal tergite 5
R3L*	Rostrum seg. 3, length	SNAT8*	Setal number, abdominal tergite 8
R2L*	Rostrum seg. 2, length	SNS*	Setal number, siphunculus
CL	Hind leg, coxa, length	SNT*	Setal number, 0.2 mm of hind tibia
CW	Hind leg, coxa, width		
FL*	Hind leg, femur, length		
FW*	Hind leg, femur, width		
TL*	Hind leg, tibia, length		
TW	Hind leg, tibia, width		
TS1W	Hind leg, tarsus seg. 1, width		

certain structures were made in an attempt to sample other components of variation. The result was a set of 52 potential morphometric characters (Table 1) for further analysis.

4 Correlation and Covariation Patterns: *Cinara nigra* (Wilson)

Descriptive statistics and correlation coefficients were calculated for each variable and pairwise combination of variables, respectively. The correlation matrix of all variables was subjected to a principal components analysis (PCA) (Foottit and Sorensen, this volume). This initial analysis showed a general-size component among the continuous measurements; however, all of the setal counts, and many of the setal length measurements, were not correlated with size. At this point three variables ($A1L$, $A1W$, CW) were eliminated from further analysis, due to relatively high variability, which is likely the result of distortion during the slide-mounting procedure.

The remaining 49 variables were measured for samples of 20 adult aphids each for 19 locations throughout the range of *C. nigra* in the mid- to northern and mountainous regions of British Columbia and Alberta. Detailed collection data is given in Foottit and Mackauer (1990).

For each sample, I calculated descriptive statistics, a matrix of correlation coefficients and a principal components analysis, and inspected the data for univariate and multivariate trends. It was apparent from this comparative analysis that the correlation and covariation of many variables was not consistent among the 19 population samples. The principal component (PC) representing size, PC I, varied from 17.1% to 36.9% of total variation and there was considerable variability in the sign and magnitude of the component scores for the setal counts and setal length measurements on the first component. Projection of the individual specimens onto the principal axes for each sample showed a consistent orientation of specimens according to overall size along the first principal axis.

A similar analysis, but of the overall patterns of variation in *C. nigra*, was carried out in order to determine if the sample size that was used ($n = 20$) was large enough to sample adequately all of the major trends in variation. Correlation coefficients between all 49 variables for all 19 samples combined ($n = 380$) were calculated and the matrix was subjected to a PCA in order to determine the strongest, overall patterns of variability. The component scores for each variable on the first three PCs are given in Table 2.

The first principal component, the general-size component (36.3% of total variation), was characterized by high positive scores for the continuous body dimensions and setal length measurements (except *SLAT5*). Some setal counts (*SNA5*, *SNGP*, *SNS*, *SNT*) showed high scores on PC I; these variables were significantly correlated with body size as shown in the correlation analysis. The second principal component (7.0% of variation) was largely a dimension influenced by decreases in the magnitude of the setal length measurements. The third principal component (3.0% of variation) consisted of shape variation, particularly in the antennal and leg dimensions, as was shown in the analysis of the smaller samples; there were also some relatively high scores for some of the setal counts. No patterns among the variable scores were evident on the remaining principal components.

The main components of variation that were shown in the PCA of the large ($n = 380$) data set were also evident in the smaller ($n = 20$) samples. It is evident that the best approach towards the morphometric characterization of a species of *Cinara* is not to emphasize a large sample, but instead to emphasize some estimate of the geographic variation, whether it is stochastic variation or a recognizable geographic pattern.

This analysis showed that although there is a major component of size-related variation in samples of *C. nigra*, there are other components of morphological variation present as well. Some of this variation is in the form of shape changes, which may or may not be allometric. Some components of variation, particularly those represented by some of the setal counts, are independent of size variation. In addition, there are strong, internal correlations among character suites of related function (e.g., those of the hind leg) that are correlated with separate components of variation, and that represent adaptation to particular modes of living. Any or all of these sources of variation may prove useful in the establishment of species boundaries.

Table 2.
Contributions of 49 variables to the first three principal components calculated from 380 specimens of *Cinara nigra*. The relative percentage of variability for components I-III is 36.2%, 7.0% and 3.5%, respectively.

Variable	PC I	PC II	PC III	Variable	PC I	PC II	PC III
BL	0.884	-0.070	-0.030	*TS1W*	0.749	-0.014	-0.229
DE	0.818	0.047	0.051	*TS1VL*	0.820	0.135	0.087
DHC	0.649	0.038	-0.083	*TS1DL*	0.576	-0.002	0.014
FRW	0.726	0.119	-0.075	*TS2L*	0.814	0.053	0.054
HL	0.575	-0.119	-0.187	*TS2W*	0.583	-0.123	-0.453
A2L	0.714	0.234	0.124	*SLH*	0.280	-0.654	0.129
A2W	0.762	0.086	-0.192	*SLA3*	0.274	-0.544	0.075
A3L	0.831	0.203	-0.036	*SLT*	0.360	-0.544	0.068
A3W	0.831	0.184	-0.039	*SLTS2*	0.254	-0.568	0.026
A4L	0.827	0.056	0.112	*SLS*	0.396	-0.590	0.079
A4W	0.741	0.188	-0.085	*SLCA*	0.348	-0.537	0.137
A5L	0.778	0.119	0.230	*SLGP*	0.546	-0.510	0.044
A5W	0.586	0.124	-0.082	*SLAT8*	0.331	-0.659	0.151
A6BL	0.365	0.010	0.300	*SLAT5*	-0.127	-0.210	-0.260
A6BW	0.540	0.182	-0.061	*SNA6SA*	0.040	0.189	0.333
A6PTL	0.190	0.276	0.137	*SNA6B*	0.122	-0.067	0.225
R5L	0.336	-0.086	0.209	*SNA5*	0.522	0.103	0.204
R4L	0.737	0.023	-0.038	*SNA2*	0.132	0.103	0.491
R3L	0.581	0.045	0.204	*SNR4*	0.052	0.023	0.350
R2L	0.508	0.104	0.224	*SNGP*	0.625	0.071	0.127
CL	0.921	0.028	-0.029	*SNAT5*	-0.019	-0.169	0.191
FL	0.915	0.228	-0.007	*SNAT8*	0.265	-0.001	0.065
FW	0.757	-0.158	-0.343	*SNS*	0.628	0.052	0.124
TL	0.847	0.286	0.035	*SNT*	0.488	-0.212	-0.163
TW	0.797	-0.068	-0.297				

5 Geographic Variation: *Cinara nigra* (Wilson)

The 19 population samples were analyzed using multiple discriminant function analysis (DFA) (Foottit and Sorensen, this volume) in order to determine further the geographic patterns among the samples. Standardized discriminant function coefficients were calculated; these represented the contributions of the variables to discrimination among the samples on each discriminant axis. Sample centroids and specimens were projected onto the first two discriminant axes (Fig. 3).

Discriminant function I (34% of variation) was largely influenced by size variation; the highest contribution to this function was made by the variable "body length" (*BL*). Fig. 3 shows that the sample centroids are aligned along this axis, which represents general-size. The length of the femur (*FL*) contributed strongly to separation along the second axis. No geographic pattern was present; most of the samples overlapped with the other samples.

Similar trends were confirmed by a UPGMA cluster analysis (Sneath and Sokal, 1973) of the Mahalanobis' generalized distances (D statistic) (Mahalanobis, 1936; Sneath and Sokal, 1973) between groups, which was calculated for all pairwise comparisons of the 19 samples. The Mahalanobis' distances were derived from the discriminant functions (DF) and can be used

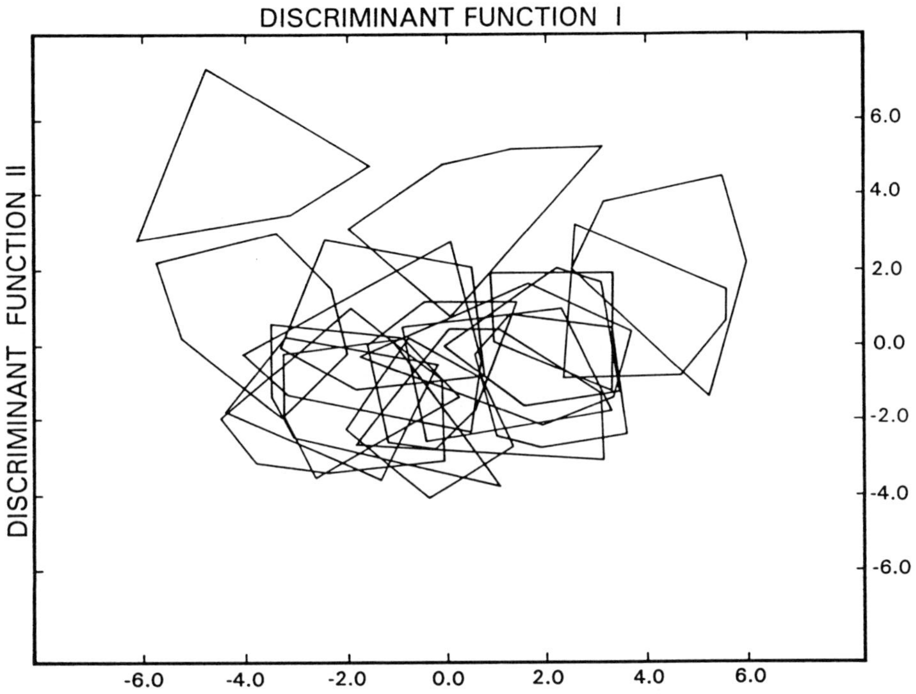

Figure 3. Nineteen population samples of *Cinara nigra* projected onto the first and second discriminant axes, based on the discriminant function analysis of 49 morphological variables. Polygons connect the outlying specimens and represent the distribution in discriminant space of each sample.

as a measure of statistical distance (dissimilarity) between any two groups. As was the result in the previous analysis, no geographic pattern in clusters of the samples was evident, except that some geographically adjacent samples were grouped together.

The lack of geographically defined clusters of the population samples (with the exception of some relatively localized sample pairs) could be due to a number of factors. However, most important is the overriding effect of size variation, which itself could be due to a number of nongenetic factors that are likely to be relatively localized. When a change in body size of an organism occurs, this in turn results in concomitant changes in many developmental and morphological relationships of an organism. Size and shape covary, so that there is a changing relationship between the two components of variation (Gould, 1966). Many of the variables, or functional groups of variables, that showed patterns of variation on the principal components subsequent to the main (size) component were probably due to allometric shape changes in response to size variation.

Other studies of the geographic variation of aphids have found a similar lack of geographic pattern. In their work on geographic variation in

Pemphigus populitransversus Riley, Sokal and Riska (1981) found a lack of demonstrable and interpretable pattern in the morphometric characters that were studied. From this work and the analysis of other *Pemphigus* species over much of the North American continent, they concluded that within a given locality, there was considerable stochastic variation from one year to the next (Sokal et al., 1980, 1991).

6 Character Reduction

On the basis of the analysis of *C. nigra* outlined above, the number of variables was reduced from 49 to 32 (Table 1). Variables that were highly correlated with other variables were eliminated, as were those that proved difficult to accurately measure or count. This gave a more manageable data set for the analysis of species differences without a great loss of information content.

The analysis of correlation and covariation patterns allowed for a thorough investigation of the relationships among the variables. Examination of the correlation coefficients of the characters and the principal components analyses of the samples revealed some important relationships among the characters. In addition, functional groups of characters, such as the antennal measurements and the hind leg dimensions, showed relationships of high internal correlation.

The reduced character set is still a relatively large one when compared to the character sets of other aphid morphometric studies. The exceptions have been that of Sokal and his colleagues (Sokal et al., 1980; Sokal and Riska, 1981), who have recommended the use of a large number of variables, and Sorensen (1983, in press), who used up to 60 and 53 morphometric traits, respectively. Sorensen (1983) concluded that among closely related species, increased PCA resolution in multivariate space could be obtained by using a reduced subset of traits that loaded dominantly in the first few PCs. Thus, it is important to include characters that sample as many areas of the body and components of variation as possible, for initial maximum information content. This is particularly important in the early stages of a biosystematic study of a group. The data set can be reduced, particularly for practical reasons, on an objective basis as the relationships among the variables become known.

The results of this study support the claim (Pimentel, 1979; Neff and Marcus, 1980) that principal components analysis is a useful dimension reducing technique. Common components of variation were revealed throughout the samples of *C. nigra* and the relationships of the functional groups of characters were shown to vary among the samples. Although the largest component of variation was size variation, there were other, some even independent, components of variation in the morphology of *C. nigra*. These results indicated that a species of aphid, even when it is characterized using techniques of multivariate analysis, has to be represented on the basis of geographical samples and on the basis of characters selected from a wide range of the holomorph. This suggests that geographically and temporally localized suites of traits may be responding separately, or differently, to locally imposed selection gradients (sensu Sorensen and Foottit, this volume; Sorensen, 1991, this volume, in press).

7 Morphometric Variation Patterns Among *Cinara* Species

There have been few studies of the morphometrics of taxonomic discrimination among species of aphids. Previous work has involved discrimination only between pairs of species or between the morphs of the same species (e.g., Blackman et al., 1977; Hand, 1986). The purpose of the next phase of this analysis was to characterize morphometric trends among a number of *Cinara* species, and to test the ability of the previously identified character set to discriminate between nine species of *Cinara* associated with *P. contorta*, *P. monticola* and *P. ponderosa*.

Samples of each putative species to be analyzed were selected after preliminary identification of a few specimens from each available field collection. The nine species selected for analysis were: *C. brevispinosa* (Gillette and Palmer), *C. contortae* Hottes, *C. medispinosa* (Gillette and Palmer), *C. murrayanae* (Gillette and Palmer), *C. nigra*, *C. parvicornis* Hottes, and *C. pergandei* (Wilson), all from *P. contorta*; *C. ponderosae* (Williams), from *P. ponderosa*; and *C. kuchea* Hottes, from *P. monticola*. Samples were selected so as to include some geographic variation in the material to be studied, and, where possible, to include differences in feeding sites among the samples of the species.

Some taxa represented in the analysis were species that are comparatively well defined, particularly *C. nigra*, *C. pergandei*, and *C. brevispinosa*. These species would serve as useful comparisons or "controls" against which the discriminatory ability of the variables could be tested. In addition, the morphometric variability patterns of these relatively well defined species would serve as reference points against which the variability patterns of the species with doubtful boundaries could be compared.

Instead of representing each species by one sample, I grouped five subsamples of five specimens each under each species. This approach allowed for the inclusion of some of the geographic heterogeneity in body size and associated features because this factor would probably have a substantial impact on the species discrimination process. The specimens were measured for the 32 morphological variables, as previously identified (Table 1). Collection data is available from the author.

Each sample was analyzed in the same manner as were the samples of *C. nigra*. A matrix of correlation coefficients was calculated for the 32 variables and subjected to a PCA. Tables of the component scores for the variables on the first three principal components are provided for three species (*C. pergandei*, Table 3; *C. kuchea*, Table 4; *C. brevispinosa*, Table 5) that best illustrate the main trends in morphometric variation among the species. Ordination plots of the first three principal axes for two species (*C. kuchea*, *C. brevispinosa*) (Fig's., 4-6) are provided and illustrate the extremes in the spatial resolution of the specimens and subsamples.

In each of the nine species, the main component was a size component (19.7% to 49.3% of variation). In general, all of the variables that were measured for all samples exhibited strong, positive contributions to this component of variation. The exception was *C. pergandei*, which had a relatively small first component of variation; many variables were negatively correlated with this axis of variation.

The second principal component (9.7% to 21.1% of variation) consisted largely of contrasts of the sign of the antennal dimensions, a reduction of the

Table 3.
Contributions of 32 morphological variables to the first three principal components calculated from 25 specimens of *Cinara pergandei*. The relative percentage of variability for components I-III is 19.7%, 14.0% and 11.1%, respectively.

Variable	PC I	PC II	PC III
BL	0.637	-0.107	0.263
FRW	0.602	0.162	-0.230
A2L	0.135	0.494	0.367
A3L	0.791	-0.066	0.253
A4L	0.396	0.775	0.191
A5L	0.551	0.719	0.229
A6BL	0.523	0.562	-0.257
A6BW	0.461	-0.123	-0.383
A6PTL	0.592	-0.168	-0.100
R5L	0.313	-0.131	-0.019
R4L	0.662	-0.112	-0.215
R3L	0.102	0.532	-0.403
R2L	-0.103	0.014	0.166
FL	0.693	-0.075	-0.277
F W	-0.188	-0.063	-0.612
TL	0.681	0.219	0.045
TS1VL	0.536	-0.011	-0.037
TS2L	0.769	0.033	0.114
SLA3	-0.429	0.417	-0.418
SLGP	0.088	0.229	-0.720
SLAT5	-0.149	0.483	-0.429
SLT	-0.238	0.630	-0.475
SNA6SA	-0.093	-0.008	0.305
SNA6B	0.420	0.377	0.015
SNA5	0.096	0.147	0.483
SNA2	-0.300	0.350	0.311
SNR4	-0.199	-0.089	-0.377
SNGP	-0.405	0.240	0.100
SNAT5	-0.567	0.479	0.257
SNAT8	-0.131	0.417	0.414
SNS	-0.137	0.609	0.438
SNT	-0.254	0.591	-0.301

rostrum segment lengths, and a reduction of the hind leg dimensions in many samples. The third principal component (7.7% to 17.3% of variation) was largely an expression of the contrasts of functional groups of characters and a trend towards a reduction in the numbers of setae.

Each species was characterized by a unique composition of the scores of the functional groups of characters. For example, variation among the antennal segments was of a number of characteristic types. In some species, such as *C.*

Table 4.
Contributions of 32 morphological variables to the first three principal components calculated from 25 specimens of *Cinara kuchea*. The relative percentage of variability for components I-III is 38.7%, 11.0% and 8.5%, respectively.

Variable	PC I	PC II	PC III
BL	0.784	-0.130	0.048
FRW	0.567	-0.183	0.262
A2L	0.876	0.072	-0.017
A3L	0.889	0.098	-0.128
A4L	0.726	0.023	-0.261
A5L	0.667	0.405	0.097
A6BL	0.763	0.283	-0.121
A6BW	-0.083	-0.309	0.358
A6PTL	-0.130	-0.673	0.180
R5L	0.418	-0.351	-0.498
R4L	0.269	-0.139	0.280
R3L	0.545	-0.531	0.018
R2L	0.444	0.498	-0.235
FL	0.948	-0.055	-0.038
FW	0.887	-0.107	0.042
TL	0.880	0.109	-0.066
TS1VL	0.774	0.012	0.258
TS2L	0.815	-0.213	0.375
SLA3	0.594	0.092	-0.653
SLGP	0.764	0.206	-0.178
SLAT5	0.642	-0.097	-0.584
SLT	0.360	-0.464	-0.454
SNA6SA	0.246	0.463	0.433
SNA6B	0.205	0.665	0.200
SNA5	0.747	0.389	0.241
SNA2	0.088	0.346	-0.107
SNR4	0.159	0.038	0.210
SNGP	0.794	-0.178	0.101
SNAT5	0.501	-0.473	0.308
SNAT8	0.348	-0.511	0.126
SNS	0.650	-0.349	0.271
SNT	0.497	0.264	0.497

pergandei and *C. contortae*, the entire set of antennal segment measurements increased with overall size, but in others, such as *C. brevispinosa* and *C. kuchea*, some antennal segments decreased in relation to the size factor. On the second component, the proximal antennal segments decreased in relation to the sixth antennal segment in some species, but the opposite occurred in other species. Although the response of the rostrum to overall size variation was consistent in all species except *C. pergandei* (Table 3), the segments of the

Table 5.
Contributions of 32 morphological variables to the first three principal components calculated from 25 specimens of *Cinara brevispinosa*. The relative percentage of variability for components I-III is 27.4%, 19.8% and 12.2%, respectively.

Variable	PC I	PC II	PC III
BL	0.827	-0.475	0.039
FRW	0.818	-0.406	0.018
A2L	0.565	-0.033	-0.350
A3L	0.852	0.120	0.230
A4L	0.031	0.896	0.062
A5L	-0.042	0.943	0.102
A6BL	0.450	-0.337	-0.627
A6BW	-0.278	-0.071	-0.471
A6PTL	0.014	0.869	0.273
R5L	0.196	0.146	-0.540
R4L	0.377	-0.022	-0.621
R3L	0.729	0.298	-0.269
R2L	0.631	-0.332	0.253
FL	0.910	0.294	0.135
F W	0.635	0.155	-0.294
TL	0.807	0.377	0.116
TS1VL	0.508	0.666	0.213
TS2L	0.636	0.551	-0.177
SLA3	0.238	0.172	0.591
SLGP	0.738	-0.161	0.230
SLAT5	0.691	0.382	0.016
SLT	0.622	0.040	0.137
SNA6SA	-0.031	0.258	0.058
SNA6B	0.172	0.374	-0.317
SNA5	-0.112	0.817	0.106
SNA2	-0.100	0.223	-0.592
SNR4	-0.041	0.489	-0.507
SNGP	0.423	-0.094	-0.219
SNAT5	0.536	-0.555	-0.295
SNAT8	0.530	-0.462	-0.071
SNS	0.326	-0.206	0.618
SNT	-0.213	0.463	-0.583

rostrum exhibited a complex pattern of variation on the second and third components. Some species, such as *C. murrayanae*, were characterized by negative loadings on all segments. In other species there was a mixture of responses among the segments of the rostrum, indicating that each segment has a unique pattern of variation.

The dimensions of the hind leg were also marked by complex variation patterns. In most species a complete positive response to overall size by all

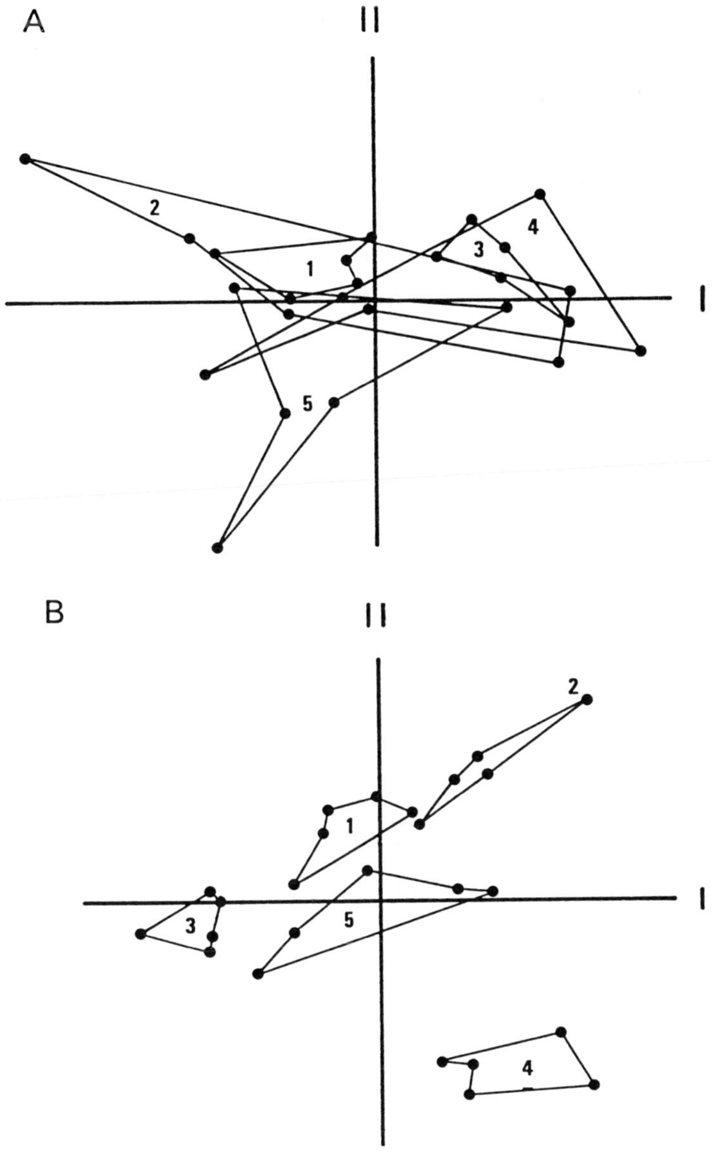

Figure 4. Diagrams showing the principal component ordination of 25 specimens of *Cinara kuchea* (A) and 25 specimens of *Cinara brevispinosa* (B) projected onto principal axes I and II, based on the analysis of 32 morphological variables (Tables 4 and 5, respectively). Polygons connect the specimens of each subsample; subsamples are numbered 1-5.

hind leg dimensions was evident. As was the case with the antennal and rostrum segments, the scores on the second and third components shown by the segments of the tarsus were varied.

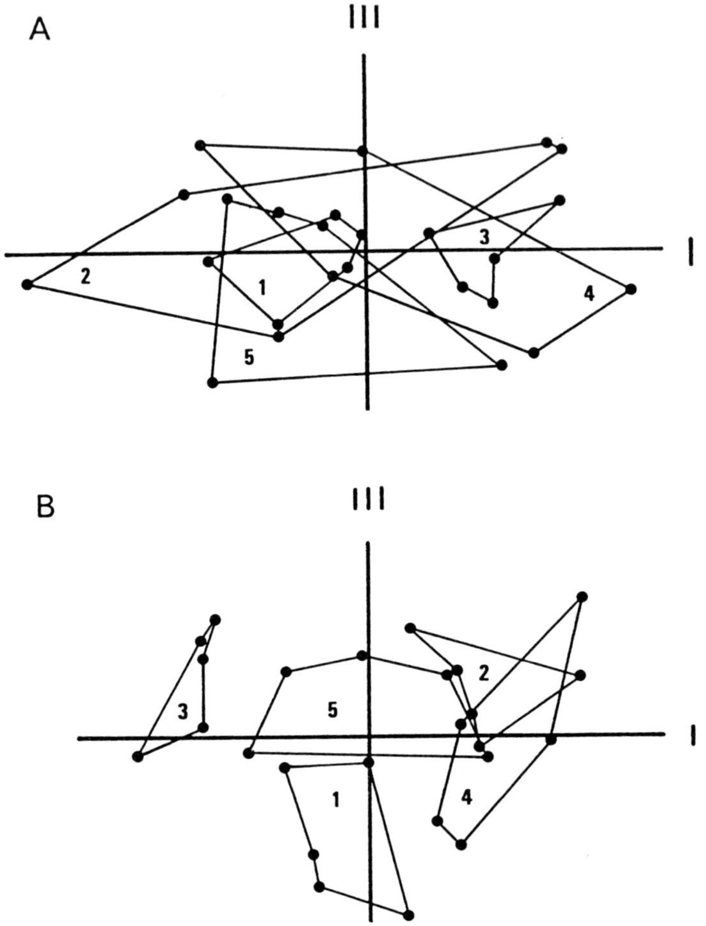

Figure 5. Diagrams showing the principal component ordination of 25 specimens of *Cinara kuchea* (A) and 25 specimens of *Cinara brevispinosa* (B) projected onto principal axes I and III, based on the analysis of 32 morphological variables (Tables 4 and 5, respectively). Polygons connect the specimens of each subsample; subsamples are numbered 1-5.

With the exception of *C. pergandei* (Table 3), the setal length and setal count variables showed a positive correlation with overall size on the first component. These characters showed a varied pattern of variation on the second and third components. However, where these variables are associated with functional areas of the aphid body, they showed trends as a group. For example, variables involving the number of setae on the abdominal area responded as a group in some species such as *C. kuchea* and *C. brevispinosa*. The setal numbers on the various antennal segments responded (decreases or increases) as a group in many species, on both the second and third components.

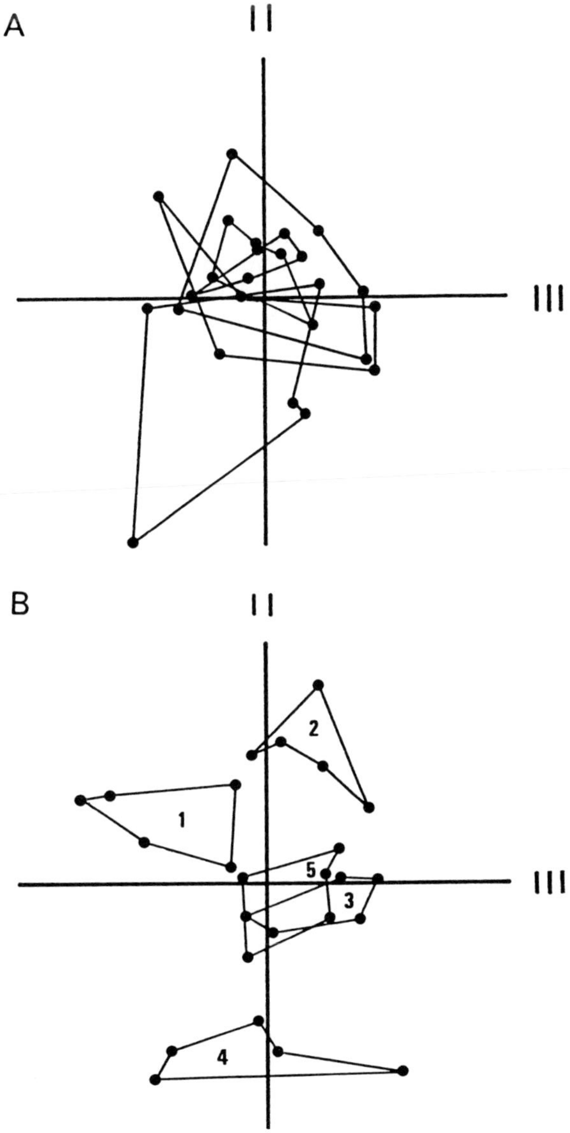

Figure 6. Diagrams showing the principal component ordination of 25 specimens of *Cinara kuchea* (A) and 25 specimens of *Cinara brevispinosa* (B) projected onto principal axes II and III, based on the analysis of 32 morphological variables (Tables 4 and 5, respectively). Polygons connect the specimens of each subsample; subsamples are numbered 1-5.

There was no consistent separation of the subsamples on the basis of geographic position or differences in feeding site. The majority of the subsamples were separated on the basis of overall size. Some species, such as

C. brevispinosa, showed an almost complete separation of the subsamples along all three principal axes of variation. These species were characterized by relatively large second and third principal components (Table 5), indicating that these species have a relatively larger proportion of morphometric variation that is not completely related to size; these species may be useful for the study of geographic variation. Other species, such as *C. kuchea*, showed considerable overlap of their subsamples (Fig's. 4-6), even along the size component axis. This indicates that morphological variation in these species is heavily influenced by factors affecting size and that these species are relatively size-variable from location to location, as was *C. nigra*.

Each of the three species for which PCA data is presented showed unique morphometric characteristics. *Cinara brevispinosa* (Table 5; Fig's. 4-6) was characterized by a relatively moderate size component (PC I, 27.4% of variation). The continuous measurements, including the setal length measurements, showed high, positive contributions to this component, with the exception of the dimensions of antennal segments V and VI. The setal counts on the antennal segments showed weak to negative loadings on this axis, indicating that variation in the antennal variables is partially independent of the general-size factor in this species. Subsamples were not only orientated along PC I according to overall size but they were also orientated along PC II and PC III. These components showed numerous contrasts of sign and magnitude of loadings within functional groups of variables. There are marked trends with respect to the setal counts on the abdomen (*SNGP*, *SNAT5*, *SNAT8*, *SNS*) on PC II and the rostrum dimensions on PC III.

Cinara kuchea exhibited a moderate size component (PC I, 38.7% of variation) with positive contributions from all variables except *A6BW* and *A6PTL*. The second component consisted of changes in the dimensions of *R5L*, *R4L* and *R3L* in relation to *R2L*, a reduction in the number of setae on the abdominal structures and an increase in the number of setae on the antennal segments. PC III represented a reduction in setal length variables.

Cinara pergandei exhibited patterns of variation that were different from those of the other species. PC I accounted for a relatively small amount (19.7%) of the total variation and it did not consist of loadings of uniform sign. PC II consisted of contrasts of sign of the dimensions of the antenna, hind legs and rostrum, indicating changes in shape. The third component was strongly influenced by a decrease in all setal length measurements. In this species, some of the other relatively minor components showed distinct patterns of variation. The uniqueness of this species will be demonstrated further in the next section.

8 Morphometric Discrimination Among Samples and Species

8.1 Samples of *Cinara contortae* Hottes

Among the geographic samples of *Cinara* species, I detected some morphological variation in the taxonomic characters previously used to distinguish between the species. For example, there were difficulties in separating all samples of *C. medispinosa* and *C. murrayanae* on the basis of the descriptive literature and reference material in collections. Among the material of another species, *C. contortae*, there were samples that differed with respect to certain characteristics, including size and pigmentation, to the extent that is was questionable whether or not these were the same species. Therefore, I grouped representative material into three categories: (a) "typical' *C. contortae*, (b) "small, thin" *C. contortae*, and (c) *C. contortae* with reduced dorsal pigmentation.

The three samples of *C. contortae* were analyzed using DFA (Table 6, Fig. 7A) in order to determine if they represented discrete taxa or morphological variants of the one species. DF I (55% of variation) was largely influenced by size variation; the largest contribution to discrimination along this axis was by the variable *BL*. Other strong contributions were by those variables associated with the antenna (*A4L*, *A5L*, *A6BL* and *SLA3*), the tibia and tarsus (*TL*, *TS1VL*, *TSVL*) and the abdomen (*SLGP*, *SLAT5*, *SNAT8*). Discrimination along this axis separated the 'typical' specimens from the "small, thin" specimens.

The second discriminant function (45.0% of variation) allowed for the almost complete separation of the "typical" specimens from those characterized by "reduced pigmentation." Nearly one-half of the variables were strong contributors to discrimination along this axis; these variables came from all of the functional and operational groups of measurements, with the exception of the lack of strong contributions from the setal length measurements. The strongest contributions were provided by the antennal segment dimensions, particularly *A5L* and the setal count taken on this segment (*SNA5*), some of the hind leg dimensions, *R3L*, and some of the setal counts taken on the abdomen (*SNAT5*, *SNAT8*, *SNS*).

I calculated a correlation matrix, based on the combined specimens ($n = 75$), and subjected this to a PCA. The results of this analysis showed that nearly one half (47.2%) of the variation in these combined samples was variation related to size differences. All of the variables, with the exception of *SNT*, were found to be significantly correlated ($P = 0.01$) with size, indicating a general-size component.

In order to further investigate the relationships among the variables and samples of *C. contortae*, I analyzed the same data set using Burnaby's method for size adjustment (Burnaby, 1966; Rohlf and Bookstein, 1988), using NT-SYS (Rohlf, 1990). This method adjusts the variables to remove the contribution of the first principal component and projects the specimens onto a space that is orthogonal to a size vector. The size-adjusted data was then analyzed by using DFA and the specimens projected onto the discriminant axes (Fig. 7B). When compared to the unadjusted data ordination (Fig. 7A), it can be seen that there is considerable overlap of the samples when size is removed, indicating that much of the apparent discrimination among the samples is due to size differences.

The importance of size in discrimination between closely related groups is also shown by Daly (this volume). In fact, Sorensen (1991) suggests that because of evolutionary energy constraints among very closely related biological entities (e.g., intraspecific populations or species complexes) that share very similar genetic covariance matrices and, therefore, respond the most similarly to imposed selection, size variance may be the last and easiest variance component to be differentiated during divergence among members. He implies that shape divergence most probably occurs earlier in phylogenetic lineages, to differentiate somewhat more distantly related groups (e.g. species-

Table 6.
Standardized discriminant function coefficients for two discriminant functions calculated from 32 morphological variables measured from three samples (n = 3x25) of *C. contortae*. The relative percentage of variability for functions I and II is 55.0% and 45.0%, respectively.

Variable	DF I	DF II
BL	1.253	-0.505
FRW	0.348	-0.346
A2L	-0.203	0.143
A3L	0.163	-0.899
A4L	-1.011	0.965
A5L	-0.791	-1.754
A6BL	0.699	0.576
A6BW	-0.154	-0.471
A6PTL	0.030	-0.994
R5L	0.306	0.349
R4L	-0.028	-0.317
R3L	-0.243	1.201
R2L	0.286	0.215
FL	-0.132	0.165
F W	0.266	1.502
TL	0.521	1.565
TS1VL	0.511	0.274
TS2L	-0.534	-2.074
SLA3	0.823	0.196
SLGP	0.487	0.306
SLAT5	-0.687	-0.037
SLT	-0.190	0.118
SNA6SA	0.018	-0.390
SNA6B	-0.036	0.540
SNA5	-0.466	1.074
SNA2	0.034	-0.022
SNR4	-0.348	0.089
SNGP	-0.256	0.118
SNAT5	-0.079	-0.556
SNAT8	0.720	-0.914
SNS	-0.015	-0.549
SNT	0.182	0.489

groups, subgenera, genera, etc.) that have less similar genetic covariance matrices. If this is so, then the removal of size-related variance may be inappropriate, or at least problematic, during phylogenetic analyses, particularly when dealing with phylogenetic hypotheses for species level taxa.

From the above analyses, I concluded that all three samples represented the taxon *C. contortae*. Although there was separation of the samples, all of the variables that discriminated the most between the samples were strongly

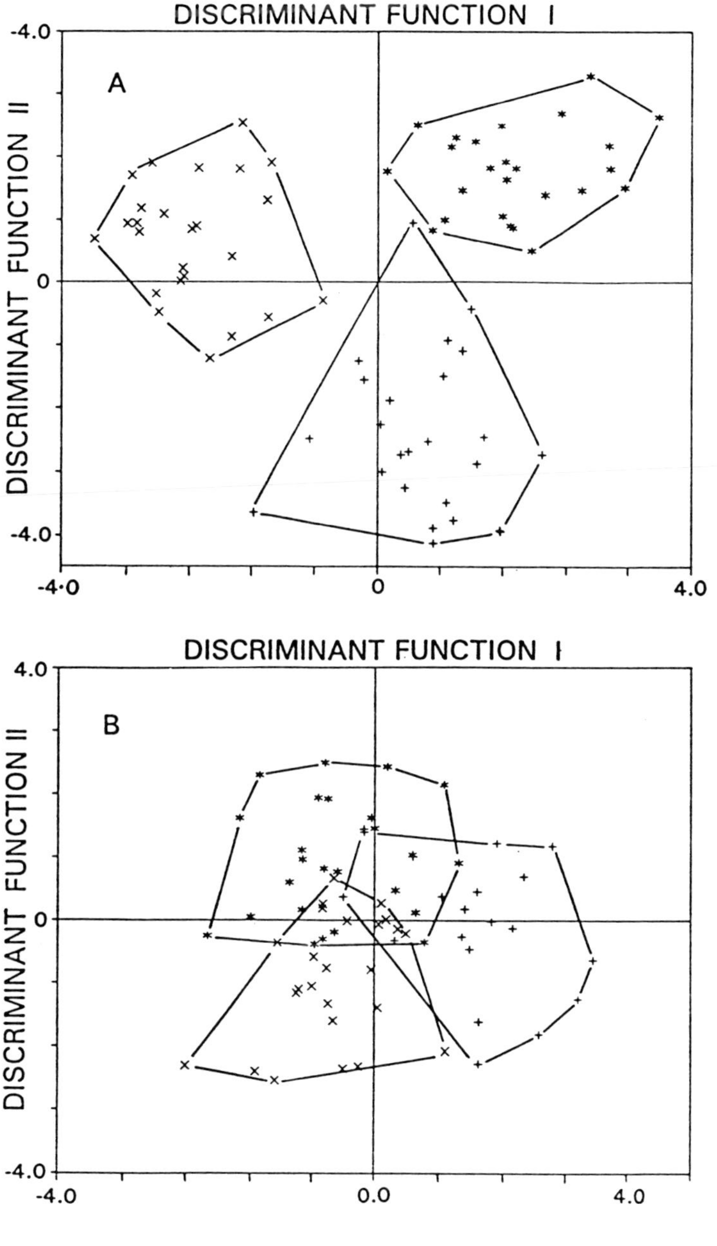

Figure 7. Specimens of three samples (n = 3x25) of *C. contortae* projected onto the first and second discriminant axes. (A) Based on the discriminant function analysis of 32 morphological variables (Table 6). (B) Based on the analysis of 22 continuous morphological variables, adjusted for size variation by Burnaby's method (Burnaby, 1966). * = *C. contortae* "typical"; x = *C. contortae* "small, thin"; + = *C. contortae* "reduced pigmentation."

influenced by size variation. Neither the "small, thin" sample nor the "reduced pigmentation" samples showed character states that were unique.

8.2 Species of *Cinara*

Samples representing the nine species of *Cinara* that were studied were analyzed using DFA (Table 7, Fig. 8A). The only clear pattern to arise from this analysis is the separation of *C. pergandei* from the other samples. This species has been placed in the subgenus *Cinarella* Hille Ris Lambers (Hille Ris Lambers, 1948; Eastop, 1972, 1976) on the basis of the presence of stalked eyes in the apterae and elongate tarsal segments. It is clear from this analysis that *C. pergandei* has unique morphometric characteristics when compared to the other species of *Cinara*. In addition to the tarsal segment character *TS1VL*, this species was separated from the other species by strong contributions from *FRW*, *A6BL*, *A6PTL*, *R4L*, *R3L*, *FL* and *SNAT5* (Table 7).

The presence of *C. pergandei* in the above analysis so altered the covariance matrix that there was no separation of the other *Cinara* species; the analysis simply determined differences between *C. pergandei* and all of the other species as a group. This global-view distortion of local differentiation among the more closely related *Cinara* reflects the local-global perspective transition problems discussed by Sorensen (this volume). Therefore, *C. pergandei* was removed and the remaining eight species were analyzed by DFA (Table 8, Fig. 8B).

The species *C. kuchea*, *C. medispinosa* and *C. murrayanae* were separated from the other species along the first discriminant axis. Separation of the species along this axis was the result of contributions to discrimination from a number of variables. The main contributors to this separation were *SLAT5*, changes in setal numbers (*SNAT8*, *SNS*, *SNT*), and dimensional changes in the sixth antennal segment (*A6BW*, *A6PTL*), the rostrum (*R4L*) and the hind leg (*FL*, *TL*). Separation of the species along the second discriminant axis was largely due to changes in the lengths of the rostrum segments (*R5L*, *R4L*, *R3L*), overall aphid size (*BL*) and dimensional changes (*A2L*, *TS1VL*) and setal length and setal count changes. No single character served to discriminate between all of the species, in fact separation along the first and second axis was the result of contributions from at least half of the variables in the data set.

These analyses showed that the range of characters that was selected was important to the successful establishment of the boundaries between the species that were studied. The polythetic nature (Sneath and Sokal, 1973) of the *Cinara* species was also evident. In all of the analyses, no single character served to completely discriminate among the species. Often, for any one analysis, it was found that as many as one half of the 32 characters were shown to be strong contributors to discrimination.

9 Higher Categories

The aphid family Lachnidae and particularly the tribe Cinarini Börner is a useful test for the application of ordination analysis to the solution of taxonomic problems because there has been considerable difference of opinion among taxonomists about not only the discrimination of taxa at the subspecies,

Table 7.
Standardized discriminant function coefficients for the first two discriminant functions calculated from 32 morphological variables measured from 9 species of *Cinara* (*C. pergandei* included). The relative percentage of variability for functions I and II is 68.6% and 13.8%, respectively.

Variable	DF I	DF II
BL	-0.017	0.062
FRW	0.593	-0.025
A2L	-0.078	0.072
A3L	0.110	0.299
A4L	-0.294	-0.223
A5L	-0.008	-0.037
A6BL	0.702	-0.127
A6BW	-0.176	0.011
A6PTL	0.542	-0.077
R5L	0.330	-0.231
R4L	-0.440	-0.400
R3L	-0.509	-0.204
R2L	-0.156	-0.014
FL	-0.407	0.386
FW	0.147	-0.119
TL	-0.427	-0.342
TS1VL	0.803	-0.252
TS2L	0.044	0.135
SLA3	0.180	0.211
SLGP	-0.099	0.029
SLAT5	0.065	0.713
SLT	0.229	-0.162
SNA6SA	0.130	-0.131
SNA6B	-0.385	0.382
SNA5	-0.316	-0.070
SNA2	0.044	0.272
SNR4	0.031	0.327
SNGP	0.114	-0.484
SNAT5	0.484	-0.037
SNAT8	-0.072	0.454
SNS	0.259	0.086
SNT	-0.111	0.340

species, subgenus, and genus levels but also of the phylogenetic placement of the Lachnidae, as a whole, within the Aphidoidea. Evolutionary trends in the Aphidoidea are difficult to determine because of uncertainties about character polarity. Some morphological features, in fact , may be plesiomorphic, but it is apparent that some morphological character states have reoccurred during the evolution of aphids. Eastop (1973) has suggested that aphids, with a

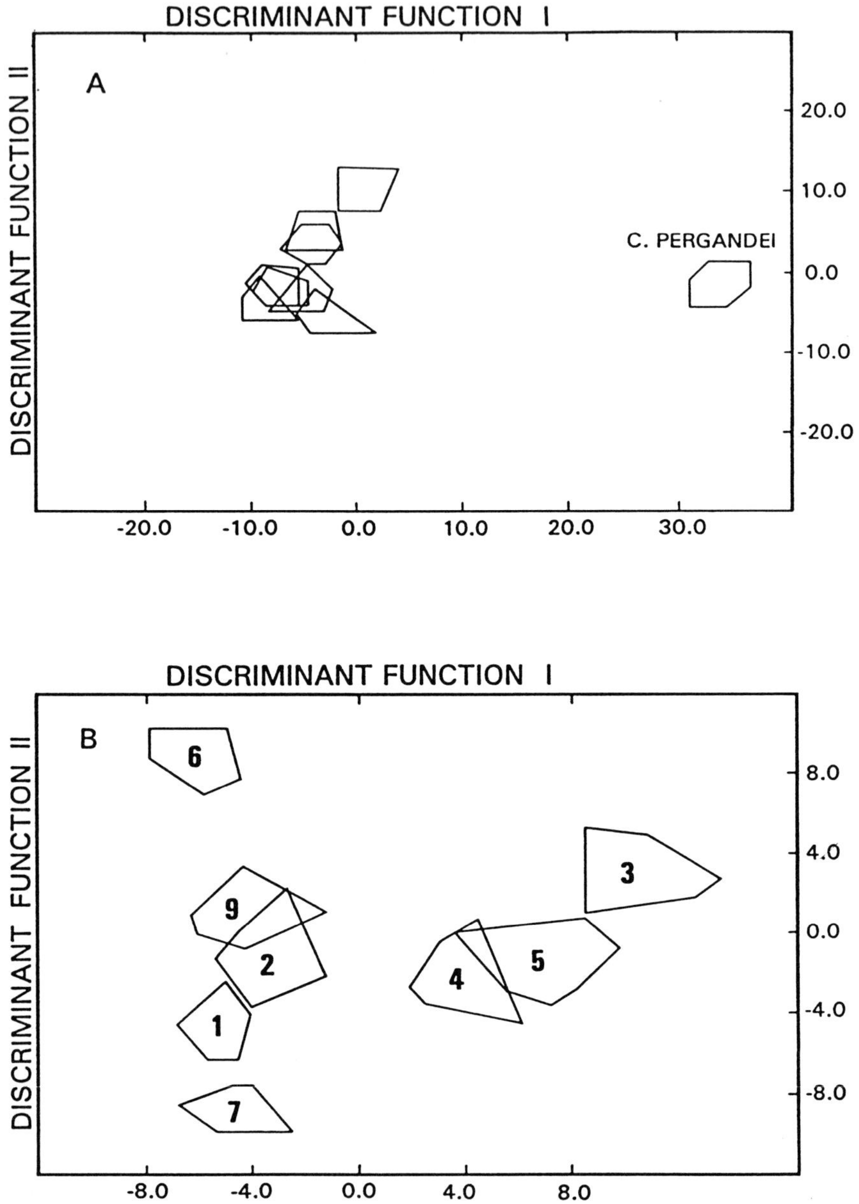

Figure 8. (A) Samples of 9 species of *Cinara* projected onto the first and second discriminant function axes, based on the analysis of 32 morphological variables (Table 7). (B) Samples of 8 species of *Cinara* projected onto the first and second discriminant function axes, based on the analysis of 32 morphological variables (Table 8). Specimens of each species are enclosed by a polygon. 1 = *C. brevispinosa*; 2 = *C. contortae*; 3 = *C. kuchea*; 4 = *C. medispinosa*; 5 = *C. murrayanae*; 6 = *C. nigra*; 7 = *C. parvicornis*; 8 = *C. pergandei*; 9 = *C. ponderosae*.

Table 8.
Standardized discriminant function coefficients for the first two discriminant functions calculated from 32 morphological variables measured for 8 species of *Cinara* (*C. pergandei* excluded). The relative percentage of variability for functions I and II is 52.5% and 25.6%, respectively.

Variable	DF I	DF II
BL	-0.189	0.525
FRW	-0.373	-0.106
A2L	0.238	0.474
A3L	0.234	-0.181
A4L	0.137	-0.161
A5L	0.173	-0.154
A6BL	0.060	-0.217
A6BW	0.295	-0.099
A6PTL	-0.320	-0.134
R5L	0.063	-0.473
R4L	-0.297	-0.750
R3L	-0.011	0.815
R2L	0.182	0.090
FL	-0.415	0.253
F W	-0.129	-0.166
TL	-0.402	-0.165
TS1VL	0.041	-0.536
TS2L	0.225	0.179
SLA3	-0.026	0.793
SLGP	0.001	-0.311
SLAT5	1.059	-0.476
SLT	0.012	0.283
SNA6SA	-0.138	0.328
SNA6B	0.165	0.307
SNA5	0.142	0.253
SNA2	-0.068	-0.008
SNR4	0.218	0.357
SNGP	0.048	0.087
SNAT5	0.061	-0.178
SNAT8	-0.353	-0.164
SNS	0.604	0.108
SNT	-0.369	-0.250

succession of morphologically distinct, parthenogenetic, generations, likely possess genetic mechanisms that result in the appearance of characteristics that may have been evolutionarily dormant for a period of time. It is evident that some aphid groups show a trend towards progressive neoteny so that relatively young aphid groups have lost some advanced adult characteristics (Richards, 1965). Many examples of convergent evolution are evident

(Heie, 1967) and parallel evolution may have resulted in a mosaic distribution of apomorphic characters (Heie, 1980). These processes could explain some of the notorious difficulties involved in the study of aphid taxonomy.

A number of European aphid taxonomists, particularly C. Börner (1952) have erected new genera and subgenera within the tribe Cinarini or have adopted these taxonomic subdivisions. Some North American taxonomists consider Börner's generic subdivisions to be more useful as subgeneric categories (Hottes, 1960). Others have maintained that application of Börner's system to the North American species of *Cinara* would result in their being divided into more taxonomically manageable groups. However, in their *Survey of the World's Aphids*, Eastop and Hille Ris Lambers (1976) placed all of the genera and subgenera of Börner and others into synonymy with *Cinara* Curtis.

There are some obvious morphological differences between the genera and subgenera in Börner's classification of the Cinarini. The host plant relationships of the genera and subgenera are generally consistent. Some of the characters used to separate the subgenera, such as the setal lengths, rostrum lengths and hind leg dimensions, have been shown by the present study to be variable in some species of *Cinara*.

There is enough morphological and ecological evidence to warrant a second look at Börner's system; perhaps modification of the species groups that he originally delineated, and the levels in the Linnaean hierarchy at which he set them, should be attempted. Many of the problems in applying Börner's system occur when one attempts to rely on univariate differences between the taxa. Analysis of the species groups within the Cinarini by methods of ordination analysis would enable the determination of the internal variability of taxonomic characters within the genera and subgenera, assessment of the discriminatory ability of these characters, and allow for the detection of patterns of variation and gaps in this variation that would enable the objective construction of a balanced taxonomic hierarchy. These methods would provide a more objective approach to the determination of character polarity in the phylogenetic reconstruction of the group.

10 **Conclusions**

A number of potentially useful taxonomic characters were identified, such as the dimensions of the sixth antennal segment, the measurements of the segments of the rostrum, and many of the setal counts. Many of these characters were shown to be of taxonomic value at the species level. The use of a multivariate morphometric approach in this study allowed for the quantitative assessment of taxonomic discrimination and for the development of a clearer definition of the species boundaries than has been previously available for *Cinara*.

On the basis of this study, it is evident that a wide range of morphological characters from all areas of the aphid body is necessary in order to establish species boundaries. In addition, geographic sampling is necessary in order to include the spatial variation inherent in these insects.

The taxonomist's judgement is still an integral part of the process of classification. However, methods of multivariate morphometrics are useful in simply sorting the large amounts of taxonomic information inherent in studies of species and generic relationships into manageable patterns. Because aphids are notorious for their morphological variability, it stands to reason that the use of this variability as taxonomic data will prove worthwhile. This is particularly evident because the recognition of the degree of internal variation and the extent of the gaps in this character variability is the essence of the process of establishing taxa.

11 Acknowledgments

I would like to thank Mr. E. Maw for assistance with some of the computational aspects of this work and Dr. J. Sorensen for many useful discussions about the applications of ordination methods in taxonomic analysis.

12 References

Blackman, R.L., Eastop, V.F. and Hills, M., 1977. Morphological and cytological separation of *Amphorophora* Buckton (Homoptera: Aphididae) feeding on European raspberry and blackberry (*Rubus* spp.). *Bulletin of Entomological Research*, 67: 286-296.

Börner, C., 1952. Europae Centralis Aphides. Die Blattlause Mitteleuropas. Namen, Synonyme, Wirtspflanzen, Generationszyklen. *Mitteilungen der Thåringischen Botanischen Gesellschaft*, 3(1): 1-259.

Bradley, G.A., 1959. Feeding sites of aphids of the genus *Cinara* Curtis (Homoptera: Aphididae) in northwestern Ontario. *Canadian Entomologist*, 91: 670-671.

Bradley, G.A., 1961. A Study of the Systematics and Biology of Aphids of the Genus *Cinara* Curtis in Canada. Ph.D. Thesis, McGill University, Montreal, 110 pp.

Burnaby, T.P., 1966. Growth-invariant discriminant functions and generalized distances. *Biometrics*, 22: 96-110.

Daly, H.V., (this volume). A statistical and empirical evaluation of some morphological variables of honey bee classification.

Dunn, G. and Everitt, B.S., 1982. An Introduction to Mathematical Taxonomy. Cambridge University Press, Cambridge, U.K., 152 pp.

Eastop, V.F., 1972. A taxonomic review of the species of *Cinara* Curtis occurring in Britain (Hemiptera: Aphididae). *Bulletin of the British Museum Natural History (Entomology)*, 27: 103-186.

Eastop, V.F., 1973. Deductions from the present day host plants of aphids and related insects. In: H.F. Van Emden (Editor), Insect/Plant Relationships. *Symposia of the Royal Entomological Society of London*, 6, pp. 157-178.

Eastop, V.F., 1976. A review of *Cinara* subgenus *Cinarella* (Hemiptera: Aphididae). *Bulletin of the British Museum Natural History (Entomology)*, 35: 1-23.

Eastop, V.F. and Hille Ris Lambers, D., 1976. Survey of the World's Aphids. Dr. W. Junk bv Publishers, The Hague, 573 pp.

Fedde, G.F., 1965. The Pine Aphid Complex of the Genus *Cinara* in the Appalachian Highlands Region of South Carolina. M.Sc. Thesis,

Clemson University, Clemson, South Carolina, 143 pp.

Fedde, G.F., 1967. Comparative Analyses of Systematic Procedures and Biological Characteristics Relating to Pine Aphids of the Genus *Cinara* of Western South Carolina. Ph.D. Thesis, Clemson University, Clemson, South Carolina, 155 pp.

Foottit, R.G. and Sorensen, J.T., (this volume). Ordination methods: their contrast to clustering and cladistic techniques.

Foottit, R.G. and Mackauer, M., 1990. Morphometric variation within and between populations of the pine aphid, *Cinara nigra* (Wilson) (Homoptera: Aphidoidea: Lachnidae), in western North America. *Canadian Journal of Zoology*, 68: 1410-1419.

Gould, S.J., 1966. Allometry and size in ontogeny and phylogeny. *Biological Reviews of the Cambridge Philosophical Society*, 41: 587-640.

Hand, S.C., 1986. The use of multivariate morphometric methods in the separation of alate morphs of the rose-grain aphid, *Metopolophium dirhodum*. *Annals of Applied Biology*, 109: 19-31.

Heie, O., 1967. Studies on fossil aphids (Homoptera: Aphidoidea). *Spolia Zoologica Musei Hauniensis*, 26: 1-274.

Heie, O., 1980. The Aphidoidea (Hemiptera) of Fennoscandia and Denmark. I. General Part. The Families Mindaridae, Hormaphididae, Thelaxidae, Anoeciidae, and Pemphigidae. Fauna Entomologica Scandinavica, Volume 9. Scandinavian Science Press Ltd., Klampenborg, Denmark, 236 pp.

Hille Ris Lambers, D., 1948. On Palestine aphids, with descriptions of new subgenera and new species (Homoptera, Aphididae). *Transactions of the Royal Entomological Society, London*, 99: 269-289.

Hottes, F.C., 1955. Three new subspecies and figures of five previously unfigured species of *Cinara* (Aphidae). *Proceedings of the Biological Society of Washington*, 68: 101-104.

Hottes, F.C., 1960. Notes on and key to the species of *Cinara* (family Aphidae) living on *Pinus edulis*. *Proceedings Biological Society of Washington*, 73: 199-214.

Kidd, N.A.C., Smith, S.D.J., Lewis, G.B. and Carter, C.I., 1990. Interactions between host-plant chemistry and the population dynamics of conifer aphids. In: A.D. Watt, S.R. Leather, M.D. Hunter and N.A.C. Kidd (Editors), Population Dynamics of Forest Insects. Intercept, Andover, Hampshire, U.K., pp. 183-193.

Mahalanobis, P.C., 1936. On the generalized distance in statistics. *Proceedings of the National Institute of Science, India*, 2: 49-55.

Mirov, N.T., 1967. The Genus Pinus. The Ronald Press Company, New York, 602 pp.

Neff, N.A. and Marcus, L.F., 1980. A Survey of Multivariate Methods for Systematics. American Museum of Natural History, New York, 221 pp.

Oestlund, O.W., 1942. Systema Aphididae. A Guide to the Phylogeny of the Aphids or Plant Lice. Part One, The Lachnea. Augustana Book Concern, Rock Island, Illinois, 78 pp.

Pepper, J.O. and Tissot, A.N., 1973. Pine feeding species of *Cinara* in the eastern United States (Homoptera: Aphididae). *Florida Agriculture Experiment Station Monograph Series*, 3, 159 pp.

Pimentel, R.A., 1979. Morphometrics, The Multivariate Analysis of Biological Data. Kendall/Hunt Publishing Company, Dubuque, Iowa, 276 pp.

Reyment, R.A., Blackith, R.E. and Campbell, N.A., 1984. Multivariate Morphometrics, (Second Edition). Academic Press, New York, 233 pp.

Richards, W.R., 1965. The Callaphidini of Canada (Homoptera: Aphididae). *Memoirs of the Entomological Society of Canada*, 44, 149 pp.

Rohlf, F.J., 1990. NTSYS-PC. Numerical Taxonomy and Multivariate Analysis System, (Version 1.60). Exeter Publishing Ltd, Setauket, New York.

Rohlf, F.J. and Bookstein, F.L., 1988. A comment on shearing as a method for "size correction." *Systematic Zoology*, 36: 356-367.

Sneath, P.H.A. and Sokal, R.R., 1973. Numerical Taxonomy, The Principles and Practice of Numerical Classification. W.H. Freeman and Company, New York, 573 pp.

Sokal, R.R., Bird, J. and Riska, B., 1980. Geographic variation in *Pemphigus populicaulis* (Insecta: Aphididae) in eastern North America. *Biological Journal of the Linnean Society*, 14: 163-200.

Sokal, R.R. and Riska, B., 1981. Geographic variation in *Pemphigus populitransversus* (Insecta: Aphididae). *Biological Journal of the Linnean Society*, 15: 201-233.

Sokal, R.R., Unnasch, R.S. and Thomson, B.A., 1991. *Pemphigus* revisited: changes in geographic variation but constancy in variability and covariation. *Evolution*, 45: 1585-1605.

Sorensen, J.T., 1983. Cladistic and Phenetic Analysis of *Essigella* Aphids: Systematics and Phylogeny in Relation to their Pinaceae Host Plants (Homoptera: Aphididae, Lachninae). Ph.D. Thesis, University of California, Berkeley (unpublished), 606 pp.

Sorensen, J.T., 1991. Phylogenetic character responses for shape component variance during the multivariate evolution of eulachnine aphids: redescription of *Pseudessigella* Hille Ris Lambers (Homoptera: Aphididae: Lachninae). *Pan-Pacific Entomologist*, 67: 28-54.

Sorensen, J.T., (in press). Deciphering biological groupings in the *Essigella knowltoni* complex (Homoptera: Aphidinea: Lachnidae), and comparison to variation in *Pinus contorta* (Coniferae: Pinaceae). *Entomologica Generalis*, 17 (1).

Sorensen, J.T., (this volume). The use of discriminant function analysis for estimation of phylogeny: partitioning, perspective and problems.

Voegtlin, D.J., 1976. A Biosystematic Study of the *Cinara* spp. (Homoptera: Aphididae) of the Conifers of the Westside Sierra Forests. Ph.D. Thesis, University of California, Berkeley, 208 pp.

Wheeler, N.C. and Guries, R.P., 1982a. Population structure, genic diversity, and morphological variation in *Pinus contorta* Dougl. *Canadian Journal of Forest Research*, 12: 595-606.

Wheeler, N.C. and Guries, R.P., 1982b. Biogeography of lodgepole pine. *Canadian Journal of Botany*, 60: 1805-1814.

Wool, D., 1977. Genetic and environmental components of morphological variation in gall-forming aphids (Homoptera, Aphididae, Fordinae) in relation to climate. *Journal of Animal Ecology*, 46: 875-889.

Wool, D. and Manheim, O., 1983. The effects of environment subdivision on morphological variation in the "cauliflower" galls of the aphid *Slavum wertheimae* (Homoptera, Aphididae, Fordinae). *Israel Journal of Entomology*, 17: 95-104.

USING ORDINATION METHODS WITH GEOGRAPHIC INFORMATION: SPECIES RESOLUTION IN A PARTIALLY SYMPATRIC COMPLEX OF NEOTROPICAL *TRAMEA* DRAGONFLIES (ODONATA: LIBELLULIDAE)

Rosser W. Garrison

Los Angeles County Agricultural Commissioner's Office / Weights and Measures, 3400 La Madera Avenue, El Monte, California 91732, U.S.A.

Abstract

Two forms of the large libellulid dragonfly *Tramea binotata* are generally distributed over southern Florida, Mexico, the Antilles, and south through southern South America. A red morph, called *Tramea insularis*, known from allopatric populations in peninsular Florida, the Bahamas and Cuba, is similar to a more wide ranging black morph, known from Mexico, south through South America. The black and red morphs are sympatric only in Puerto Rico, Jamaica and Hispaniola. Analysis of seven mensural characters using principal components analysis (PCA) and discriminant function analysis (DFA) revealed differences in relation to geography, which support treating the two forms as separate species. PCA showed differing patterns of covariance for the allopatric OTUs in attribute space, suggesting differentiation among the character suites for the traits analyzed. In DFA, allopatric populations, used as "knowns," were employed to assign individuals, as "unknowns," from the areas of sympatry, with a 68% rate of correct classification in relation to color phena. A PCA on a large sample of Puerto Rican specimens from one locality resulted in a bipolar cluster, consistent with their assignment to red or black phenotypes.

1 Introduction

Multivariate methods provide the investigator with a variety of means for detecting and analyzing the overall patterns of variation between and within clusters of taxa. Among their chief advantages are the detection of independent sources of variation, the recognition of highly correlated characters, and the ability to classify questionably determined specimens according to an a priori data base (e.g., Daly and Balling, 1978; Garrison, 1984). Other inferentional uses of some multivariate techniques have been their application in quantitative genetics by detecting coadapted gene complexes (Sorensen and Foottit, this volume), and to cladistics by sequestering closely related taxa that are cladistically derived (Sorensen, this volume).

Odonata, or dragonflies, lend themselves to studies using multivariate techniques. They are large, primitive insects, whose rich network of wing veins provides a multitude of easily measured characters. Their chief disadvantage arises in having no assurance that taxonomic identifications represent valid species. Many species of dragonflies are wide ranging, aerial predators, whose biologies do not lend themselves to breeding in the laboratory (but, for an exception, see Leong, 1989). As with so many other organisms, differential morphology is used for evaluating species status.

Numerical analyses of dragonfly measurement data sets has almost always involved univariate techniques (Huggins, 1927; Bennefield, 1965; Dumont and Dumont, 1969; Yablokov et al., 1970; Maibach 1986). Such methods are relatively easy to use, and replication of various tests applied to suites of characters has often been used to detect differences. The chief disadvantage of these methods, however, is their inability to detect highly correlated characters. The determination of a series of highly significant character state differences may be the result of one or a few genes, or simply due to a common growth factor. Multivariate methods can help by partitioning, or at least facilitating the researcher's recognition of the effects of these two factors.

Phenetic methods in odonate systematics have lagged far behind that of other arthropod groups. I am familiar with only three papers which apply multivariate techniques to dragonflies. Garrison (1976) analyzed geographic variation of a wide ranging Nearctic libellulid, *Libellula luctuosa* Burmeister, and evaluated certain wing vein characters, as well as a liberal number of coded (qualitative) character states. A high concordance of results was achieved by subjecting the data set to principal components analysis (PCA), discriminant function analysis (DFA), and hierarchical clustering techniques. The results allowed speculation on distributional patterns and geographical differentiation of this species. A more practical application of multivariate techniques was given by Garrison (1984), using discriminant analysis to separate sibling species of immature damselflies. Larvae were reared to adults, which were positively identifiable to species, and the measurements taken from their exuviae were used as identified samples for the immatures. A second set of specimens, most of which were also reared so that positive identifications could be made, was treated as "unknowns" and classified according to the data from the "known" sample. About 85% of the unknowns were correctly allocated among five taxa. Garrison also provided an algorithmic "key," based upon DFA, which investigators unfamiliar with odonate systematics could follow for identification of questionable specimens.

Leong (1989) and Leong and Hafernik (in press) provide excellent examples of PCA and DFA, that were used to analyze morphological variation, seasonal variation and hybridization between two species of San Francisco Bay area damselflies. Sympatric and allopatric samples of two sister species, *Ischnura denticollis* (Burmeister) and *I. gemina* (Kennedy) were collected at several sites. Using various body and wing measurements, Leong detected allometric shifts in some size parameters over different seasons. Field-caught specimens in sympatric areas were considered hybrids because they were morphologically intermediate, and they were analyzed and compared with hybrids produced in the laboratory. Both PCA and DFA indicated that the field-caught intermediates were similar to the laboratory-reared hybrids; both were phenotypically intermediate between parental types. However, DFA was more sensitive than was PCA in detecting differences that were based on allometric patterns of size.

Here, I use multigroup principal components analysis and discriminant function analysis in an attempt to interpret species resolution among two partially sympatric dragonfly morphs of *Tramea binotata* Rambur and *T. insularis* Hagen. A brief summary of the status of these species is warranted. *Tramea binotata* is a black morph dragonfly that is widespread over central Mexico, south through all of northern and central South America; the name is the oldest of seven available for the taxon. The name *Tramea insularis* has been applied to a predominantly reddish orange, but seemingly morphologically identical, morph that is known from peninsular Florida, the Bahamas, and Cuba. Although these color morphs are largely allopatric, they occur sympatrically on three islands of the Antilles: Puerto Rico, Hispaniola, and Jamaica. Body coloration appears to be stable in both morphs, although young *T. binotata* are initially orange when teneral but become black at maturity. Both morphs are characterized by having a red (*T. insularis*) or black (*T. binotata*) hind wing anal band (Fig. 1), which varies in width, both between and within populations. Although systematists have generally been able to distinguish the phena based on body and wing coloration, I examined patterns of variation using seven continuous measured variables (Table 1, Figs. 1 and 2).

2 Methods and Materials

Two data sets were employed in this study. The first consisted of 110 males (too few females were available for comparable analysis) from six allopatric areas. My objective criterion for the assignment to groups was based solely on locality: 41 red morphs were examined from south Florida ($n = 15$), the Bahamas ($n = 18$), and Cuba ($n = 8$); and 69 black morphs were examined from Belize ($n = 30$), Panama ($n = 9$), and Venezuela ($n = 30$); these localities represented the sample groups FLA, BAH, CUB, BEL, PAN, VEN, respectively, for the analyses in which a priori groupings were required.

I assumed that each locality-based group shared a common gene pool. All three localities for black *T. binotata* were composed of samples collected from the same locality or (for Panama) within a few kilometers of one another. The small sample from Cuba, representing specimens from different parts of that island, probably represented individuals from different populations.

The second (= "unknown") data set consisted of 66 males from Puerto Rico ($n = 55$), the Dominican Republic ($n = 10$), and Jamaica ($n = 1$). The Puerto Rican sample consisted of 39 red (*T. insularis*) and 16 black (*T. binotata*) morphs from the east and west sides of Laguna Tortuguero in northern Puerto Rico. Three of the 10 Dominican Republic specimens were *T. insularis*, and

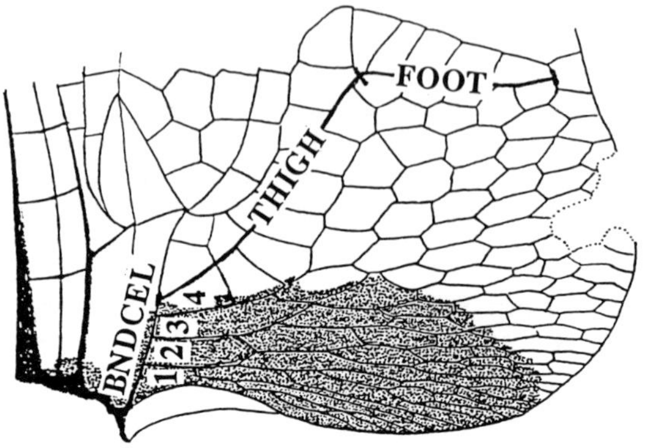

Figure 1. Pictorial depiction of characters *BNDCEL*, *THIGH*, and *FOOT* from right hind wing of *Tramea binotata*. *BNDCEL* in this figure is assigned a value of 4.0.

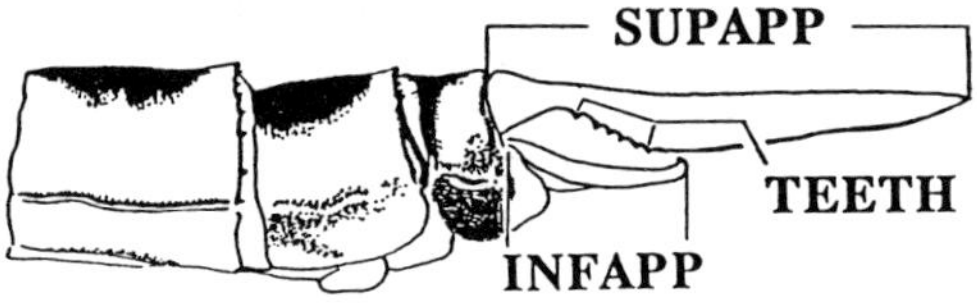

Figure 2. Pictorial depiction of characters *SUPAPP*, *INFAPP*, and *TEETH* from left lateral view of caudal appendages of male *Tramea insularis*.

both morphs were collected in different parts of the island. The single Jamaican specimen was a black *T. binotata*.

The first data set was analyzed using multigroup PCA and DFA programs (Pimentel and Smith, 1986). The 55 Puerto Rican specimens of the second data set were also analyzed using a single group PCA. The second "unknown" data set was classified using the group centroids of the first data set as reference. Analyses were R-mode and, because data measurements were mixed mode, they were transformed to z-scores for the PCA. All measurements were taken from perfect specimens; there were no missing values.

3 Results

3.1 Multigroup Principal Components Analysis

3.1.1 Allopatric Data Set

The six data groups revealed, not surprisingly, an inequality of group centroids (F = 7.40, 35 and 414 df, P = 0.00). Factor loadings for all seven

Table 1.
Morphometric Characters.

No.	Abbreviation	Description
1	*BNDCEL*	Width of hind wing band measured in cells or fractions of cells from the basal paranal cell distally to, and past, the basal cells in the anal loop. Because there are only three paranal cells and three cells at base of the anal loop in the hind wing, measurements are equally scaled for all specimens (Fig. 1).
2	*THIGH*	Length (mm) of the midrib of the anal loop from the cubitus to the ankle (Fig. 1).
3	*FOOT*	Length (mm) of the midrib of the anal loop from the ankle to the tip of the toe (Fig. 1).
4	*SUPAPP*	Length (mm) of the superior appendage (Fig. 2).
5	*INFAPP*	Length (mm) of the inferior appendage (Fig. 2).
6	*TEETH*	Number of teeth (denticles) on the ventral side of the superior appendage (Fig. 2)
7	*H W*	Hind wing length (mm).

characters for the first, second, and third axes for each group PCA and for the total (within) group PCA are given in Tables 2-4. Within-group PCA encompasses variation common to all six localities. Character weighting was different for each group. For example, the first axis for FLA, CUB, BEL was primarily concerned with size; but in each axis, *BNDCEL* reacted antagonistically to all other characters. Individuals from FLA and CUB were, on average, smaller than those of BEL; width of the hind wing band (*BNDCEL*) in FLA and CUB samples increased, whereas values for most other characters decreased. The opposite was true for the BEL sample. For these general-size components, all characters contributed significantly (i.e., $P > 0.25$), except for *FOOT* (CUB) and *TEETH* (BEL).

Only character loadings for BEL individuals show any similarity to the overall within PCA (Table 2), which indicates that the Belize sample is average with respect to all six groups. Total percentage of variation explained for the first component ranged from a 22.8% (Venezuela) to 44.7% (Cuba). Size variation accounted for the greatest component of the Cuban examples relative to all other groups. I believe this was due to the small sample size and because they were collected over different parts of the island. Of all six samples, the Cuban individuals showed the greatest variation in size, but not in overall morphology.

Interpretation of eigenvectors 2 (Table 3) and 3 (Table 4) is difficult. Each explains an average of 20-30% (component 2) to 10-18% (component 3) of the total variation. The high factor loadings, attributed to one or few characters for many samples, suggest allometric patterns of growth. Character loadings for most localities seem to indicate a complex interaction among characters, which differ within each group. The small sample size from some localities may have caused spurious results. As for example, tests of sphericity for all components (i.e., testing whether each orthogonal component is arbitrarily placed) were not significant for the FLA and PAN samples. For component 1,

Table 2.
Principal components analysis: eigenvector 1 loadings for allopatric data set.

Character	FLA	BAH	CUB	BEL	PAN	VEN	WITHIN
	Group						
BNDCEL	0.31	- 0.46	0.33	- 0.02	0.31	0.30	0.13
THIGH	- 0.21	- 0.28	- 0.46	0.50	- 0.50	0.55	- 0.51
FOOT	- 0.26	0.13	- 0.14	0.43	-0.28	0.44	0.35
SUPAPP	- 0.56	- 0.51	- 0.30	0.38	0.11	- 0.05	- 0.39
INFAPP	0.40	0.57	- 0.50	0.31	- 0.06	0.11	- 0.37
TEETH	- 0.33	- 0.31	- 0.37	0.09	- 0.45	- 0.51	0.16
H W	- 0.47	0.11	- 0.44	0.57	- 0.60	0.37	- 0.54
% variation	35.5	36.6	44.7	36.4	30.3	22.8	29.5

Table 3.
Principal components analysis: eigenvector 2 loadings for allopatric data set.

Character	FLA	BAH	CUB	BEL	PAN	VEN	WITHIN
	Group						
BNDCEL	0.25	0.18	0.18	0.68	0.43	- 0.59	0.56
THIGH	0.61	- 0.59	- 0.11	- 0.14	0.09	0.28	0.18
FOOT	- 0.59	- 0.36	0.57	0.30	0.42	0.14	0.43
SUPAPP	0.01	- 0.00	0.49	0.11	0.70	0.33	- 0.16
INFAPP	- 0.27	0.08	- 0.20	- 0.08	0.32	0.55	- 0.16
TEETH	0.10	0.26	0.46	- 0.64	0.21	0.31	- 0.64
H W	0.37	- 0.65	0.36	0.00	0.04	0.20	0.10
% variation	19.4	24.6	31.9	21.2	25.3	21.5	18.8

there is similarity in importance of characters for BEL individuals compared to the within PCA for components 2 and 3. For example, *BNDCEL* increased as the number of *TEETH* decreased (PC-2), and *FOOT* reacted positively at the expense of *INFAPP* (PC-3).

A further comparison of each locality PCA with the overall PCA is accomplished by evaluating the angular differences between-group PCA axes relative to the respective within-group axes. Angular comparisons of about 20° or less have been considered to represent overall similarity between PCAs (Blackith, 1965). First component scores for BEL were most similar to the within-group PCA (Table 2); Table 5 substantiates this observation. The angular difference for the first component between BEL and the first component is only about 9.7°. The second and third axes show values of 25.5° and 33.2°, respectively, but these are much less than for any similar comparisons with any other group. Angular differences are most pronounced with BAH and PAN; consequently, these samples are most poorly typified in the within-group PCA.

Table 4.
Principal components analysis: eigenvector 3 loadings for allopatric data set.

	Group						
Character	FLA	BAH	CUB	BEL	PAN	VEN	WITHIN
BNDCEL	0.29	- 0.49	0.85	- 0.18	0.05	- 0.08	- 0.36
THIGH	- 0.35	- 0.11	0.35	- 0.06	0.08	- 0.13	0.07
FOOT	0.18	- 0.60	- 0.18	0.44	0.59	0.13	0.33
SUPAPP	0.02	0.30	- 0.03	- 0.05	0.05	- 0.65	- 0.48
INFAPP	- 0.48	- 0.22	- 0.07	- 0.85	- 0.77	- 0.10	- 0.61
TEETH	0.71	0.50	0.22	0.12	- 0.23	0.40	0.24
H W	0.14	0.07	0.25	0.20	0.06	0.60	0.32
% variation	15.4	18.1	10.8	13.4	16.2	17.8	14.6

Table 5.
Comparison of angles between principal components for 6 locality groups versus pooled (within) dispersion matrix. Angles >20° (*) from within principal components should be considered to be significantly different.

	Group					
Vector	FLA	BAH	CUB	BEL	PAN	VEN
1	25.6°*	64.2°*	23.4°*	- 9.7°	39.8°*	- 57.8°*
2	89.7°*	- 86.1°*	52.2°*	27.5°*	83.6°*	- 57.4°*
3	64.8°*	83.9°*	- 81.4°*	33.2°*	56.8°*	43.9°*
4	- 72.0°*	- 82.8°*	- 88.9°*	- 54.3°*	- 80.3°*	61.9°*
5	- 82.2°*	31.6°*	- 75.6°*	72.1°*	80.7°*	58.3°*
6	66.9°*	- 58.2°*	- 80.4°*	- 48.7°*	- 83.1°*	71.8°*
7	51.0°*	- 80.6°*	47.1°*	- 13.4°	- 74.7°*	42.9°*

Figs. 3-4 show within-group PCA scattergrams for all *Tramea binotata* (black and gray symbols) and *T. insularis* (white symbols). Important character vectors are also shown. *THIGH* and *HW* are primarily explained by PC-1, *BNDCEL*, *FOOT*, and *TEETH* by PC-2. A strong polarization between PAN and VEN *T. binotata* and all *T. insularis* is apparent in Fig. 3. Most *T. binotata* occur in the quadrants 1 and 4, and most *T. insularis* occur in quadrant 2 and the upper part of quadrant 3. A small overlap exists with eight *T. binotata* and two *T. insularis*. One VEN *T. binotata* was well into quadrant 2; it was the smallest specimen (*HW* = 34 mm) of its series. The next smallest specimen from this locality had a *HW* of 37 mm. One Floridian *T. insularis* occurred with *T. binotata* because it had the greatest number of *TEETH* (10 ventral teeth); all other specimens from FLA, BAH, and CUB had no more than nine ventral teeth. On the other hand, 10 specimens from BEL, PAN, and VEN had 10 teeth, six had 11, and three had 12 teeth. One specimen from BAH occurred in the lower right area of quadrant 3, and had a uniquely small

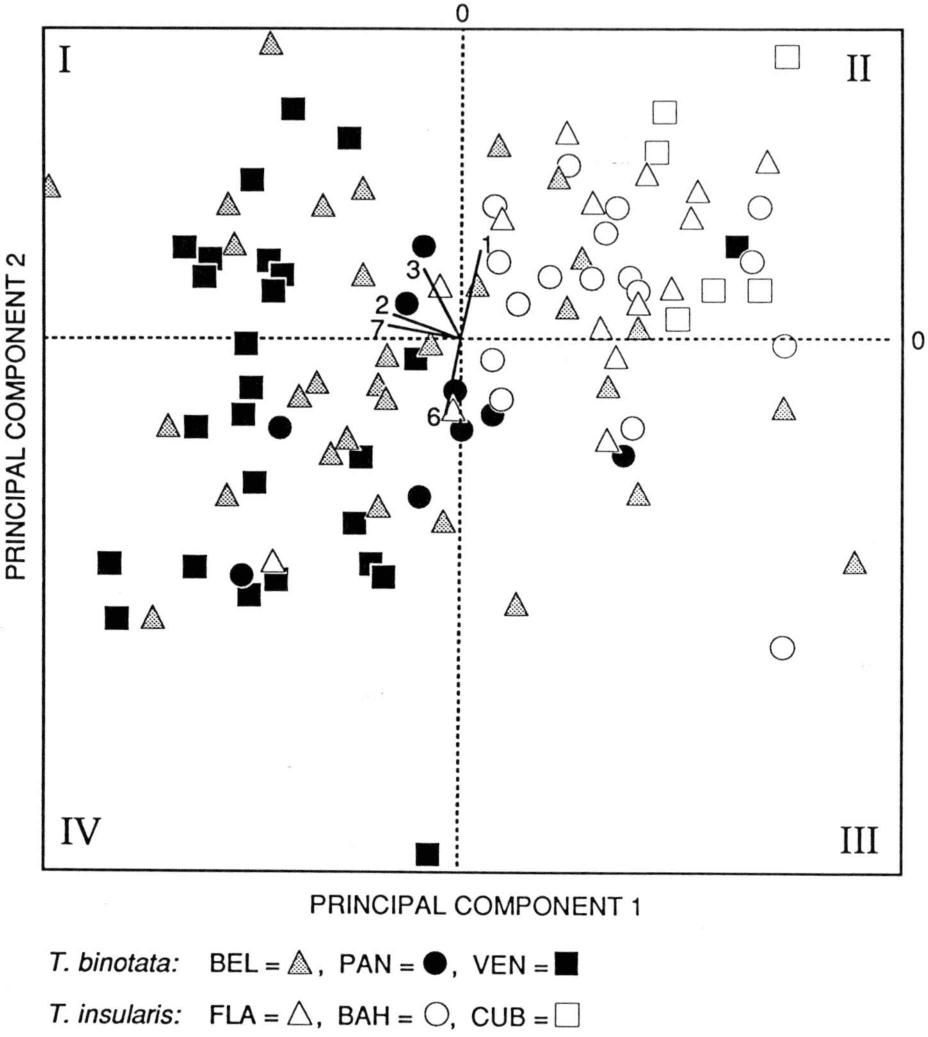

Figure 3. Scattergram of within-group PCA of six *Tramea* locality data set along PC-1 and PC-2. Localities for three *T. binotata* (black morphs) and three *T. insularis* (red morphs). Character vectors: 1 = *BNDCELL*, 2 = *THIGH*, 3 = *FOOT*, 6 = *TEETH*, 7 = *HW*.

FOOT (3.52 mm); all other specimens from the locality had that variable in excess of 4.00 mm.

The best discriminatory characters for these taxa were chiefly *BNDCEL* and *TEETH*. Populations of *T. insularis*, in general, tend to have a wider hind wing band and fewer teeth on the ventral side of the superior appendage. PAN *T. binotata* became allied more closely with *T. insularis* because, on average, they possess a broad hind wing band ($\bar{x} = 4.1$) typical of *T. insularis* (CUB $\bar{x}$ = 4.7, BAH $\bar{x} = 4.3$, FLA $\bar{x} = 4.0$).

In Fig. 3, gray shows BEL *T. binotata*; two important points are evident: they are (1) intermediate to all other localities, and (2) the least clustered of any group. BEL material has the widest morphological variability of any locality.

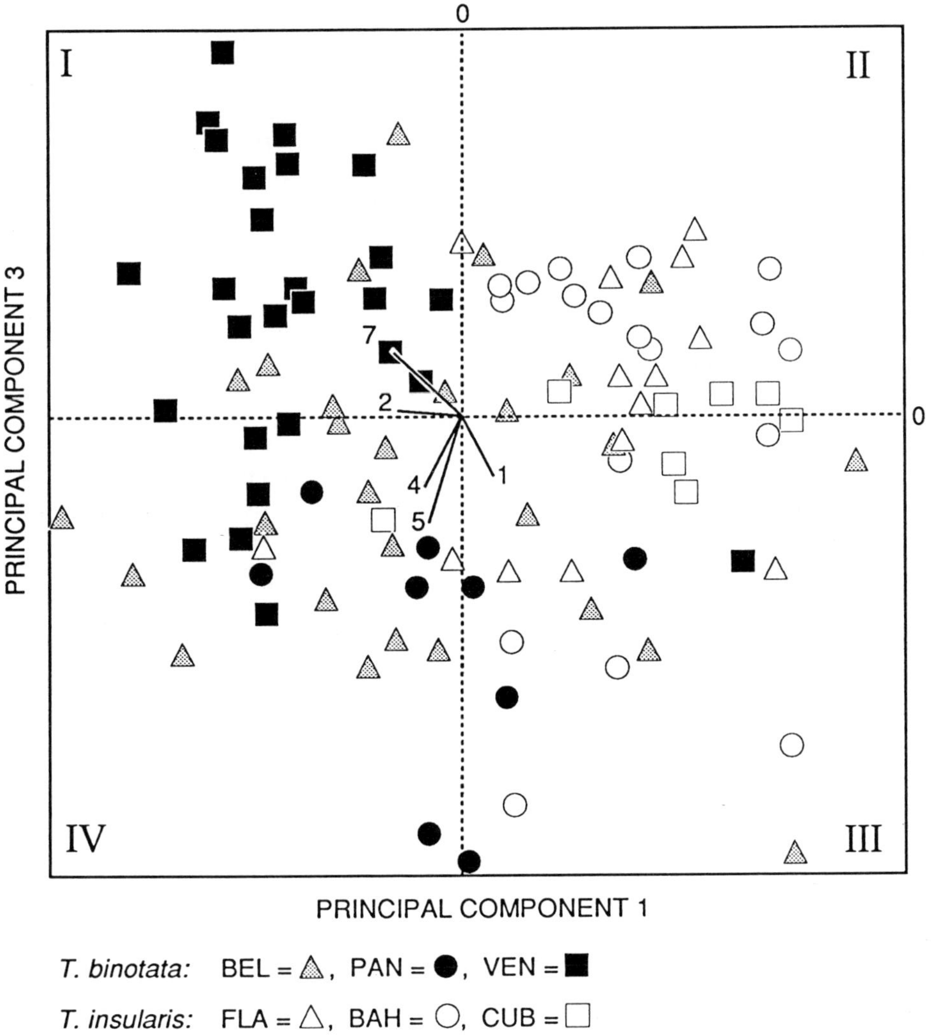

Figure 4. Scattergram of within-group PCA of six *Tramea* locality data set along PC-2 and PC-3. Character vectors: 1 = *BNDCELL*, 2 = *THIGH*, 4 = *SUPPAPP*, 5 = *INFAPP*, 7 = *HW*. See Figure 3 for further explanation.

Fig. 4 reveals similar patterns of variation to those expressed on components 1 and 2. The most important characters allied with PC-3 are *BNDCEL*, *SUPAPP*, and *INFAPP*. PAN specimens segregate from most VEN *T. binotata* on the basis of a wider hind wing band. Separation between these two localities is not apparent on PC-2 alone, despite the important factor loading (0.56) for that character. *BNDCEL*, on PC-2, is primarily concerned with the differentiation between *T. binotata* and *T. insularis*. As in Fig. 3, the distribution of *T. binotata* from BEL on PC-1 and PC-3 (Fig. 4) shows an intermediacy between two *T. insularis* and the remainder of *T. binotata* localities. Similar distributions in multivariate space were observed for specimens from FLA and BAH. Five specimens from FLA occupy areas in quadrants 3 and 4; three specimens from the BAH occur at the lower part of quadrant 3. All of these specimens had a longer *INFAPP* than other members of their groups. Similarly, one CUB specimen in quadrant 4 had a longer *INFAPP* than any other specimen from that island.

3.1.2 Sympatric Data Set

A single PCA of the 55 males from Puerto Rico is depicted in Figs. 5 and 6. Although I could discriminate between these *T. insularis* and *T. binotata* by color, I analyzed them as a single locality sample, because both taxa were sympatric in the Laguna Tortuguero region, and because the possibility of gene flow existed among them. The percentage of total variation explained by the first three components was 27.6, 22.3, and 16.3%, respectively. PC-1 can be considered a general-size component, because eigenvalues for all characters except *TEETH* were negative. Important sources of variation that were explained by PC-1 were those for *FOOT*, *SUPAPP*, and *HW*. *TEETH*, and especially *BNDCEL*, were major components of variation that were explained by PC-2 (Fig. 5). Assignment of red (*T. insularis*) and black (*T. binotata*) phenotypes reveals a strong polarization between the two groups. The PCA, based on the seven characters, corroborates my a priori assessment of the two sympatric forms. *Tramea insularis*, characterized by a relatively wider *BNDCEL* and fewer *TEETH*, occurs primarily in quadrants 3 and 4, and tends to form a more cohesive group than does *T. binotata*. The latter forms a less aggregated assemblage in quadrants 1 and 2.

Segregation between *T. insularis* and *T. binotata* is much less in evidence on PC-3 (Fig. 6). The principal components of variation on PC-3 are *THIGH*, *SUPAPP*, and *INFAPP*, but little difference between the two color morphs is manifested using these characters. *Tramea insularis* displays more variability for these characters than does *T. binotata*.

3.2 Discriminant Function Analysis

3.2.1 Allopatric Data Set

Classification of the first allopatric data (from six localities) is shown in Table 6. My a priori assessment was to lump the red populations (FLA, BAH, CUB) as *T. insularis*, and the black populations as *T. binotata*. Geisser classification probabilities show that 81 individuals (74%) were correctly

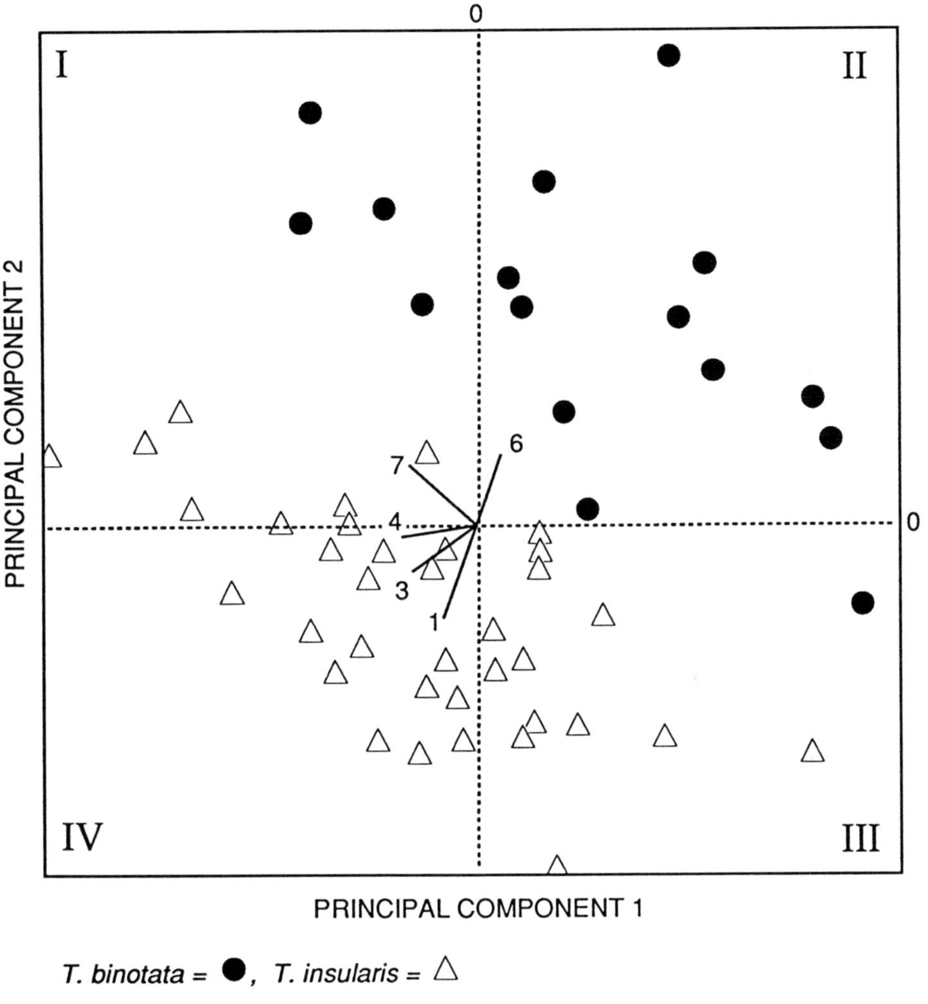

Figure 5. Within-group PCA of Puerto Rican sympatric data set along PC-1 and PC-3. All specimens are from Laguna Tortugue-ro. *T. binotata* are black phenotypes, *T. insularis* are red phenotypes. Character vectors: 1 = *BNDCELL*, 3 = *FOOT*, 4 = *SUPPAPP*, 6 = *TEETH*, 7 = *HW*.

allocated to each parental group; 29 individuals (26%) were misclassified. Most errors involved misclassification into geographically surrounding groups. That is, four FLA males were classified as BAH and one as CUB; only two were allocated to the BEL sample of *T. binotata*. Ninety-nine (90%) of 110 specimens were correctly classified according to my a priori assignment into *T. insularis* and *T. binotata*. Most misclassified individuals were erroneously classified as *T. binotata* from BEL. The BEL sample was, according to PCA, the most heterogeneous of the six groups, and it was intermediate among all

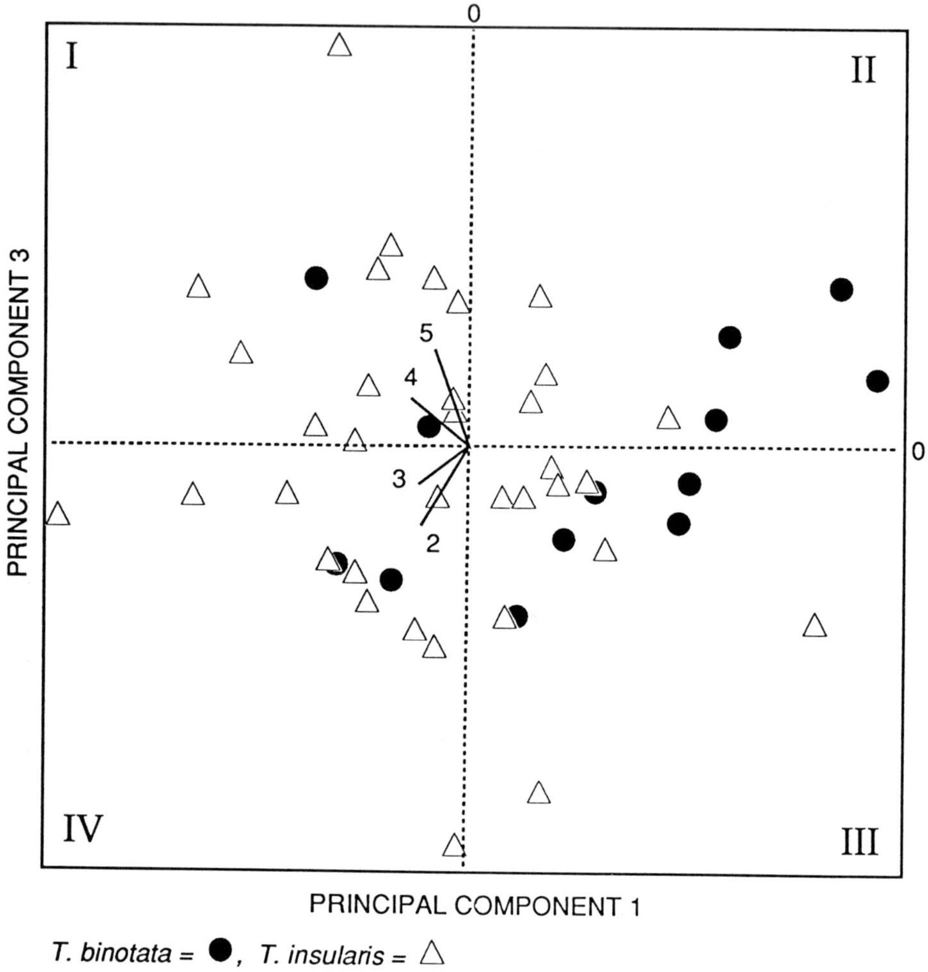

Figure 6. Within-group PCA of Puerto Rican sympatric data set along PC-1 and PC-3. Character vectors: 2 = *THIGH*, 3 = *FOOT*, 4 = *SUPPAPP*, 5 = *INFAPP*. See Figure 5 for further explanation.

groups. Not surprisingly, the incorrectly placed specimens from the other five localities were most often classified there. In Fig. 7, the centroids and 95% confidence radii are shown, and it is evident that the FLA, BAH, and CUB centroids clump together, with overlap among these three groups. The allocation of red phenotypes to respective localities was prone to error, due to the proximity of all centroids. Of these three populations, the centroid of the FLA group is closest to any centroid for a black population; in this case, BEL. The centroids for VEN and PAN populations are well apart from any other centroid. The first two canonical axes (Table 7) were responsible for most

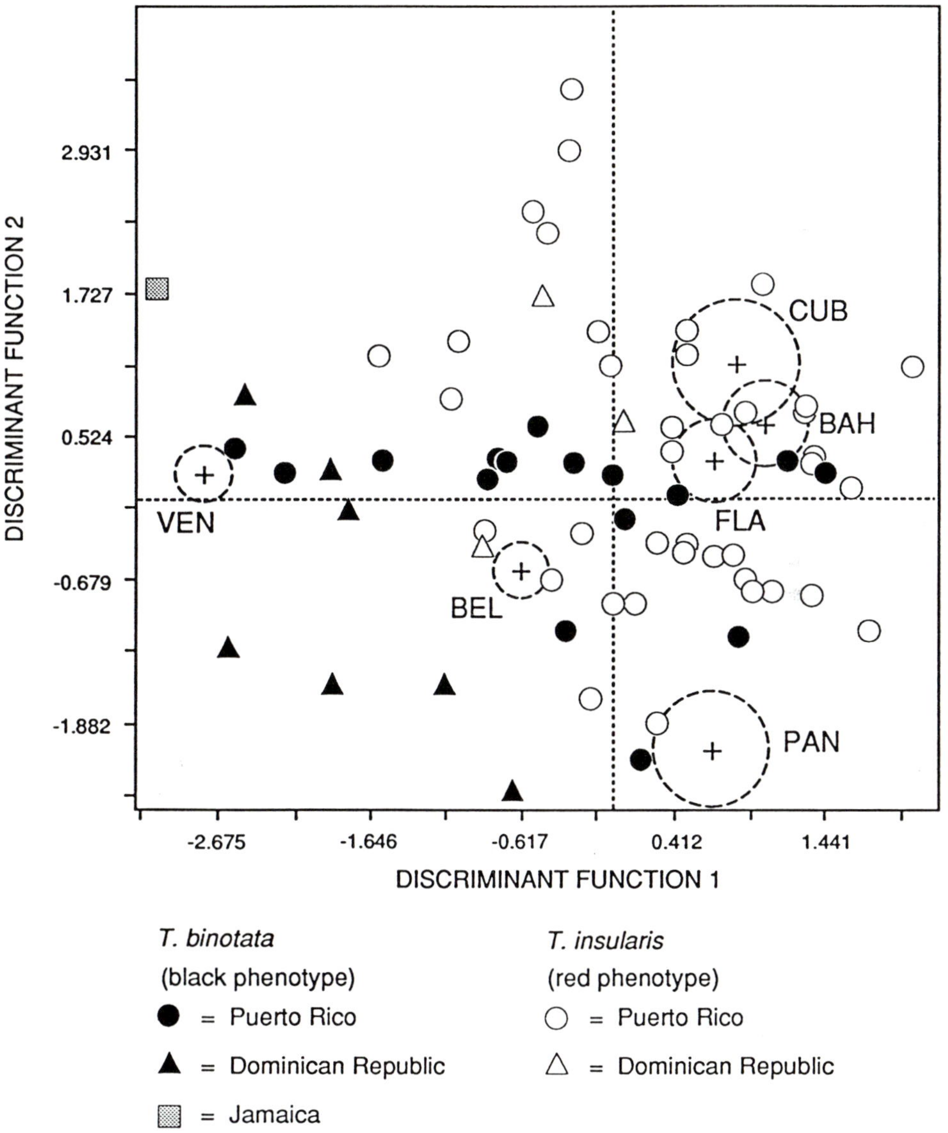

Figure 7. Discriminant function scattergram along first two axes of sympatric *Tramea* data set. Dashed circles are 95% confidence radii of six allopatric data set, and + represents their centroid.

(88.5%) of the discrimination among the six localities. *TEETH* and *HW* were of no importance in discriminating among the six groups; whereas, *BNDCEL* played only a minor role as a discriminatory character along the first

Table 6.
Discriminant function analysis: classification table of allopatric groups by locality. By locality: 81 individuals (74%) correctly classified; 29 individuals (26%) misclassified. By species: 99 individuals (90%) correctly classified; 11 individuals (10%) misclassified.

		Group					
		Red Phena (*T. insularis*)			Black Phena (*T. binotata*)		
Locality	*n*	FLA	BAH	CUB	BEL	PAN	VEN
FLA	15	8	4	1	2	--	--
BAH	18	3	13	--	1	1	--
CUB	8	--	1	5	2	--	--
BEL	30	1	3	--	22	3	1
PAN	9	--	--	--	2	7	--
VEN	30	--	--	1	3	--	26

Table 7.
Discriminant function analysis: standardized canonical vector coefficients and the percentage of variation explained by each canonical variate for the allopatric data set.

Character	CV-1	CV-2	CV-3
BNDCEL	1.01	- 0.11	0.61
THIGH	- 1.67	- 0.52	2.96
FOOT	- 1.03	0.87	- 2.06
SUPAPP	2.01	0.32	0.91
INFAPP	2.54	- 9.29	- 3.77
TEETH	0.02	- 0.13	- 0.02
H W	0.10	0.04	0.28
% variation	69.1	19.4	8.07

Table 8.
Classification of Puerto Rican (n = 55), Hispaniolan (n = 10) and Jamaican (n = 1) *Tramea* to the respective allopatric localities listed in Table 6. Arbitrarily defined phenotypes (red versus black) are given below.

			Group					
			Red Phena (*T. insularis*)			Black Phena (*T. binotata*)		
Locality	Color	*n*	FLA	BAH	CUB	BEL	PAN	VEN
Puerto Rico	red	39	13	10	3	10	--	--
	black	16	6	--	--	7	12	--
Hispaniola	red	3	--	1	--	2	--	--
	black	7	--	--	--	3	1	3
Jamaica	black	1	--	--	--	--	1	--

canonical axis. Important discriminatory characters on axis 1 were *SUPAPP* and *INFAPP*. Males of VEN and, to a lesser extent, BEL had a relatively longer *SUPAPP* and *INFAPP* compared with *T. insularis* populations. The second canonical variate is totally dominated by *INFAPP*, which was the only discriminatory character along that axis.

Canonical variates 1 and 2 show an important allometric, or shape, difference among the samples from BEL, PAN and VEN. *INFAPP* is longest in PAN ($\bar{x}$ = 1.99 mm) and *SUPAPP* is shortest ($\bar{x}$ = 4.73 mm), followed by BEL (*INFAPP* $\bar{x}$ = 1.83 mm, *SUPAPP* $\bar{x}$ = 4.75 mm). However, the opposite relationship is found among VEN males (*INFAPP* $\bar{x}$ = 1.78 mm, *SUPAPP* $\bar{x}$ = 4.92 mm).

3.2.2 Sympatric Data Set

Classification of the "unknown" data set to the respective allopatric samples is shown in Table 8 and Fig. 7. I arbitrarily classified sympatric individuals as red (*T. insularis*) or black (*T. binotata*). Of 39 red Puerto Rican males, 26 (66.7%) were classified to one of the three allopatric red localities (FLA, BAH, CUB). Ten others classified as BEL specimens. One Hispaniolan *T. insularis* was allocated to BAH, and the two others to BEL.

Results of the classification of black (*T. binotata*) morphs are of more interest. Of 16 *T. binotata* from Puerto Rico, 10 of 16 (63%) were classified into one of the black allopatric localities. Seven of the 10 were classified as BEL males. Six of 16 (38% were classified as FLA *T. insularis*. These six individuals had the smallest *THIGH* measurements of the 16 Puerto Rican *T. binotata*. A seasonal and/or geographic component may have played a role in the segregation between the six misclassified and 10 correctly classified males. Four misclassified males were collected on 7 January, another on 13 January, and one on 18 December. Collection dates for correctly classified specimens were as follows (number in parentheses = *n*): 7 June (1), 21 June (1), 27 July (1), 15 Sept. (2), 18 Dec. (4), and date unknown (1). Although eight of the 10 correctly classified *T. binotata* came from the eastern side of Laguna Tortuguero, the two others came from the western side. Contrasting with this are five of six misclassified *T. binotata* which came from the western side of Laguna Tortuguero.

3.3 Classification of the Puerto Rican Data Set

Because the PCA on the 55 Puerto Rican males yields a bipolar cluster, consistent with my assignment of groups into *T. insularis* and *T. binotata* (Fig. 5), I subjected these data to DFA. All but three *T. binotata* were correctly classified, resulting in a 95% correct classification. As with the previous DFA for the six localities, *TEETH* and *HW* were of virtually no importance in discriminating between Puerto Rican *T. insularis* and *T. binotata*. Instead, the best discriminatory characters were *FOOT* and *BNDCEL*. *Tramea insularis* generally had a larger *FOOT* ($\bar{x}$ = 4.58 mm) than did *T. binotata*

($\bar{x}$ = 4.28 mm), but a wider *BNDCEL* characterized *T. insularis* ($\bar{x}$ = 3.9 mm) from *T. binotata* ($\bar{x}$ = 2.9 mm).

4 Discussion

The results of this study indicate that the red and black phenotypes of *Tramea* probably represent two separate species, *T. insularis* and *T. binotata*, respectively. The multivariate methods used here corroborate a priori assessments that were made based on color. Another possibility exists: that only one species, with a balanced polymorphism, exists in Puerto Rico, Jamaica and Hispaniola. I find this hypothesis less credible, because I believe that black and red morphs should occur sympatrically over broad geographic areas; the mechanisms that would maintain a polymorphism on only three islands is unclear. Few characters, other than overall color, have traditionally been used to separate these taxa. Recently, odonatists have detected two differences in the color pattern of the abdomen. In *T. binotata*, the tergite of abdominal segments 8-10 are entirely black, whereas inverted black triangles occupy the same area in *T. insularis*. The ventral margin of the tergites in *T. binotata* is chiefly dark, but this area is pale in *T. insularis*. However, I have found intermediate conditions existing in *T. binotata* and *T. insularis*. Additionally, juvenile *T. binotata* are orange and become entirely black at maturity, making these patterns difficult to detect. To date, no morphological differences have been detected to separate the two.

Although principal components analysis is not meant to be used as a discriminatory technique, results of a PCA on the 55 Puerto Rican males (Fig. 5) clearly show a strong polarization into two groups that is consistent with my a priori assessment. The results here do not support the conclusion that the red and black morphs represent a single species, because if such were the case, I would expect less differentiation among the seven morphological characters than occurs. Furthermore, a Geisser classification of the 55 Puerto Rican males, when subjectively classified into red or black phenotypes, leads to a 95% correct classification. Classification of the same 55 individuals, plus the one Jamaican and 10 Hispaniolan males, did not result in a 100% placement of red and black morphs to respective red and black allopatric localities. Unequal sample sizes can distort such classification because those samples with a greater dispersion tend to "capture" more individuals than others with a smaller dispersion. The misclassifications from Table 8 are, however, easily explainable. It is not surprising that 10 Puerto Rican *T. insularis* were classified as BEL *T. binotata*, because the PCA showed these latter individuals to be average with respect to all other allopatric localities. The misclassification of six Puerto Rican *T. binotata* as FLA *T. insularis* might at first seem puzzling, but the Floridian *T. insularis* are closest phenotypically to *T. binotata* (Fig. 7).

The results of these analyses indicate that populations of *T. binotata* are more different from one another than are comparable allopatric populations of *T. insularis*. PCA and DFA showed virtually no differences among Floridian, Bahaman, or Cuban individuals. The DFA of the Puerto Rican *T. insularis* shows that they possess characters that are somewhat intermediate between those of individuals from Florida and Belize. The Venezuelan *T. binotata* are well differentiated from all other groups, and the same is true of *T. binotata*.

Although Puerto Rican *T. insularis* and *T. binotata* can be diagnosed by some of the characters evaluated here, their phenotypic dissimilarity to Venezuelan individuals indicates that the origin of Antillean *T. binotata* probably was Central American, perhaps in the Yucatan Peninsula. Less clear is the possible origin of *T. insularis*. Because the major area of distribution of *T. insularis* includes southern Florida, the Bahamas, and Cuba, it is reasonable to assume that the species dispersed to Hispaniola, Jamaica, and Puerto Rico.

Without additional data, more precise conclusions on the history of these two species cannot be ascertained, because of the inadequacy of the current data set. Even though the sample sizes evaluated here were small, the results indicate complex patterns of variation that may be easier to interpret when larger series become available.

5 Acknowledgments

Earlier drafts of this paper were criticized by Drs. Thomas W. Donnelly (State University of New York, Binghamton), Sidney W. Dunkle (Dept. of Plant Industry, Gainesville, Florida), and John T. Sorensen (California Dept. of Food and Agriculture, Sacramento). Dr. Sorensen also edited this paper and helped during the evolution of the manuscript. My wife, Jo, edited and typed the manuscript. To all of these people, I extend my sincere thanks.

6 References

Bennefield, B.L., 1965. A taxonomic study of the subgenus *Ladona* (Odonata: Libellulidae). *University of Kansas Science Bulletin*, 45: 361-396.

Blackith, R.E., 1965. Morphometrics. In: T.H. Waterman and H.J. Morowitz (Editors), Theoretical and Mathematical Biology. Blaisdell Publishing Company, New York, pp. 225-249.

Daly, H.V. and Balling, S.S., 1978. Identification of Africanized honey bees in the Western Hemisphere by discriminant analysis. *Journal of the Kansas Entomological Society*, 51: 857-869.

Dumont, H.J. and Dumont, S., 1969. A biometrical analysis of the dragonfly, *Ischnura elegans* (Van der Linden) with a special reference to its chloride-tolerance and generation number. *Biologisch Jaarboek Dodonaea*, 37: 50-60.

Garrison, R.W., 1976. Multivariate analysis of geographic variation in *Libellula luctuosa* Burmeister (Odonata: Libellulidae). *Pan-Pacific Entomologist*, 52: 181-203.

Garrison, R.W., 1984. Revision of the genus *Enallagma* of the United States west of the Rocky Mountains and identification of certain larvae by discriminant analysis (Odonata: Coenagrionidae). *University of California Publications in Entomology, Berkeley*, 105: 1-129.

Huggins, J.R., 1927. Variations in size of *Calopteryx maculata* and a proposed new species. *Transactions of the American Entomological Society*, 52: 355-364.

Leong, J.M., 1989. Morphological variation and hybridization in two damselflies, *Ischnura denticollis* and *Ischnura gemina* (Odonata: Coenagrionidae). M.S. Thesis. San Francisco State University, San Francisco, (unpublished), 84 pp.

Leong, J.M. and Hafernik, J.E. Jr., (in press). Seasonal variation in allopatric populations of *Ischnura denticollis* (Burmeister) and *Ischnura gemina* (Kennedy) (Odonata: Coenagrionidae. *Pan-Pacific Entomologist*.

Maibach, A., 1986. Revision systematique du genre *Calopteryx* Leach (Odonata, Zygoptera) pour l'Europe occidentale. II. Analyses morphologiques et synthese. *Mitteilungen der Schweizerischen Entomologischen Gesellschaft*, 59: 389-406.

Pimentel, R.A. and Smith, J.D., 1986. BIOΣTAT II, a Multivariate Statistical Toolbox. Sigma Soft, Placenta, California.

Sorensen, J.T., (this volume). The use of discriminant function analysis for estimation of phylogeny: partitioning, perspective and problems.

Sorensen, J.T. and Foottit, R.G., (this volume). The evolutionary quantitative genetic rationales for use of ordination analyses in systematics: phylogenetic implications.

Yablokov, A.V., Eatin, V.J. and Pritikina, L.N., 1970. Variability of wing venation of the dragonfly. *Beiträge zur Entomologie*, 20: 503-526.

VARIATION IN POPULATIONS OF LEAFHOPPERS AND PLANTHOPPERS (AUCHENORRHYNCHA): BIOTYPES AND BIOLOGICAL SPECIES

Michael F. Claridge and Malcolm C. Gillham

School of Pure and Applied Biology, University of Wales, Cardiff CF1 3PT, Wales, U.K.

Abstract

Leafhoppers and planthoppers are particularly well suited to biotaxonomic studies: they are all herbivores characterized by particular patterns of host plant utilization, and sympatric species usually maintain reproductive isolation by means of acoustic signals that are easily analyzed. They are also well suited for morphometric analysis and have been the subjects for both canonical variates and principal components analyses. We examine case studies where such analyses, coupled with biological information, have made possible the discrimination of both biotypes and biological species, and the identification of environmentally induced variation.

1 Introduction

The major task of the taxonomist is to order the diverse array of living organisms into recognizable and manageable groups, the fundamental unit of which is the species. The delimitation, description, and naming of species on the basis of morphology has proceeded systematically since the days of Linnaeus. However, the use of discontinuities in morphology alone as evidence for taxonomic status has severe limitations; for example, where the sexes, life history stages and polymorphic offspring of an organism show greater dissimilarities than do different species. The species, therefore, cannot be regarded merely as a series of individuals showing continuous morphological variation, but must be defined in a biological context.

Most biologists today accept the "biological species concept" of Mayr (1942, 1963) as the most useful for sexually reproducing organisms. This defines species as groups of actually or potentially interbreeding natural populations which are reproductively isolated from other such groups. The essential characteristic separating biological species is, therefore, reproductive isolation and the demonstration of such isolation is the main task of biological taxonomy (Claridge, 1988). The widespread acceptance of the biological species concept further questions the validity of the "morphospecies", because, by definition, species may exist that show very little or no morphological differentiation. The growing use of biochemical and behavioral techniques as measures of gene flow between sympatric populations has uncovered many such cryptic or sibling species.

The study of patterns of morphological variation is, therefore, only one approach in the attempt to find discontinuities between populations that may provide evidence for reproductive isolation. However, it does have advantages over other methods. For example, individual organisms often need not be alive, or even preserved in any special way, in order to be studied. Also, the tissues of individuals need not be destroyed during examination and thus may be preserved for reference or as nomenclatural type specimens.

The Auchenorrhyncha are particularly suited to biotaxonomic studies. Premating isolation between sympatric species is usually achieved mainly by specific acoustic signals, and these are relatively easy to record and analyze (see Claridge, 1985). Discontinuities in song pattern between populations can be used as evidence of reproductive isolation, and individuals that have been assigned to species on the basis of song structure can then be examined subsequently for diagnostic morphological characters.

All species of Auchenorrhyncha, so far as known, are herbivores. Not only do the nymphs and adults generally feed on the host plant, but the eggs are usually laid within its tissues. Thus, the relations with the host are very close. In most groups so far studied, a tendency towards host plant specificity has been shown (Claridge and Reynolds, 1972; Claridge and Wilson, 1981; Booij, 1982; den Bieman, 1987; Gillham, 1991). Populations associated with different host plant species may vary in form from, at one extreme, distinct and reproductively isolated biological species, to, at the other, subpopulations of an effectively panmictic population. The controversial hypothesis has frequently been proposed that intermediate host races, or biotypes, between these two extremes may also be found in sympatry (discussion in Claridge, 1988). Certainly there is no doubt that allopatric populations may show such a continuum of genetic differentiation. In order to quantify differences between such obviously closely related forms, refined methods of estimating morphological differentiation are required.

Morphometrics, the measure and analysis of form, provides techniques with which to quantify subtle differentiation between closely related populations. Insects make ideal subjects for morphometric studies. The rigid exoskeleton enables repeatable measurements to be made more easily. Also, the adult form is usually distinct from that of the immature stages and does not normally grow further after final ecdysis, preventing the need to take variation between different adult growth stages into account. No attempt has been made here to discuss in detail the statistical basis of the wide range of morphometric methods available (for general reviews, see Neff and Marcus,

1980; Reyment et al., 1984). Instead this review is specifically concerned with the use of the more common morphometric techniques for quantifying variation within and between populations of leafhoppers and planthoppers. A number of case studies are given that illustrate the sort of taxonomic problems to which these methods can be applied.

2 Data Collection

Although a wide range of methods for collecting morphometric data are available (see Rohlf, 1990), a convenient one, and that most widely used, is to break down the overall form of an organism into a series of selected distance measurements. As in any phenetic study, the more measurements that are made on an individual, the greater the chance there is of representing it accurately. Computer technology enables a large number of measurements to be made quickly and precisely. The method most commonly used is to superimpose the character to be measured over a digitizing tablet, either as a projected image or by use of a camera lucida, and to input extreme points of measures as cartesian coordinates using an electronic cursor. The shortest distance between the points can then be calculated using the appropriate computer software.

When selecting points for measurement, it is important that they are repeatable. Specimens must be orientated such that each character is measured in the same plane and that each point is homologous, so far as possible, between all individuals. However, selection of distance measures is usually based on characters that are traditionally used in the taxonomy of the group under study. Such characters are often described by terms such as "maximum lengths" and "narrowest distances," which give no guarantee of homology. A better practice, as outlined by Bookstein et al. (1985), would be to select objectively a series of repeatable points, or "landmarks," the relative positions of which could then be analyzed.

Strauss and Bookstein (1982) have devised a system of collecting distance measurements between landmarks which, in the spirit of D'Arcy Thompson (1942), attempts to reconstruct the geometry of the form being studied. In their "truss" method, distances are measured, where possible, to form a series of diagonally intersected quadrilaterals (Fig. 1). In this way, an even coverage in a number of planes can be made on the specimen, as a series of short distances. This gives a geometrical arrangement that is sensitive to changes in shape between individuals and allows the direction of any shape changes to be identified. Although this method of data collection provides an accurate way of quantifying changes in form, it has not yet been applied to studies of the Auchenorrhyncha and, indeed, is only just beginning to be used on insects (e.g., Wetton, 1988).

3 Data Analysis

A variety of statistical techniques are available with which to analyze landmark data in morphometrics, and new ones are regularly being developed (see Rohlf, 1990). Univariate analyses have been found to be limiting in representing patterns of variation among populations when compared to

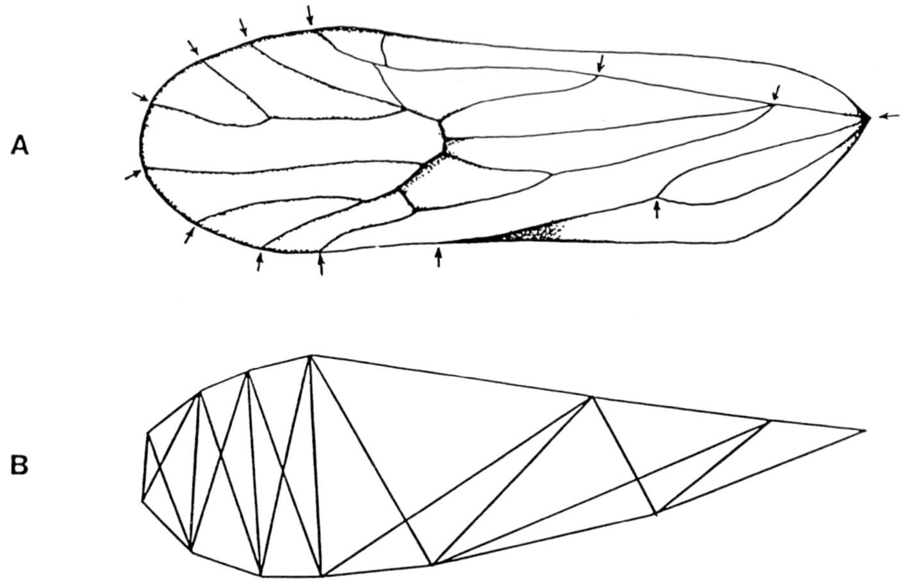

Figure 1. Forewing of the brown planthopper, *Nilaparvata lugens*. (A) Selected landmarks (homologous between individuals). (B) Distance measurements between landmarks to form a truss network, using the method of Strauss and Bookstein (1982).

multivariate techniques (Willig et al., 1986). By taking into account covariation among a number of variables, a multivariate analysis may give a more stable picture of variation than can the values of a single distance measurement.

Ordination techniques represent graphically patterns of multivariate variation among populations, in a few dimensions. This is preferable to using hierarchical techniques, such as cluster analysis, which are limited to representing variation in only one dimension and are based on overall similarity measures (Sneath and Sokal, 1973). Such a categorical method can cannot satisfactorily display the relative relationships between a number of populations, which is important in studies on geographic variation (Thorpe, 1976). It also cannot indicate the degree and significance of any overlap between populations that is based upon the covariation of variables, which is an important consideration when the status of sympatric populations are questioned. The two ordination techniques often used in systematics are canonical variates analysis [CVA] (the multiple-group ordination form of discriminant function analysis [DFA]) and principal components analysis [PCA] (James and McCulloch, 1990).

4 Canonical Variates Analysis (CVA): Patterns of Variation

CVA can be used to find the best separation that is possible between groups of individuals, with a number of distance measures. New uncorrelated axes of variation, called canonical variates, are calculated so as to maximize the variance between groups relative to the pooled variance within groups. The first canonical variate explains most of the between-group variation, and the remaining canonical variates are ranked in order of the amount of any remaining variation that they explain. The number of axes derived is dependent upon the number of groups being analyzed. The scores for individuals on the first two, or even three, canonical variates can be plotted against each other to give a scattergram. This is a convenient representation of the relative degree of discrimination between groups that includes most of the variation.

The Mahalanobis' generalized distance is usually used as a measure of the degree of discrimination between groups, and the F-statistic can be calculated as a significance value between pairs of groups. CVA, therefore, provides a statistical and graphical means by which to represent patterns of variation among populations. This is particularly useful for studies of geographical variation, but can also be used to obtain a measure of similarity among sympatric populations or closely related species. It can also be used to determine the relationship of individuals of unknown origin to known populations.

If populations cannot be separated completely, using a classification coefficient that is based on all the extracted canonical variates (or discriminant functions in the two group situation), it means the populations are morphologically inseparable, at least on the basis of the variables used. If significant differences in morphology are found, it is important to know how much of the variation is heritable and how much is the result of environmental effects, before any decisions on genetic status can be made. The influence of habitat on morphology is liable to be particularly marked in phytophagous insects that are associated with different host plant species.

Rearing eggs or early instar nymphs from different populations on the same host plant under the same laboratory conditions should reduce any environmental effects, although if significant differences are found these could still have resulted from different environmental effects on the developing eggs and young nymphs. Only by rearing a complete generation of each population under the same laboratory conditions, can any environmental effects on morphology between the populations be ruled out. Even in this case, extreme care must be taken when using individuals from long established laboratory cultured populations, because they may have changed dramatically during the process of sampling and culturing. However, such experimentation may not be practicable when studying populations that are derived from a wide geographic range, or that cannot be bred easily; therefore, decisions on the degree of inherited morphological variation that is present among populations may have to be made from the analysis of a single generation of field collected specimens. In such cases, a method of data analysis is needed that can identify the heritable components of variation.

Differences in the shapes of morphological structures between sympatric populations have long been used by taxonomists to delimit and describe species. The rationale behind this use is that size is more influenced by environmental factors, whereas shape is more likely to have a larger genetic component. Shape variation may not always be obvious, as in a discontinuity in the outline of a single structure, but may be more subtle or difficult to visualize and be spread over the form of an organism. Shape can, therefore, be interpreted, in a multivariate sense, as a factor that is present in the covariation of distance measures, which may or may not describe a single structure (Humphries et al., 1981). (A discussion of the semantics of "size" and "shape" is given by Bookstein [1989].)

In CVA, the contribution of characters to group separation can be estimated from the relative magnitudes of the coefficients of each on the axis of variation. However, CVA has limitations in representing patterns of variation of shape. Information on shape is lost in CVA, because measures which may be important in describing shape variation, but which contribute little to overall discrimination between groups, will be suppressed when adjustments are made to maximize between-group variation in the analysis (Humphries et al., 1981).

5 Principal Components Analysis (PCA): Size and Shape

PCA can be used to clarify patterns of covariation among distance measures. As in CVA, uncorrelated axes of variation, in this case principal components, are extracted as linear combinations of the original variables. Unlike CVA, however, PCA does not maximize the variation between groups compared to the variation within groups; instead, it extracts components in the order of the amount of variation that they explain among ungrouped individuals. Analysis can be made on a pooled correlation or a variance-covariance matrix.

The relative contribution of the variables to each component is given by the magnitude of their coefficients. The coefficients on PC-1, the axis which explains most of the variation, are often roughly equal in value and are the same sign. This has lead to the first principal component, calculated from logged distance measures, being interpreted as representing growth or overall size among individuals. The remaining components can then be thought of as representing residual shape variance when the variables on these components have coefficients of opposite signs and are substantially different from zero. However, it should be realized that the first principal component may also contain an unknown amount of allometrically-related shape variation, and that interpretation of the second principal component as a shape axis alone is also unwise (Mosimann and James, 1979). A number of modifications to PCA have been proposed which attempt to partition size and shape more effectively (Humphries et al., 1981; Bookstein et al,, 1985; Somers, 1986), the relative merits of which are still the subject of debate (Rohlf and Bookstein, 1987; Somers, 1989; Sundberg, 1989).

Despite these reservations, PCA provides a useful technique with which to look for the hidden patterns of variation within a group of individuals, while allowing any discontinuities shown on each component to be interpreted in terms of the covariation of distance measures. This is particularly useful when populations need to be identified in mixed samples. However, individuals which are known to be from different populations should not be pooled before analysis, because this will result in within- and between-group sources of variation being confounded (Thorpe, 1983a, 1983b). A widely available and robust procedure that takes group structure into account is a PCA performed on the pooled within-group variance-covariance matrix, known as a multiple-group principal components analysis (MGPCA) (see Thorpe, 1983a, 1983b, 1988).

MGPCA allows axes of variation between groups to be interpreted as in an ordinary PCA, although as mentioned above, a strict interpretation of size and shape axes should be avoided. However, components that explain relatively small amounts of variation are less likely to contain large amounts of overall size variation, because most size variation will have been explained already by larger (earlier) components. If two populations are, therefore, completely discriminated on these residual components, they are most probably different in shape. If these populations are sympatric, this is good evidence for reproductive isolation. The contribution of the covariation of variables to the discrimination on each component can be read from the magnitude of the coefficients on each, though again this is somewhat subjective. By taking group structure into account, the degree of between-group variation on each component can be measured statistically.

A CVA on the individual scores from all the components derived from a MGPCA will give the same ordination result as a CVA on the original data. Therefore, if the first principal component is recognized as basically representing an overall size factor, a "growth-out" analysis can then be made by inputing the scores for the individuals on all components except the first into a CVA. The removal of a general-size factor should give patterns of variation that are much less influenced by environmental factors. This would be particularly useful for studies on geographic variation in phytophagous insects when the plant species on which the populations sampled have developed is unknown.

6 Case Studies

6.1 *Nilaparvata lugens*: Biotypes and Species

Nilaparvata lugens (Stål) is the brown planthopper, a major pest of irrigated rice in much of Asia. The insect is a phloem-feeder and is restricted to cultivated and wild rices (*Oryza* species) as host plants. A major strategy for control has been the breeding of varieties of rice that are resistant to feeding by the insect.

Populations of the planthopper have been identified with distinctly different patterns of feeding virulence to rices that incorporate different genes for resistance. These populations are often termed biotypes and identified by a system of numbering (summaries in: Claridge and den Hollander, 1980; Wilson and Claridge, 1985). Differences of opinion have been expressed as to the status of these biotypes (review in: Claridge and Morgan, 1987).

Saxena and his colleagues, working at the International Rice Research Institute (IRRI), Philippines, take the view that the three biotype populations (biotypes 1, 2 and 3) maintained in culture at IRRI on cultivars incorporating respectively, no gene for resistance, *Bph1* and *bph2*, are genetically well differentiated forms (Saxena and Rueda, 1982; Saxena and Barrion, 1985). On the other hand, Claridge and colleagues claim that the biotypes at IRRI show only slight genetic differentiation and that one may be effectively converted to another by intense selection on an appropriate rice variety over as few as 10 generations (Claridge and den Hollander, 1980, 1982).

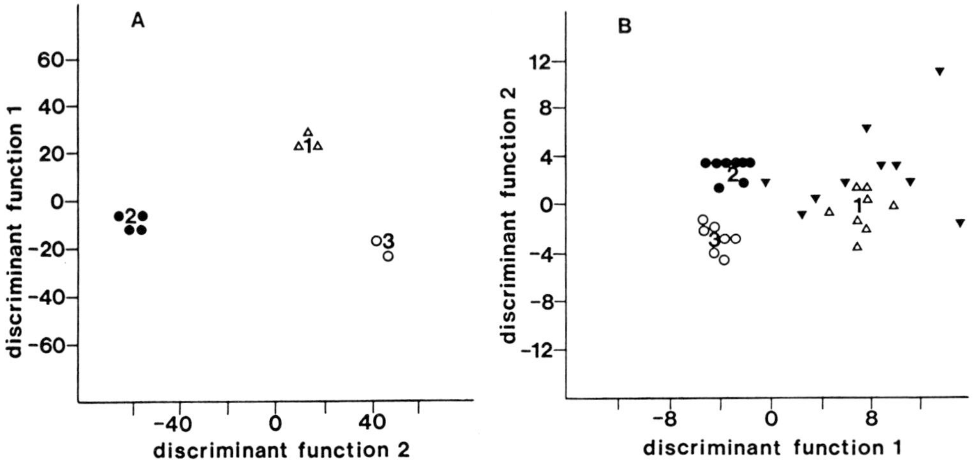

Figure 2. Canonical variate plots for *Nilaparvata lugens* from the Philippines. (A) 'Biotypes': 1 [= white triangles], 2 [= black dots], and 3 [= white dots]. (B) 'Biotypes' 1, 2 and 3 and population from *Leersia hexandra* [= upside-down black triangles]. (after Saxena and Barrion, 1985)

Saxena and Rueda (1982) made a detailed morphometric analysis of samples from the IRRI biotype cultures using 109 different measurable characters of the rostrum, legs and antennae of adult brachypterous, and macropterous males and females. Their results were analyzed using the Wilks' Lambda criterion for canonical variates analysis. In this way, they were able to show 100% discrimination between the biotype populations (Fig. 2A). They then went on, with no further evidence, to suggest that similar morphometric differences between field populations might be used to predict "biotype" status (Saxena and Barrion, 1985). Claridge et al. (1984) did a similar, but smaller scale, morphometric analysis on the same biotype cultures. They studied 10 selected characters for brachypterous males of biotypes 1, 2 and 3 (Fig. 3). The same system of CVA as used by Saxena and Rueda (1982) provided clear, but less complete, discrimination between the three populations (Fig. 4A). The less complete discrimination of the biotypes was probably due to the smaller number of characters used by Claridge et al. (1984). Thus, it is clear that the IRRI biotypes are clearly distinguishable by CVA.

In order to test the nature of the differentiation revealed by the CVA, Claridge et al. (1984) reared the three biotype populations for only one generation on the same susceptible variety of rice (TN1). The resulting adult insects were measured in the same way and subjected to CVA. The analysis showed that after rearing on the same variety, there was then very little discrimination between the biotype populations with wide overlaps between them (Fig. 4B). It was concluded that the differences that were previously demonstrated did not represent major genetic differentiation, but were primarily induced by environmental factors; these were probably associated with the relative food qualities of the different rices on which the populations

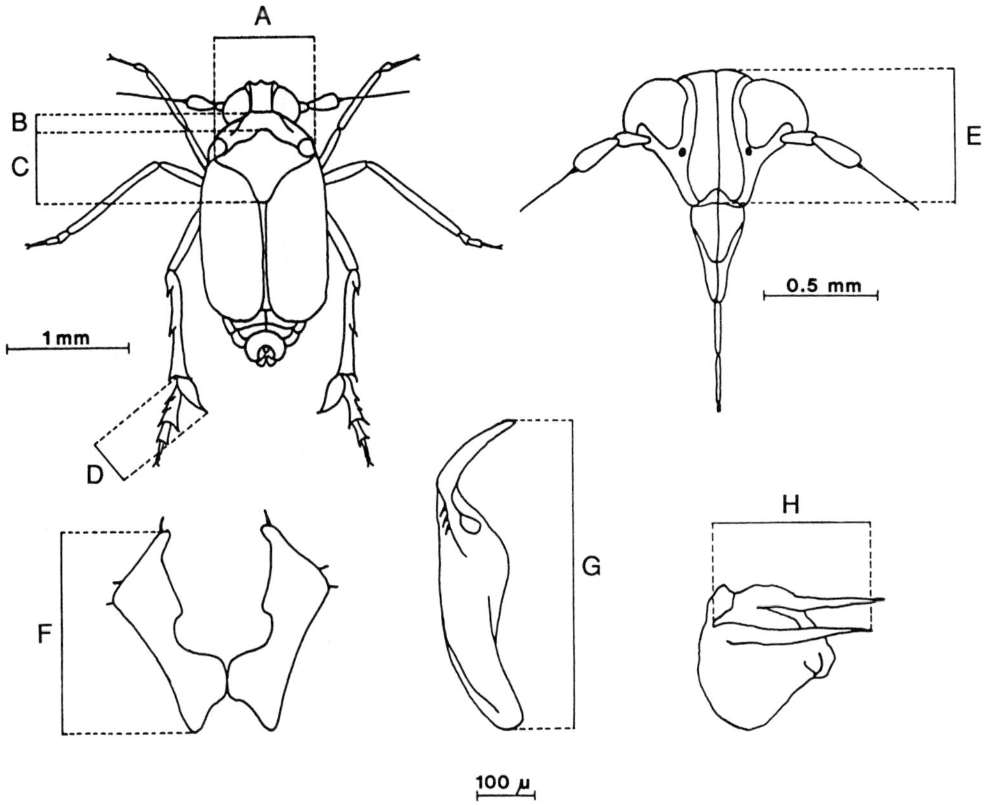

Figure 3. Adult brachypterous male of *Nilaparvata lugens* showing some of the measurements used as characters for the morphometric analysis of 'biotypes' by Claridge et al. (1984). Characters: a = scutellum width, b = pronotum length, c = scutellum length, d = length of tarsal spur, e = frons length, f = paramere length, g = aedeagus length along perimeter, h = anal tube length.

were normally maintained. Thus, the demonstration of clear differences by CVA is not of itself evidence for genetic divergence of the biotypes and certainly does not justify the extrapolation of such results to field populations.

As stated above, *N. lugens* is restricted to rices as field food plants. However, Medrano and Heinrichs (1982) first noted that populations apparently of the same species feeding on the grass *Leersia hexandra* in the Philippines. *Leersia hexandra* is a common weed of irrigation channels and reservoirs near rice paddies. Similar populations have since been found widely in Asia and Australia (Claridge et al., 1988). Saxena described the *Leersia*-associated population as "a non-virulent biotype." CVA applied to characters of the rostrum, legs and antennae showed that there was some discrimination between the populations, though less than previously demonstrated between the rice-associated biotypes.

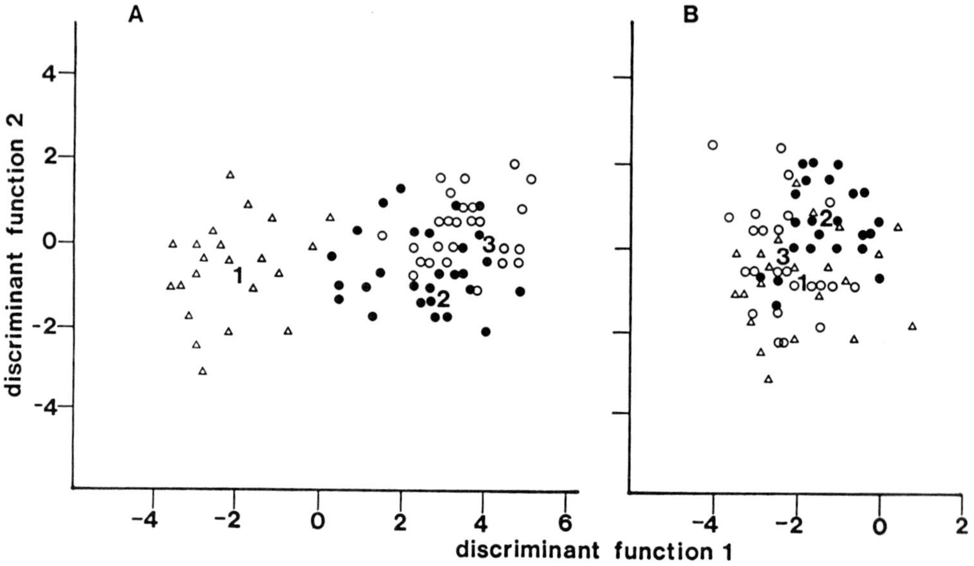

Figure 4. Canonical variate plots for 'biotypes' 1, 2 and 3 of *Nilaparvata lugens* from the Philippines. (A) Reared, respectively, on varieties TN1, Mudgo and ASD7. (B) All reared on TN1 (after Claridge et al., 1984).

Studies on sympatric *Leersia*- and rice-associated populations from different geographical regions (Claridge et al., 1985, 1988) showed them to be reproductively isolated in the field and, therefore, to be different biological species.

6.2 *Nephotettix*: Species Hybridization

Species of *Nephotettix* are known as green leafhoppers and are common in and around rice fields in Africa, Asia and Australia (Wilson and Claridge, 1985). *Nephotettix virescens* (Distant), and to a lesser extent *N. nigropictus* (Stål), are vectors of the important tungro virus disease of rice in tropical Asia. The two species are generally distinguished primarily on a basis of the pattern of dark markings on the adults (Ghauri, 1971). Sometimes in field populations morphologically intermediate individuals may be found and it has been suggested that they might represent natural interspecific hybrids (Ramakrishnan and Ghauri, 1979).

Haslam (1984), using a DFA technique, was able clearly to separate field populations of the two species (Fig. 5A). When morphologically intermediate individuals were included in the analysis, it was found that they fell within the range of variation of one or other of the two species, but not in between (Fig. 5B). However, when true interspecific hybrids were obtained in the laboratory between the same field populations, and were incorporated into the analysis, they were found to scatter between the two species on the plot of discriminant

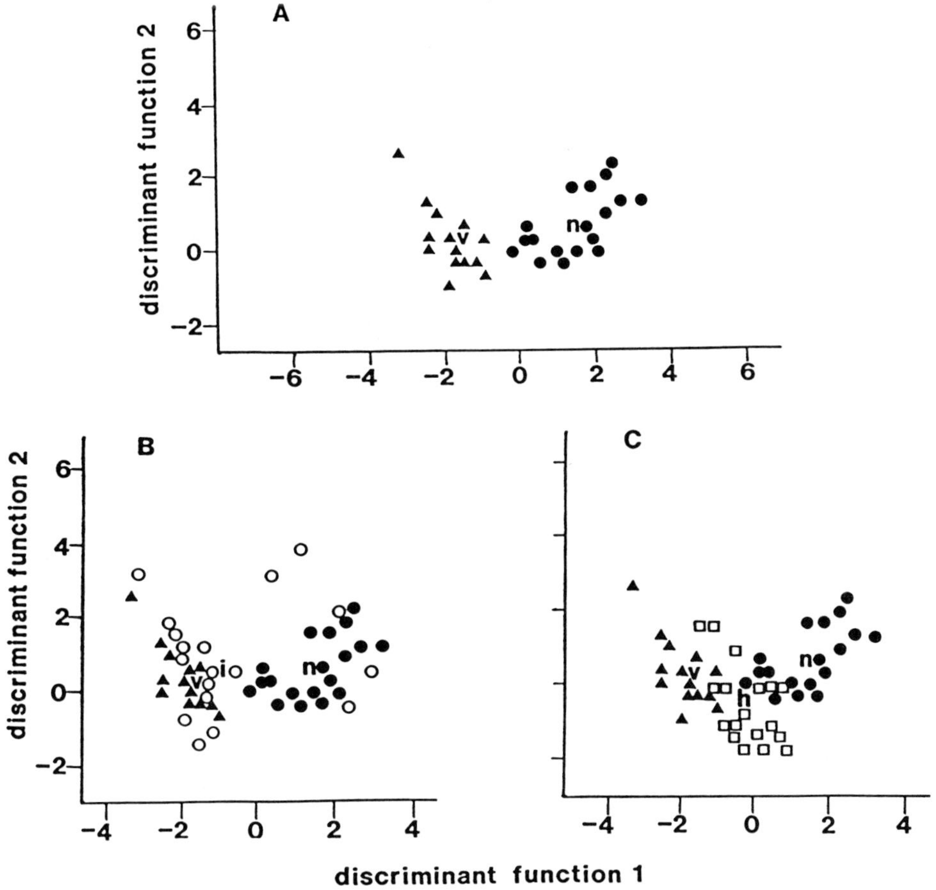

Figure 5. Canonical variate plots for *Nephotettix nigropictus* [= n, black dots] and *N. virescens* [= v, black triangles] from Sri Lanka. (A) The two species, excluding morphological intermediates. (B) The two species including morphological intermediates [= i, white dots]. (C) The two species and laboratory produced interspecific hybrids [= h, white squares]. (after Haslam, 1984)

functions 1 and 2 (Fig. 5C). True hybrids were truly intermediate. It was thus suggested that the supposed field hybrids, which were based on intermediate morphology, were not true genetic hybrids. There is then no evidence for field hybridization between *N. virescens* and *N. nigropictus*.

6.3 *Oncopsis*: Sibling Species

The genus *Oncopsis* includes a group of essentially northern Holarctic leafhoppers that feed on the phloem of broad leaved trees, especially species of

Betula, Alnus, Corylus and related genera. Species taxonomy is usually very difficult because of the generally slight differentiation of male genitalia and the development of complex color pattern polymorphisms (Claridge and Nixon, 1986).

Two very closely related species, *O. flavicollis* (L.) and *O. subangulata* (Sahl.), have generally been recognized as living, often together in mixed populations, on the two native tree birches in Britain, *Betula pendula* Roth and *B. pubescens* Ehrh. Confirmation of their species status was first obtained by recording their acoustic courtship and calling signals (Claridge and Reynolds, 1972).

Claridge and Nixon (1986) recently made CVA studies on both males and females of *O. flavicollis* and *O. subangulata* from South Wales. In both sexes very significant discriminations were obtained (Fig. 6). However, in addition, populations of *O. flavicollis* from *B. pendula* and *B. pubescens* were also significantly discriminated. Standardized discriminant function coefficients showed that for males the measurements of the internal abdominal apodemes provided the best discriminators. Indeed, further study showed that these could be used reliably to discriminate not only the species *O. flavicollis* and *O. subangulata*, but also the populations of *O. flavicollis* from *B. pendula* and from *B. pubescens*. It was thus suggested that the latter might represent also very closely allied biological species. Acoustic analysis not only confirmed this suggestion, but also indicated the presence of a further species within the *O. flavicollis* complex in eastern Britain.

6.4 *Alnetoidia alneti*: Host Plant Associated Populations

Alnetoidia alneti is an abundant, polyphagous, mesophyll-feeding leafhopper, commonly found on the leaves of a wide variety of broad leaved trees. Claridge and Wilson (1981) listed species of 13 different genera of trees as hosts in Britain. Claridge and Wilson (1976) cited some data which suggested that populations associated with two common hosts, alder (*Alnus glutinosa* (L.)) and hazel (*Corylus avellana* L.) differed in size and host preferences and could represent sibling species. Indeed, Ribaut (1936) recognized these populations as different species, based on size and pigmentation.

Gillham and Claridge (unpublished data) made a morphometric study of *A. alneti* populations collected separately from 13 different species of host trees in South Wales. CVA showed a series of overlapping ranges of variation among insects from these host plants, with the populations from alder and hazel forming extreme groups. Populations that were collected widely in Britain from these two hosts, over a period of more than ten years, were then analyzed in more detail. CVA once again discriminated clearly the two host associated clusters. All of the alder samples were grouped closer to each other than to any of the hazel samples; the same was also true for the hazel samples themselves. Thus, the effects of host plant appear clearly to be more important than either geography or time.

MGPCA on pooled within-group variance-covariance matrices from logged data gave complete separation of individuals from alder and hazel; males separated on component 1 (Fig. 7) and females separated on components 1 and 2. These axes were regarded as representing overall size, due to their

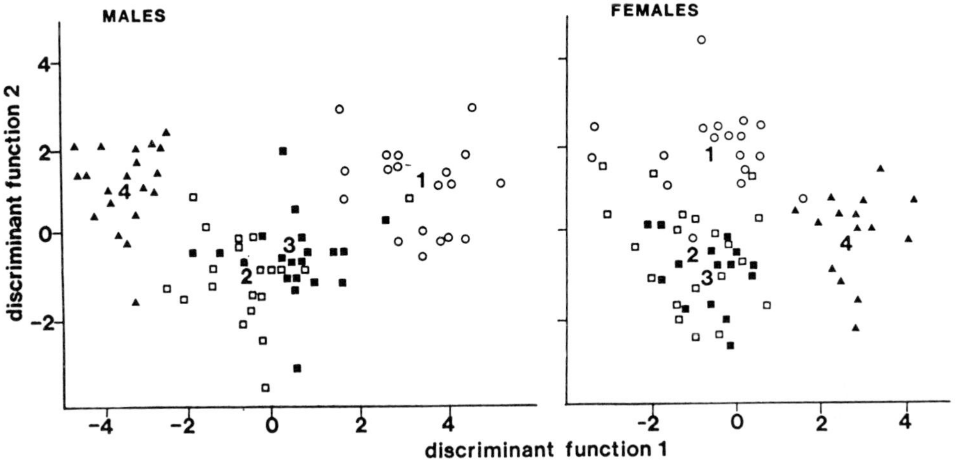

Figure 6. Canonical variate plots for males (left) and females (right) of *Oncopsis flavicollis* from *Betula pendula* near Cardiff [= 1, white squares], from *B. pubescens* near Cardiff [= 2, white squares], from *B. pubescens* in Brecon Beacons [= 3, black squares], and for *O. subangulata* from *B. pubescens* in Brecon Beacons [= 4, black triangles]. (after Claridge and Nixon, 1986)

coefficient values. Although there were significant differences on the remaining components, complete separation was not found for either sex. There was, therefore, no evidence for any morphological variation between *A. alneti* from alder and hazel that could be attributed to differences in shape. This suggested a lack of reproductive isolation. Differences, essentially due to size, were found between individuals from the hosts, and may have some genetic basis.

In order to attempt to test the nature of the morphometric differentiation between the hazel and alder populations, transfer experiments were performed. First and second instar nymphs that were collected from one hazel plant were reared separately on foliage from different individual hazel plants; additionally, nymphs from one alder plant were reared separately on the foliage from different alder plants and also on hazel foliage. The resulting adults were subjected to the same CVA procedure as above. Those reared on foliage from different individual trees, but of the same plant species, were found to overlap almost completely, whereas no overlap was found between those reared on alder and those reared on hazel. The adults, taken as nymphs from alder but reared on hazel, overlapped with those taken from and reared on hazel, but not with those taken from and reared on alder (Fig. 8).

It is clear, therefore, that in this species the nymphal food plant radically affects morphometric variability of resulting adults. There is no evidence for genetically controlled differences. Gillham (1989), in studies on the acoustic behavior of *A. alneti*, showed that there was no difference in the pattern of the call produced by males, between individuals that were collected as nymphs on alder and those that were collected as nymphs on hazel. Also, electrophoretic

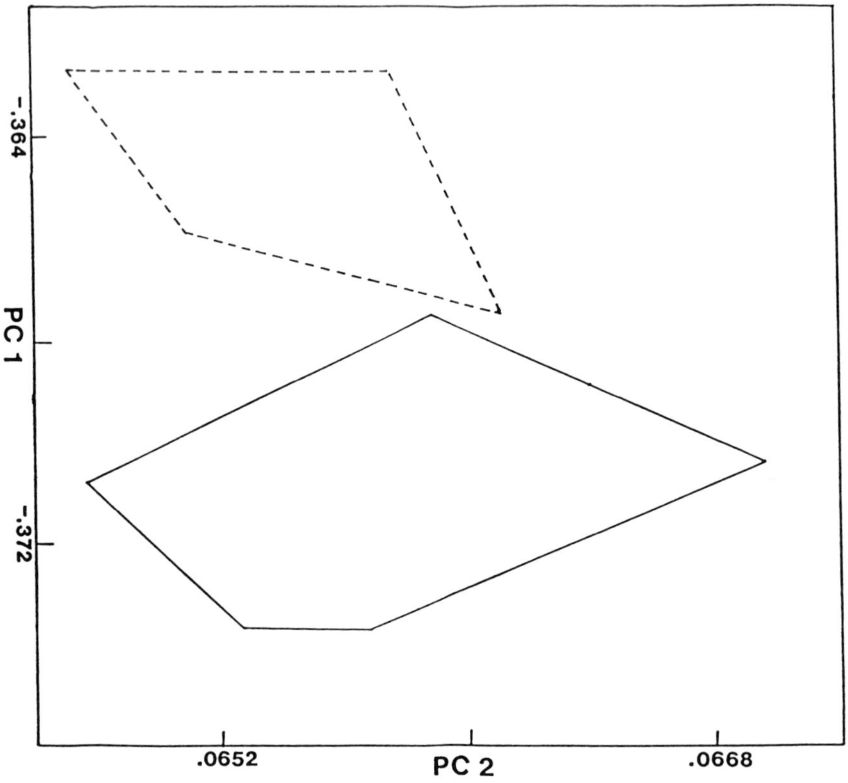

Figure 7. Multiple-group PCA of host-associated populations of *Alnetoidia alneti* in South Wales. PCA space circumscribed for adults collected on *Alnus glutinosa* [=solid line] and *Corylus avellana* [= broken line].

analysis, using three enzyme systems, failed to reveal any diagnostic loci between individuals taken from alder and hazel field populations. *Alnetoidia alneti* appears then to be a widely polyphagous species with a hierarchy of host plant suitability and preference, with alder and hazel as extremes. This may explain the more common occurrence of *A. alneti* on alder than hazel in south Wales (MCG, personal observation).

6.5 Other Studies

Space does not allow us to give details of all publications that have used ordination techniques for studying morphometric variation in leafhoppers and planthoppers. Other important investigations include those of Sorensen (1987) and Sorensen and Sawyer (1989) on the two forms of the *Macrosteles fascifrons* (Stål) complex, long known in culture in the United States. PCA and CVA techniques separate two forms which, supported by acoustic evidence (Purcell

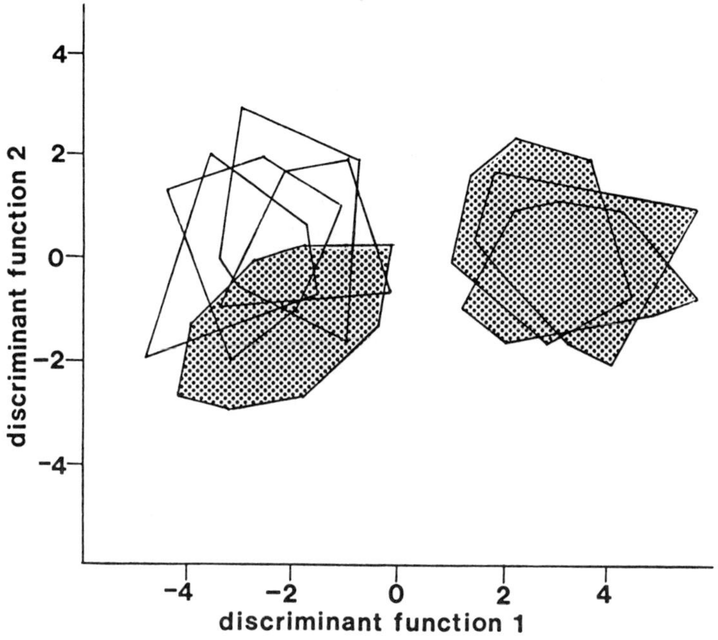

Figure 8. Canonical variate plot of space circumscribed for adults of *Alnetoidia alneti* reared on *Corylus avellana* [grouped to the left] and *Alnus glutinosa* [grouped to the right]; shaded segments represent adults derived from nymphs collected on *Alnus glutinosa*.

and Loher, 1976), should be regarded as different biological species. Further, CVA shows that these species apparently have different phenological regressions for seasonal morphological change in leg and wing traits (Sorensen and Sawyer, 1989). Further data and analyses of field populations are needed on this complex.

Siwi (1986) has begun the first major study on geographical variation on the green leafhopper, *Nephotettix virescens*, in Indonesia. She reports analyses of populations that were collected in an east to west series from Sumatra, Java, Bali, Lombok, Sulawesia and Irian Jaya. PCA techniques show discrimination between some populations, but no clear geographical pattern is obvious. Further work on these important insect pests is needed and could most usefully be combined with hybridization and acoustic studies.

7 Discussion and Conclusions

The studies briefly outlined above clearly demonstrate the value of ordination techniques in the analysis of variation between identifiable populations of leafhoppers and planthoppers. This applies equally to apparently sympatric populations that are associated with different host plants, as well as to geographically definable and presumably truly allopatric ones. The principles of analysis can, of course, be applied to other insect groups and used to compare populations of those which develop on different animal or plant hosts.

Although all studies mentioned here have used only morphological data, ordination techniques can also be applied to other characteristics, such as biochemical or behavioral traits. Indeed, it is clear that valid conclusions about the variation of natural populations can only be achieved by such a diversity of approaches. For example, Gillham (in press) used CVA and PCA on a number of variables selected from the acoustic calling signals of *Alebra* leafhoppers to show patterns of geographic, host plant and temperature related variation.

CVA and PCA methods provide powerful techniques, but in principle they are not able by themselves to determine whether or not populations are reproductively isolated. Direct observation and experiment alone can provide conclusive data. The example of *Nilaparvata lugens* described above is instructive. The rice-feeding populations, often known as biotypes, are clearly separable by CVA (Fig. 2A). However, only experimental transfer between different rice varieties demonstrated that the differences are almost entirely host plant induced (Fig. 4). On the other hand, the populations from rice and from the grass, *Leersia hexandra*, whilst they show less discrimination by CVA than the rice biotypes (Fig. 2B), are reproductively isolated and distinct biological species.

Thus, ordination techniques are enormously valuable, and will undoubtedly become even more important as new and more powerful computer systems become widely available. Hopefully the lively debates presently taking place in morphometrics will result in robust techniques becoming readily available to those interested in elucidating the taxonomic status and patterns of genetic variation in insect field populations. They must, nevertheless, be seen as tools which, only when coupled to field observation and experiment, will give us new insights into the nature of variation in insect populations.

8 Acknowledgments

We thank Panos Petrakis for reading the manuscript and for detailed discussion on ordination techniques. Studies on leafhoppers and planthoppers at Cardiff have been funded in part by the Overseas Development Administration (ODA) and the Natural Environmental Research Council (NERC), for whose support we are deeply indebted.

9 References

Booij, C.J.H., 1982. Biosystematics of the *Muellerianella* complex (Homoptera, Delphacidae): hostplants, habitats and phenology. *Ecological Entomology*, 7: 9-18.

Bookstein, F.L., 1989. "Size and shape": a comment on semantics. *Systematic Zoology*, 38: 173-180.

Bookstein, F.L., Chernoff, B., Edler, R.L., Humphries, J.M., Smith, G.R. and Strauss, R.E., 1985. Morphometrics in Evolutionary Biology. Special

Publication, 15. Academy of Natural Sciences, Philadelphia, 277 pp.

Claridge, M.F., 1985. Acoustic behavior of leafhoppers and planthoppers: species problems and speciation. In: L.R. Nault and J.G. Rodriguez (Editors), The Leafhoppers and Planthoppers. John Wiley and Sons, New York, pp. 103-125.

Claridge, M.F., 1988. Species concepts and speciation in parasites. In: D.L. Hawksworth (Editor), Prospects in Systematics. Clarendon Press, Oxford, U.K., pp. 92-111.

Claridge, M.F. and den Hollander, J., 1980. The 'biotypes' of the rice brown planthopper, *Nilaparvata lugens*. *Entomologia Experimentalis et Applicata*, 27: 23-30.

Claridge, M.F. and den Hollander, J., 1982. Virulence to rice cultivars and selection for virulence of the brown planthopper, *Nilaparvata lugens*. *Entomologia Experimentalis et Applicata*, 32: 213-221.

Claridge, M.F., den Hollander, J. and Haslam, D., 1984. The significance of morphometric and fecundity differences between the 'biotypes' of the brown planthopper, *Nilaparvata lugens*. *Entomologia Experimentalis et Applicata*, 36: 107-114.

Claridge, M.F., den Hollander, J. and Morgan, J.C., 1985. The status of weed-associated populations of the brown planthopper, *Nilaparvata lugens* (Stål): host race or biological species? *Zoological Journal of the Linnean Society*, 84: 77-90.

Claridge, M.F., den Hollander, J. and Morgan, J.C., 1988. Variation in hostplant relations and courtship signals of weed-associated populations of the brown planthopper, *Nilaparvata lugens* (Stål), from Australia and Asia: a test of the recognition species concept. *Biological Journal of the Linnean Society*, 35: 79-93.

Claridge, M.F. and Morgan, J.C., 1987. The brown planthopper, *Nilaparvata lugens* (Stål), and some related species: biotaxonomic approach. In: M.R. Wilson and L.R. Nault (Editors), Proceedings of 2nd International Workshop on Leafhoppers and Planthoppers of Economic Importance, held at Brigham Young University, Provo, Utah USA, 28th July - 1st August 1986. CAB International Institute of Entomology, London, pp. 19-32.

Claridge, M.F. and Nixon, G.A., 1986. *Oncopsis flavicollis* (L.) associated with tree birches (*Betula*): a complex of biological species or a host plant utilization polymorphism? *Biological Journal of the Linnean Society*, 27: 381-397.

Claridge, M.F. and Reynolds, W.J., 1972. Host plant specificity, oviposition behavior and egg parasitism in some woodland leafhoppers of the genus *Oncopsis* (Hemiptera: Cicadellidae). *Transactions of the Royal Entomological Society of London*, 124: 146-166.

Claridge, M.F. and Wilson, M.R., 1976. Diversity and distribution patterns of some mesophyll-feeding leafhoppers of temperate woodland canopy. 1976. *Ecological Entomology*, 1: 231-250.

Claridge, M.F. and Wilson, M.R., 1981. Host plant associations, diversity and species-area relationships of mesophyll-feeding leafhoppers of trees and shrubs in Britain. *Ecological Entomology*, 6: 217-238.

den Bieman, C.F.M., 1987. Host plant relations in the planthopper genus *Ribautodelphax* (Homoptera, Delphacidae). *Ecological Entomology*, 12: 163-172.

Ghauri, M.S.K., 1971. Revision of the genus *Nephotettix* Matsumura (Hemiptera: Cicadelloidea: Euscelidae) based on the type material. *Bulletin of Entomological Research*, 60: 481-512.

Gillham, M.C., 1989. Biosystematic Studies on some Mesophyll-feeding Leafhoppers Associated with Trees and Shrubs. Ph.D. Thesis. University of Wales, Cardiff, (unpublished), 257 pp.

Gillham, M.C., 1991. Polymorphism, taxonomy and host plant associations in *Alebra* leafhoppers (Homoptera: Cicadellidae). *Journal of Natural History*, 25: 233-255.

Gillham, M.C., (in press). Variation in acoustic signals within and among leafhopper species of the genus *Alebra* (Homoptera: Cicadellidae). *Biological Journal of the Linnean Society*.

Haslam, D., 1984. Taxonomic Studies on Some Planthopper and Leafhopper Pests of Rice. M.Sc Thesis. University of Wales, Cardiff, (unpublished), 176 pp.

Humphries, J.M., Bookstein, F.L., Chernoff, B., Smith, G.R., Elder, R.L. and Poss, S.G., 1981. Multivariate discrimination by shape in relation to size. *Systematic Zoology*, 30: 291-308.

James, F.C. and McCulloch, C.E., 1990. Multivariate analysis in ecology and systematics: panacea or Pandora's box? *Annual Review of Ecology and Systematics*, 21: 129-166.

Mayr, E., 1942. Systematics and the Origin of Species from the Viewpoint of a Zoologist. Columbia University Press, New York, 334 pp.

Mayr, E., 1963. Animal Species and Evolution. Oxford University Press, Oxford, U.K., 797 pp.

Medrano, F.G. and Heinrichs, E.A., 1982. *Leersia hexandra* as weed host for the brown planthopper. *International Rice Research Newsletter*, 7: 15-16.

Mosimann, J.E. and James, F.C., 1979. New statistical methods for allometry with application to Florida red-winged blackbirds. *Evolution*, 33: 444-459.

Neff, N.A. and Marcus, L.F., 1980. A Survey of Multivariate Methods for Systematics. American Museum of Natural History, New York, 221 pp.

Purcell, A.H. and Loher, W., 1976. Acoustical and mating behavior of two taxa in the *Macrosteles fascifrons* species complex. *Annals of the Entomological Society of America*, 69: 513-518.

Ramakrishnan, U. and Ghauri, M.S.K., 1979. Probable hybrids of *Nephotettix virescens* (Distant) and *N. nigropictus* (Stål) (Hemiptera: Cicadellidae) from Sabah, Malaysia. *Bulletin of Entomological Research*, 69: 357-361.

Reyment, R.A., Blackith, R.E. and Campbell, N.A., 1984. Multivariate Morphometrics, (Second Edition). Academic Press, New York, 233 pp.

Ribaut, H., 1936. Fauna de France, 31: Homopteres Auchenorrhynques, 1 (Typhlocybidae). Paul Chevalier, Paris, 231 pp.

Rohlf, J.F., 1990. Morphometrics. *Annual Review of Ecology and Systematics*, 21: 299-316.

Rohlf, F.J. and Bookstein, F.L., 1987. A comment on shearing as a method of "size correction." *Systematic Zoology*, 36: 356-367.

Saxena, R.C. and Barrion, A.A., 1985. Biotypes of the brown planthopper *Nilaparvata lugens* (Stål) and strategies in development or host plant resistance. *Insect Science and its Application*, 6: 271-289.

Saxena, R.C. and Rueda, L.M., 1982. Morphological variations among three biotypes of the brown planthopper *Nilaparvata lugens* in the Philippines. *Insect Science and its Application*, 3: 193-210.

Siwi, S.S., 1986. Variation in morphological characteristics of *Nephotettix virescens* (Distant) (Homoptera, Euscelidae) from different islands in Indonesia. *Contributions to the Central Research Institute for Food Crops, Bogor (Indonesia)*, 75: 1-26.

Sneath, P.H.A. and Sokal, R.R., 1973. Numerical Taxonomy, the Principles and Practice of Numerical Classification. W.H. Freeman and Company, New York, 573 pp.

Somers, K.M., 1986. Multivariate allometry and removal of size with principal components analysis. *Systematic Zoology*, 35: 359-368.

Somers, K.M., 1989. Allometry, isometry and shape in principal components analysis. *Systematic Zoology*, 38: 169-173.

Sorensen, J.T., 1987. The multivariate evolution and taxonomic analysis of leafhopper biotypes and species complexes: use of character correlations and quantitative genetic methods. In: M.R. Wilson and L.R. Nault (Editors), Proceedings of 2nd International Workshop on Leafhoppers and Planthoppers of Economic Importance, Brigham Young University, Provo, Utah USA, 28th July - 1st August 1986. CAB International Institute of Entomology, London, pp. 217-234.

Sorensen, J.T. and Sawyer, S.M., 1989. Assessing the multivariate evolutionary responses of phenological differentiation for sibling species: biosystematics in the *Macrosteles fascifrons* complex (Homoptera: Cicadellidae). *Annals of the Entomological Society of America*, 82: 250-261.

Strauss, R.E. and Bookstein, F.L., 1982. The truss: body form reconstructions in morphometrics. *Systematic Zoology*, 31: 113-135.

Sundberg, P., 1989. Shape and size-constrained principal components analysis. *Systematic Zoology*, 38: 166-168.

Thompson, D'Arcy W., 1942. On Growth and Form, (Second Edition). Cambridge University Press, Cambridge, U.K., 1116 pp.

Thorpe, R.S., 1976. Biometric analysis of geographic variation and racial affinities. *Biological Reviews*, 51: 407-452.

Thorpe, R.S., 1983a. A review of the numerical methods for recognising and analysing racial differentiation. In: J. Felsenstein (Editor), Numerical Taxonomy, NATO Advanced Study Institute, Series G, Number 1. Springer-Verlag, New York, pp. 404-423.

Thorpe, R.S., 1983b. A biometric study of the effects of growth on the analysis of geographic variation: tooth number in the green geckos (Reptilia: *Phelsuma*). *Journal of Zoology, London*, 201: 13-26.

Thorpe, R.S., 1988. Multiple group principle component analysis and population differentiation. *Journal of Zoology, London*, 216: 37-40.

Wetton, M.N., 1988. A morphometric analysis of mandibular teeth in *Folsomia* (Collembola: Isotomidae). *Zoological Journal of the Linnean Society*, 94: 99-109.

Willig, M.R., Owen, R.D. and Colbert, R.L., 1986. Assessment of morphometric variation in natural populations: the inadequacy of the univariate approach. *Systematic Zoology*, 35: 195-203.

Wilson, M.R. and Claridge, M.F., 1985. The leafhopper and planthopper faunas of rice fields. In: L.R. Nault and J.G. Rodriguez (Editors), The Leafhoppers and Planthoppers. John Wiley and Sons, New York, pp. 381-404.

THE USE OF ORDINATION TECHNIQUES TO DISCRIMINATE WITHIN PEST APHID SPECIES COMPLEXES

R.L. Blackman

Department of Entomology, British Museum (Natural History), Cromwell Road, London, SW7 5BD, U.K.

Abstract

Multivariate techniques are particularly useful for systematic analyses of aphids, especially when clonal samples are used as the groups in canonical variates analysis. The effects of environmental variables, such as temperature, can be studied on single genotypes in the laboratory, providing information that helps greatly in the interpretation of results. Discrimination between closely related taxa is usually good, but may be confounded by the considerable phenotypic plasticity shown by aphids, which seems to be particularly great in the case of permanently parthenogenetic (apomictic) populations. This chapter addresses the coordinated use of discriminant analysis and karyotypic studies in problems of intraspecific variation and species recognition within aphid pest complexes.

1 Introduction

Aphid species of agricultural importance tend to be highly variable organisms, and often pose problems for taxonomists. Populations of what is nominally a single species may differ in their host plant relationships, life cycle and/or morphology, and these differences can give rise to doubts about whether it is correct to regard all these populations as one species, or whether a complex of closely related or sibling species might, in fact, be involved.

A related problem is that pest species often have close relatives that look very like them but feed specifically on noneconomic plants. Even when it can be shown that a pest and non-pest are distinct, reproductively isolated species with different biologies, it may be difficult to find reliable morphological characters to discriminate between them.

The aphid taxonomist clears his specimens and mounts them permanently on microscope slides, with the body almost flattened and the appendages displayed in a single plane of focus. Morphometric measurements are thus readily obtainable and, in many aphid genera, morphometric characters provide the principal or only means of discriminating between closely related species. Tables of comparative morphometric data have been used for many years in aphid species descriptions and revisionary works. Differences in general body size are traditionally compensated for by use of simple ratios or bivariate plots that include one strongly size-related character (Eastop, 1985). Functions involving as many as six variables have been empirically derived by some taxonomists for use in difficult species separations (e.g., Stroyan, 1982). It is evident, therefore, that multivariate methods are exceptionally well suited to the needs, as well as the traditions, of aphid taxonomists.

Throughout spring and summer, aphids reproduce parthenogenetically and form highly structured populations consisting of one or more lineages. Aphid parthenogenesis is an ameiotic (apomictic) form of thelytoky (Blackman, 1978) so that, rare mutational events apart, all individuals of a single parthenogenetic lineage have the same genetic constitution. This is a great advantage for multivariate work because it means that, with a little extra work, aphids collected in the field can be cloned in the insectary. A series of individuals from a single clone can then be mounted and measured to provide quantitative data describing the phenotype much more adequately than could measurement of a single specimen. Such data can provide the basis for a statistical morphological comparison of different clones, or of the same clone reared in several different environments. If the existence of more than one taxon is suspected, principal components or principal coordinate analysis can be carried out using mean values for a series of clonal samples of each putative taxon. Alternatively, the clonal samples can be used as the groups in a canonical variates analysis (CVA).

CVA seems particularly appropriate for the multivariate analysis of aphid morphometric data. Its great advantage is that it maximizes between-group differences; but this, of course, means that the data have to be organized into predefined groups. In aphids, each group can comprise individuals of a single parthenogenetic lineage, so that the analysis can be entirely objective, with no prior assumptions about the interrelationships of individuals, populations or taxa. The limitations that make data unsuitable for CVA can usually be discounted. Within-group data is unlikely to be heterogeneous (although it is still wise to look out for deviations from homogeneity) and, provided that the taxa included in the analysis are closely related, morphometric characters are unlikely to be heteroscedastic (much more variable in some groups than in others). Experience has shown that with clonal samples, small groups of about 10 specimens can provide reliable data for analysis.

Some morphometric characters are highly correlated within clonal samples, so that much of the information that such characters provide is redundant in the analysis of intergroup relationships. CVA, apart from being generally robust to the inclusion or exclusion of characters, would discount such redundant information anyway, so it continues to work well when the data set is pruned of highly correlated characters. After some preliminary work has been done, it is, therefore, possible to obtain meaningful results with small numbers of characters, reducing the time for data acquisition considerably.

If it is not possible to rear clonal samples for analysis, then it is justifiable to make the assumption that aphids collected from the same colony on the same day can be treated as a valid group for the purposes of CVA. This in no way prejudges the subsequent ordination of the groups. A test for within-group heterogeneity is, however, desirable with such field-collected material.

The main purpose of this paper is to review some examples of the application of CVA to aphid morphometric data, and to show how this method of analysis provides a powerful tool for the recognition and discrimination of closely related taxa, even in cases where such taxa exhibit considerable phenotypic plasticity.

2 Separation of *Rubus* Aphids: *Amphorophora rubi* (Kalt.) and *A. idaei* Börner

Until 50 years ago, all large green aphids with swollen siphunculi colonizing *Rubus* species in Europe were regarded as one species, *A. rubi*. Then, Börner (1939) decided on the basis of host plant transfers, color in life and certain morphological differences, that the aphids on *R. idaeus* (European raspberry), which he called *A. idaei*, were distinct from those on *R. caesius* (European dewberry) and *R. fruticosus* agg. (blackberry, brambles), both of which were colonized by what he regarded as true *A. rubi*. For many years, however, Börner's separation was not widely accepted, mainly because the morphological differences that he claimed to exist between the two taxa did not stand up to more extensive sampling (Hille Ris Lambers, 1949; Hill, 1953). Briggs (1959) regarded Börner's aphid from *R. caesius* and *R. fruticosus* as a "strain" of *A. rubi*, treating it as equivalent to other *Amphorophora* strains that are important as vectors of raspberry viruses. This left unanswered the question of whether the brambles that grow so profusely in the wild provide a natural reservoir of aphids carrying raspberry viruses.

Firm evidence in support of Börner's conclusion, that two species were involved, came from cytological observations that British populations of *Amphorophora* on *R. idaei* have a karyotype with $2n$(female) = 18, whereas those from *R. fruticosus* have $2n$(female) = 20 (Blackman et al., 1977). To test whether the taxa could be separated morphometrically, data for eight characters were obtained from a large number of samples. The characters selected were those that were previously found to be useful in species discrimination in this genus of aphids. All were continuous variables except one, caudal hair number, chosen because Börner had used it in his separation of *A. rubi* and *A. idaei*. The samples used ranged from clones reared under controlled conditions to field-collected samples from various European countries.

As an initial experiment, a CVA was done on 10 clones, four of *A. idaei* (karyotyped as $2n$ = 18) and six of *A. rubi* (karyotyped as $2n$ = 20). All originated from collections at separate localities in southeast England and were reared on excised leaves of their respective host plants at a constant temperature of 15°C and 12 h photoperiod. Mean scores on the first canonical variate (CV-1) for these clonal samples were widely separated between taxa (Fig. 1). This result shows the discriminatory powers of CVA, once the complicating influences of the environment are minimized and the data primarily reflect genetic differences between samples.

When all available samples were included in the analysis, a very different result was obtained (Fig. 2). CV-1 still discriminated between the taxa, but with a much greater range of scores and hardly any separation between the two groups, so that the lack of overlap between them seems merely fortuitous.

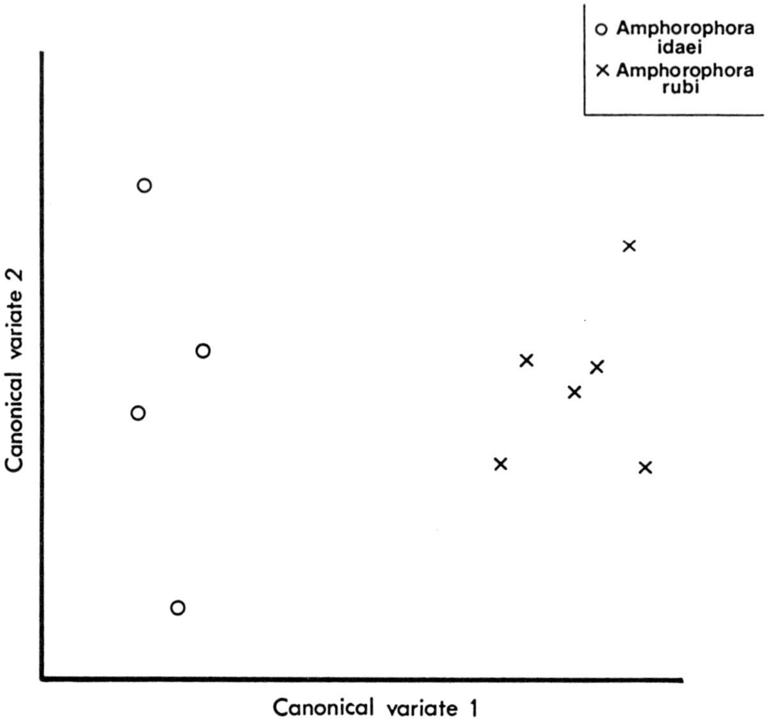

Figure 1. Plots of the scores on the first two canonical variates, based on eight characters, for clonal samples of apterae of *Amphorophora rubi* and *A. idaei* reared on excised leaves of their respective host plants in controlled conditions.

Without the information from karyotype and host plant, one would not be able to ascertain a discriminating value on CV-1 between the two taxa.

Although *A. rubi* does not colonize *R. idaeus* in the field, it will feed when caged on excised leaves of this plant in the laboratory. A clone of *A. rubi* reared on *R. idaeus* for five generations still grouped with *A. rubi*, showing that the morphological differences between taxa could not simply be an effect of the host plant. Five clones of *A. rubi* were reared in the dark at 6°C. These gave very different scores on both CV-1 and CV-2 from the same clones reared at 15°C (Fig. 2). The direction and magnitude of the differences were inconsistent, possibly because the 6°C treatment was rather extreme and caused wide variation in the physiological condition of the host plant. CV-1 can thus be strongly affected by environmental conditions, yet still discriminate between taxa.

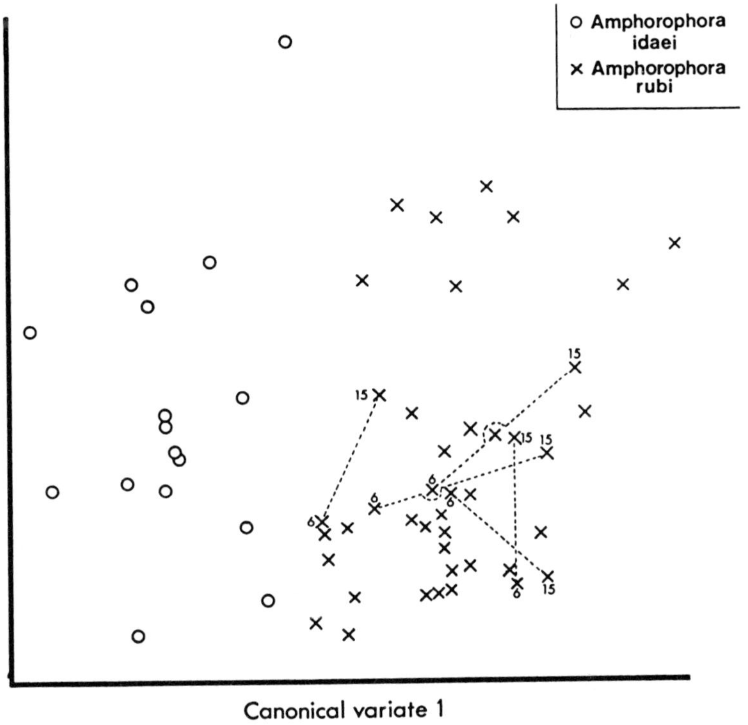

Figure 2. Plots of the scores on the first two canonical variates, based on eight characters, for reared and field-collected samples of apterae of *Amphorophora rubi* and *A. idaei*; the samples were reared on excised leaves of their respective host plants in controlled conditions (broken lines join samples of the same clone reared under different conditions).

3 Discrimination in the Green Peach Aphid: *Myzus persicae* Species Complex

Myzus persicae (Sulzer) is well known as probably the most cosmopolitan and polyphagous aphid species, but for many years there has been uncertainty about whether all the populations placed under this name were in reality a single taxonomic entity. Recently, discriminant analysis was used in conjunction with karyological studies to provide evidence that two forms (*M. antirrhinii* Macchiati and *M. nicotianae* Blackman) should be separated from *M. persicae* s. str. (Blackman and Paterson, 1986; Blackman, 1987).

In aphid taxonomy, it is important to take account of the life cycle, because this will influence the nature and degree of genetic isolation between taxa. *Myzus persicae*, itself, shows life cycle variation; it reproduces continuously by parthenogenesis in all parts of the world where winters are mild or nonexistent, yet it is nevertheless essentially a bisexual species, retaining the capability to be holocyclic (i.e., to produce sexual morphs and cold-resistant fertilized eggs on its primary host, peach [*Prunus persica*]) wherever the climate favors this method of overwintering (Blackman, 1974). Thus, populations of *M. persicae* are at least potentially interbreeding, even on a world scale. *Myzus antirrhinii* is permanently parthenogenetic, and probably originated from within ancestral *M. persicae* stock by permanent loss of the sexual phase. It can be recognized karyologically, because it has $2n$ = 13 or 14 instead of the normal *M. persicae* $2n$ = 12, apparently due to dissociations of either one or two autosomes.

When *M. nicotianae* was first shown to be distinct, it too was believed to be permanently parthenogenetic, but populations have now been found in Greece that go through a sexual phase on peach like *M. persicae*. The mechanism for reproductive isolation between *M. persicae* and *M. nicotianae* in such populations is not yet known.

Myzus persicae is also liable to be confused, particularly in Europe, with several closely related taxa in the *Myzus* subgenus *Nectarosiphon* that are more easily recognized by their biology and host plant relationships than by the morphology of slide-mounted specimens. *Myzus myosotidis*, for example, looks almost identical to *M. persicae*, but lives all the year around on water forget-me-not, *Myosotis palustris*, producing oviparae and apterous males on this plant in autumn at a time when holocyclic populations of *M. persicae* produce gynoparae (winged parthenogenetic females) and alate males that migrate back to peach. Before considering the use of CVA to distinguish hitherto unrecognized taxa, how well does it discriminate between the known species in the group?

A CVA for apterae, based on 14 morphometric variables, provided a range of CV-1 scores for "*persicae*" (including, at this stage, *M. antirrhinii* and *M. nicotianae*) that was separate from those of other species of *Myzus* (*Nectarosiphon*) (Fig. 3). A feature of this result was that CV-1, accounting for 39.2% of the total variance, seemed virtually independent of rearing temperature; samples of the same clone reared at 10°C and 20°C had almost identical scores on CV-1. In contrast, CV-2, accounting for a further 16.3% of the variance in the data, was strongly temperature-dependent and could not, therefore, contribute usefully to the separation of taxa, in particular splitting samples of *M. certus* into two clusters. Similarly, CV-3 and CV-4, accounting together for an additional 18.5% of total variance, seemed also to be concerned more with environmental than with genetic variance components and, therefore, did not improve the separation of taxa.

CV-5, however, with just 6.5% of the total variance, grouped non-"*persicae*" samples excellently according to expected taxonomic relationships (Fig. 4). The compactness of the clusters in a CV-1/CV-5 plot and the closeness of clonal samples of *M. persicae* reared at different temperatures, indicates that environmental components contribute minimally to CV-1 and CV-5, and that the 45.8% of the total variance accounted for by these two vectors was predominantly genetic.

Extraction of the first five vectors in this analysis still left about 20% of variance unaccounted for. Nevertheless, the best procedure for examination of populations hitherto lumped as *M. persicae* was to reanalyze the data omitting the taxa already resolved.

Fig. 5 shows the mean scores on the first three canonical variates for samples of apterae of 48 *M. persicae*-group samples of known karyotype from Europe and America. Aphids with $2n$ = 12 (circles) formed two distinct clusters, comprising *M. persicae*, itself, with low scores on CV-1 and CV-3,

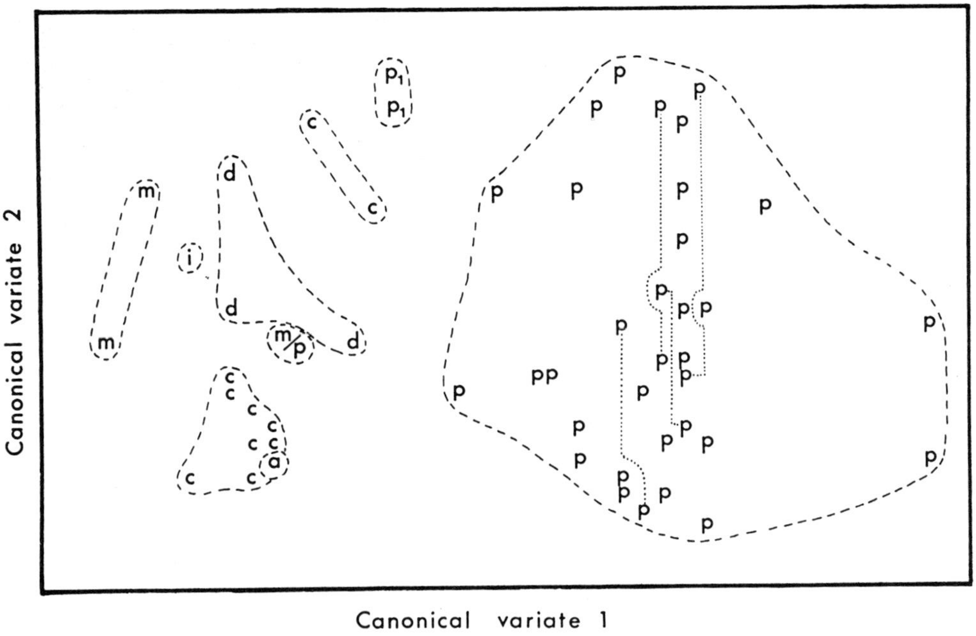

Figure 3. Plot of CV-1 versus CV-2 for a CVA of apterae of *Myzus* subgenus *Nectarosiphon*, based on 14 characters. Dotted lines join samples of the same clone reared at two different temperatures. a = *M. ajugae*, c = *M. certus*, d = *M. dianthicola*, i = *M. icelandicus*, m = *M. myosotidis*, m/p = hybrid clone between *M. myosotidis* and *M. persicae*, p and p_1 = *M. persicae* (including *M. antirrhinii* and *M. nicotianae*). (From Blackman and Paterson, 1986)

and *M. nicotianae* (from America) with high scores on CV-1 and CV-2. Both clusters included samples with and without the *A1,3* chromosomal translocation that is associated with a high level of insecticidal resistance. Samples of *M. antirrhinii* with $2n$ = 13 or 14 also formed two discrete clusters according to their CV-1 and CV-3 scores, with North American (diamonds) and European samples (squares), as well as 13- and 14-chromosome karyotypes, in both clusters. However, included in this analysis were samples of three clones of European *M. antirrhinii* that were reared at two different temperatures, 10°C and 20°C. Because the specimens reared at 10°C fall into one cluster and those reared at 20°C into the other, no phylogenetic significance can be attributed to these clusters.

The best way to look for heterogeneity among putative *M. antirrhinii* is to run a CVA exclusively on all available samples from Europe and North America with $2n$ = 13 or 14. When this was done (Fig. 6A), there were no clearly defined clusters. CV-1 and CV-2 here accounted respectively for 48.7% and 14.5% of the total variance, and were both strongly dependent on the environment, as also was CV-3 (12.5%). In fact, almost all the variation in the data seemed to be attributable to the combined effects of temperature and host

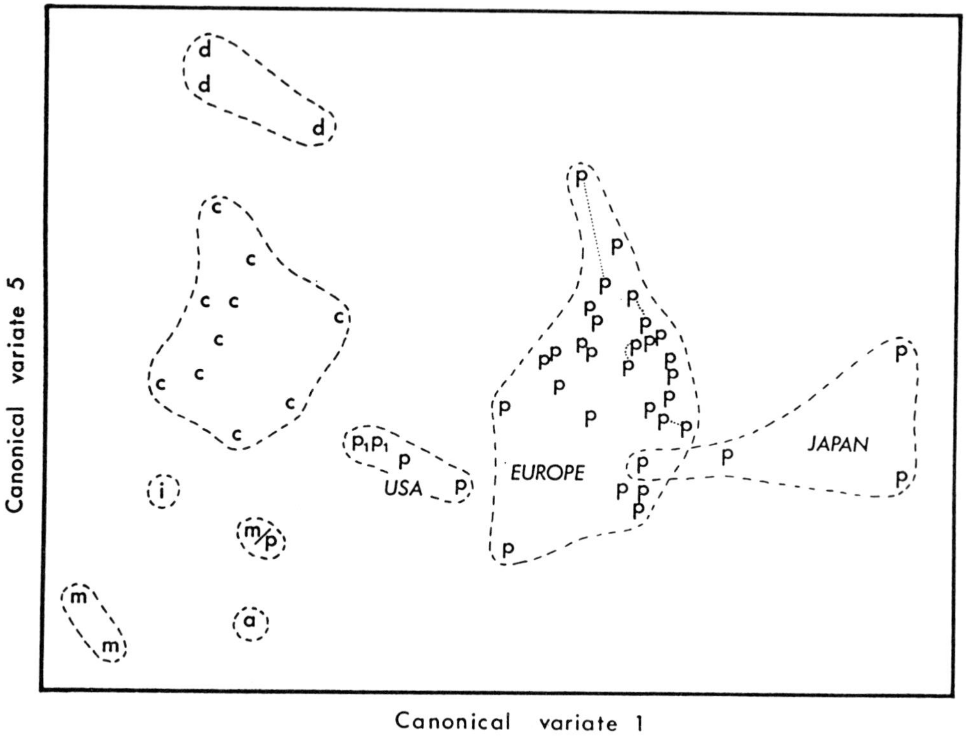

Figure 4. Plot of CV-1 versus CV-5 for a CVA of apterae of *Myzus* subgenus *Nectarosiphon*, based on 14 characters. Dotted lines join samples of the same clone reared at two different temperatures. a = *M. ajugae*, c = *M. certus*, d = *M. dianthicola*, i = *M. icelandicus*, m = *M. myosotidis*, m/p = hybrid clone between *M. myosotidis* and *M. persicae*, p and p_1 = *M. persicae* (including *M. antirrhinii* and *M. nicotianae*). (From Blackman and Paterson, 1986)

plant, indicating that all the data could be ascribed to a single taxon (i.e., *M. antirrhinii*).

Running the same data again, but with 19 samples of *M. persicae* also included, was informative (Fig. 6B). CV-1 was again very dependent on the environment, and had very similar scores for the *M. antirrhinii* samples as when these samples were analyzed alone. However, CV-1 here accounted for a reduced proportion of the total variance (33.9%).

CV-2 had changed completely; its 17.6% of variance clearly had a major genetic component providing discrimination between *M. persicae* and *M. antirrhinii* with no significant temperature effects, although the host plant still had a major influence. CV-3 in this analysis (14.8% of total variance) seemed to be equivalent to CV-2 in the analysis of *M. antirrhinii* samples only. This example shows how CVA can achieve discrimination even when there are major environmental influences on morphology.

Separations of *M. persicae* from *M. nicotianae* (Fig. 7), and of *M. nicotianae* from *M. antirrhinii* (Fig. 8), were more straightforward. Mean scores on CV-1

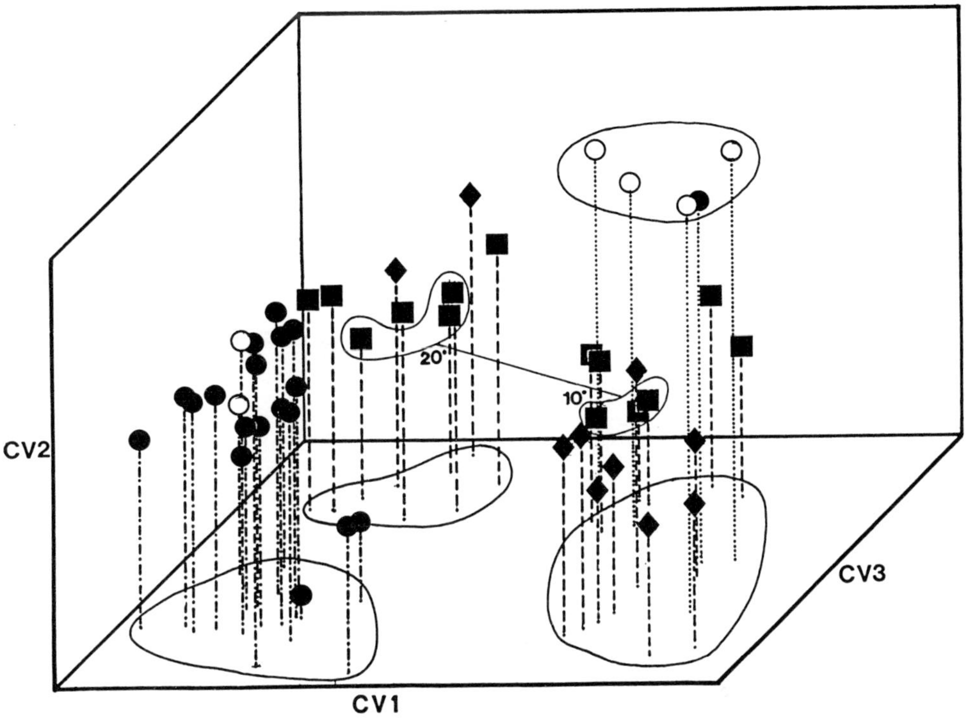

Figure 5. Three-dimensional plot of the first three CVs in a CVA based on 14 characters of samples of apterae of the *Myzus persicae* group of known karyotype from Europe and North America. Samples with $2n$ = 12 (open circles, homozygous; solid circles, translocated) group as *M. persicae* and *M. nicotianae* (see text). Samples of *M. antirrhinii* with $2n$ = 13 and 14 (squares = European; diamonds = North American) include three clones reared at two temperatures.

were discriminatory in each case, except that three samples of *M. nicotianae* containing very small individuals did not separate satisfactorily from *M. antirrhinii*. The CV for the *y*-axis was chosen as that least influenced by temperature, as indicated by samples of the same clones reared at 10°C and 20°C. In the CVA comparing *M. nicotianae* and *M. antirrhinii* (Fig. 8), CV-1 and CV-5 scores for *M. antirrhinii* formed a remarkably good cluster, considering the problems of phenotypic variability encountered with this taxon in other analyses.

4 Discrimination in the Corn Leaf Aphid Group: *Rhopalosiphum maidis*

Variation in the corn leaf aphid, *R. maidis*, needs to be interpreted in the context of this aphid's lack of sexual reproduction. No functional sexual phase has ever been described and, therefore, populations of *R. maidis* on cereals,

A

B

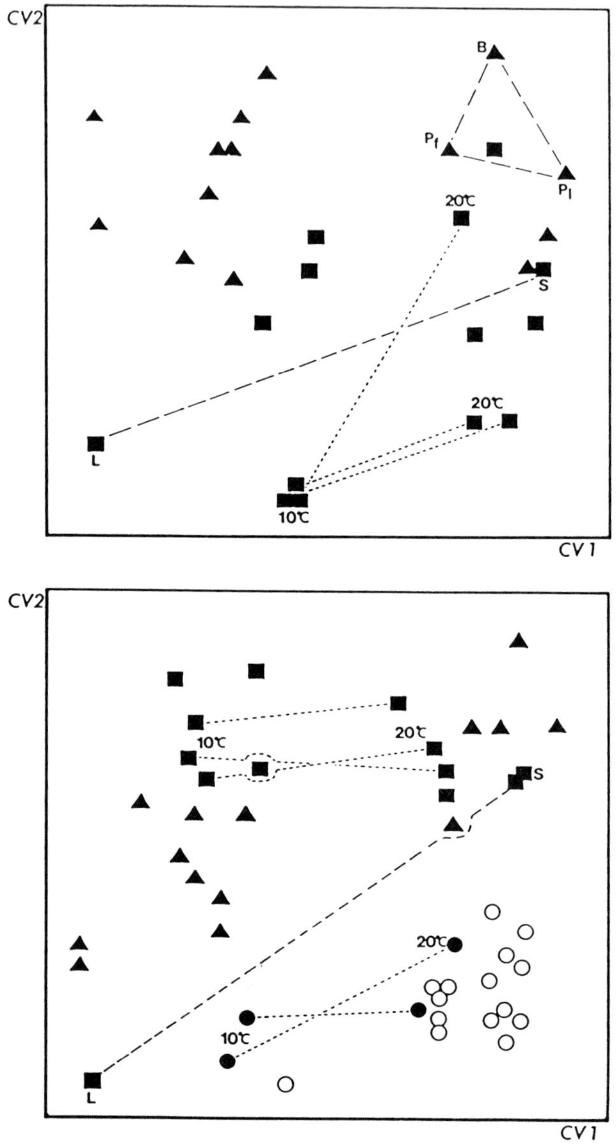

Figure 6. CVA based on 14 characters of apterae of samples of *M. antirrhinii* from Europe (squares) and North America (triangles). (A) Plot of CV-1 versus CV-2 for 28 samples of *M. antirrhinii* alone. (B) Plot of CV-1 versus CV-2 for the same *M. antirrhinii* samples run with 19 samples of *M. persicae*. Dotted lines join samples of the same clone reared at two temperatures, broken lines join samples from the same population reared on different host plants. L = field-collected on *Lonicera nitida*, S = laboratory-reared on *Solanum tuberosum*, P_f = field-collected on *Pittosporum eugenioides*, P_l = laboratory reared on *P. eugenioides*, B = laboratory-reared on *Brassica juncea*. (From Blackman, 1987)

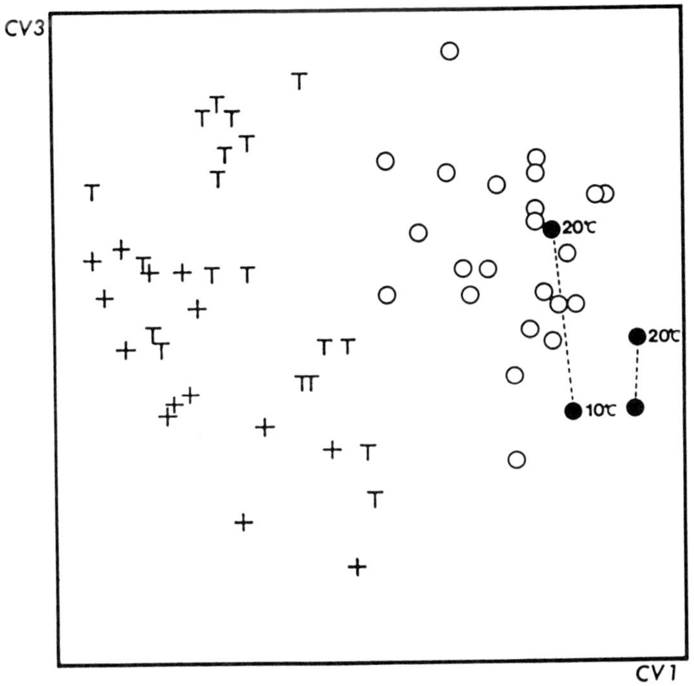

Figure 7. CVA based on 14 characters of apterae comparing samples of *M. persicae* (circles) and *M. nicotianae* (+ = North America, T = Old World countries). Broken lines join samples of the same clone reared at different temperatures. (From Blackman, 1987).

world-wide, comprise an indefinite number of isolated, parthenogenetic lineages.

Some of these lineages can be recognized by their different karyotypes, which are to some extent associated with particular host plants (Brown and Blackman, 1988). In the northern hemisphere, *R. maidis* populations on barley and eupanicoid grasses, such as *Echinochloa crusgalli* (barnyard grass), have $2n = 10$, whereas populations on *Zea mays* and *Sorghum* spp. usually have $2n = 8$. The $2n = 8$ karyotype resembles that of several other species of *Rhopalosiphum*, such as *R. padi* (L.) and *R. rufiabdominalis* (Sasaki), and is probably the primitive form, with $2n = 10$ derived from it by two autosomal dissociations.

To look for morphological differences between these forms, 10 morphometric characters were measured on 70 samples of apterae (802 specimens) and 60 samples of alatae (643 specimens) from all parts of the world. When specimens were grouped according to karyotype and the overall means of each of the 10 variables were compared between karyotypes, several characters showed significant differences between 8-chromosome and 10-chromosome aphids. For example, both apterae and alatae with $2n = 10$ had on average a longer terminal process to the antenna, a longer third antennal

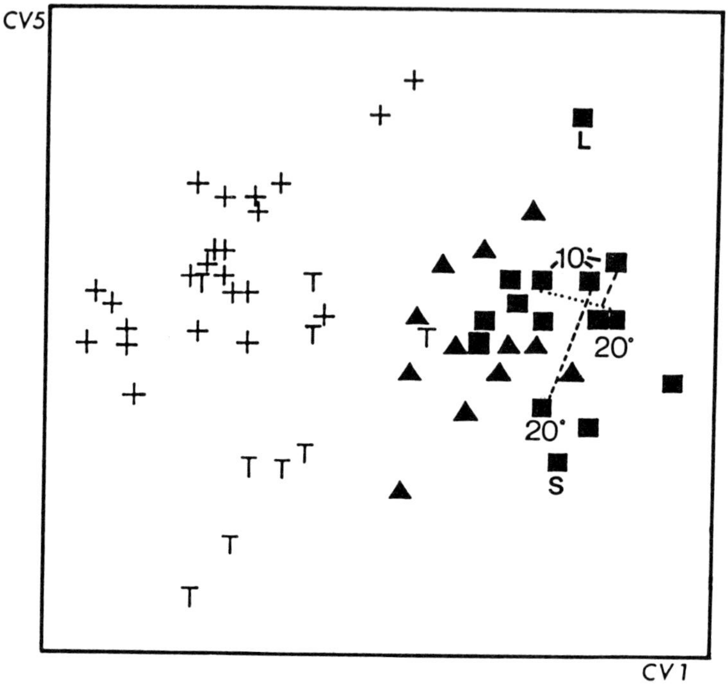

Figure 8. CVA based on 14 characters of apterae comparing samples of *M. nicotianae* and *M. antirrhinii* (squares = Europe, triangles = North America,). Broken lines join samples of the same clone reared at different temperatures. L = field-collected on *Lonicera*, S = the same clone laboratory-reared on potato (cf., Fig. 6). (From Blackman, 1987).

segment and a shorter second hind tarsal segment than those with $2n = 8$. However, individual measurements showed a very broad range of overlap, even when general-size was taken into account, so that it was impossible to find any simple, reliable discriminants.

CVA based on all 10 characters also failed to achieve a complete separation of the 8-chromosome and 10-chromosome forms of *R. maidis* (Fig. 9), and attempts to find linear discriminant functions based on up to 10 characters have so far been unsuccessful. Functions giving over 80% correct classification with the original data set have completely failed with new data.

These problems with the corn leaf aphid seem to have two basic causes that are likely to apply also to other permanently parthenogenetic aphid species. First, the phenotype can be greatly modified by the environment, as shown by the effects of rearing a single clone of $2n = 10$ *R. maidis* at different temperatures (Fig. 9). In particular, there appears to be a strong interaction between environmental and genetic components of variation, so that no CV is independent of the environmental conditions experienced during development. Such phenotypic plasticity was also encountered with *Myzus antirrhinii*, and seems to be a feature of the genome of permanently parthenogenetic species. It

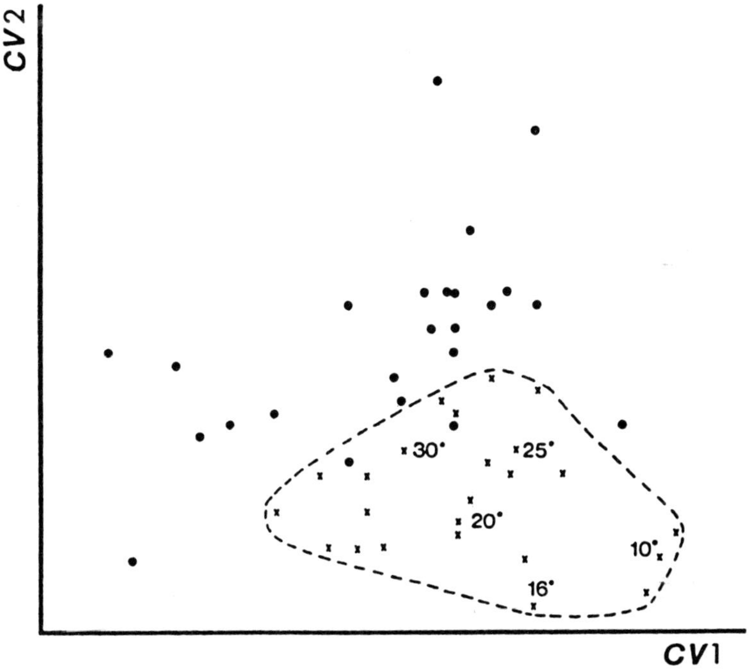

Figure 9. Plot of the mean scores on CV-1 and CV-2 for samples of alatae of *Rhopalosiphum maidis* karyotyped as $2n = 10$ (crosses) and $2n = 8$ (black circles). (From Brown and Blackman, 1988).

may be associated with the high level of heterozygosity found in old parthenogenetic lineages, perhaps coupled with the selection of epistatic mechanisms that may help to maintain adaptability in the absence of genetic recombination (Atchley, 1977; Parker, 1979; Lynch and Gabriel, 1983).

Secondly, although it seems probable that $2n = 10$ *R. maidis* is a single parthenogenetic lineage, the number of separate lineages with $2n = 8$ is unknown. The $2n = 10$ lineage just happens to be recognizable by its karyotype and host plant association. Populations of $2n = 8$ *R. maidis* on *Zea* and *Sorghum*, for example, may comprise separate lineages, or there may also be other $2n = 8$ lineages with different host plant affinities or more restricted geographical distributions. Although the overall means of certain characters differ significantly between 8- and 10-chromosome aphids, this could be due to the presence of one or two widely distributed clones with $2n = 8$ that are morphologically distinct from the 10-chromosome form. If other, less common genotypes with a greater morphological resemblance to the 10-chromosome form are also present in the data set, then these could be blurring the distinction at the level of individual aphids or samples.

Likewise, CVA should achieve separation of a small number of genetically different lineages, each represented by a similar number of samples, but is likely to fail if an unknown number of other lineages, represented by one or a few samples, is also included in the data base.

More work clearly needs to be done, particularly to gain better knowledge of the genotypic structure of corn leaf aphid populations using more precise methods for the characterization of individual parthenogenetic lineages, such as DNA fingerprinting and enzyme electrophoresis. Karyotype information has provided only a partial insight into the problem, but, nevertheless, it has indicated that the dissection of the *R. maidis* group into a number of recognizable entities is possible. If other lineages within the group transpire to have clear-cut differences in host plant, such as those shown by $2n = 10$ and $2n = 8$ forms in northwest U.S.A. (Blackman et al., 1990), then it is important to find ways to recognize them, and multivariate methods will continue to play a part in this process.

5 Conclusions

Experience with use of CVA on aphids prompts the following conclusions:

(1) CVA can give a good representation of taxonomic relationships within a genus, even when it contains species of very similar morphology, and where only small numbers of samples of some of the species are available.

(2) For particularly difficult species separations, the resolving power of the method can be greatly increased by excluding all other taxa and running a CVA with approximately equal numbers of samples of the two putative species.

(3) Information on life cycle (i.e., breeding system), host plant and karyotype can help considerably in the interpretation of data. (Enzyme or DNA data could also provide very useful additional information.)

(4) Samples of a few clones of one or more taxa reared in different conditions of host plant and/or temperature provide very useful information to ensure that the effects of environment on variation in the study material are fully taken into account, and to show which CVs are least dependent on the environment (i.e., have the greatest genetic component).

(5) A high proportion of the total variance in field-collected aphids is invariably due to environmental factors, which may, therefore, contribute greatly to the first few canonical variates. Contrary to experience with many other organisms, taxonomically relevant variation may be included in vectors that account for less than 10% of the total variance. Although not so far tried, it might even be worth looking beyond the fifth CV for variation with a strong genetic component.

(6) Particular problems may arise in the analysis of permanently parthenogenetic aphid populations. These may be due partly to phenotypic plasticity, associated with a strong interaction between environmental and genetic components of variance, and partly to lack of information about the number of separate parthenogenetic lineages represented in the data base.

6 References

Atchley, W.R., 1977. Biological variability in the parthenogenetic grasshopper *Warramaba virgo* (Key) and its sexual relatives. 1. The eastern Australian populations. *Evolution*, 31: 782-799.

Blackman, R.L., 1974. Life cycle variation of *Myzus persicae* (Sulz.) (Hom., Aphididae) in different parts of the world, in relation to genotype and environment. *Bulletin of Entomological Research*, 63: 595-607.

Blackman, R.L., 1978. Early development of the parthenogenetic egg in three species of aphids. *International Journal of Insect Embryology and Morphology*, 7: 33-44.

Blackman, R.L., 1987. Morphological discrimination of a tobacco-feeding form from *Myzus persicae* (Sulzer) (Hemiptera: Aphididae), and a key to New World *Myzus* (*Nectarosiphon*) species. *Bulletin of Entomological Research*, 77: 713-730.

Blackman, R.L., Eastop, V.F. and Hills, M., 1977. Morphological and cytological separation of *Amphorophora* Buckton (Homoptera: Aphididae) feeding on European raspberry and blackberry (*Rubus* spp.). *Bulletin of Entomological Research*, 67: 285-296.

Blackman, R.L., Halbert, S.E. and Carroll, T.W., 1990. Association between karyotype and host plant in corn leaf aphid (Homoptera: Aphididae) in the northwestern United States. *Evironmental Entomology*, 19: 609-611.

Blackman, R.L. and Paterson, A.J.C., 1986. Separation of *Myzus* (*Nectarosiphon*) *antirrhinii* (Macchiati) from *Myzus* (*N.*) *persicae* (Sulzer) and related species in Europe (Homoptera: Aphididae). *Systematic Entomology*, 11: 267-276.

Börner, C., 1939. Neue Gattungen und Arten der mitteleuropischen Aphidenfauna. *Arbeiten ber Physiologische und Angewandte Entomologie*, 6: 75-83.

Briggs, J.B., 1959. Three new strains of *Amphorophora rubi* (Kalt.) on cultivated raspberries in England. *Bulletin of Entomological Research*, 50: 81-87.

Brown, P.A. and Blackman, R.L., 1988. Karyotype variation in the corn leaf aphid, *Rhopalosiphum maidis* (Fitch), species complex (Hemiptera: Aphididae) in relation to host-plant and morphology. *Bulletin of Entomological Research*, 78: 351-363.

Eastop, V.F., 1985. The acquisition and processing of taxonomic data. In: H. Szelegiewicz (Editor), Evolution and Biosystematics of Aphids. Ossolineum, Wroclaw, Poland, pp. 245-270.

Hill, A.R., 1953. Aphids associated with *Rubus* species in Scotland. *Entomologist's Monthly Magazine*, 89: 298-303.

Hille Ris Lambers, D., 1949. Contributions to a monograph of the Aphididae of Europe. IV. The genera *Aulacorthum* Mordv., 1914; *Microlophium* Mordv., 1914; *Hyalopteroides* Theob., 1916; *Idiopterus* Davis, 1909; *Pentalonia* Coquerel, 1859; *Amphorophora* Buckton, 1876; *Wahlgreniella* Nov. Gen.; *Megoura* Buckton, 1876; *Megourella* Nov. Gen.; *Hyperomyzus* Börner, 1933; *Nasonovia* Mordv., 1914. *Temminckia*, 8: 182-324.

Lynch, M. and Gabriel, W., 1983. Phenotypic evolution and parthenogenesis. *The American Naturalist*, 122: 745-764.

Parker, E.D. Jr, 1979. Phenotypic consequences of parthenogenesis in *Cnemidophorus* lizards. II. Similarity of *C. tesselatus* to its sexual parental species. *Evolution*, 33: 1167-1179.

Stroyan, H.L.G., 1982. Revisionary notes on the genus *Metopolophium* Mordvilko 1914, with keys to European species and descriptions of two new taxa. *Zoological Journal of the Linnaean Society of London*, 75: 91-140.

MORPHOLOGICAL VARIATION IN AN ECTOPARASITE: PARTITIONING ECOLOGICAL AND EVOLUTIONARY INFLUENCES

M.A. Houck

Department of Biological Sciences, Texas Tech University, Lubbock, Texas 79409, U.S.A.

Abstract

Forty-two morphological characters from 95 females of the ectoparasitic mite *Chrysomelobia labidomerae* were evaluated using multivariate morphometric techniques. Mites were collected from nine Mexican and two United States locations. These collections were from five species of chrysomelid beetles (hosts). Effects of locality and host on morphological variation were partitioned using principal components analysis and size-free discriminant function analysis. Character allometries were calculated across the form. Vector correlations were used to examine patterns of character covariation. Clustering (UPGMA) of Mahalanobis' distances, by locality and host, were used to determine phenetic similarities among populations. These procedures are discussed in light of their potential for interpreting parasite-host coevolution.

1 Introduction

The parasitic mite *Chrysomelobia* is found throughout the Western Hemisphere, and it is thought that the genus contains only a single New World species, *Chrysomelobia labidomerae* Eickwort (Acari: Podapolipidae). The estimated range of the mite extends from Central America to southern Canada (Eickwort, 1975; Eickwort and Eickwort, 1986; Drummond et al., 1989), where it is an obligate ectoparasite of chrysomelid beetles. These mites occur under the beetle elytra, where they attach to abdominal tissues and feed on hemolymph. All developmental stages of *C. labidomerae* occur on chrysomelid beetles; specifically, chrysomelids within the tribe Doryphorini (= Zygogrammini) (Drummond et al., 1985).

1.1 The Host

The Doryphorini is characterized as containing taxa that: (1) are monophagous (or restricted-oligophagous) herbivores (Jacques, 1988); (2) congregate in dense localized aggregations; and (3) may express bright coloration and elaborate chemical defenses (Deroe and Pasteels, 1977, 1982; Pasteels and Daloze, 1977; Daloze and Pasteels, 1979; Pasteels et al., 1982, 1983, 1984; Daloze et al., 1986; Dettner, 1987). Thus far, *C. labidomerae* has been found in association with only two of the five genera within the Doryphorini: *Leptinotarsa* and *Labidomera*. Specimens of *C. labidomerae* collected from *Leptinotarsa* and *Labidomera* are taxonomically indistinguishable (Eickwort and Eickwort, 1986).

Mexico is the historical zone of origin and radiation of species of *Leptinotarsa* (Jolivet, 1986) and as the genus has expanded northward species richness has declined. There are presently 29 described Mexican species, 12 species found in the U.S., and one in Canada (Jacques, 1988). Though the mite is distributed throughout Mexico with some degree of frequency on *L. decemlineata* (Say), it has also been found in association with *L. undecemlineata* Stål, *L. signaticollis* Stål, and *L. cacica* Stål. There is question of the taxonomic validity of *L. signaticollis*, which may be synonymous with *L. undecemlineata* (Weiman, 1910; Tower, 1918; Hsiao and Hsiao, 1983).

Species of *Leptinotarsa* that have radiated into the U.S. are restricted to the Southwest, and the race of *L. decemlineata* occurring in the U.S. has never been found to be parasitized by *C. labidomerae* (Drummond et al., 1984; Drummond et al., 1985). Throughout most of the continental U.S., *C. labidomerae* is, instead, found in association with *Labidomera clivicollis clivicollis* (Kirby) (milkweed leaf beetle) (Drummond et al., 1984). *Labidomera,* the sister-group of *Leptinotarsa* (Hsiao and Hsiao, 1983; Jacques, 1988), is a specialist on the Asclepiadaceae.

Leptinotarsa and *Labidomera* are not well studied, with the exception of one species, *Leptinotarsa decemlineata*, the Colorado potato beetle, an important agricultural pest in the Western Hemisphere. This is the only species of *Leptinotarsa* whose range extends beyond the New World, having been introduced to Europe and Asia through international commerce.

Recent interest in the ecology of plant-herbivore interactions has significantly advanced our understanding of the proliferation of species of the

Doryphorini, and it is now thought that the radiation observed in *Leptinotarsa* can be correlated to adaptations to the Solanaceae (e.g., *L. decemlineata, L. undecemlineata and L. signaticollis*) and the Compositae (e.g., *L. cacica*) (Hsiao, 1981, 1986; Hsiao and Hsiao, 1983). Microhabitat selection among species of *Leptinotarsa* may also have played an important role in speciation and resource-partitioning within the genus (Hsiao, 1986).

Gregarious behavior (associated with oligophagy) and microhabitat specialization in *Leptinotarsa* contribute to increased inbreeding and potential endemism within local beetle populations, and have secondarily imposed inbreeding constraints on the mites. Endemism in *Leptinotarsa* has resulted in morphologically distinct species with no known species complexes, sibling species, or interspecific hybrids (Hsiao, 1988).

1.2 The Parasite

Nymphal mites typically parasitize the same beetle upon which they were deposited as eggs. Only adult females disperse to colonize new hosts, and transfer of mites occurs during chrysomelid copulation (Baker and Eickwort, 1975). The frequency of beetle mating (hence mite host-transfer) is correlated with beetle age and ambient temperature (Drummond et al., 1989). Accidental displacement of parasites does occur, however, and individuals that are dislodged from their hosts account for some colonization of alternate host-species.

Dislodged mites may survive for 7.6 ± 2.3 days without a host, and those individuals which are successful at invading an unfamiliar host-taxon experience reduced survival rates (Drummond et al., 1989). The frequency of effective invasion is a function of the encounter rate of the various beetle species, and is related to the degree of herbivore-plant fidelity. Interspecific contagion is improbable among strict monophagous hosts.

Coupling of parasite transmission to host reproductive cycle has significantly limited the potential dispersal of the mite and curiously linked its distribution to the mate-choice of beetles. A positive consequence of this is presumed to be a more effective host accommodation.

Because the mite is an obligate parasite, its success has depended upon prudent responsiveness to biological changes occurring in the beetle as shifts in primary host-plants have occurred. Because it is an external parasite of an ectotherm, its survival also depends upon its level of tolerance and accommodation to ambient environmental conditions.

1.3 The Problem

The analyses described here were applied to augment a research program investigating the coevolution of beetles and their associated mites. Such interactions are common between mites and beetles. The interactions grade from situational and ephemeral (i.e., phoretic), to obligate parasitism.

In general, ecological and behavioral correlations have been central to studies of such interspecific interactions, and the literature contains many detailed accounts of bizarre interactions. Morphological arguments are less often invoked, usually concentrate on features that are undeniably specific to the coassociates, and generally focus on one character to define the interaction. Frequently cited morphological correlations include the tongue lengths of avian and insect pollinators with flower corolla length, and tissue thickness of ovipositional sites with ovipositor length of wasps.

When extracting information from observed interactions, the morphology of a taxon is considered to some extent a historical record of the successful adjustments to selection that have been made to negotiate the coevolutionary contract between the species. However, morphological patterns are not exclusively structured by species interactions but are the product of many selective forces, the most important of these being accommodations to the physical environment.

This study explores the potential of ordination techniques for partitioning the morphological consequences of interspecific (host) influences, from physical (locality) influences. It is hypothesized here, that: (1) *C. labidomerae* that infest different host species (occurring within a common geographic location) should exhibit discernable differences in morphological adaptation related to host-specificity; and (2) *C. labidomerae* mites from any single host species should exhibit geographic clines in morphological adaptation (across the range of that host) in response to habitat. The *Chrysomelobia*-chrysomelid interaction is used here as a methodological example that can be generalized to other tightly coupled (presumed coevolved) associations. A suite of techniques is examined that have value in evolutionary studies of phylogenetic systematics and speciation. The application of discriminant function analysis (DFA), as applied here, has apparently not been suggested or explored in the literature previously. The work focuses on applied techniques and methodology, with less emphasis on biological content, which will be developed elsewhere.

2 Materials and Methods

2.1 Source of Samples: Habitat Contrasts

Ninety-five females of *Chrysomelobia labidomerae* were collected from nine disjunct populations in Mexico and two in the U.S. (Table 1). Five populations were collected from *Leptinotarsa decemlineata* at different localities in the state of Morelos, two were from *L. signaticollis* in the states of Mexico and Puebla, and two were from *L. cacica* and *L. undecemlineata* from the same site in the state of Veracruz. Three of the collection sites in Morelos (Apatlaco, Temilpa, and Las Estecas) were very similar in plant diversity. Moyotepec is the driest of the five Morelos sites. Canon de Lobo, Morelos, is in an ecological zone of transition between the wetter region to the north and the drier regions to the south. Ixtapan, Mexico is an area of higher elevation, greater precipitation, and earlier rainy season than are the Morelos locations, and it is considerably cooler. Atlixco, Puebla, is situated on a plateau, similar to the Morelos sites but it is drier and higher, with much less lush vegetation. The two sites in Veracruz are in tropical rain forests and thus they are warmer and wetter than all other Mexican sites. The two Veracruz sites are separated by a distance of only 2 km and differ approximately 15 m in altitude. In the United States, collections were made in Ithaca, New York, and in the Midwest

Table 1.
Localities and hosts for *Chrysomelobia labidomerae* collected from Mexico and the United States. Letters and numbers in parentheses represent the labels for the scattergrams of localities and hosts in the PCA and DFA plots.

Locality	Alt (m)	Chrysomelid Hosts	Number
Morelos, Mexico			
(A) Apatlaco	1500	(1) *Leptinotarsa decemlineata*	10
(B) Moyotepec	1330	*Leptinotarsa decemlineata*	8
(C) Canon de Lobo	1470	*Leptinotarsa decemlineata*	10
(D) Temilpa	1170	*Leptinotarsa decemlineata*	9
(E) Las Estecas	1170	*Leptinotarsa decemlineata*	7
Mexico, Mexico			
(F) Ixtapan	2170	(2) *Leptinotarsa signaticollis*	10
Puebla, Mexico			
(G) Atlixco	1930	*Leptinotarsa signaticollis*	7
Veracruz, Mexico			
(H) Chocaman	1315	(3) *Leptinotarsa undecemlineata*	9
Chocaman	1330	(4) *Leptinotarsa cacica*	10
United States			
(I) Ithaca, New York		(5) *Labidomera c. clivicollis*	10
(J) Midwest (= Iowa, Indiana, Michigan, Minnesota)		*Labidomera c. clivicollis*	5

from *Labidomera clivicollis clivicollis*. The U.S. collections represent a maximum contrast to the Mexican sites in habitat, and differ in host genus.

2.2 **Protocol for Data Collection**

Mites were removed from the beetles, cleared in lactophenol, and mounted in Hoyer's medium (by the methods of OConnor and Houck, 1991). Camera-lucida drawings were produced using a Wild M-20 phase-contrast microscope. From these drawings, anatomical landmarks and helping points (sensu Bookstein et al., 1985) were electronically converted to Cartesian coordinates using a digitizing pad (Houston Instrument Hipad DT-114) and an IBM PC/AT microcomputer. Euclidean distances (scaled in μm; Table 2), to serve as morphometric characters, were computed among selected pairs of points. Distances were converted to natural logarithms to equalize variances, preserve

Table 2.
Definitions of labels and abbreviations for characters used in the multivariate analyses of female *Chrysomelobia labidomerae*. These are applied uniformly throughout the text and figures. Labels (letters) represent characters plotted in the vector plots. Ventral setal lengths were not measured as characters because of their uniform small size.

Labels			
Dorsum		Venter	
Label	Character	Label	Character
a	Circumference of gnathosoma	*a*	Body length
b	Length of setae *hui*	*b*	Gnathosomal length
c	Length of setae *do*	*c*	Opisthosomal width
d	Length of setae *lui*	*d*	Body circumference
e	Length of setae *pch*	*e*	Distance between setae *pm*
f	Length of setae *vi*	*f*	Distance between setae *cxI*
g	Distance between setae *pch*	*g*	Distance between setae *cxII*
h	Distance between setae *vi*	*h*	Distance between setae *pri*
i	Distance between setae *sc*	*i*	Distance between setae *cxIII*
j	Distance between setae *hui*	*j*	Distance between setae *cxIV*
k	Distance between setae *hue*	*k*	Distance between setae *c*
l	Distance between setae *do*	*l*	Distance between gnath. & setae *pm*
m	Distance between setae *lui*	*m*	Distance between setae *pm* & *cxI*
n	Distance between gnath. & setae *pch*	*n*	Distance between setae *cxI* & *cxII*
o	Distance between setae *pch* & *vi*	*o*	Distance between setae *cxII* & *pri*
p	Distance between setae *vi* & *sc*	*p*	Distance between setae *cxIII* & *pri*
q	Distance between setae *sc* & *hui*	*q*	Distance between setae *cxIII* & *cxIV*
r	Distance between setae *hui* & *hue*	*r*	Distance between setae *cxIV* & *c*
s	Distance between setae *hue* & *do*	*s*	Distance between setae *c* & anus
t	Distance between setae *do* & *lui*	*t*	Sternal apodeme (*ap. sa.*) length
u	Distance between setae *lui* & anus	*u*	Sejugal apodeme (*ap. sj.*) width

Abbreviations			
abbrev.	character	abbrev.	character
ap. sa.	sternal apodeme	*lui*	internal lumbar setae
ap. sj.	sejugal apodeme	*pch*	dorsal gnathosomal setae
c	caudal setae	*pdsI*	prodorsal shield setae I
cxI	setae of coxal segment, leg I	*pdsII*	prodorsal shield setae II
cxII	setae of coxal segment, leg II	*pm*	ventral gnathosomal setae
cxIII	setae of coxal segment, leg III	*pri*	internal presternal setae
cxIV	setae of coxal segment, leg IV	*sc*	scapular setae
do	dorsal setae	*vi*	internal vertical setae
hui	internal humeral setae		

(linearize) allometric relationships among characters (Bookstein et al., 1985), and produce a scale-invariant covariance matrix (Jolicoeur, 1963a).

Characters were chosen for their repeatability of measurement, minimal distortion due to mounting, and equitable distribution across the form. "Pseudo-landmarks" marking the boundary of the form were avoided because of membrane distortion due to engorgement or specimen mounting. Dorsal and ventral tergites or sclerites, and associated setae, were the primary structures from which mensural characters were derived. Character distances from bilateral setal pairs were averaged. The character set included 21 dorsal and 21 ventral characters of females of *C. labidomerae* (Table 2).

2.3 Statistical Procedures

All ordination analyses were done on an IBM PC/AT microcomputer using the SAS-PC statistical package (SAS Institute, 1985). Body size of the 95 female mites was examined both as a conventional univariate character (total length, in μm) and as a latent multivariate variable (general-size: Bookstein, 1989) contributing to total character variance. In both cases, ANOVA procedures were applied to analyze the influence of locality and host on body size. The Tukey studentized range test on all main-effect means was used to determine statistical differences in size, among groups, at the overall level of $\alpha = 0.05$. Although size variation plays a central role in ontogenetic and phylogenetic interpretations of character differences within and among groups (Houck et al., 1990), differences in form due to evolutionary size shifts must be distinguished from concomitant allometric changes to accurately assess phylogenetic associations (Rohlf and Bookstein, 1987).

I applied two multivariate statistical procedures to the data that offer complementary interpretations of variation and discrimination within the character space. The first was principal components analysis (PCA), based on the covariance matrix of log-transformed data. Use of a covariance matrix is appropriate for morphometric data because it is an uncentered matrix, containing information on a relative scale, that can be applied to character data measured in comparable metric units. PCA is an exploratory multivariate statistical technique that determines, by an orthogonal sequence of linear combinations of variables (eigenvectors), the combination of variables that best explains the greatest proportion of variation in the data set (Digby and Kempton, 1987). This is done without prior assignment of individuals (or samples) to categorical groups. Character variation was examined by performing the PCA only once, but plotting the scores as scattergrams with two different sets of labels: one labeled by locality, and one by host.

The second multivariate procedure used was discriminant function analysis (DFA), an optimization technique that extracts from the character set the linear combinations of variables that best separates (discriminates) predefined groups. It maximizes F-ratios for which the groups differ by the greatest multiple of their pooled within-group variance. Because size may be responsible for (or contribute to) discrimination in some unpredictable way, size (as a latent variable) was regressed out of the data set on a character-by-character basis. This allowed biological patterns in form to be interpreted as size-free shape contrasts ("uncorrelated with some size measure but covarying with physical scale": Bookstein, 1989), independent of the allometric consequences of size variation. Logistically, it was first determined that there

Table 3.
Principal components analysis, character loadings, and allometric coefficients for dorsal and ventral traits of *Chrysomelobia labidomerae* females.

Dorsal Traits				Ventral Traits			
Trait Code	PC-1 loading	PC-2 loading	Allometric Coefficients	Trait Code	PC- 1 loading	PC- 2 loading	Allometric Coefficients
a	0.643	0.332	1.30	*a*	0.897	- 0.270	1.15
b	0.360	- 0.145	0.74	*b*	0.637	0.390	1.08
c	0.497	- 0.158	1.04	*c*	0.785	0.245	1.34
d	0.269	0.182	0.53	*d*	0.913	- 0.136	1.57
e	0.583	0.458	1.19	*e*	0.495	0.474	0.81
f	0.361	0.397	0.73	*f*	0.504	0.319	0.88
g	0.497	0.613	0.99	*g*	0.565	0.514	0.99
h	0.406	0.379	0.81	*h*	0.465	0.345	0.84
i	0.513	0.495	1.03	*i*	0.568	0.106	0.98
j	0.413	0.465	0.84	*j*	0.630	0.319	1.07
k	0.511	0.349	1.04	*k*	0.452	- 0.052	0.78
l	0.449	0.463	0.90	*l*	0.514	0.482	0.87
m	0.627	0.291	1.27	*m*	0.553	0.430	0.94
n	0.553	0.473	1.11	*n*	0.485	0.415	0.82
o	0.414	0.042	0.82	*o*	0.677	- 0.388	1.17
p	0.505	0.572	1.02	*p*	0.490	0.208	0.84
q	0.548	- 0.176	1.12	*q*	0.520	0.463	0.92
r	0.403	0.203	0.82	*r*	0.630	- 0.659	1.10
s	0.518	- 0.507	1.10	*s*	0.365	- 0.034	0.62
t	0.616	- 0.569	1.28	*t*	0.432	- 0.050	0.76
u	0.662	0.166	1.35	*u*	0.641	0.480	1.09

was a significant size-vector present in the data. When the specimens analyzed vary significantly in some measure of body size, the first eigenvector of the within-group variance-covariance matrix of the log-transformed character distances usually represents a general-size factor (Jolicoeur, 1963b). Subsequent eigenvectors of the matrix usually represent shape components. Existence of such a general-size factor was established from the PCA (Table 3). A general-size vector is typically represented by loadings that are unidirectional in sign and similar in magnitude. The size factor can also be determined from univariate plots (Figs. 1A, 1B). Once the first principal component (PC-1) was determined to be a size factor, each log-transformed character was regressed upon it. This procedure removes general-size on PC-1, but may not remove all size effects. Other treatments of size can be found in Burnaby (1966), Bookstein et al. (1985), and Rohlf and Bookstein (1987); Bookstein (1989) discusses the terms "size" and "shape" in detail. The size-free residuals were then calculated and a DFA was subsequently performed by group, independently by locality and by host, on the residuals. Scores were plotted as scattergrams. Discriminant character loadings were estimated by computing the linear correlations between the discriminant function scores

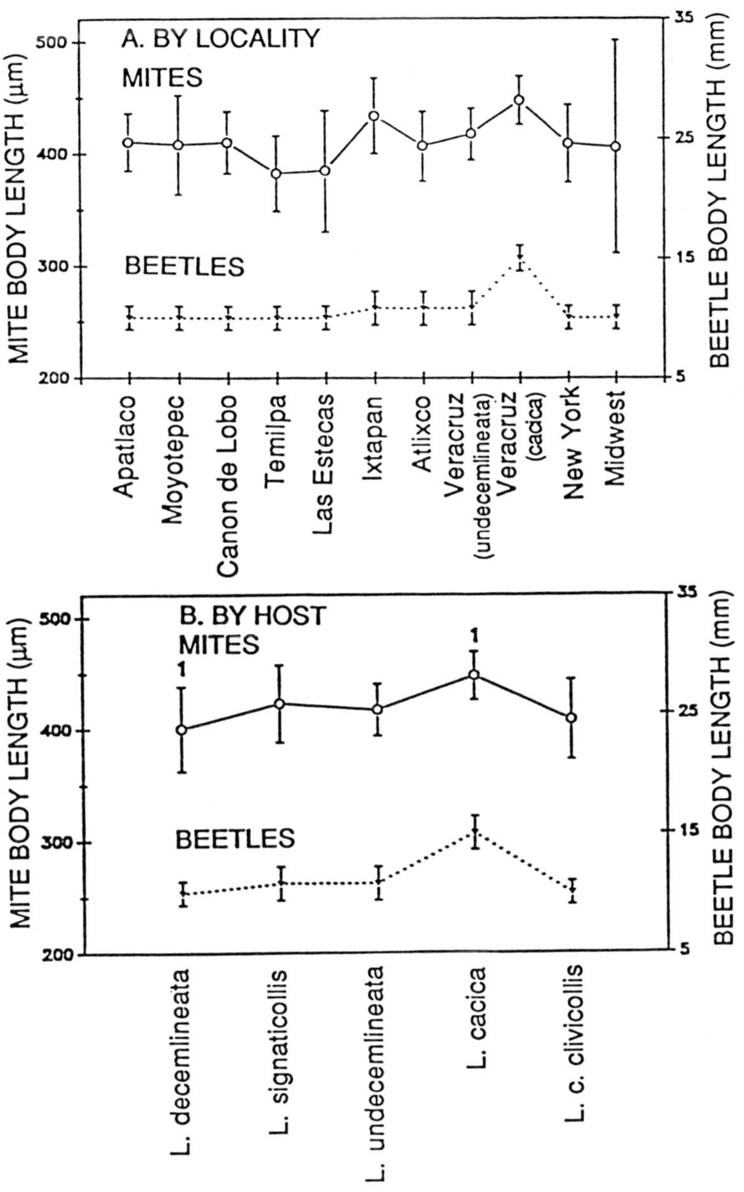

Figure 1. Mean body size (μm ± SD) of female *Chrysomelobia labidomerae* pooled by: (A) Locality: from the ANOVA it was determined that there were no between-group statistical differences in size when localities were examined in pairwise combinations. Mean body size (mm ± SD) of the corresponding host species is given for perspective. (B) Host: a statistical difference in body size occurred (α = 0.05) only between mites collected from *Leptinotarsa decemlineata* and *L. cacica* (indicated by the number "1"). Mean body size (mm) of the corresponding host species is given for perspective.

and the character residuals. The correlations were plotted as vector diagrams (Wright, 1954, 1968). Vector analysis was done in this manner because discriminant function coefficients (weights) are not inherently biologically interpretable with respect to shape (Bookstein et al., 1985), whereas biological interpretations can be made based on vector correlations.

PCA and DFA provide a contrast in character analysis. PCA addresses the question of patterns of morphological variation within and among groups. This is a different question in both tone and fact from that of discrimination among groups. For example, if within-group variance is high compared to among-group variance, then PCA may not separate groups. Even if mean differences among group centroids are statistically significant, the groups may overlap in multivariate space with no obvious discontinuities. In contrast, if within-group variance is low and among-group variation is high, then patterns of separation may occur. DFA, however, maximizes among-group differences. PCA and DFA applied in tandem provide information for the understanding of patterns of variation, as well as for distinguishing groups; they should not be considered to be alternatives. In both applications, however, it is important to distinguish evolutionarily derived influences from passive allometric influences.

Static allometric coefficients among groups ("evolutionary" allometries of Lande, 1985) were scaled with respect to the general-size factor. Loadings (coefficients of variables) on PC-1 were scaled to a mean of +1.0 and interpreted as multivariate allometric coefficients on general-size. Allometric coefficients not significantly different from +1.0 indicate isometry, whereas those greater than +1.0 indicate positive allometry and those less than +1.0 indicate negative allometry.

Mahalanobis' distances were calculated using the primary, size-free, discriminant function (DF), and a clustering algorithm (UPGMA: Sneath and Sokal, 1973) was used to determine phenetic similarities, by locality and host. Resulting dendrograms were interpreted as degree of separation of gene pools but were not interpretable as statements of phylogenetic relationships because (lacking extrinsic polarity) they do not distinguish primitive from derived similarity. Mahalanobis' distances represent the distances among the group centroids, in the DFA clusters. Centroid analysis positions the "average" individual in DFA space for a cluster. In some cases, it is instructive to plot the centroids ± one standard deviation for a fuller interpretation of structure in a data set (Delfinado-Baker and Houck, 1989). The distances were calculated for the primary size-free discriminant functions using the residuals of the PCA. Although Mahalanobis' distances can be chosen to represent any subspace of discriminant axes (e.g., DF-1 - DF-3), in this study the higher components (> DF-4) usually contained minimal biological information.

The phenogram for the mites was contrasted with a constructed consensus cladogram for *Leptinotarsa*. A character matrix for the beetles was extracted from Jacques (1988), and character states were ordered where appropriate; this matrix contained 49 characters for 27 species of *Leptinotarsa*. Characters and taxa are given in Appendices 1A and 1B, respectively; the matrix used is given in Appendix 2. Trees were constructed from this data using PAUP (Swofford, 1985) with local branch swapping, and were rooted at the midpoint of greatest patristic distance. The best four, equally parsimonious trees (consistency index, CI = 0.443), were used to generate the consensus tree using Swofford's program CONTREE.

3 Results and Discussion

3.1 Patterns of Morphological Variation

3.1.1 Variation in Body Length

Overall, locality accounted for 25.2% of the total variation in body length among samples of mites (ANOVA; P = 0.004). Body length among females, sampled across localities, ranged from 382.4 ± 33.6 μm (mean ± SD, format hereafter) to 433.9 ± 33.8 μm with no significantly different pairwise comparisons of means and no clear geographical cline in size (Fig. 1A). Mites from the most northern collection (New York) were intermediate in body length (409.1 ± 34.7 μm).

To examine the relationship between body size of *Chrysomelobia* and their beetle hosts, mite size was plotted and compared with the recorded mean body size (mm) for species of *Leptinotarsa* (Jacques, 1988) and *Labidomera c. clivicollis* (Dillon and Dillion, 1972). Mite size was somewhat responsive to local variations in habitat, with mites from Ixtapan and Veracruz having the largest mean size. Small sample size of the Midwest mites (n = 5) resulted in the largest sampling error. Aside from the sampling error of the Midwest mites, mites collected from beetles from Las Estecas had the greatest within-group population variance.

Among-sample analysis of host species effects (Fig. 1B) accounted for approximately the same amount of variation in body length (r^2 = 0.222, P = 0.0004). Mites from *Leptinotarsa cacica*, which are approximately 50% larger than the other beetle taxa, were the largest (447.4 ± 21.8 μm), but only statistically larger (by 12%) than those from *L. decemlineata* (400.3 ± 37.7 μm). Mite body size paralleled the mean body size of the hosts, indicating host-parasite congruence in this character, which has been found in other ectoparasitic mites (Delfinado-Baker and Houck, 1989).

3.1.2 Variation: Principal Components Analysis

3.1.2.1 Locality

In the PCA, the first eigenvector accounted for 29% of total variation in the data set. The magnitude and sign-distribution of the loadings indicated that PC-1 was a moderate size vector (Table 3). The remainder of the variation was partitioned across the remaining components, with PC-2 representing 13% of the residual variance. The PCA, labeled by locality, resulted in complete overlap of the samples in character space (Fig. 2), due to significant within-group variation. No additional separation occurred on subsequent components.

3.1.2.2 Host

A scatterplot of PC-1 versus PC-2, with samples labeled by host, shows overlap in variation among all groups with the exception of mites collected from *L. cacica* (Fig. 3) suggesting a significant host effect on variation. This

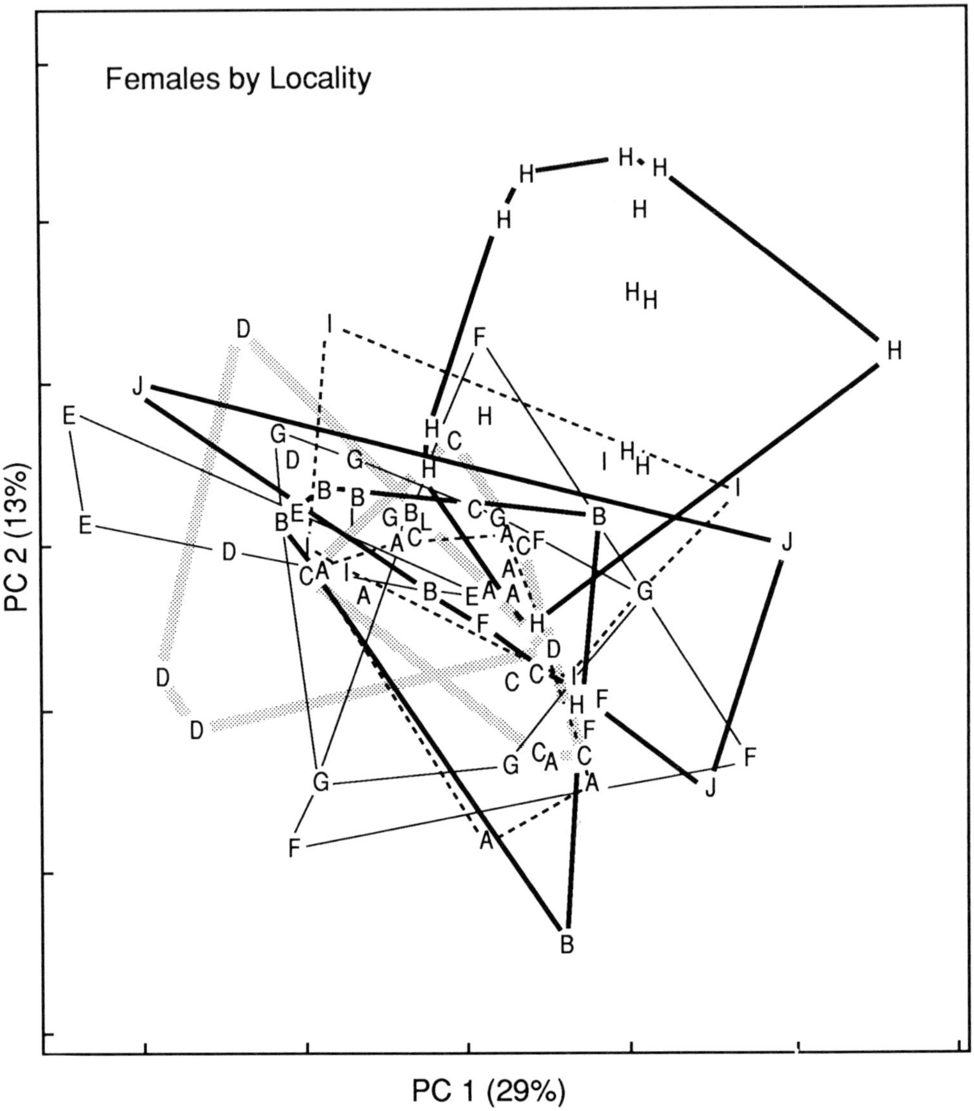

Figure 2. Scatterplots of the first two principal components of female *Chrysomelobia labidomerae* by locality. Localities: A = Apatlaco (Morelos); B = Moyotepec (Morelos); C = Canon de Lobo (Morelos); D = Temilpa (Morelos); E = Las Estecas (Morelos); F = Ixtapan (Mexico); G = Atlixco (Puebla); H = Chocaman (Veracruz); I = Ithaca, New York; J = Midwest (Iowa, Indiana, Michigan, Minnesota).

sample of mites also had the largest mean body length (447.4 ± 21.8 μm) in the univariate analysis. In the PCA plot, *L. cacica* had the largest mean general-size factor and also separated on the first shape axis (PC-2). No further

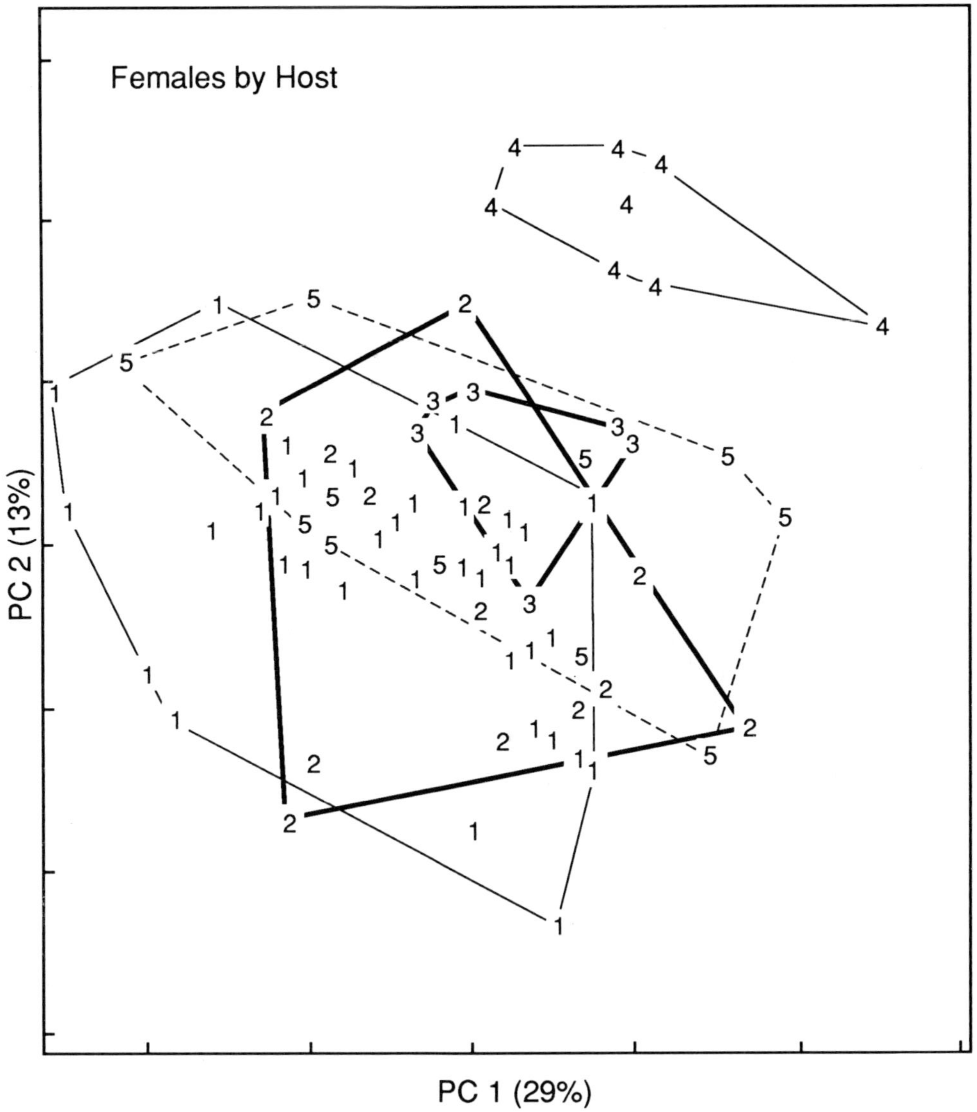

Figure 3. Scatterplots of the first two principal components of female *Chrysomelobia labidomerae* by host. Hosts: 1 = *Leptinotarsa decemlineata*; 2 = *L. signaticollis*; 3 = *L. undecemlineata*; 4 = *L. cacica*; 5 = *Labidomera c. clivicollis.*

separation of groups occurred on subsequent components.

A significant host effect is further supported because other mites from Veracruz (collected from *L. undecemlineata*) cluster with mites from the other Mexican states (and the U.S.), rather than with those collected in Veracruz from *L. cacica*.

3.2 Allometric Relationships

3.2.1 Dorsum

Allometric coefficients of dorsal and ventral characters (Table 3) are portrayed in Figs. 4A and 4B. Positively allometric responses to size variation are relatively evenly distributed longitudinally across the dorsal idiosoma, with the largest relative increase occurring in the posterior dorsal sector of the body. Lateral expansion, is nearly isometric except for the enlargement of tergite 3, which is positively allometric and significant. The gnathosome also lengthens isometrically at the position of setae *pch*, but its circumference is positively allometric (+1.30). Dorsal setal lengths are, in general, either isometric with general-size or negatively allometric (Table 3). The exception is the pair of setae on the gnathosome (*pch*), which become relatively longer (+1.19) with an increase in size.

3.2.2 Venter

Patterns of ventral allometric coefficients are consistent with those of the dorsum (Fig. 4B), with near isometry across the idiosoma. Only the very posterior tip of the venter significantly deviates towards negative allometry, compensating for the positive allometric shifts in the dorsum. Functionally, the lateral displacement of the dorsal caudal setae is correlated with a narrowing of the setal distances (setal pair *c*) on the corresponding ventral surface.

There is also a disproportionate increase in circumference in the structure housing the feeding apparatus (the gnathosome). Ventral allometric coefficients indicate that an increase in general-size in females is accommodated by a widening of the sejugal area. Note, however, that this measurement has the largest potential for associated error because, even though the sclerites of the body are somewhat fixed in form, it is not landmark-based.

3.3 Patterns of Morphological Discrimination

3.3.1 Locality

The primary size-corrected discriminant function, based on groupings by locality, accounted for 32% of total variation in the data (Table 4). Primary discrimination occurred between females from Las Estecas and all other Mexican localities (Fig. 5). Discrimination due to dorsal character shifts (Fig. 6A) of the mites from Las Estecas is due primarily to their relatively larger gnathosome, wider opisthosomal (mid) region, and longer *do* setae. The wider body is correlated with an increased separation of all dorsal setal pairs. Ventral characters (Fig. 6B) contributing to discrimination parallel the wider dorsal shields in the propodosomal area and in tergites 1 and 2, but not in tergite 3. There is also an increase longitudinal distance between legs I and II and between legs III and IV in mites from Las Estecas.

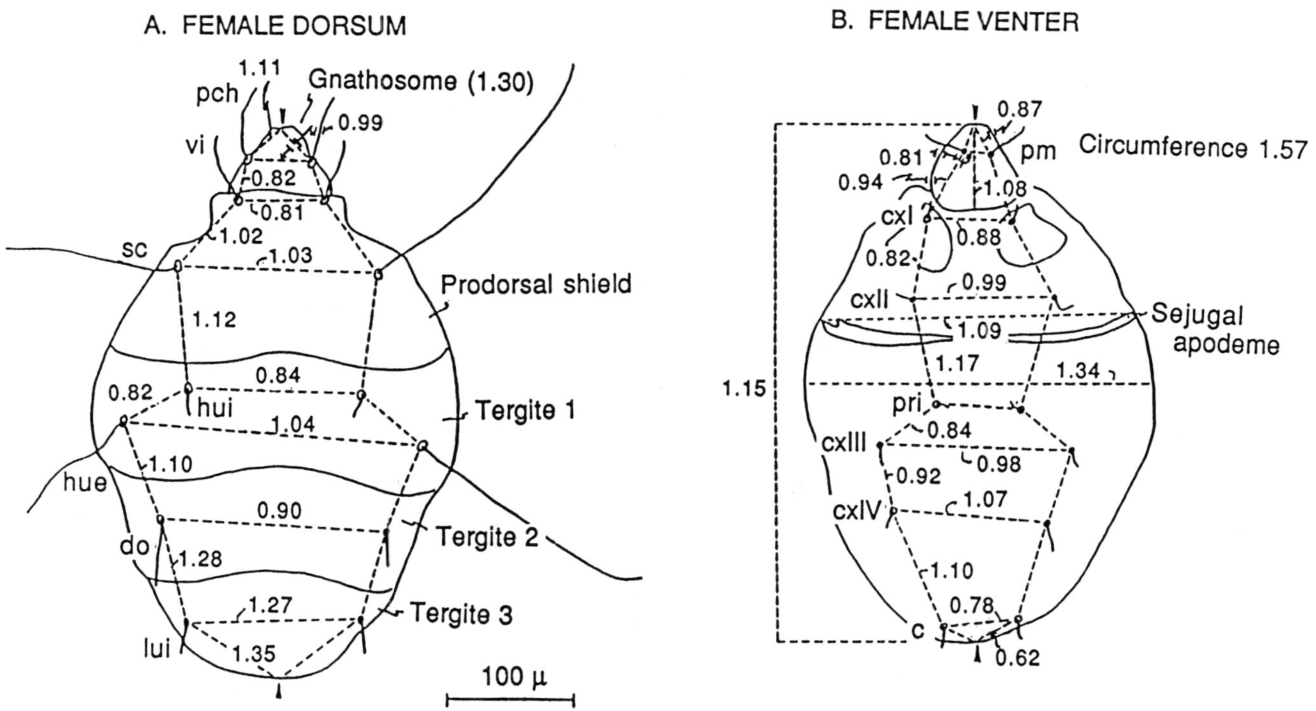

Figure 4. Evolutionary allometries, across the form, for females of *Chrysomelobia labidomerae*: (A) Dorsal; (B) Ventral.

Table 4.
Attributable percentages of total variance that are accounted for by the first three discriminant functions for dorsal and ventral characters of *Chrysomelobia labidomerae* females.

% of Variance Attributable to	DF-1	DF-2	DF-3	Total for DFs-1 to -3
Locality	32 %	24 %	12 %	68 %
Host	41 %	32 %	13 %	86 %

Secondary discrimination (DF-2 = 24%) separated the U.S. clusters from the Mexican groups, with the mites collected from the Midwest intermediate in the magnitude of shape change between the New York mites and the Mexican mites. The discriminating dorsal characters for the U.S. groups include a relatively elongated and wider gnathosome, a shorter distance between setae of tergites 2 and 3, and narrower intersetal distances in tergite 1.

3.3.2 **Host**

Discriminant functions 1 and 2 represented 41% and 32%, respectively, of total variation in the size-corrected data. The parasites from the U.S. host, *L. c. clivicollis*, could be completely discriminated from clusters representing the four Mexican hosts (Fig. 7). Mites from the Mexican hosts, *L. cacica* and *L. decemlineata*, could not be uniquely distinguished from one another on the first two (or subsequent) discriminant functions, and those from *L. undecemlineata* and *L. signaticollis* could not be discriminated from one another.

There is, however, a general cline in shape variation (DF-2) among the Mexican clusters, representing a host-related shift in shape change. *Chrysomelobia* from *L. undecemlineata* (Veracruz) are completely separable from mites from *L. cacica* (Veracruz) on DF-2. Mites from *L. decemlineata* (Morelos) were more similar to those from *L. cacica* (Veracruz) than to any other populations, including those collected within the same Mexican state.

In general, the Mexican populations have significantly larger posterior body regions and elongated gnathosoma in relation to mites of U.S. hosts. Dorsal characters contributing to the discrimination of U.S. from Mexican mites include a wider, more elongate dorsum (especially posterior to tergite 1) in Mexican mites and a shortening of setae *lui* in U.S. mites (Fig. 8A). Ventral discriminating characters of U.S. mites include a longer gnathosoma, a reduction in the distance between setae *pri*, and an increased distance from setal pair *c* to the anus (Fig. 8B).

Discrimination of mites collected from *L. decemlineata* and *L. cacica* (both from Veracruz) was due primarily to a greater separation of the setal pair *pm* on the tip of the gnathosome and wider dorsal tergites 2 and 3 in *L. cacica*.

3.4 Clustering of Mahalanobis' Distances and Phenetic Similarities Among Populations

Mahalanobis' distances indicate the *C. labidomerae* from the U.S. may be considered as a discrete morphotype, distinct from all Mexican populations (Fig. 9A). Selection or drift, due to geographic separation, and host differences both contribute to the morphological variation. This conclusion, derived from the morphological data, is corroborated by independent behavioral observations of Drummond et al. (1984), who determined that two biotypes of *C. labidomerae* exist in the Western Hemisphere: a Mexican biotype, which parasites *L. decemlineata*, and a U.S. biotype, which cannot parasite *L. decemlineata*. Mites in this study were collected from *Labidomera* and not *L. decemlineata*, because the natural range of *L. decemlineata* does not extend into the northern U.S. There has not been a recent evolutionary association between the northern mite race and *L. decemlineata*. Separation in time has acted not in terms of reduced efficiency of parasitization, but through absolute loss of the ability to exploit a host from the previous zone of origin.

Mites associated with *L. decemlineata* in Morelos are most similar to mites collected from *L. cacica* in Veracruz (Fig. 9B), indicating that they may have shared a common gene pool that has dissociated relatively recently, conserving morphological integrity across habitats of eastern coastal Mexico and interior Mexico. A cladistic interpretation of *Leptinotarsa* has not been published. However, before his death, Jacques (1988) presented a thorough treatment of *Leptinotarsa*, including much information on discrete character states. His untimely death prevented any potential cladistic treatment of the data. I have extracted the character state analysis from Jacques' (1988) work and have subsequently generated a consensus cladogram (see methods) based upon his morphological data (Fig. 10). Although currently, a cladistic analysis using molecular data (T. Hsiao, personal communication) is underway, its results are unavailable for inclusion here; when finished, it will be of great value in determining how valid my cladogram, based on Jacques' characters, is.

Because the *Leptinotarsa* cladogram (Fig. 10) represents a tentative phylogenetic history of these beetles, independent of inferences developed from mite morphology, it can be used to test the conclusions developed from the *Chrysomelobia* phenogram. The two important conclusions from cladistic analysis that are relevant to the current study are that: (1) *L. cacica* occurs on a distant branch of the network from *L. decemlineata*, and (2) among the species in question, *L. decemlineata* is most closely aligned to *L. undecemlineata* (*Leptinotarsa signaticollis* was not distinguished from *L. undecemlineata* in Jacques' treatise).

Mites collected from *L. cacica* and *L. decemlineata* share a considerable amount of morphological similarity, as shown by the size-corrected multivariate discrimination comparisons (Fig. 7), despite the apparent influence that having come from different hosts (Fig. 3). A parsimonious explanation for the patterns of morphological relationships among the mites, relative to that of the historical relationships of the beetles involves the following scenario. (1) Mites colonized ancestral beetles prior to the speciation events that resulted in the diversion of *L. cacica* and *L. decemlineata* lineages. The fact that *Labidomera*, the sister-group of *Leptinotarsa*, also harbors the mite would suggest that this host-parasite interaction may have occurred prior to the primary node of the *Leptinotarsa* clade. (2) Habitat dissimilarity and

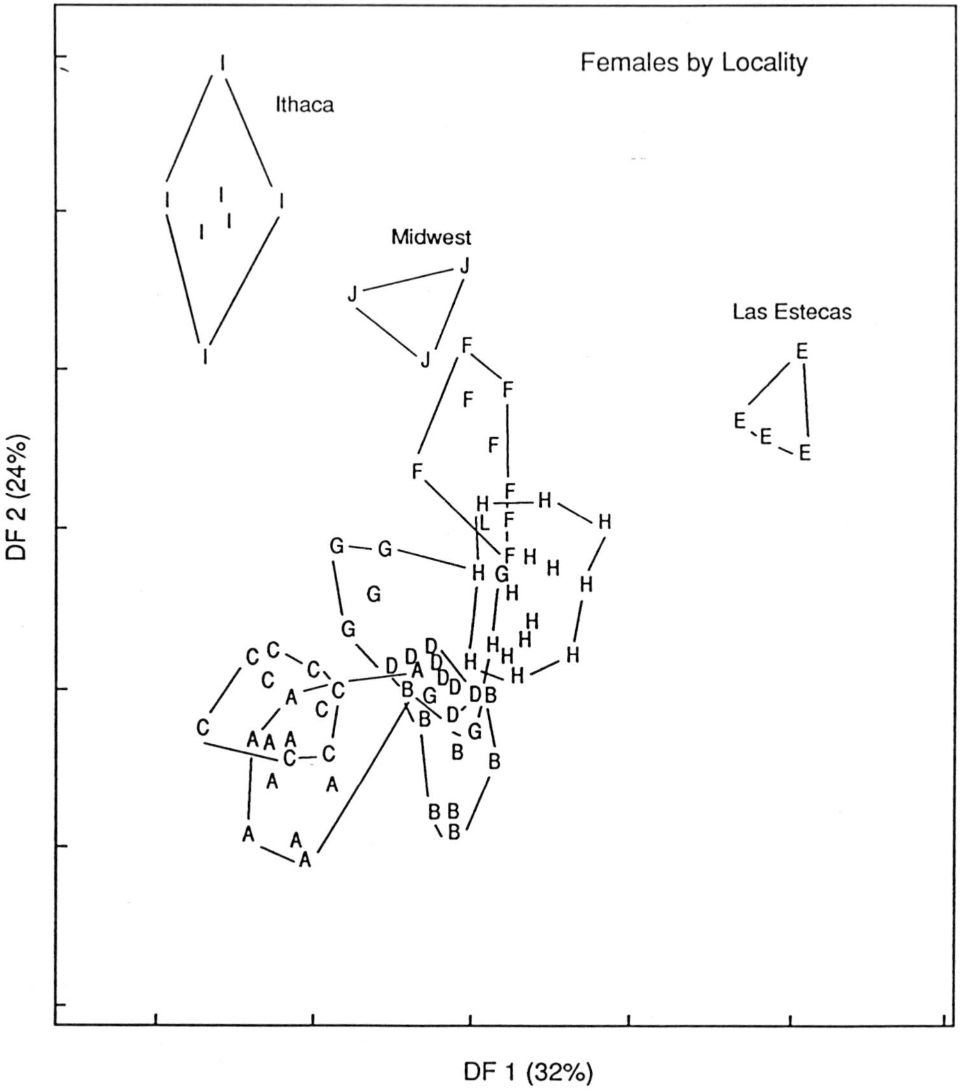

Figure 5. Discriminant function analysis of character variation in *Chrysomelobia labidomerae* females by locality. Localities: A = Apatlaco (Morelos); B = Moyotepec (Morelos); C = Canon de Lobo (Morelos); D = Temilpa (Morelos); E = Las Estecas (Morelos); F = Ixtapan (Mexico); G = Atlixco, (Puebla); H = Chocaman (Veracruz); I = Ithaca, New York; J = Midwest (Iowa, Indiana, Michigan, Minnesota).

geographic separation accentuated morphological differences, but to a lesser extent. (3) The northern dispersal of *C. labidomerae*, concomitant with the extension of the host range, may have resulted in attenuation of the host-parasite relationship. This may have occurred because the mites did not

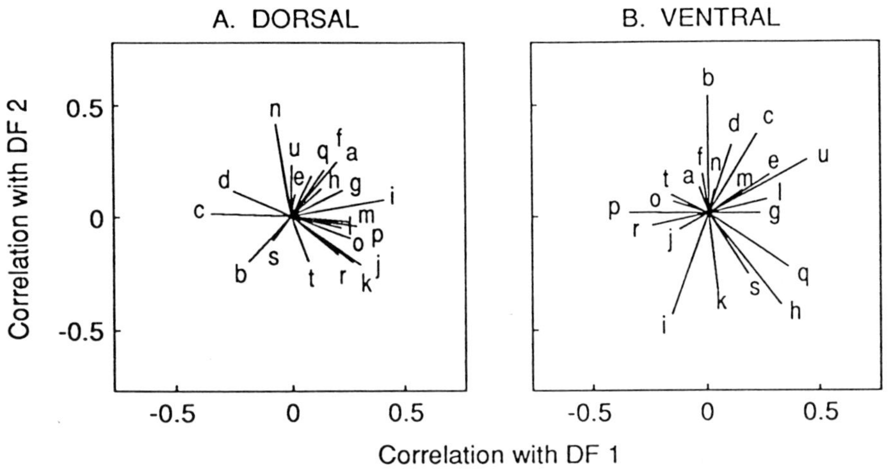

Figure. 6. Vector analysis of character correlations of DF-1 and DF-2 for the females by locality data set (Fig. 5). (A) Character correlations of DF-1 and DF-2 for the dorsal character set. (B) Character correlations of DF-1 and DF-2 for ventral character set. The complement of label names is given in Table 2; only those characters with a significant DF correlation are given in the figure. To interpret the vector plots, the polar coordinates and direction of the vectors indicated those characters that most contribute to discrimination among character expression (analyzed by locality or host) among the groups in complementary polar coordinates of the DF plots. For example, dorsal characters *a*, *b*, *f*, *g*, *q*, *h*, and *s* (Fig. 3) distinguish mites of Las Estecas from all other Mexican mites. The length of the vector from point of origin (the 0.0 position = no correlation) is proportional to the relative magnitude of contribution of that character to group separation. An evenly spaced floret of vectors would represent independence in character evolution. Notice that in Fig. 6A above, characters *a*, *f*, *g*, and *h*, however, represent a suite of anterior characters which respond as a complex. Character *g* is a prodorsal shield character,and *s* is a character representing change between tergites 1 and 2. In Fig. 2, mites from Las Estecas are represented by a reduction of the length of setae *hui* (vector *b*) and the distance between setae *hue* and *do* (vector *s*), as compared to all other Mexican mites;

survive at the northern extent of the host range, or because of genetic drift and speciation of mite-free subsets of the parent populations.

Mites associated with *L. undecemlineata* in Veracruz have shared similarity with mites from *L. signaticollis* in the interior areas of Mexico, and less morphological communality with sympatric mites from different hosts (Fig. 9B). Mites from *L. signaticollis* and *L. undecemlineata* were also not discriminated from one another in the previous DFA. This is especially interesting because these hosts can interbreed (Logan et al., 1987), which may indicate a recent, incomplete dichotomy within the host genus. Despite a probable recency of the separation of *L. signaticollis* and L. *undecemlineata*, however, the two taxa show some significant morphological shifts (Node *B* of Fig. 9B).

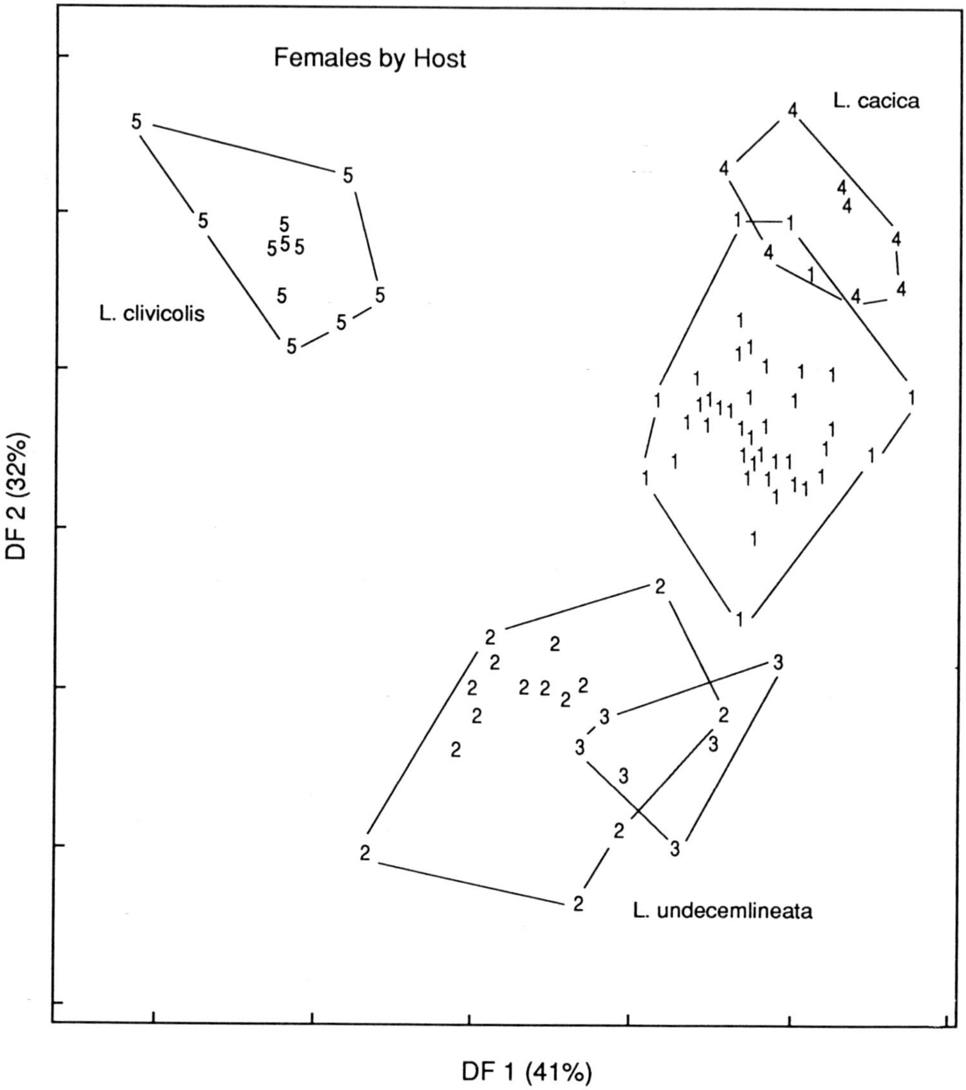

Figure 7. Discriminant function analysis of character variation in *Chrysomelobia labidomerae* females by host. Hosts: 1 = *Leptinotarsa decemlineata*; 2 = *L. signaticollis*; 3 = *L. undecemlineata*; 4 = *L. cacica*; 5 = *Labidomera c. clivicollis.*

No obvious ecological characteristics (parameters) clearly distinguish Las Estecas from the other Morelos sites. However, mites collected from Las Estecas show significant morphological divergence from all other populations collected from Morelos (Fig. 9A). This divergence may reflect a vestige population expressing morphological independence, due to habitat isolation

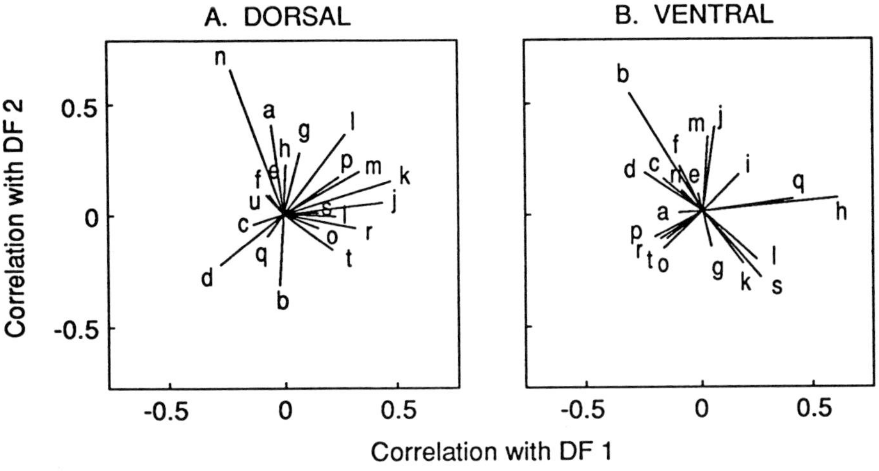

Figure 8. Vector analysis of character correlations with DF-1 and DF-2 for females by host data set (Fig. 7). (A) Character correlations with DF-1 and DF-2 for dorsal character set. (B) Character correlations with DF-1 and DF-2 for ventral character set. For Figs. 8A and 8B, the complement of label names are given in Table 2; only those characters with a significant DF correlation are given in the figures; see Fig. 6 for additional explanation.

(drift) or physical differences in habitat. Clearly, elements associated with habitat must be considered contributory. It is curious that these mites also had the highest within-group variance in size, indicating a general pattern of increased population variation.

4 Conclusions

(1) Inspection of the univariate measure of size (body length) indicates a lack of among-sample geographic clinal variation in females of *C. labidomerae*. The ecological generalization of increasing body size with increasing latitude, "Bergmann's rule," does not apply here. Mites collected from *L. cacica* were the largest (447.4 ± 21.8 μm) and significantly larger than those from *L. decemlineata* (400.3 ± 5.9 μm). Within-sample variation in mean-length was greatest when body length was partitioned by locality and less when considering the effect of host-species on body length. Mites collected from beetles from Las Estecas had the greatest within-group population variance. Mite body size was correlated with the body size of beetles from which they were collected.

(2) PCA indicated that a moderate size factor existed in the data. Within-group (population) variation was extensive, and among-group separation in multivariate space was minimal. Character variance indicated no major trends due to locality or host within any of the first three PCA axes. Female morphological variation overlapped in mites from all host groups except one, *L. cacica*, which also represented the population with the largest general-size factor of the five groups.

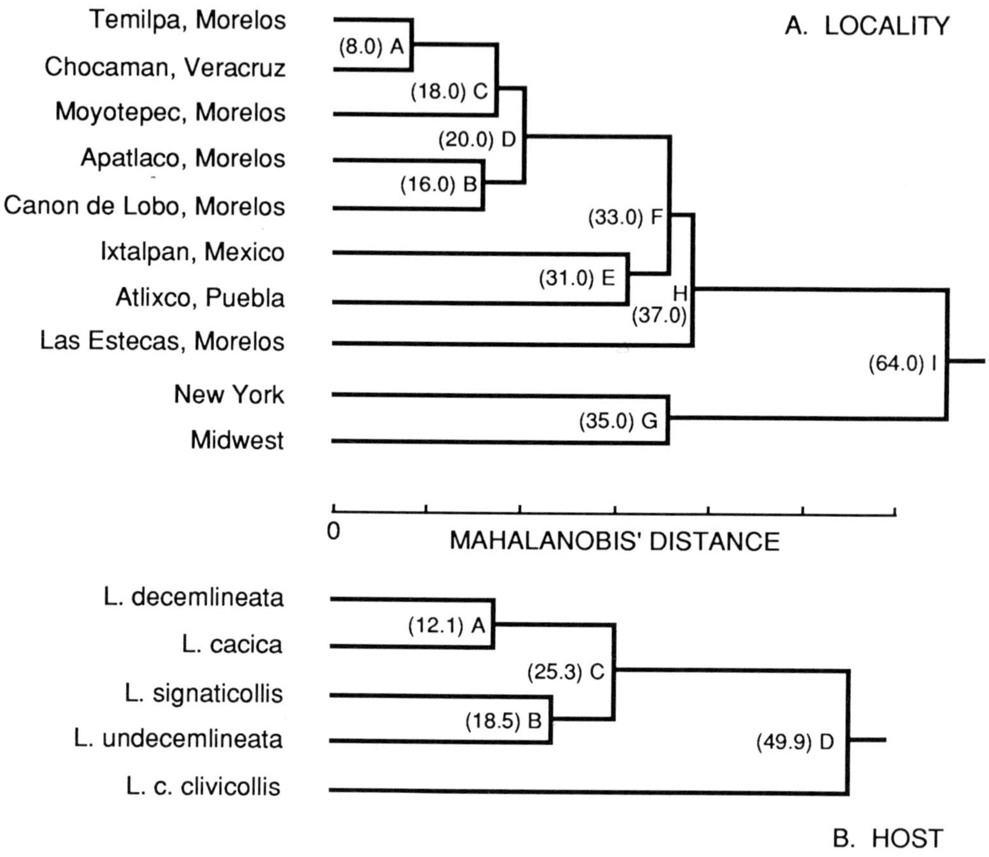

Figure 9. Mahalanobis' distances for females of *Chrysomelobia labidomerae* portrayed as dendrograms of relationships by locality and host. (A) Locality; (B) Host.

(3) DFA indicated that some environmental component present at one Mexican site, Las Estecas, has significantly influenced mite morphology. This component may have been in the form of physical parameters or geographical isolation.

(4) DFA also indicated that there was clear discrimination of U.S. populations of mites from Mexican populations, at the level of host genus. When morphological variation attributable to the first three discriminant functions was pooled, host species effects (86%) accounted for more variation in morphology than locality effects (68%).

(5) This study demonstrates the value of multivariate analysis in distinguishing cryptic taxa that other taxonomic methods have not resolved, and the vector analysis allows the examination of individual characters responsible for observed differences. *Chrysomelobia* mites in the Western Hemisphere were previously considered to be monospecific. Conspicuous structural differences, such as those between *C. labidomerae* and its only

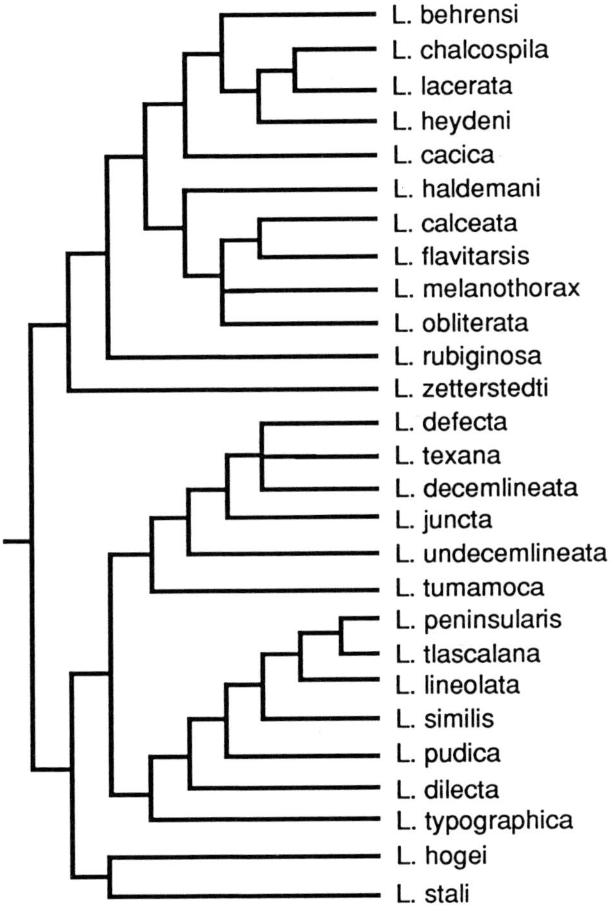

Figure 10. Consensus tree of 27 species of *Leptinotarsa* based on four equally parsimonious trees (CI = 0.443). Character matrix extracted from Jacques (1988), see Appendix 2.

described congener, *C. mahunkai* Regenfuss from Europe (Regenfuss, 1968; Eickwort, 1975), are not present in these populations.

(6) Patterns of morphological variation in the parasite do not, however, indicate an "expected" resolution of one mite species per host species of beetle, throughout its the range. For instance, mites on *L. decemlineata* from Las Estecas are as different from other populations on *L. decemlineata* as they are from populations on other hosts.

(7) Though more concentration of work is necessary to fully interpret these host-parasite relationships, the most likely hypothesis at this time is that *Chrysomelobia* was a successful parasite of *Leptinotarsa* prior to the major speciation events of the host clade, and that geographic restriction of the mite, due to plant-herbivore specificity, has acted to produce a mosaic of

morphological variation in mite morphology. The host-parasite relationship cannot be examined as a unit without considering the host plants as a contributory force. This may be very important to the interpreting of "habitat" by *Chrysomelobia*. A tri-level assessment (plant, herbivore, and parasite) of interactions is called for by the structure of this biological system. In this study, the plant host species were not part of the multivariate analysis. Clearly, future studies would benefit from incorporating such information.

5 Relevance of Ordination Procedures to Coevolution Studies

The process of evolution creates novel changes in organisms manifested in two ways (Carson and Templeton, 1984); it produces adaptive characters (morphological shifts) and it produces reproductive isolation. Morphological shifts in *Chrysomelobia* are clearly responsive to two prominent evolutionary forces: habitat related and host related. Geographic (and perhaps reproductive) isolation may have occurred secondarily in *Chrysomelobia* as a natural consequence of reduced gene-flow in its herbivorous hosts as the hosts developed resource specialization (monophagy). Monophagy of the host resulted in disjunct populations of mites. Reduced panmixis and genetic drift favored interruption of morphological homogeneity within the species as inbreeding influences accumulated across time.

Although these circumstances provide an appropriate arena in which coevolution (Ehrlich and Raven, 1964) may become a functional organizing factor within an established ecological community, detecting coevolution is far more complex than the interpretation of evolutionary processes within a single species. The available methodological tools for defining coevolution have historically been few. A common way of empirically defining coevolution between associated organisms is by inferential correlation, where researchers observe correlated patterns of occurrence and interaction between two species, both temporally and spatially, and define this relationship to be coevolved.

Mitter and Brooks (1983) and Brooks (1979, 1985) argue that the degree of historical coevolution of two groups of taxa may be assessed by examining independently generated phylogenetic hypotheses (cladograms) for historical congruence. They reason that if taxa have had a long history of intimate association, then the historical patterns represented in their clades should be concordant. This is a cladistic interpretation of Fahrenholz's rule (Eichler, 1948). To the extent that populations of parasites have faithfully tracked vicariance and dispersal events of their hosts, maintaining species associations, the patterns of divergence of host and parasite will be congruent. If the host-parasite interaction is weak and more than one host is available, however, then a resource-tracking model may apply (Kethley and Johnston, 1975). Parallel cladogenesis and resource-tracking represent extremes of a continuum based on dispersal opportunities and habitat choices available to a parasite (Timm, 1983). Environmental parameters (nonhost accommodation) can influence pathways of cladogenesis or resource-tracking in reticulate ways and any ecological interpretation must be coupled with phylogenetic inference (Ross, 1972; Boucot, 1983) to fully interpret phenotypic shifts in form. Unfortunately, there is currently no way to integrate environmental influences into cladistic analyses or the resource model.

Multivariate morphometric tools offer an auxiliary course in the study of coevolution that can enhance interpretations derived with other methodologies. A mosaic pattern of morphological change within a lineage is to be expected in ectoparasites, and a consensus of evolutionary response among all characters should never be anticipated. Various functional units across the morphological form may respond at different rates and magnitudes to either environmental constraints or host accommodation, or more probably with some interaction of both. Assessing the information content of morphological variation becomes difficult if our perspective is limited to one methodological ideology. It is helpful to partition the interpretation of the morphology into two levels: the pattern of variation among individuals (PCA and DFA) within and among populations, and variation of various functional components of a morphological form (vector analysis).

In an attempt to partition host and habitat effects on the general patterns of variation (weighted discrimination), or on individual characters of interest (vector analysis), the multivariate ordination approach is appropriate because: (1) It extends conclusions beyond those available with qualitative generalism or inferential correlation and can provide quantitative descriptors of natural groups. (2) It allows interpretation of within-group variation relative to among-group variation in a multidimensional space, and provides an opportunity to partition environmental effects from those of host-parasite coevolution. (3) It focuses on variation within and among groups (the currency of natural selection), which can be interpreted in reference to information on cladistic branching patterns (the outcome of natural selection). A variance shift represents a slower change and a more evolutionarily conservative parameter for a population than does a shift in mean character value (Roughgarden, 1972; Case, 1982). A shift in a character-rich multivariate space is likely to be more informative about the morphological integrity of populations, than any mean-shift in a univariate analysis. (4) It is not dependent on availability of fully resolved clades, and subsets of the clade may be legitimately examined if the host-parasite association is not a general attribute of the full clade. In the case of *Chrysomelobia*, parasitization could only be examined with a subset of the host possibilities partly due to limitation of available specimens and partly because the mite is not associated with all members of the monophyletic clade. (5) Vector analysis can be examined for correlated character suites and thus allows interpretation of functional character complexes. (6) It can separate the incidence of an evolutionary increase in size, as a selected univariate character, from ecological size shifts. It can also partition evolutionary shape change from shape change passively responding to an increase in size. This provides an alternate hypothesis (allometry) to test for evolutionary size-shape change in organisms.

Multivariate approaches to the study of host-parasite associations are rare and underexplored. I offer here one possible approach which is not without flaw. For instance, the portrayal of multicharacter variance patterns in a two-dimensional space is a snap-shot in evolutionary time. Variance structure is influenced by sampling error as well as dynamic changes in populations. The inclusion of additional data may give a different snap-shot, and the stability of the variance structure must be tested, not assumed. Standard errors of parameters and loadings can be estimated (e.g., using a jackknife procedure; Sokal and Rohlf, 1981) to determine how influential any component of the analysis is on the established variance structure.

I believe that the exploratory procedures outlined here are a potential first step in a projected two-step process. In this study, morphological variation in the ectoparasites has been partitioned and interpreted, by host and locality, in an multivariate context. The next logical step is to evaluate, in like manner, the morphology of the hosts from which these individuals were recovered. This will provide a covariance configuration of the hosts, as reflected by association with the ectoparasites. Before coevolution can be adequately demonstrated, it is necessary to show that both organisms of the association are responsive to their interaction. Multivariate host and parasite configurations could subsequently be superimposed. One could translate and rotate the interactive (nonenvironmental) orthogonal axes and interpret the degree of congruence (Procrustes rotation: see Digby and Kempton, 1987). Though this has not yet been empirically completed for the *Chrysomelobia*-chrysomelid association, it is the next logical exploratory step.

6 Acknowledgments

I thank Richard E. Strauss, Byron Alexander and George C. Eickwort for advice and comments on this work which significantly improved its technical content and clarity. George C. Eickwort initiated my interest in this project. I thank Byron Alexander for digitization and mounting of the mites. Francis A. Drummond provided beetles from which the mites were collected, and the detailed Mexican site descriptions. Daniel Brooks, Howell Daly, Michael Donoghue, David Dussourd, William Fink, Catherine Tauber and James Woolley offered comments which significantly improved an original version of this manuscript. This work was partially supported by National Science Foundation grant BSR-8307711 and BARD grant IS-1397-87.

7 References

Baker, T.C. and Eickwort, G.C., 1975. Development and bionomics of *Chrysomelobia labidomerae* (Acari: Tarsonemina; Podopolipidae), a parasite of the milkweed leaf beetle (Coleoptera: Chrysomelidae). *Canadian Entomologist*, 107: 627-638.

Bookstein, F.L., 1989. "Size and shape": a comment on semantics. *Systematic Zoology*, 38: 173-180.

Bookstein, F.L., Chernoff, B., Elder, R.L., Humphries, J.M., Smith, G R. and Strauss, R.E., 1985. Morphometrics in Evolutionary Biology. Special Publication, 15. Academy of Natural Sciences, Philadelphia, 277 pp.

Boucot, A.J., 1983. Does evolution take place in an ecological vacuum? *Journal of Paleontology*, 57: 1-29.

Brooks, D.R., 1979. Testing the context and extent of host-parasite coevolution. *Systematic Zoology*, 28: 299-307.

Brooks, D.R., 1985. Historical ecology: a new approach to studying the evolution of ecological associations. *Annals of the Missouri Botanical Garden*, 72: 660-680.

Burnaby, T.P., 1966. Growth-invariant discriminant functions and generalized distances. *Biometrics*, 22: 96-110.

Carson, H.L. and Templeton, A.R., 1984. Genetic revolutions in relation to speciation phenomena: the founding of new populations. *Annual Review of Ecology and Systematics*, 15: 97-131.

Case, T., 1982. Coevolution in resource-limited communities. *Theoretical Population Biology*, 21: 69-91.

Daloze, D., Braekman, J.C. and Pasteels, J.M., 1986. A toxic dipeptide from the defense glands of the Colorado beetle. *Science*, 233: 221-223.

Daloze, D. and Pasteels, J.M., 1979. Production of cardiac glycosides by chrysomelid beetles and larvae. *Journal of Chemical Ecology*, 5: 63-76.

Delfinado-Baker, M. and Houck, M.A., 1989. Geographic variation in *Varroa jacobsoni* (Acari: Varroidae): application of multivariate morphometric techniques. *Apidologie*, 20: 345-358.

Deroe, C. and Pasteels, J.M., 1977. Defensive mechanisms against predation in the Colorado beetle (*Leptinotarsa decemlineata,* Say). *Archives de Biologie*, 88: 289-304.

Deroe, C. and Pasteels, J.M., 1982. Distribution of adult defense glands in chrysomelids (Coleoptera: Chrysomelidae) and its significance in the evolution of defense mechanisms within the family. *Journal of Chemical Ecology*, 8: 67-82.

Dettner, K., 1987. Chemosystematics and evolution of beetle chemical defenses. *Annual Review of Entomology*, 32: 17-48.

Digby, P.G.N. and Kempton, R.A., 1987. Multivariate Analysis of Ecological Communities. Chapman and Hall, London, 206 pp.

Dillon, E.S. and Dillion, L.S., 1972. A Manual of Common Beetles of Eastern North America, Volume II. Dover Publications Inc., New York, 884 pp.

Drummond, F.A., Casagrande, R.A., Chauvin, R., Hsiao, T.H., Lashomb, J.H., Logan, P.A. and Atkinson, T.H., 1984. Distribution and new host records of a race of *Chrysomelobia labidomerae* Eickwort (Acari: Tarsonemina: Podapolipidae) attacking the Colorado potato beetle in Mexico. *International Journal of Acarology*, 10: 179-180.

Drummond, F.A., Casagrande, R.A. and Logan, P.A., 1989. Population dynamics of *Chrysomelobia labidomerae* Eickwort, a parasite of the Colorado potato beetle. *International Journal of Acarology*, 15: 31-45.

Drummond, F.A., Logan, P.A., Casagrande, R.A. and Gregson, F.A., 1985. Host specificity tests of *Chrysomelobia labidomerae,* a mite parasitic on the Colorado potato beetle. *International Journal of Acarology*, 11: 169-172.

Ehrlich, P.R. and Raven, P.H., 1964. Butterflies and plants: a study in coevolution. *Evolution*, 18: 586-608.

Eichler, W., 1948. Some rules ectoparasitism. *Annual Magazine of Natural History, Series 12*, 1: 588-598.

Eickwort, G.C., 1975. A new species of *Chrysomelobia* (Acari: Tarsonemina; Podapolipidae) from North America and the taxonomic position of the genus. *Canadian Entomologist*, 107: 613-626.

Eickwort, R.C. and Eickwort, G.C., 1986. Effects of parasitism by the mite *Chrysomelobia labidomerae* (Acari: Podapolipidae) on the longevity and fecundity of its host beetle, *Labidomera clivicollis* (Coleoptera: Chrysomelidae). *International Journal of Acarology*, 12: 223-227.

Houck, M.A., Gauthier, J.A. and Strauss, R.E., 1990. Allometric scaling in the earliest fossil bird, *Archaeopteryx lithographica*. *Science*, 246: 195-198.

Hsiao, T., 1981. Ecophysiological adaptations among geographic populations of the Colorado potato beetle in North America. In: J.H. Lashomb and R. Casagrande (Editors), Advances in Potato Pest Management. Hutchinson Ross Publishing Company, Stroudsburg, Pennsylvania, pp. 69-85.

Hsiao, T., 1986. Specificity of certain chrysomelid beetles for Solanaceae. In: W.G. D'Arcy (Editor), Solanaceae: Biology and Systematics, Second International Symposium. Columbia University Press, New York, pp. 345-377.

Hsiao, T., 1988. Host specificity, seasonality and bionomics of *Leptinotarsa* beetles. In: P. Jolivet, E. Petitipierre and T.H. Hsiao (Editors), Biology of Chrysomelidae. Kluwer Academic Publishers, Dordrecht, Netherlands, pp. 581-599.

Hsiao, T. and Hsiao, C., 1983. Chromosomal analysis of *Leptinotarsa* and *Labidomera* species (Coleoptera: Chrysomelidae). *Genetics*, 60: 139-150.

Jacques, R.L., 1988. The Potato Beetles: the Genus *Leptinotarsa* in North America (Coleoptera: Chrysomelidae). E.J. Brill Publishing Company, New York, 144 pp.

Jolicoeur, P., 1963a. The multivariate generalization of the allometry equation. *Biometrics*, 19: 497-499.

Jolicoeur, P., 1963b. The degree of generality of robustness in *Martes americana*. *Growth*, 27: 1-27.

Jolivet, P., 1986. Insects and Plants: Parallel Evolution and Adaptations. Flora and Fauna Handbook, 2. E.J. Brill Publishing Company, New York, 197 pp.

Kethley, J.B. and Johnson, D.E., 1975. Resource tracking patterns in bird and mammal ectoparasites. *Miscellaneous Publications of the Entomological Society of America*, 9: 231-236.

Lande, R., 1985. Genetic and evolutionary aspects of allometry. In: W.L. Jungers (Editor), Size and Scaling in Primate Biology. Plenum Press, New York, pp. 21-33.

Logan, P.A., Casagrande, R.A., Hsiao, T.H. and Drummond, F.A., 1987. Collections of natural enemies of *Leptinotarsa decemlineata* (Coleoptera: Chrysomelidae) in Mexico, 1980-1985. *Entomophaga*, 32: 249-254.

Mitter, C. and Brooks, D.R., 1983. Phylogenetic aspects of coevolution. In: D.J. Futuyma and M. Slatkin (Editors), Coevolution. Sinauer Associates Inc., Sunderlund, Massachusetts, pp. 65-98.

OConnor, B.M. and Houck, M.A, 1991. Preparation of mite specimens. In: D. Rosen (Editor), Armored Scale Insects, Their Biology, Natural Enemies and Control, Volume B. Elsevier Science Publishers, Amsterdam, pp. 243-245.

Pasteels, J.M. and Daloze, D., 1977. Cardiac glycosides in the defense secretion of chrysomelid beetles: evidence for their production by the insects. *Science*, 197: 70-72.

Pasteels, J.M., Braekman, J.C. and Daloze, D., 1982. Chemical defense in chrysomelid larvae and adults. *Tetrahedron*, 38: 1891-1897.

Pasteels, J.M., Rowell-Rohier, M., Braekman, J.C. and Dupont, A., 1983. Salicin from host plant as precursor of salicylaldehyde in defensive secretion of chrysomeline larvae. *Physiological Entomology*, 8: 307-314.

Pasteels, J.M., Rowell-Rohier, M., Braekman, J.C. and Daloze, D., 1984. Chemical defenses in leaf beetles and their larvae: the ecological, evolutionary and taxonomic significance. *Biochemical Systematics and*

Ecology, 12: 395-406.
Regenfuss, H., 1968. Untersuchungen zur morphologie, systematik und okologie der Podapolipidae (Acarina, Tarsonemini). *Zeitschrift fur Wissenschafliche Zoologie*, 177: 183-282.
Rohlf, F.J. and Bookstein., F.L., 1987. A comment on shearing as a method for "size correction." *Systematic Zoology*, 36: 356-367.
Ross, H.H., 1972. An uncertainty principle in ecological evolution. In: R.T. Allen and F.C. James (Editors), A Symposium on Ecosystematics. Occasional Paper, 4. University of Arkansas Museum, Fayetteville, Arkansas, pp. 133-157.
Roughgarden, J., 1972. Evolution of niche width. *The American Naturalist*, 106: 683-718.
SAS Institute Inc., 1985. SAS/STAT: Guide for Personal Computers, (Version 6 Edition). SAS Institute Inc., Cary, North Carolina, 429 pp.
Sokal, R.R. and Rohlf, F.J., 1981. Biometry, (Second Edition). W.H. Freeman and Company, New York, 859 pp.
Sneath, P.H.A. and Sokal, R.R., 1973. Numerical Taxonomy, the Principles and Practice of Numerical Classification. W.H. Freeman and Company, New York, 573 pp.
Swofford, D.L., 1985. PAUP, Phylogenetic Analysis Using Parsimony, (Version 2.4). D.L. Swofford, Illinois Natural History Survey, Urbana, Illinois.
Timm, R.M., 1983. Fahrenholz's rule and resource tracking: a study of host-parasite coevolution. In: M.H. Nitecki (Editor), Coevolution. University of Chicago Press, Chicago, pp. 225-265.
Tower, W.L., 1918. The mechanism of evolution in *Leptinotarsa*. Publication 263. Carnegie Institution, Washington, D.C., 384 pp.
Wieman, H.L., 1910. A study in the germ cells of *Leptinotarsa signaticollis*. *Journal of Morphology*, 21: 135-215.
Wright, S., 1954. The interpretation of multivariate systems. In: O. Kempthorne, T.A. Bancroft, J.W. Gowen and J.L. Lush (Editors), Statistics and Mathematics in Biology. Iowa State College Press, Ames, Iowa, pp. 11-33.
Wright, S., 1968. Evolution and the Genetics of Populations, Volume I. Genetic and Biometric Foundations. University of Chicago Press, Chicago, 469 pp.

8 Appendices

Appendix 1A.
Character descriptions for species of *Leptinotarsa* Stål. The numeric codes for character states appear in square brackets after their state descriptor, alternative state descriptor terms or comments appear in curved brackets.

I. Head:
  A) Color:
    1) Bronze [0]; metallic blue/green [2]; black [3]; yellow {flavous} [4]; red {rufescent} [5]; brown {testaceous} [6]; brassy {geneous} [7].
  B) Punctations, type:
    2) Fine [0]; course [1]; variable [2]; none [3].
  C) Punctations, regularity:
    3) Regular [0]; irregular [1].
  D) Punctations, nature:
    4) Dense [0]; sparse [1]; in rows [2]; differ at margins [3]; moderate [4]; smooth {default if not given} [5]; differ at spots [6]; differ at eyes [7].
  E) Interocular distance (mm):
    5) 1.4-1.5 [0]; 1.6-1.7 [1]; 1.8-1.9 [2]; 2.0-2.1 [3].
  F) Head width (mm):
    6) 1.6-1.8 [0]; 1.9-2.1 [1]; 2.2-2.4 [2]; 2.5-2.7 [3]; 2.8-3.0 [4].
  G) Maculation:
    7) Immaculate [0]; maculate [1].
  H) Maculation, color:
    8) Bronze [0]; metallic blue/green [2]; black [3]; yellow {flavous} [4]; red {rufescent} [5]; brown {testaceous} [6]; brassy {geneous} [7].

II. Thorax:
  A) Color:
    9) Bronze [0]; metallic blue/green [2]; black [3]; yellow {flavous} [4]; red {rufescent} [5]; brown {testaceous} [6]; brassy {geneous} [7].
  B) Punctations, type:
    10) Fine [0]; course [1]; variable [2]; none [3].
  C) Punctations, regularity:
    11) Regular [0]; irregular [1].
  D) Punctations, nature:
    12) Dense [0]; sparse [1]; in rows [2]; differ at margins [3]; moderate [4]; smooth {default if not given} [5]; differ at spots [6]; differ at eyes [7].
  E) Maculation:
    13) Immaculate [0]; maculate [1].
  F) Maculation, color:
    14) Bronze [0]; metallic blue/green [2]; black [3]; yellow {flavous} [4]; red {rufescent} [5]; brown {testaceous} [6]; brassy {geneous} [7].

III. Pronotum:
  A) Color:
    15) Bronze [0]; metallic blue/green [2]; black [3]; yellow {flavous} [4]; red {rufescent} [5]; brown {testaceous} [6]; brassy {geneous} [7].
  B) Punctations, type:
    16) Fine [0]; course [1]; variable [2]; none [3].
  C) Punctations, regularity:
    17) Regular [0]; irregular [1].
  D) Punctations, nature:
    18) Dense [0]; sparse [1]; in rows [2]; differ at margins [3]; moderate [4]; smooth {default if not given} [5]; differ at spots [6]; differ at eyes [7].

E) Maculation:
   19) Immaculate [0]; maculate [1].
F) Maculation, color:
   20) Bronze [0]; metallic blue/green [2]; black [3]; yellow {flavous} [4]; red {rufescent} [5]; brown {testaceous} [6]; brassy {geneous} [7].
G) Length (mm):
   21) 1.5-1.8 [0]; 1.9-2.2 [1]; 2.3-2.6 [2]; 2.7-3.1 [3].
H) Width (mm):
   22) 3.84-4.0 [0]; 4.1-4.5 [1]; 4.6-5.0 [2]; 5.1-5.5 [3]; 5.6-6.0 [4]; 6.1- 6.5 [5]; 6.6-7.0 [6]; 7.1-7.6 [7].

IV. Scutellum:
A) Color:
   23) Bronze [0]; metallic blue/green [2]; black [3]; yellow {flavous} [4]; red {rufescent} [5]; brown {testaceous} [6]; brassy {geneous} [7].
B) Punctations, type:
   24) Fine [0]; course [1]; variable [2]; none [3].
C) Punctations, regularity:
   25) Regular [0]; irregular [1].
D) Punctations, nature:
   26) Dense [0]; sparse [1]; in rows [2]; vary at margins [3]; moderate [4]; smooth {default if not given} [5]; vary at spots [6]; vary at eyes [7].

Appendix 1B.
List of species of *Leptinotarsa*, and their abbreviations, included in the cladistic analysis. (Note: *Leptinotarsa boucardi* Achard, *L. dohrini* Jacoby, *L. novemlineata* Stål, and *L. virqulata* Achard were omitted from the cladistic analysis because these species were unavailable for examination.)

Abbrev.	Taxon	Abbrev.	Taxon
BEHR	*L. behrensi* Harold	MELA	*L. melanothorax* (Stål)
CACI	*L. cacica* Stål	OBLI	*L. obliterata* (Chevorlat)
CALC	*L. calceata* Stål	PENI	*L. penisularis* (Horn)
CHAL	*L. chalcospila* Stål	PUDI	*L. pudica* Stål
DECE	*L. decemlineata* (Say)	RUBI	*L. rubiginosa* (Rogers)
DEFE	*L. defecta* (Stål)	SIMI	*L. similis* Bowditch
DILE	*L. dilecta* Stål	STAL	*L. stali* Jacoby
FLAV	*L. flavitarsis* (Guerin-Meneville)	TEXA	*L. texana* Schaeffer
HALD	*L. haldemani* (Rogers)	TLAS	*L. tlascalana* (Stål)
HEYD	*L. heydeni* Stål	TUM	*L. tumamaca* Tower
HOGE	*L. hogei* Jacoby	TYPO	*L. typographica* Jacoby
JUNC	*L. juncta* (Germar)	UNDE	*L. undecemlineata* (Stål)
LACE	*L. lacerata* Stål	ZETT	*L. zetterstedti* Stål
LINE	*L. lineolata* Stål		

Appendix 2.
Morphological character states used for the cladistic analysis of *Leptinotarsa* species (from Jacques, 1988). Characters 1-4, 7-20, 23-34, and 37-49 are unordered. The definitions of characters and taxa listed are given in Appendices 1A, 1B.

Taxon	Characters (1-49)
BEHR	0000330000130000130036000012010000541001001000103
HALD	2001210020110020110022330532012000213000030003000
CACI	2001320022130022130025200014113000643001021002100
RUBI	5004220050130050130023330515012000225000050005000
DEFE	4100001342131342131311400404112123104130040014010
TEXA	4214111340131340131312330504102143114130040004000
TUM	5117010050131350131311630514116153105000150005000
UNDE	3214211662131362131312330504116153313000230203002
JUNC	5214321350111350110323330504106153225130051315103
DECE	4214111340131340131312430504106153214130040014010
CHAL	7000440070130070130037700016100183653000230003000
LACE	3004330031130031130026311313112154643001031003100
HEYD	3000330030130030130026300113112184443001021002100
CALC	2014210023050022140013630534112146212001121004110
MELA	3011110030010030010022330544112146223000030003000
PENI	5217000062130062130000601134102136005001150005000
TLAS	6117000060130060130000630533102124006000160006000
ZETT	5017220050130050130013530505112134325000150005000
TYPO	5217111350130050130002530514116193115000150005000
DILE	2004110020130020130002230514116150112000000000000
OBLI	3217320070010070000024730534112130323000230003000
PUDI	3210000031130031130012330504112146212000020002000
HOGE	7000120071130071130012730517116164217000270007000
STAL	7000120070130070130022330513112134113000070007000
LINE	2100000020130020130000230514114146007000060006000
FLAV	2214220023050023050013230504112133322000020014010
SIMI	2217000020130020130001230506114192102000020002000

DISCRIMINANT ANALYSIS OF YEAR CLASSES OF PERIODICAL CICADA BASED ON WING MORPHOMETRIC DATA ENHANCED BY MOLECULAR INFORMATION

Chris Simon

Department of Ecology and Evolutionary Biology, University of Connecticut, U-43, Storrs, Connecticut 06269, U.S.A.

Abstract

Discriminant analysis has been used previously to demonstrate that year-classes (broods) of periodical cicadas, *Magicicada* sp., can be separated based upon morphometric measurements of their wing veins. The proteins and mitochondrial DNA of these insects has also been studied. Three morphologically distinct species of 17-year cicadas exist, which each have a 13-year sibling species. In one of these sibling pairs (*M. septendecim* [17-year] versus *M. tredecim* [13-year]), discriminant analysis had previously indicated that two populations of 13-year broods masqueraded as 17-year broods by occupying the discriminant space of the latter. Recent molecular evidence indicates that these 13-year populations are actually derived from the 17-year lineage rather than other 13-year cicadas, and that discriminant analysis correctly displayed the evolutionary and genetic relationships involved. This article shows how molecular data, especially that of mitochondrial DNA, can be used to aid in the interpretation of morphometric data.

1 Introduction

Although groups of populations may overlap with respect to many characters viewed singly, the same groups may be entirely distinct when characters are view jointly (Jolicoeur, 1959). Discriminant function analysis (Lachenbruch, 1975) separates groups by giving less weight to characters with high within-group relative to among-group variances. Discriminant function analysis can be used as an identification procedure to assign individuals to preexisting groups, or as used in this chapter, as an ordination procedure to help visualize groups in a space which is reshaped so that among-group differences are maximized.

Discriminant analysis was used to show that year-classes (broods) of periodical cicadas could be differentiated based on measurements of wing veins (Simon, 1979a, 1983a). Studies of proteins (Simon, 1979a, 1979b; Archie et al., 1985) and later studies of mitochondrial DNA (mtDNA) (Simon, 1988; Martin and Simon, 1988, 1990) examined molecular genetic differentiation in periodical cicadas. This chapter demonstrates how these molecular data, especially the mtDNA data, can be used to enhance the interpretation of the discriminant analysis of periodical cicada wing morphometric data.

Periodical cicadas were chosen for study because individuals within broods are reproductively isolated from individuals belonging to other broods. In this respect, broods can be analogized with incipient species, and can be used to examine the morphological and molecular differentiation that is associated with the early stages of genetic isolation. In addition to the presence of discrete year-classes or broods, the life history of periodical cicadas is further complicated by the existence of two life cycles, one of 13 years and one of 17 years. Three broods of 13-year cicadas and twelve broods of 17-year cicadas exist at present.

There are three morphologically distinct species of 17-year cicada, each of which has a 13-year sibling species. The three sibling pairs are (17-year followed by 13-year): *Magicicada septendecim* and *M. tredecim* ("Decim" siblings); *M. cassini* and *M. tredecassini* ("Cassini" siblings); and *M. septendecula* and *M. tredecula* ("Decula" siblings). Most of the 17-year broods and all of the 13-year broods contain all three morphologically distinct species.

The purpose of my original discriminant analysis (Simon, 1979a, 1983a) was to determine whether there are morphological differences between 13- and 17-year Decim siblings, and whether morphological differences could be detected between populations of Decim siblings belonging to two temporally and spatially adjacent 17-year broods (thought to be very closely related). Thirteen-year Cassini siblings were included for comparison. Although previous authors were not able to find any characters to clearly distinguish between life cycles or year-classes, it was predicted that modern methods of morphometric analysis would be able to do so.

2 Morphometrics

Morphometric analyses were performed on two year-classes of *M. septendecim* (Brood XIII and Brood XIV), one year-class of *M. tredecim* (Brood XXIII) and one year-class of *M. tredecassini* (also part of Brood XXIII). Figs. 1 and 2 show the distributions of Broods XIII, XIV, XXIII and X (Brood X is relevant to later discussions). Morphological measurements of 48 wing dimensions for 80-100 individuals per year-class were made using a digitizer. Data points were input directly into a computer and stored. Table 1 lists localities of samples and sample sizes of measured individuals. Fig. 3 shows the 48 wing characters measured.

A two level nested analysis of variance (Sokal and Rohlf, 1974) was performed for each of the 48 wing vein characters for the Decim species. Broods were found to differ significantly for 79% of the 48 characters, although populations differed for only 60% of the characters. The variation was partitioned as follows. The majority of the variation was within populations (average 72.5% for the 48 characters). Cicadas at any one locality varied a

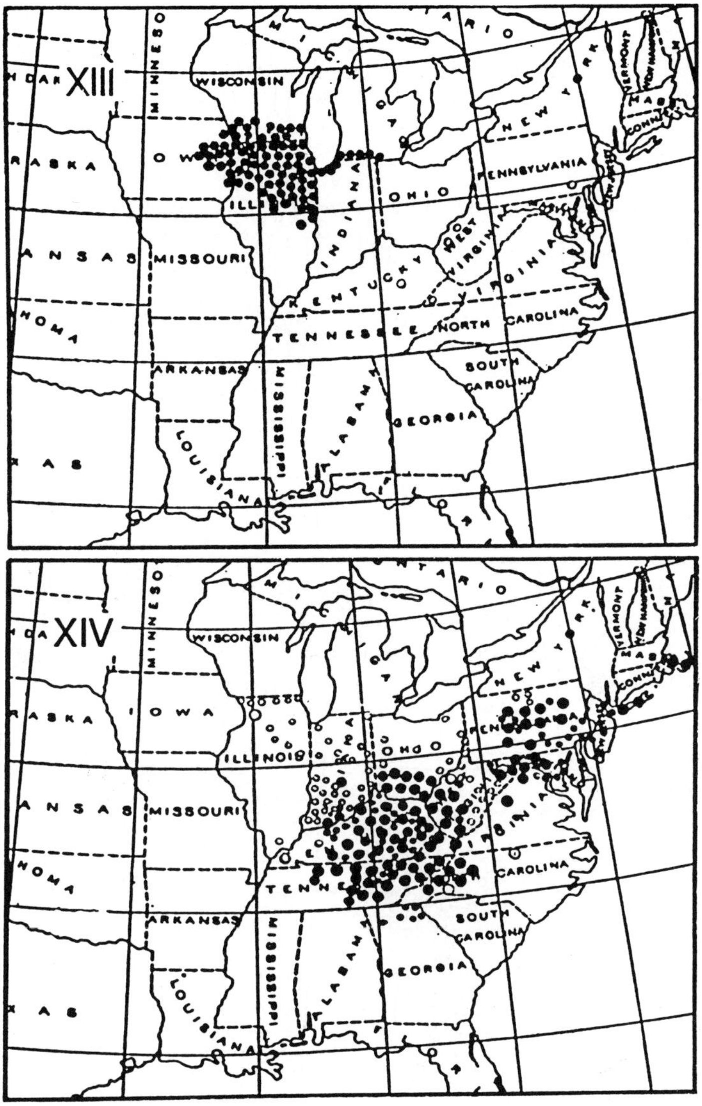

Figure 1. Distribution of 17-year Broods XIII and XIV. Maps here and Fig. 2 are updated from Marlatt (1907) with small black dots representing smaller populations and large black dots representing larger populations. Black dots with white centers represent populations that have become extinct or for which there are unclear records. Broods XIII, XIV were included in the morphometric analyses.

great deal in size and shape of the wing. Of the remaining variation, 9.3% was among populations and 18.2% was among broods. Thus, 17-year broods that are no older than the last glacial maximum (~18,000 years), and 13- versus

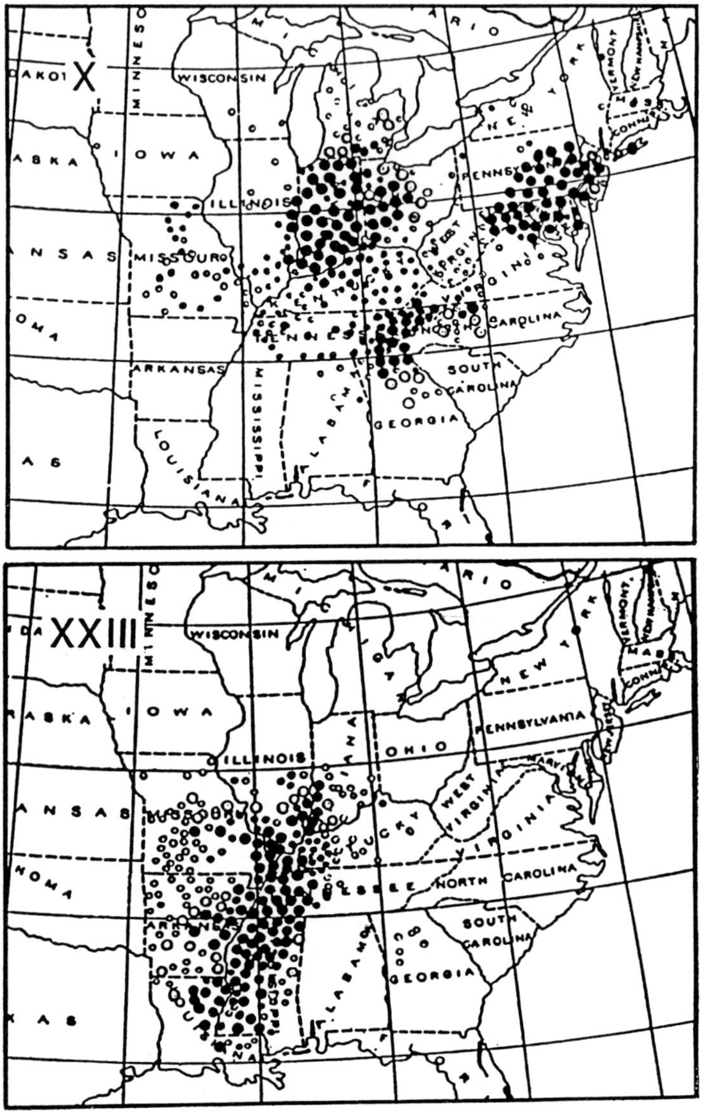

Figure 2. Distribution of 17-year Brood X and 13-year Brood XXIII (Decim, Cassini, and Decula siblings combined). Small black dots represent smaller populations and large black dots represent larger populations. Black dots with white centers represent populations that have become extinct or for which there are unclear records. Brood XXIII was included in the morphometric analyses; 17-year Brood X is shown because it is the year-class from which "13-year cicadas" in Missouri, Illinois, and Indiana are thought to be derived.

17-year broods that are probably no older that the Pleistocene in origin, have developed morphological differences.

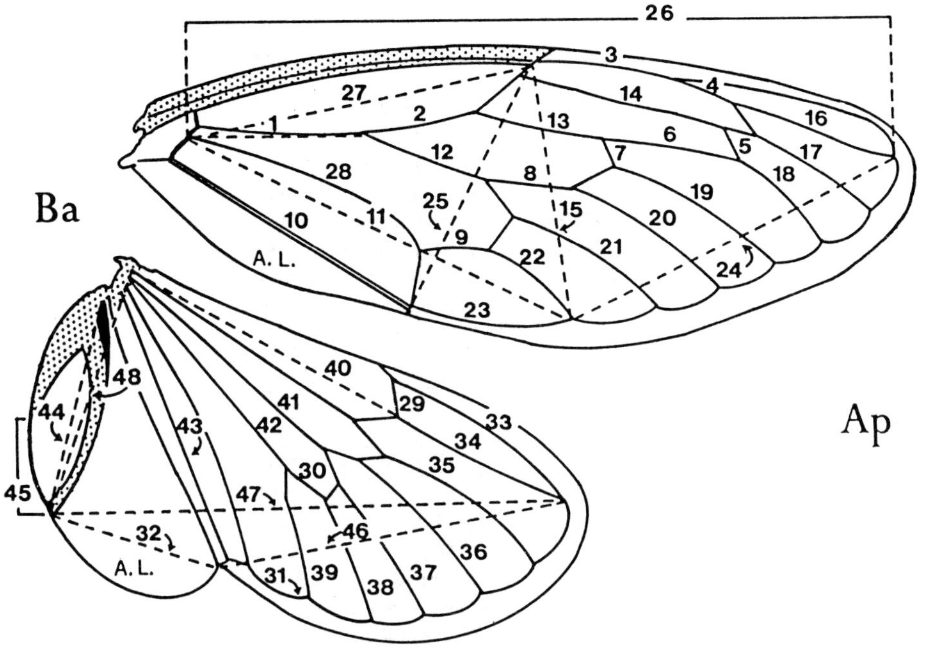

Figure 3. The 49 wing characters measured. The fore wing is drawn above the hind wing. The costal (anterior) margin of the fore wing is stippled. The basal portion of both wings (labeled Ba) is closest to the insect's body. The apical portion of the wing is labeled Ap. Character numbers are written just above or to the right of the vein that they identify unless otherwise indicated. Dashed lines describe length and width characters. A.L. is the abbreviation used for the anal lobe in both wings. Characters 11 and 43 measure the diagonal width of the anal lobes of the fore and hind wings, respectively. (From Simon, 1983a)

Because of the large amount of among-brood variation and the small amount of among-population variation, it seemed likely that broods of periodical cicadas could be differentiated but that populations within broods could not; discriminant analysis upheld this expectation. Discriminant analyses were performed on wing morphometric data (Fig. 3) for the populations listed in Table 1 using programs written by Bruce Riska (see Riska, 1981). The first three discriminant functions accounted for 79% of the total discrimination, and seven of the 22 possible discriminant functions were significant at or above the 5% level.

Figs. 4 and 5 show two dimensional plots of the population centroids and 95% confidence ellipses about these bivariate means. The area encompassed by each ellipse has a 95% probability of including the true populational mean. The size of the ellipse depends on population sample size and the tilt of the ellipse depends on the covariances along each discriminant axis. The shape of the ellipse depends on the amount of within-population variation along each discriminant axis.

Significance tests associated with discriminant function analysis assume that within-population variation (measured by the within-groups variance-

Table 1.
Localities and sample sizes for morphometric analysis of periodical cicada wing veins.

Population Number	Year-Class	*Magicauda* Species	State	County	Sample Size
1	XIV	*septendecim*	North Carolina	Buncombe-1	10
2	XIV	*septendecim*	Maryland	Washington	10
3	XIV	*septendecim*	North Carolina	Buncombe-2	10
4	XIV	*septendecim*	Pennsylvania	Adams	10
5	XIV	*septendecim*	Pennsylvania	Berks	10
6	XIV	*septendecim*	New York	Suffolk-1	10
7	XIV	*septendecim*	New York	Suffolk-2	10
8	XIV	*septendecim*	Massachusetts	Barusta-1	10
				Total Decim XIV	100
9	XXIII	*tredecim*	Louisiana	Ouachita Par.	10
10	XXIII	*tredecim*	Tennessee	Carrol	10
11	XXIII	*tredecim*	Missouri	Perry	10
12	XXIII	*tredecim*	Indiana	Knox	10
13	XXIII	*tredecim*	Kentucky	Marshall	10
14	XXIII	*tredecim*	Mississippi	Hinds	10
15	XXIII	*tredecim*	Mississippi	Holmes	10
16	XXIII	*tredecim*	Mississippi	Warren	10
				Total Decim XXIII	100
17	XIII	*septendecim*	Illinois	Will	80
18	XIII	*septendecim*	Illinois	Laporte	20
				Total Decim XIII	100
19	XXIII	*tredecassini*	Indiana	Gibson	20
20	XXIII	*tredecassini*	Tennessee	Shelby	20
21	XXIII	*tredecassini*	Tennessee	Lake	20
22	XXIII	*tredecassini*	Illinois	Union-1	20
23	XXIII	*tredecassini*	Illinois	Union-2	20
				Total Cassini XXIII	100

covariance matrix) is equivalent for each population. Therefore, all confidence contours, once standardized, should be circular. The elliptical contours in Figs. 4 and 5 indicate that the assumption of equal covariance structure is violated in this case. Tests of equality of variances confirmed this fact (see Simon, 1979a, 1983a). The violation of this assumption, however, does not preclude the use of discriminant function analysis as a descriptive tool. Despite these unequal covariance structures, populations were lumped into their respective broods with little overlap.

Two of the year-classes could not be completely separated along at least one axis. Two populations of Brood XXIII (13-year, Decim) overlapped with one population of Brood XIV (17-year, Decim) when viewed on axis 1 versus 2 and

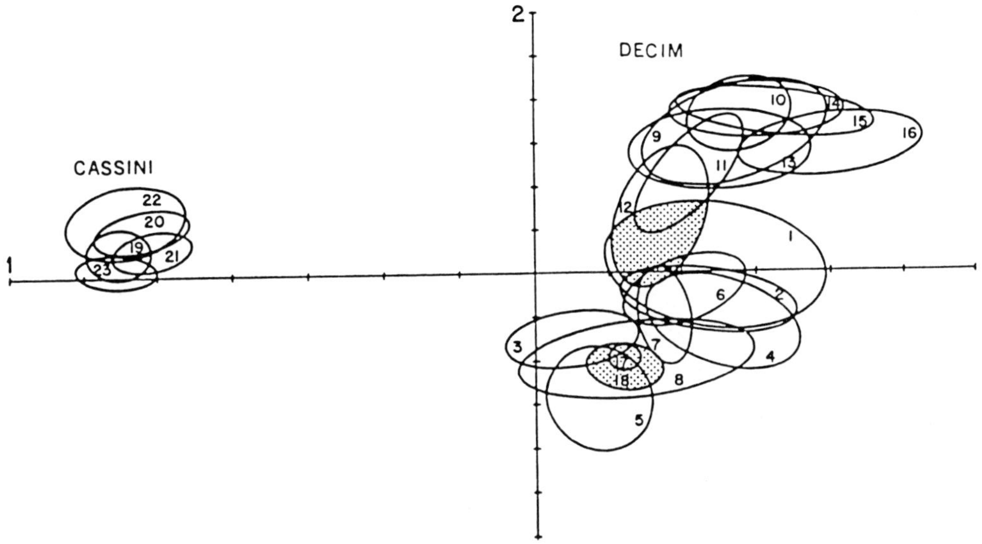

Figure 4. Ninety-five percent confidence ellipses about population centroids plotted on discriminant axes 1 and 2. Population localities and sample sizes are given in Table 1. Note populations 11 and 12 of 13-year Decim siblings of Brood XXIII overlap with 17-year Decim siblings of Brood XIV in this figure. The axes are labeled in Mahalanobis' distance units: one tic mark = one distance unit. The scale on the ordinate is compressed relative to that of the abscissa making the ellipses appear less circular then they actually are. (From Simon, 1983a)

on axis 2 versus 3 (Fig. 5). When viewed in light of molecular data, this overlap is not unexpected.

3 Molecular Differentiation

Martin and Simon (1988, 1990) found that throughout most of their ranges, the 13- versus 17-year Decim siblings represented two distinct mitochondrial DNA lineages with a 2.6% sequence divergence measured by restriction enzymes (50 sites). Neither the Cassini nor the Decula siblings showed restriction site differences between 13- versus 17-year species.

Given the ability to recognize distinct 13- and 17-year Decim lineages, one question of interest was whether there has been gene flow between these life cycle forms in regions where they have historically overlapped. At present, there are only two local populations in which 13- and 17-year Decim siblings are known to occur in the same trees; both of these are in southeastern Iowa (Simon, 1988). In the past, there was extensive overlap of broods of these two life cycles in Missouri, Illinois, northern Arkansas, and western Indiana; however, cicadas with 17-year life cycles no longer occur in the overlap zone.

Martin and Simon (1988, 1990) examined 13-year Decim populations for evidence of hybridization between 13- and 17-year Decim lineages in areas of

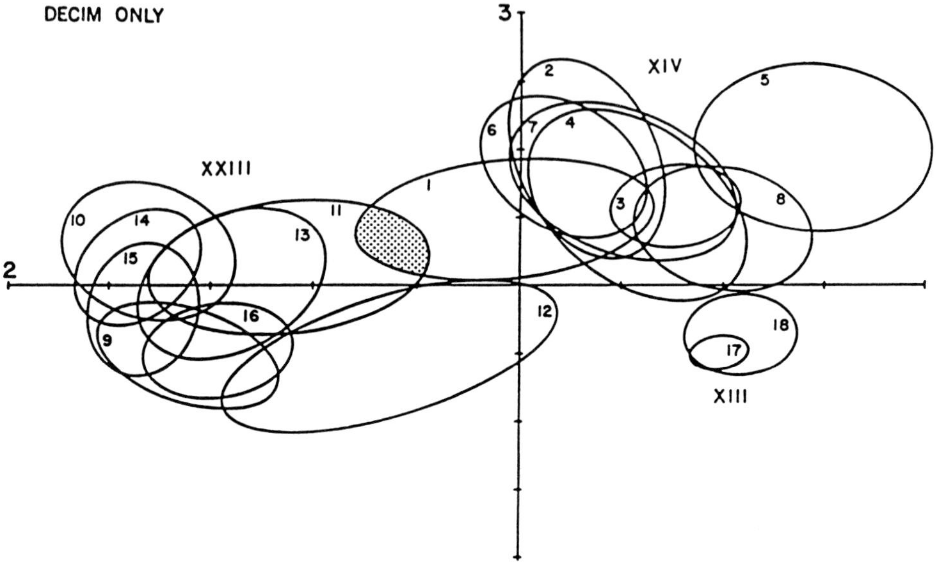

Figure 5. Ninety-five percent confidence ellipses about population centroids plotted on discriminant axes 2 and 3. Population localities and sample sizes are given in Table 1. The Cassini populations are left out of this drawing (but not left out of the analysis) because they lie directly behind the main cluster (but well behind the plane of the picture if the third dimension were to be drawn) and would only confuse the drawing. Note especially that Decim siblings of 17-year Broods XIII and XIV do not overlap here, although populations 11 and 12 of 13-year Decim siblings of Brood XXIII overlap with 17-year Decim siblings of Brood XIV here and in Fig. 4 The axes are labeled in Mahalanobis' distance units: one tic mark = one distance unit. The scale on the ordinate is compressed relative to that of the abscissa making the ellipses appear less circular then they actually are. Also note that the scale here is smaller than for Fig. 4 because of the greater length of axis 1. (From Simon, 1983a)

former overlap. Instead of finding evidence for gene flow, they came to the surprising conclusion that the 13-year Decim siblings inhabiting these midwestern areas were not 13-year cicadas at all, but rather are 17-year cicadas masquerading as 13-year cicadas. Genetic and morphological evidence, including mitochondrial DNA restriction site polymorphisms (50 sites examined), an allozyme locus and an abdominal color polymorphism, identified these 13-year Decim as belonging to the 17-year Decim lineage. Historical evidence identified their origin as 17-year Brood X. These Brood X cicadas had apparently switched life cycles by shortening their development by four years. Fig. 2 shows the present and former distributions of Brood X.

The identification of a life cycle switching event in periodical cicadas was not altogether unexpected. Various authors (Lloyd and Dybas, 1966; Lloyd and White, 1976; Simon and Lloyd, 1982; Lloyd et al., 1983; Kritsky, 1988) had found circumstantial evidence for such events. Martin and Simon's (1988, 1990) study provided the first clear genetic evidence for switching and was the first demonstration that this phenomenon had occurred on a massive scale. Their results also demonstrated that outside of this anomalous area, 13- and 17-year lineages were distinct and well-differentiated, indicating that frequent switching was not occurring in these areas.

The mechanism for the four year life cycle switch has not been identified but White and Lloyd's (1975) study of nymphal development suggests a partial explanation. In both life cycles, all but four weeks is spent as underground nymphs (five instars). The difference in life cycle length between 13- and 17-year cicadas is attributed to a four year dormancy period in the second instar of 17-year nymphs (White and Lloyd, 1975). The loss of this dormancy period could change a 17- to a 13-year cicada. Lloyd and White (1976) suggest that nymphal crowding may trigger four year accelerations in development.

4 Combination of Molecular and Morphometric Data

In light of the new molecular findings, the first question that comes to mind is whether my original discriminant analysis (Simon, 1979a, 1983a) of 13- versus 17-year Decim siblings included any masquerading populations from northern Arkansas, Missouri, Illinois, or Indiana. Any 13-year populations sampled from this area would be expected to cluster in discriminant space with the 17-year cicadas rather than with other 13-year populations. In fact, two 13-year populations were collected from this area in my earlier studies, population 11 (Perry County, Missouri) and population 12 (Knox County, Indiana); and, as expected, these were the two populations that overlapped their 17-year siblings in those analyses. This result suggests that the periodical cicadas from populations 11 and 12 were actually 17-year cicadas that had switched their life cycle to 13-years. DNA sequences for individuals from these particular localities have not yet been examined; this work is in progress.

Only in hindsight was the significance of the overlap of populations 11 and 12 apparent in the previous discriminant analysis of morphometric data. This is because differentiation in these recently evolved entities is slight, making the groups difficult to detect; therefore, their partial overlap did not cause concern. The fact that populations 11 and 12 did not cluster entirely within the 17-year Brood XIV populations in discriminant space, but rather overlapped with the Decim populations of both Broods XXIII and XIV, can now be explained in light of Martin and Simon's (1988, 1990) evidence that these anomalous populations originated not from 17-year Brood XIV but rather from 17-year Brood X. Morphometric analyses of Brood X Decim would have to be undertaken to test this hypothesis.

5 Describing the Usefulness of Individual Morphometric Characters for Discrimination of Year-Classes

In order to determine which characters were most important in separating broods, Simon (1983a) calculated discriminant functions using broods instead of populations as the a priori groups. All three of the discriminant functions were highly significant. Plots of the brood mean scores and character coefficients are shown in Fig. 6 for discriminant axes 1 and 3 and in Fig. 7 for discriminant axes 1 and 3. The insets show characters represented as vectors

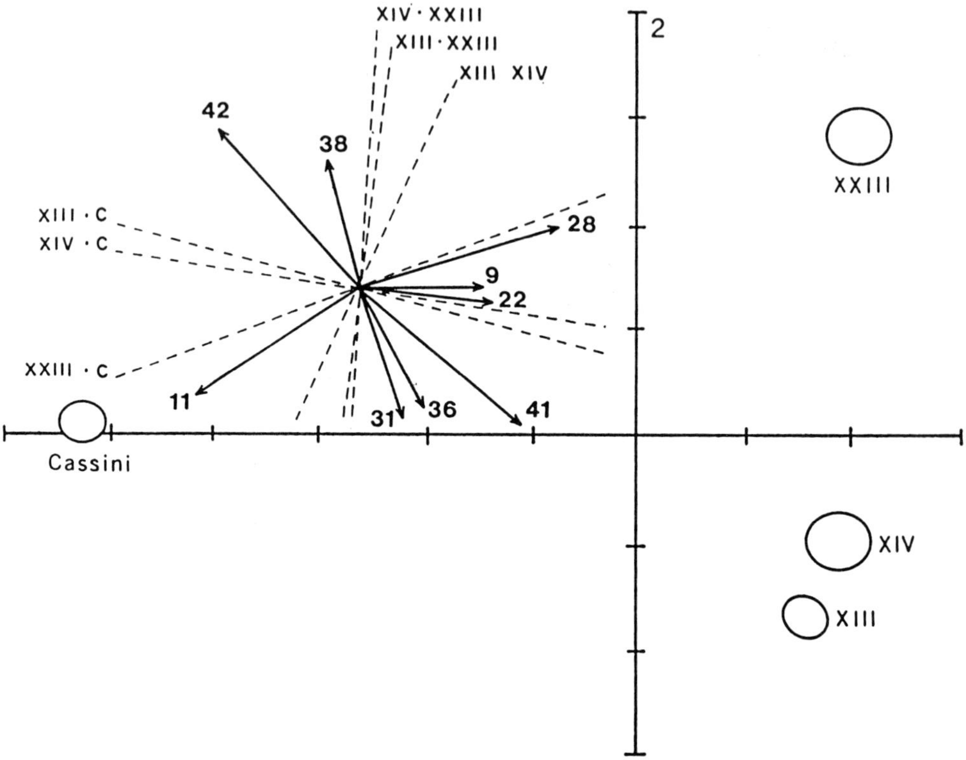

Figure 6. Ninety-five percent confidence ellipses about brood centroids on discriminant axes 1 and 2. Labeled vectors represent characters most important in separating groups. Dashed lines labeled with brood numbers are parallel to imaginary lines connecting the centers of each pair of ellipses and represent the direction of variation of each group from each of the other groups. Only those characters with discriminant weights greater than 40 (an arbitrary cutoff) are shown; other characters would be less important discriminators. Refer to Fig. 3 for character number identification; scales are the same as in Fig. 4. (From Simon, 1983a)

that join the bivariate plot of their coefficient on each axis with the origin (a similar illustrative technique was used by Jolicoeur [1959]). Only those characters that are most important in discriminating groups are shown. Refer to the wing diagram (Fig. 3) to match wing veins with their character numbers in Figs. 6 and 7.

The character vectors that are longest and most parallel to an imaginary line connecting any two brood centroids (the ellipse centers) represent the characters that are the most important in separating those two broods in those two dimensions. In the insets in Figs. 6 and 7, the labeled, dashed lines are parallel to these imaginary lines and can be used as a reference to choose more easily the characters most important in each separation. For each pair of broods, a different combination of characters is important for discrimination. Additional broods might best be separated using a different subset of the

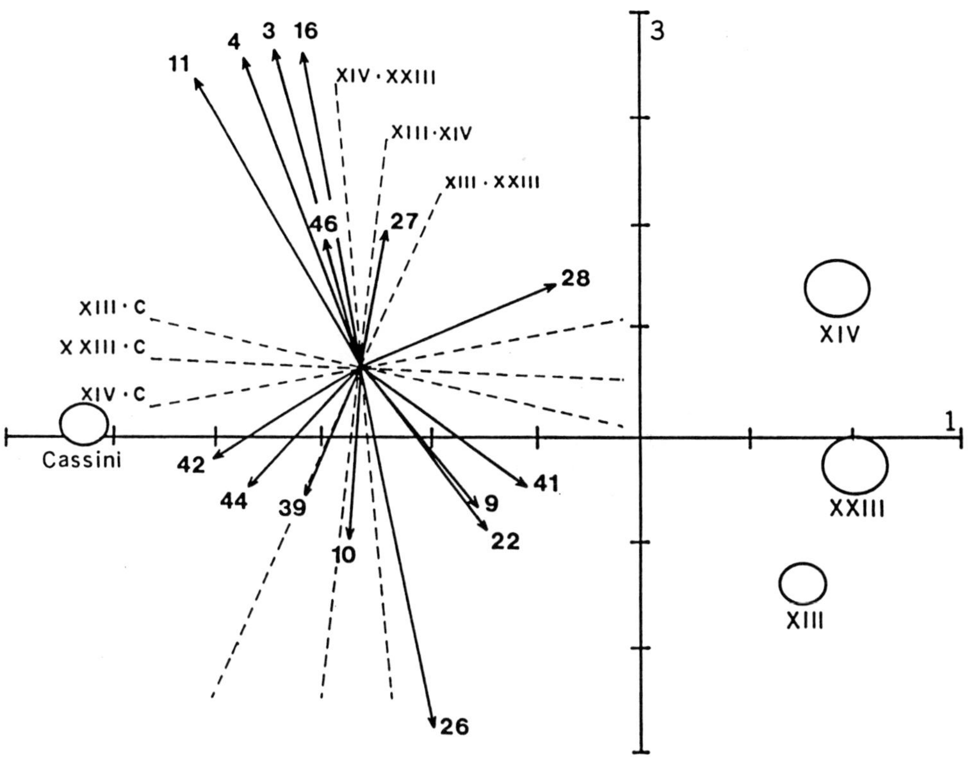

Figure 7. Ninety-five percent confidence ellipses about brood centroids on discriminant axes 1 and 3. Labeled vectors represent characters most important in separating groups. Dashed lines labeled with brood numbers are parallel to imaginary lines connecting the centers of each pair of ellipses and represent the direction of variation of each group from each of the other groups. Only those characters with discriminant weights greater than 40 (an arbitrary cutoff) are shown; other characters would be less important discriminators. Refer to Fig. 3 for character number identification; scales are the same as in Fig. 4. (From Simon, 1983a)

characters; therefore, it would be unwise to discard any characters which have not been examined among all groups under consideration.

6 Phylogenetic Analysis Using Morphometric Data

The usefulness of discriminant analysis of morphometric data for separating groups of morphologically similar insect taxa has been demonstrated many times (e.g., Simon, 1983a; Pungerl, 1986; Inayatullah et al., 1987; Wetton, 1987), but the usefulness of discriminant analysis transformations of morphometric data for phylogenetic analysis is controversial. Phylogenetic analyses for the periodical cicadas, using morphometric data that was coded according to discriminant scores, were not attempted. This was because, as Crespi (this volume) points out, there is no a priori reason to suspect that taxa will evolve to maximize among-group differences for all characters. Nevertheless, other authors (Schluter, 1984; Sorensen, 1987) have produced biologically reasonable phylogenies based on data that were transformed using ordination procedures which adjust characters according to their genetic variances and covariances. The rationale for these multivariate transformations is based on Lande's (1979) model for the evolution of the mean phenotype in populations.

Untransformed morphometric data from wing veins appears to be of little use in the construction of phylogenetic hypotheses for year-classes of periodical cicadas. A phylogenetic tree based on untransformed morphological characters was constructed for four year-classes of 17-year Decim (Broods I, II, XIV, and XIII) and two year-classes of 13-year Decim (Broods XIX and XXIII) using 13-year Cassini (Brood XXIII) as an outgroup (Simon, unpublished data). Morphometric characters were coded using homogeneous subset coding (Simon, 1983b; Archie, 1985), which is designed to capture all the statistically significant differences among characters. These coded data were analyzed phylogenetically using a distance Wagner procedure (Farris, 1972). The results of the phylogenetic analysis linked taxa according to size, an immediately suspicious result. Thirteen-year cicadas were most derived, and 17-year broods were arranged in descending order, as: Broods I + II, XIV, and XIII.

7 Concluding Remarks

Allozymic data for these same year-classes has revealed only limited information (Archie et al., 1985). Year-classes can be grouped only into eastern 17-year cicadas (including Broods, I, II, V-X, XIII and XIV), western 17-year cicadas (Broods III and IV) and central 13-year cicadas (Broods XIX, XXII, and XXIII). Earlier phylogenetic analysis (Simon, 1979a, 1979b), which examined only five year-classes (17-year Broods XIII and XIV, 13-year Broods XIX, XXII, and XXIII), suggested that 13-year cicadas were more derived than 17's, but this result was questioned by later studies (Archie et al., 1985), which showed that all broods of periodical cicadas are allozymically very similar. In addition, analyses of mitochondrial DNA suggest that 13-year cicadas are older than 17-year cicadas (Martin and Simon, 1990).

Preliminary phylogenetic studies using restriction enzyme digests of mtDNA and sequence data from Domain III of the mitochondrial 12S ribosomal region have little more resolution than allozymic data. None of the three types of data can be used to distinguish between year-classes, but all can be used to distinguish between 13- and 17-year Decim siblings and provide a phylogenetic estimate for the relationships between them and the Cassini and Decula siblings (Fig. 8). Neither the mtDNA data nor the allozyme data previously examined could distinguish between 13- and 17-year Cassini siblings, or between 13- and 17-year Decula siblings. A definitive phylogeny of year-classes is obviously not provided by any of these data.

Although no differences in nucleotide sequence among 17-year broods has been found in our limited survey to date, mitochondrial DNA sequence data do have the potential to be useful in distinguishing among closely related year-classes. There are regions of the mitochondrial DNA that are predicted to

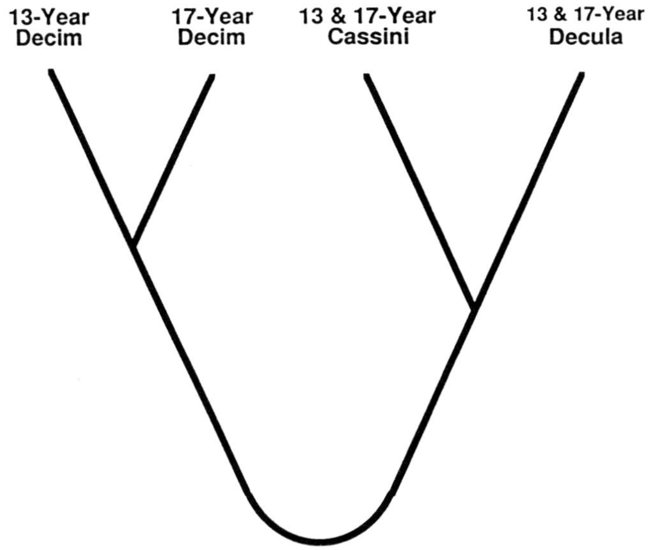

Figure 8. Phylogenetic relationships among *Magicicada septendecim*, *M. tredecim*, *M. cassini*, *M. tredecassini*, *M. septendecula*, and *M. tredecula* based on mtDNA restriction site polymorphisms (50 sites examined) and on nucleotide sequences from a 350 base-pair region of Domain III of the mitochondrial small ribosomal gene.

evolve faster than the average for the entire mtDNA molecule (measured using restriction enzymes). These include unconstrained nucleotides in: (1) the control region (A+T-rich region); (2) third positions of codons belonging to 4-codon families of protein coding genes; and (3) nonstructurally important segments of ribosomal genes (Simon, 1991). These regions are currently being explored to test their phylogenetic usefulness.

Discriminant function analysis of morphometric data from the wings of periodical cicadas is useful for recognizing reproductively isolated year-classes. Another useful aspect of this analysis is that individual characters that are important in distinguishing groups can be identified. Molecular data can be combined with morphometric data to aid in interpretation. For example, what appeared to be an imperfection in the separation of 13- and 17-year populations by discriminant function analysis can now be explained as the result of the misidentification of two populations. This misidentification is understandable in light of the fact that midwestern 17-year periodical cicadas (identified by DNA, allozyme and coloration pattern characters) have switched their life cycle and are masquerading as 13-year cicadas.

The phylogenetic usefulness of morphometric data transformed via ordination procedures remains controversial. To resolve the controversy, phylogenetic hypotheses need to be constructed using both transformed and untransformed data and evaluated in light of information from other kinds of data (e.g., DNA sequences).

8 Acknowledgments

This research was supported by NSF Grants BSR-88-22710 (C. Simon), BSR-85-09164 (C. Simon), DEB-78-10710 (M. Lloyd), DEB-81-07038 (M. Lloyd), and BMS-20522 (R. Koehn); an NSF Postdoctoral Fellowship Grant, and grants from Sigma Xi and the Theodore Roosevelt Memorial Fund are also acknowledged. Acknowledgments for support from many individuals are given in my papers cited below. I thank J. Archie, C. McIntosh, and M. Whiting for comments on and/or assistance with the manuscript.

9 References

Archie, J.W., 1985. Methods for coding variable morphological features for numerical taxonomic analysis. *Systematic Zoology*, 34: 326-345.

Archie, J.W., Simon, C. and Wartenburg, D., 1985. Geographical patterns and population structure in periodical cicadas based on spatial analysis of allozyme frequencies. *Evolution*, 39: 1261-1274.

Crespi, B.J., (this volume). Natural selection and morphometrics.

Farris, J.S., 1972. Inferring phylogenetic trees from distance matrices. *The American Naturalist*, 106: 645-668.

Inayatullah, C., Webster, J.A. and Fargo, W.S., 1987. Morphometric variation in the alates of greenbug (Homoptera: Aphididae) biotypes. *Annals of the Entomological Society of America*, 80: 306-311.

Jolicoeur, P., 1959. Multivariate geographic variation in the wolf *Canus lupus* L. *Evolution*, 13: 283-299.

Kritsky, G., 1988. The 1987 emergence of the periodical cicada (Brood X) in Ohio (Homoptera: Cicadidae). *Ohio Journal of Science*, 88: 168-170.

Lachenbruch, P.A., 1975. Discriminant Analysis. Hafner Press, New York, 128 pp.

Lande, R., 1979. Quantitative genetic analysis of multivariate evolution, applied to brain: body size allometry. *Evolution*, 33: 402-416.

Lloyd, M. and Dybas, H., 1966. The periodical cicada problem. II. Evolution. *Evolution*, 20: 466-505.

Lloyd, M., Kritsky, G. and Simon, C., 1983. A simple Mendelian model for 13- and 17-year life cycles of periodical cicadas, with historical evidence of hybridization between them. *Evolution*, 37: 1162-1180.

Lloyd, M. and White, J., 1976. Sympatry of periodical cicada broods and the hypothetical 4-year acceleration. *Evolution*, 30: 786-801.

Marlatt, C.L., 1907. The periodical cicada. *Bulletin of the United States Department of Agriculture, Division of Entomology, n.s.*, 18: 52-58.

Martin, A. and Simon, C., 1988. Anomalous distribution of nuclear and mitochondrial DNA markers in periodical cicadas. *Nature*, 336: 237-239.

Martin, A. and Simon, C., 1990. Differing levels of among-population divergence in the mitochondrial DNA of 13- versus 17-year periodical cicadas. *Evolution*, 44: 1066-1080.

Pungerl, N.B., 1986. Morphometric and electrophoretic study of *Aphidius* species (Hymenoptera: Aphidiidae) reared from a variety of aphid hosts. *Systematic Entomology*, 11: 327-354.

Riska, B., 1981. Morphological variation in the horseshoe crab *Limulus polyphemus*. *Evolution*, 35: 647-658.

Schluter, D., 1984. Morphological and phylogenetic relations among the Darwin's finches. *Evolution*, 38: 921-930.

Simon, C., 1979a. Evolutionary Relationships Among the 13- and 17-year Periodical Cicadas. Ph.D. Thesis. State University of New York at Stony Brook, Stony Brook, New York (unpublished), 280 pp.

Simon, C., 1979b. Evolution of periodical cicadas. I. Phylogenetic inferences based on allozymic data. *Systematic Zoology*, 28: 22-39.

Simon, C., 1983a. Morphological differentiation in wing venation among broods of 13- and 17-year periodical cicadas. *Evolution*, 37: 104-115.

Simon, C., 1983b. A new coding procedure for morphometric data with an example from periodical cicada wing veins. In: J. Felsenstein (Editor), Numerical Taxonomy, NATO Advanced Study Institute, Series G, Number 1. Springer-Verlag, New York, pp. 378-382.

Simon, C., 1988. Evolution of 13- and 17-year periodical cicadas (Homoptera: Cicadidae: Magicicada). *Bulletin of the Entomological Society of America*, 34: 163-176.

Simon, C., 1991. Molecular systematics at the species boundary: exploiting conserved and variable regions of the mitochondrial genome of animals via direct sequencing from amplified DNA. In: G.M. Hewitt, A.W.B. Johnston and J.P.W. Young (Editors), Molecular Techniques in Taxonomy. NATO Advanced Studies Institute, Series H, Number 57. Springer-Verlag, New York, pp. 33-71.

Simon, C. and Lloyd, M., 1982. Disjunct populations of periodical cicadas: relicts or evidence for polyphyly. *Journal of the New York Entomological Society*, 90: 275-301.

Sokal, R. and Rohlf, F.J., 1974. Biometry, (Second Edition). W.H. Freeman and Company, New York, 859 pp.

Sorensen, J.T., 1987. Multivariate statistical approach to deduction of phylogeny within *Essigiella* (sic) (Aphididae: Lachninae). In: J. Holman, J. Pelikan, A.G.F. Dixon and L. Weismann (Editors), Population Structure, Genetics and Taxonomy of Aphids and Thysanoptera, Proceedings of International Symposia, held at Smolenice Czechoslovakia, September 9-14, 1985. SPB Academic Publishing, The Hague, pp. 243-260.

Wetton, M.N., 1987. Morphological variation in British *Folsomia quadrioculata* Tullberg (Collembola: Isotomiade): a multivariate study. *Systematic Entomology*, 12: 257-270.

White, J. and Lloyd, M., 1975. Growth rates of 17- and 13-year periodical cicadas. *American Midland Naturalist*, 94: 127-143.

PHYLOGENETIC ANALYSIS OF QUANTITATIVE AND QUALITATIVE CHARACTERS OF GALL-INDUCING MIDGES AND THE HISTORICAL RELATION TO THEIR HOSTS: *DASINEURA* (DIPTERA: CECIDOMYIIDAE) ON ROSACEAE AND MONOCOTS

J.C. Roskam[1] and M. Zandee[2]

[1] Department of Population Biology, University of Leiden, Schelpenkade 14a, 2313 ZT Leiden, Netherlands; [2] Institute of Theoretical Biology, University of Leiden, Kaiserstraat 63, 2311 GP Leiden, Netherlands

Abstract

Qualitative and quantitative morphological characters, and qualitative bionomical data, were assessed for 96 western Palaearctic species of gall midges in *Dasineura* Rondani. We compared qualitative data cladograms, generated using group compatibility (CAFCA) and/or Wagner parsimony analysis (HENNIG-86), with quantitative data dendrograms constructed using single-linkage clustering of pairwise Euclidean distances between species that were generated in principal components analysis (PCA). The analyses involved both different semaphoronts separately, and a combined data set with a complete set of available semaphoronts and characters. We propose a phylogeny for the latter with polytomies that obviously either share synapomorphies for species' feeding mode (inquilinism), or host plant association. Among a set of species associated with Rosaceae, all species on *Filipendula* appear monophyletic, but those on Maloideae appear polyphyletic. Among a set of combined species on monocots, gymnosperms and ferns, sister-groups are absent for those on Poaceae and Cyperaceae; but on Poaceae, congruence exists among generated cladograms. PCAs show many phenetic similarities for species on the same host family or genus. No congruence exists between PCA scattergrams for semaphoronts, nor between dendrograms, versus cladograms for species on Rosaceae or monocots. Therefore, 'gestalt' grouping of species that share a host taxon may reflect homoplasy, and may be poly- or paraphyletic.

1 Introduction

Gall-inducing midges, as obligate plant parasites that manipulate their host plants to evoke galls, have very intensive long term ecological bonds with their host plants (Roskam, 1992). Therefore, the phylogeny of this group might reflect, to a major extent, the reciprocal, coevolutionary processes among gall midges and their associated host plants. Many small but morphologically well defined gall midge genera have species that are associated with hosts which are closely related phylogenetically (Roskam, 1985). This suggests that the host plant associations of larger genera might also allow phylogenetic interpretations of relationships. Reciprocal illumination (sensu Hennig, 1966) could be a major tool for investigating the phylogeny of a group of organisms that is associated with another group which has a better known phylogeny.

There are pit-falls in the definition of the term coevolution, however. The effects of the host plants predominate in the reciprocal associations between plants and their gall-making parasites, forcing the parasites to evolve and adapt accordingly. Jermy (1976) coined the term sequential evolution for this concept. As in many coevolutionary processes, the resulting phylogenetic patterns of the associates are not strictly parallel. Shifts to structurally and chemically similar, but not necessarily phylogenetically close, host plants may obscure expected parallel traits among the phylogenies of associates. One might, therefore, expect a more diffuse, rather than a strict, congruent pattern as result of coevolutionary processes (see many references in Futuyma and Slatkin, 1983; Nitecki, 1983). If closely related gall midge species should be present on phylogenetically close host plant taxa, one would expect clusters of such midge species within host plant genera or families.

The taxonomy and phylogeny of this well known group of insects are still poorly understood because of their trophic type. For the major part of their life-cycle, the midges live encapsulated in galls and are subjected, therefore, to parallel selective forces. Homoplasies thus could be expected to cause considerable noise in phylogenetic studies of such a group. Clear-cut, qualitative morphological traits that are diagnostic in adults, as well as immatures, are scarce. Quantitative assessments of their relationships, in particular by using multivariate techniques (Sylvén and Carlbäcker, 1981), remained almost unexplored. Many biological traits are heritable, as are morphological characters, and could similarly be used in phylogenetic studies (Sylvén, 1975). Host plant associations are such traits, as are the kind of host plant tissue that is galled, the complexity of the gall structure, the kind of aggregation of the larvae, and the place where the midges pupate. Biochemical diagnostic techniques are less appropriate for exploratory analyses of the group, at least at a larger scale, because until now fresh material was required but was (and remains) often unavailable.

Dasineura, the largest genus of the tribe Oligotrophini, is one of the largest cosmopolitan genera of gall midges. It has about 260 species described for the Palaearctic (Skuhravá, 1986; *Rabdophaga* Westwood excluded) and 110 Nearctic species (Gagné, 1989). A few more species have been described for India and Japan (Mani, 1973; Yukawa, 1971), but these areas remain poorly known. In general, we adopt Skuhravá's (1986) taxonomy. However, the transfer of some species to other genera is necessary.

Until now, *Dasineura* has been considered a 'catch all' genus (Gagné, 1989). It is one of the most species-rich entities in the gall-inducing Cecidomyiidae. An analysis of some patterns of evolutionary radiation for the genus is, therefore, of paramount importance for an understanding of the phylogeny of this major group of phytophages. Because of the problematic nature of diagnostic characters for *Dasineura*, we do not have a firm opinion about its generic limits. The type-species of *Dasineura*, *D. obscura* Rondani, was caught in flight; its type is presumably lost. Unfortunately, Rondani's description, in 1840, fits a large group of gall midges having a curved R_5 vein that reaches the wing margin distinctly anteriad of the wing tip. The ovipositor is soft, telescopic, and its cerci are fused. The postabdomen of the male is devoid of diagnostic traits, except for more or less reduced pubescence on the dorsal surface of the gonostyli. Most species induce simple galls on vegetative plant tissues, but a minority evoke galls in flowers or fruit capsules. The larvae generally live gregariously and drop to the ground for pupation. If species induce more complex galls, pupation occurs within these galls. Some species live in the inflorescences of grasses and sedges without causing any apparent damage, but others are inquilinous on galls induced by other gall midge species. Although most plant families have species that are galled by *Dasineura*, the Rosaceae shows the greatest number.

We analyze the phylogeny of some groups of *Dasineura* species in connection with the phylogeny of their host plants. We will do so on three levels of host plant phylogeny. First, we determine whether *Dasineura* reflects its host plant phylogeny at the family level. Second, we derive a cladogram for the species associated with Rosaceae to see if it shows congruence with a host phylogeny. Also, we focus on the *Dasineura* species that are associated with some less commonly colonized host plant groups, such as monocots, gymnosperms and ferns. The results of the cladistic analyses, based on qualitative characters, are compared with dendrograms of *Dasineura* species that are based on the multivariate analyses of quantitative characters. This comparison may allow an otherwise difficult phylogenetic interpretation of quantitative characters (Pimentel and Riggins, 1987).

2 Material and Methods

The species associated with Rosaceae, Poaceae, Cyperaceae, gymnosperms and ferns that were available for analysis are listed in Table 1. A complete list of the 96 species used for comparison of midges and host patterns at the plant family level is available on request. All material used in this study was slide-mounted following Sylvén (1975) and Roskam (1977). Because morphology, particularly of the adult, might be easily influenced by differences in rearing techniques, the material of the Stelter collection was preferred; that collection has the most extensive holdings. Other collections were used for measurements of third instar, mature larvae.

The qualitative characters used in the cladistic analyses are presented in Table 2. Quantitative characters are listed for adults in Table 3, and for larvae in Table 4. The terminology used follows Gagné (1981) and Roskam (1977) for adults and larvae, respectively. All measurements were taken with an ocular micrometer using phase contrast microscopy.

We initially analyzed 10 males, females and larvae, each, per species to determine intraspecific variation. An example of the coefficients of variation for one species, *D. mali*, is shown in Tables 3 and 4. Because all characters appeared to be normally distributed, we used only three specimens per species to achieve a fairly stable estimate of the mean value for each character; for some species, we used less, however, due to a lack of specimens.

Table 1.
List of midge species examined. Nomenclature follows Skuhravá (1986). Host plant records and gall numbers follow Buhr (1964-65); if gall numbers are absent, the descriptions refer to the original publications. MA = male, FE = female, LA = larva. Collection source as collection of: HB = H.F. Barnes (British Museum [Natural History], London), WN = W.C. Nijveldt, JR = J.C. Roskam (WN, JR in Zoological Museum, Amsterdam), ES = E. Sylvén (Swedish Museum of Natural History, Stockholm), HS = H. Stelter (2251 Post Broderstorf, Germany), MS = M. Skuhravá (Czechoslovakia Academy of Natural Sciences, Prague); * = only 1 or 2 examined. Collection sites listed only as abbreviated countries: GB = Great Britain, D = Germany, NL = Netherlands, S = Spain, A = Austria.

			Collection Source			Collection Site		
Midge	Host Plant(s)	Gall No.	MA	FE	LV	MA	FE	LV
D. abietiperda (Henschel, 1880)	*Picea* spp.	4781	HB*	HB*	--	GB	GB	--
D. alopecuri (Reuter, 1895)	*Alopecurus, Agropyron* spp.	seed	--	HS	HB	--	D	GB
D. crataegi (Winnertz, 1853)	*Crataegus* spp.	2072	WN	WN	JR	NL	NL	NL
D. dactylidis (Metcalfe, 1933)	*Dactylis glomerata* L.	seed	HB	HB	HB	GB	GB	GB
D. engstfeldi (Rübsaamen, 1889)	*Filipendula ulmaria* (L.) Maxim.	2744 -54	HS	HS	HB	D	D	GB
D. festucae (Barnes, 1939)	*Festuca rubra* L.	seed	HB*	HB*	HB	GB	GB	GB
D. filicina (Kieffer, 1889)	*Pteridium aquilinum* Kuhn	5290	HS	HS	JR	D	D	GB
D. filipendula (Kieffer, 1889)	*Filipendula vulgaris* Moench	2756	--	--	ES	--	--	S
D. graminis (Felt, 1908)	*Agrostis vulgaris* With.	seed	HB*	HB*	HB	GB	GB	GB
D. harrisoni (Bagnal, 1912)	*Filipendula ulmaria* (L.) Maxim.	2725 -38	HS	HS	JR	D	D	NL
D. inflata Stelter, 1986	*Carex otrubae* Podp.	utricle	HS	HS	--	D	D	--
D. koesterbecki Stelter, 1986	*Carex acutiformis* Ehrh.	utricle	HS	HS	--	D	D	--
D. mali (Kieffer, 1904)	*Malus sylvestris* Mill.	4116	WN	WN	WN	NL	NL	NL
D. minungula Stelter, 1986	*Carex paniculata* L.	utricle	HS	HS	--	D	D	--
D. oxyacanthae (Rübsaamen, 1914)	*Crataegus* spp.	2103	HS	HS	ES	D	D	S
D. plicatrix (H. Loew, 1850)	*Rubus* spp.	5907	WN	WN	WN	NL	NL	NL
D. poae (Mühle, 1957)	*Poa pratensis* L.	seed ?	HB*	HS	ES	D	D	S
D. potentillae (Wachtl, 1885)	*Potentilla argentea* L.	5146	HS	HS	--	D	D	--
D. pteridicola (Kieffer, 1901)	*Dryopteris, Pteridium* spp.	2416, 5291	HS	HS*	JR	D	D	GB
D. pustulans (Rübsaamen, 1889)	*Filipendula ulmaria* (L.) Maxim.	2741	HS	HS	JR	D	D	NL
D. pyri (Bouché, 1847)	*Pyrus communis* L.	5342	HS	HS	WN	D	D	NL
D. sodalis (F. Löw, 1877)	ex galls *D. tortrix*	inquiline	HS	HS	WN	D	D	NL
D. tortrix (F. Löw, 1877)	*Prunus spinosa* L.	5229 -32	WN	WN	WN	NL	NL	NL
D. triseti (Barnes, 1939)	*Trisetum flavescens* (L.) P.B.	7230	HB	HB	HB	A	A	A
D. ulmaria (Bremi, 1847)	*Filipendula ulmaria* (L.) Maxim.	2737 -53	WN	WN	WN	NL	NL	NL

Table 2.
Qualitative characters and character states.

No.	Character/(type)	State
	(Bionomy)	
1	Feeding mode	gall-inducing/ phytophagous/ inquiline
2	Plant tissue	leaf/ shoot / inflorescence/ flower/ fruit
3	Aggregation	gregarious/ solitary
4	Hibernation	soil/ both/ gall
	(Larva)	
5	Spatula width, µm	15-24 / 25-34 / etc
6	Spatula shape	9 classes with acute/ blunt/ rounded lobes, ditto incision, completely or incompletely developed manubrium
7	Anal papillae	small/ big/ fused
	(Male)	
8	Ratio R_5/le wing	0.95-0.99/ 0.90-0.94/ 0.85-0.89/ 0.80-0.84
9	Relation aedeagus/cerci	a=c/ a<c/ a>c/ a>>c
10	Shape mediobasal lobe	rectangular/ thick triangular/slender triangular
11	Shape gonocoxites	oblique/ angulate/ carinate
12	Shape gonostyli	6 classes, with curved/ parallel shape and attenuate/ constricted/ blunt top
13	Pubescence dorsal surface gonostyli	5 classes
14	Pubescence ventral surface gonostyli	5 classes
	(Female)	
15	Shape cercus	subglobose/ ovate/ oblong/ elongate/ cylindrical/ conical/acute
16	Pubescence cercus	multitricous patches/ ditrichous patches/ transverse rows of microtrichia/ striate patches/scattered microtrichia

We used principal components analysis (PCA) (Pimentel, 1979) to describe the patterns of variation among standardized quantitative (continuous) characters. Separate analyses were conducted for male, female and larval semaphoronts; a joint analysis for all three semaphoronts was also executed using a more restrictive number of species (54) and characters (55).

Attaching a phylogenetic interpretation to the PCA results is troublesome because the data are a mixture of plesiomorphic and apomorphic conditions. Discriminating plesiomorphic from apomorphic information from continuous characters by out-group comparison is problematic. In most studies, continuous character variation is broken into discrete states by one of several possible procedures (Simon, 1983; Archie, 1985). We choose a different approach and tried to impose a phylogenetic interpretation on the patterns of

Table 3.
Quantitative characters for *adults* (males and females). For characters with double numbers, the values for both sexes were used in the linked matrix; for single numbered characters, only the values for males were used in the linked matrix, although values for both sexes were available. CoV = coefficients of variation for *D. mali* (n = 10); Abbrev. = abbreviation used in Fig. 1. Quantitative characters 35-56 are larval characters, see Table 4.

Character	Description	Abbrev.	CoVmale	CoVfemale
1	head: height	*he head*	6.40	8.74
2	tentorium: height	*he tent*	5.30	5.87
3	first palp segment: length	*palp1*	16.33	15.06
4	fourth palp segment: length	*palp4*	12.12	9.66
5 & 30	antenna: number of flagellomeres	*nr flm*	3.32	3.43
6 & 31	antenna: length	*le ant*	8.90	9.44
7 & 32	third flagellomere: length node	*le flm*	8.22	7.24
8 & 33	third flagellomere: width node	*wi flm*	7.44	7.03
9	third flagellomere: length male stalk	*st flm*	9.31	--
10 & 34	third flagellomere: length longest seta	*se flm*	15.34	13.90
11	metafemur: length	*fem*	11.98	12.17
12	metatibia: length	*tib*	15.09	13.49
13	second metatarsus: length	*tars2*	11.29	6.97
14	fifth metatarsus: length	*tars5*	8.16	8.25
15	wing: length	*le wing*	8.01	9.70
16	wing: width	*wi wing*	9.37	10.50
17	wing: length R_1	R_1	8.90	10.71
18	wing: length R_5	R_5	7.41	10.63
19	wing: length Cu_{1a}	Cu_{1a}	10.18	13.90
20	wing: length Cu_{1b}	Cu_{1b}	8.47	12.69
21	wing: length Cu_2	Cu_2	15.05	14.79
22	gonocoxite: length	*le gcox*	6.14	--
23	gonocoxite: width	*wi gcox*	9.10	--
24	gonostylus: length	*gst*	3.69	--
25	cercus: length incision	*cerc*	10.04	--
26	cercus: length	*le cerc*	--	10.78
27	cercus: height	*he cerc*	--	9.85
28	hypoproct: length	*hpr*	--	7.26
29	mediobasal lobe gonocoxite: length	*mbl*	3.58	--

variation in continuous characters by looking for congruence between dendrograms derived from the PCA, on the one hand, and cladograms depicting phylogenetic relations based on qualitative characters, on the other. Two different approaches to phylogeny reconstruction were used here: group-compatibility as implemented in CAFCA (Zandee, 1987, 1989; Zandee and Geesink, 1987), and Wagner parsimony analysis as implemented in HENNIG-86 (Farris, 1988). CAFCA analyses were run using the 'partial monothetic sets' (PMS) option. In HENNIG-86, the options 'hennig' and 'bb*' were used. Under the PMS option, the concept of parsimony in CAFCA is

Table 4.
Quantitative characters for *larvae*. CoV = coefficients of variation for *D. mali* (*n* = 10); Abbrev. = abbreviation used in Fig. 1; *a1* = first abdominal segment; *a8* = eighth abdominal segment; *as* = anal segment; *av* = anterior ventral seta; *d2* = second dorsal seta; *ip* = interior pleural seta; *pi* = inferior pleural seta; *ps*, =superior pleura seta; *pv* = posterior ventral seta, *th3* = third thoracic segment; *tl* = longest terminal seta; *ts* = shortest terminal seta. Quantitative characters 1-35 are adult characters, see Table 3.

Character	Description	Abbrev	CoVlarva
35	body: length	*le body*	6.23
36	body: width	*wi body*	6.30
37	head capsule: width	*wi head*	9.21
38	head capsule between antenna: width	*wi ant*	10.71
39	spatula: length	*le sp*	8.36
40	spatula: width	*wi sp*	7.68
41	spatula: height lobes	*he lobe*	10.06
42	spatula: width manubrium	*wi man*	19.28
43	verrucae *a1*: number	*a1 verr*	8.97
44	spinules *a1*: number/unit measurement	*a1 spin*	7.21
45	spinules *a1*: height	*he spin*	11.10
46	stigma *a8*: diameter	*a8 stg*	8.60
47	seta *th3-d2*: length	*th3d2*	8.29
48	seta *th3-ps*: length	*th3ps*	9.81
49	seta *th3-ip*: length	*th3ip*	6.15
50	seta *a8-d2*: length	*a8d2*	5.62
51	seta *a8-ps*: length	*a8ps*	5.11
52	seta *a8-pi*: length	*a8pi*	6.24
53	seta *a8-av*: length	*a8av*	15.70
54	seta *a8-pv*: length	*a8pv*	11.23
55	seta *as-tl*: length	*astl*	9.43
56	seta *as-ts*: length	*asts*	10.20

more restricted than in Wagner parsimony because all groups in a cladogram are bound to possess at least one unique character state, at the cost of extra steps if necessary. The data matrix used for the reconstruction of host plant phylogeny was derived under assumption 'zero' (Brooks, 1981; Zandee and Roos, 1987; Wiley, 1987) from a particular cladogram for *Dasineura* and the host plant associations of *Dasineura* and analyzed using CAFCA.

3 Results and Discussion

3.1 Patterns Among Host Plant Families

All quantitative characters for all semaphoronts (males, females [Table 3], and mature larvae [Table 4]) of 54 species were linked in one data matrix that underwent PCA. The first principal axis (41% of the variation) mainly represents aspects of general-size because all loading coefficients, except one dealing with larval spinules, are positive. The functional relation of the characters is more distinct when we look at the second and third principal axes (Fig. 1). Characters of the male postabdomen have positive loadings on the second principal axis, whereas those representing the larval body and spatula load most heavily on the third principal axis. Male wing and hind leg characters score in the first quadrant. Setal characters of the larvae score in quadrant II, whereas characters of male head, and female antenna and ovipositor, score in quadrant IV.

The host plant family associations of the midge species are labeled in Fig. 2. Between species, from most plant families that are represented by only two or three midge species, the pairwise Euclidean distances in the first three dimensions (60% of total variation) are usually short; these distances are shown in brackets below. Examples are: *D. fastidiosa* and *D. interbracta*, both inquilines of *Semudobia* spp. on *Betula* [2.32]; *D. affinis* and *D. violae* on *Viola* [3.27]; *D. epilobii* and *D. kiefferi* on *Epilobium* [3.31]; *D. gentneri* and *D. gleditchiae* [2.27], and *D. gentneri* and *D. trifolii* [3.47] on Fabaceae; *D. hygrophila* and *D. galii* on Rubiaceae [3.89]; *D. myosotidis* and *D. symphyti* on Boraginaceae [2.72]; *D. dactylidis* and *D. festucae* on Poaceae [3.16]. These short Euclidean distances are expressions of morphological similarity and may indicate radiation events within the plant taxa involved. On the other hand, pairs like *D. tiliae* and *D. thomasianae* on *Tilia* [8.35], and *D. filicina* and *D. pteridicola* on *Pteridium* [4.77] are mutually distant, which might indicate independent colonization. Host plant families with many midge associates, such as Rosaceae, have their species scattered over the multivariate space defined and spanned by all *Dasineura*.

In the absence of polarization of the continuous variables, the morphological similarities between these pairs of species may be based at least in part on plesiomorphy; therefore, their evolutionary relationships must remain conjecture. Using corroborative evidence, however, it seems appropriate to conclude, through reciprocal illumination, that unique colonization events and subsequent radiations may have occurred for such midge groups, if plant cladograms demonstrate sister-group relationships for their hosts.

Using HENNIG-86 on qualitative data (Table 2), we found more than 100 cladograms with a minimum length of 214 steps (consistency index, CI = 0.26) for *Dasineura*. The midges associated with Poaceae form a monophyletic group, as do *D. plicatrix*, *D. crataegi* and *D. mali* on Rosaceae. Three species on *Filipendula*, *D. engstfeldi*, *D. harrisoni* and *D. pustulans* are placed in one clade also. However, these species share their clade with three species associated with other plant families. If we reconstruct the phylogenetic relationships among plant families from the midge phylogeny, by viewing host species as 'areas of endemism' for gall midges (Brooks, 1981; Zandee and Roos, 1987; Wiley, 1987), we obtain two CAFCA cladograms with a length of 151 steps (CI = 0.68). Both should be rejected because they completely disagree with current opinions about angiosperm classification (Takhtajan, 1980), although this classification is not strictly phylogenetic in the Hennigian sense. Alternative schemes (e.g., Dahlgren and Bremer, 1985), however, do not include most of the host plant families relevant for this study.

For the same qualitative data on the midge species, the single most parsimonious cladogram found using CAFCA gives 306 steps (CI = 0.19) in

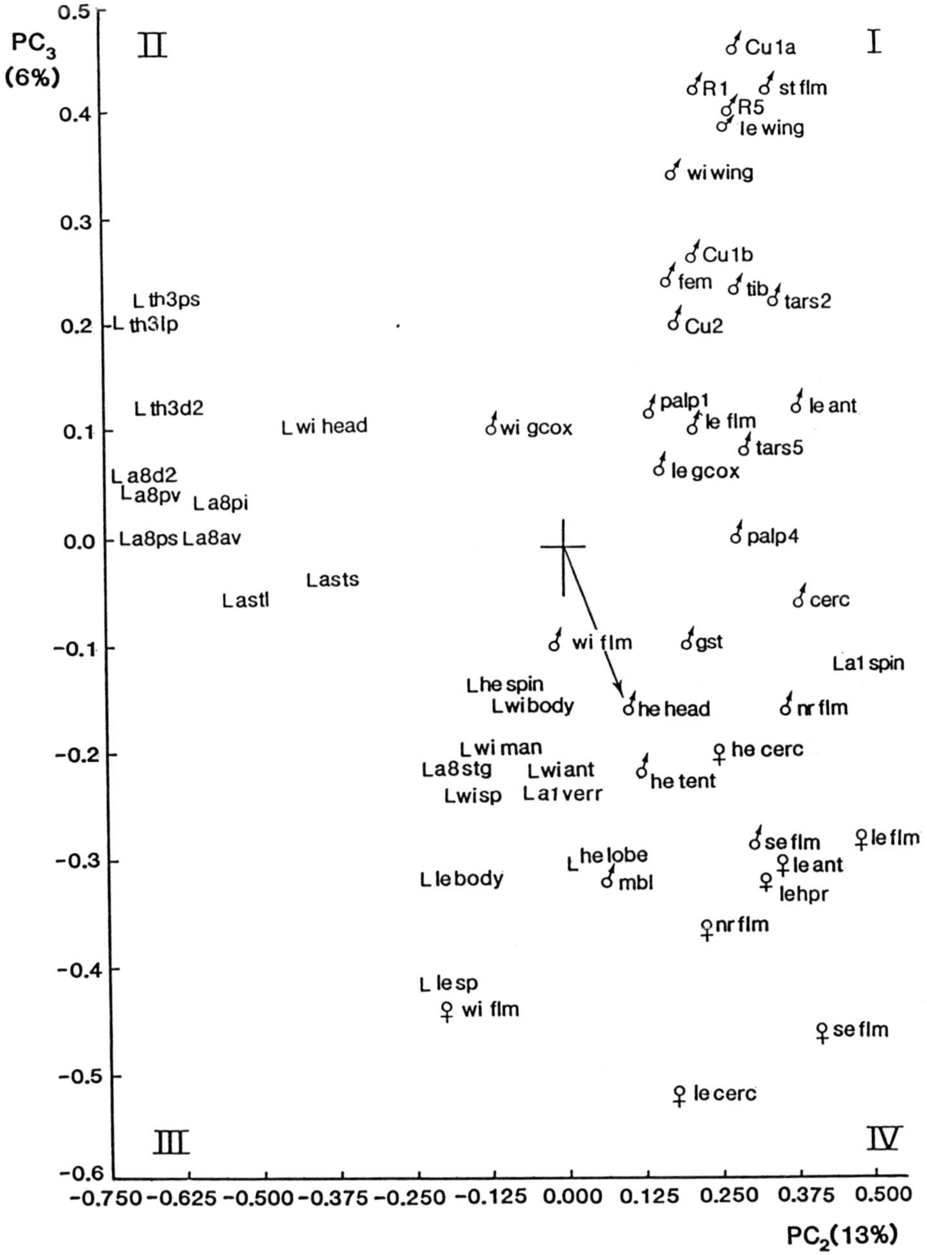

Figure 1. *Dasineura* character vectors for PCA axes 2 versus 3 on the linked data matrix. Only the vector for height of the head ("he head") is shown [arrow]; the tips of the other vectors occur at the position (score) of the characters specified. See Tables 3 and 4 for abbreviations.

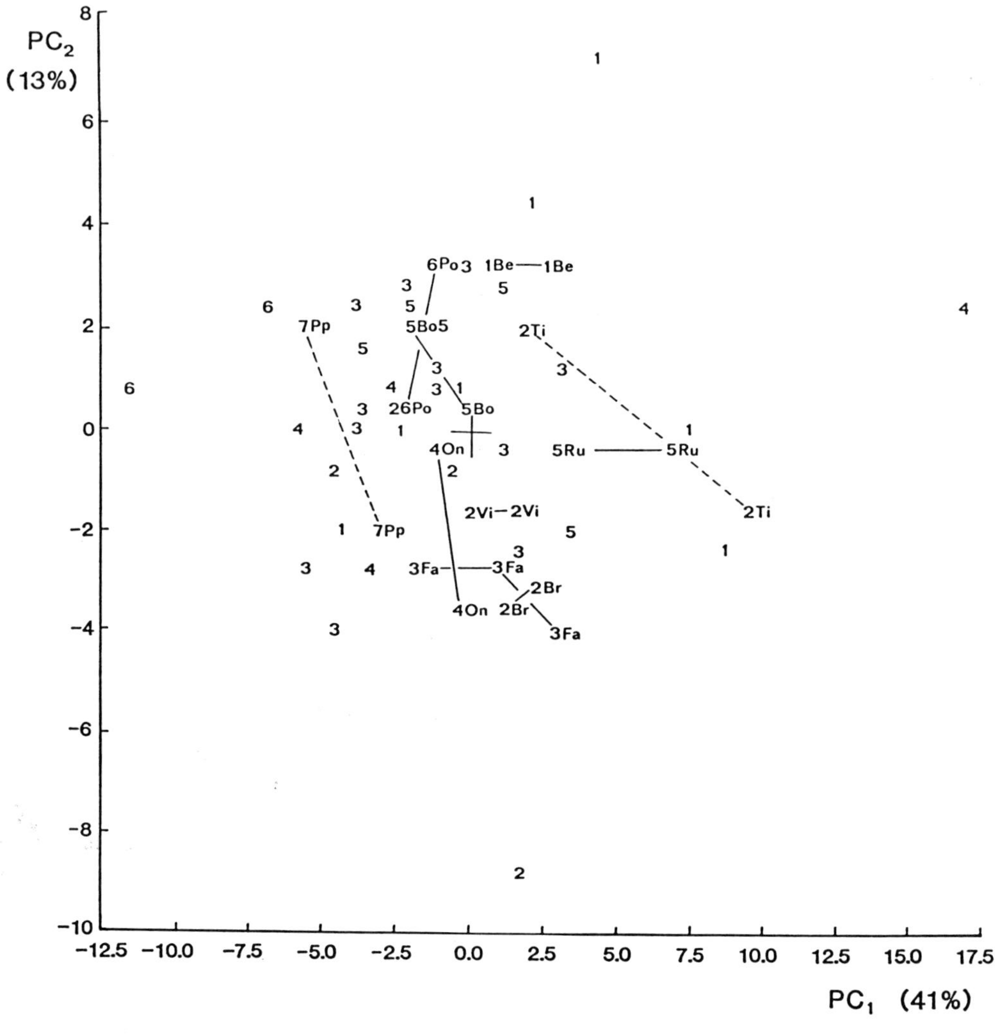

Figure 2. Plot of PCA axes 1 versus 2 for the linked data matrix showing the host plant associations of midge species. Numbers refer to host plant orders, letters to plant families according to the underlining in family names; 1 = Ranunculidae (Ranunculaceae, Papaveraceae), Hamamelidae (Ulmaceae, Urticaceae, Betulaceae), Caryophyllidae (Polygonaceae, Caryophyllaceae); 2 = Dilleniidae (Brassicaceae, Hypericaceae, Salicaceae, Tiliaceae, Violaceae); 3 = Rosidae p.p. (Fabaceae, Rosaceae, Saxifragaceae); 4 = Rosidae p.p. (Aceraceae, Apiaceae, Oleaceae, Onagraceae, Rhamnaceae); 5 = Asteridae (Asteraceae, Boraginaceae, Capriforiaceae, Gentianaceae, Lamiaceae, Rubiaceae); 6 = Commelinidae (Poaceae); 7 = ferns (Polypodiaceae). Drawn lines connect midge species associated with a same plant family; the Euclidean distance between these species is considered as *short* [< 4]. Dashed lines connect midge species separated by *long* [> 4] distances.

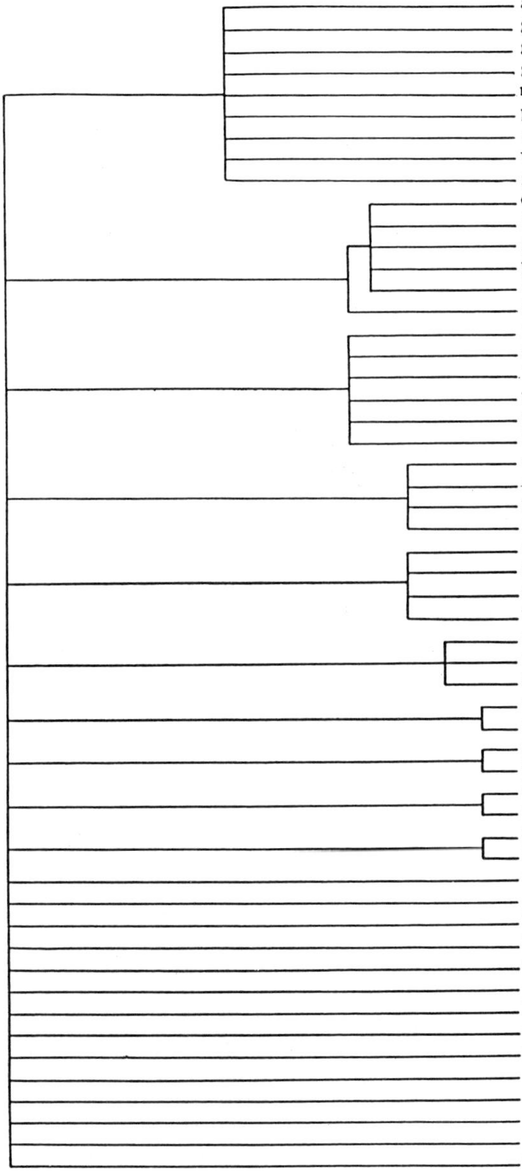

GALL MIDGE SPECIES	HOST PLANT FAMILY	HOST PLANT SUBCLASS
aparines	Rubiaceae	Ast
galiicola	Rubiaceae	Ast
gentianae	Gentianaceae	Ast
glechomae	Lamiaceae	Ast
myosotidis	Boraginaceae	Ast
periclymeni	Caprifoliaceae	Ast
symphyti	Boraginaceae	Ast
virgaeaureae	Asteraceae	Ast
galii	Rubiaceae	Ast
dactylidis	Poaceae	Com
festucae	Poaceae	Com
graminis	Poaceae	Com
poae	Poaceae	Com
triseti	Poaceae	Com
filicina	Polypodiaceae	fern
crataegi	Rosaceae	Ros
frangulae	Rhamnaceae	Ros
lotharingi	Caryophyllaceae	Car
mali	Rosaceae	Ros
plicatrix	Rosaceae	Ros
tortrix	Rosaceae	Ros
alpestris	Brassicaceae	Dil
gentneri	Fabaceae	Ros
pyri	Rosaceae	Ros
trifolii	Fabaceae	Ros
affinis	Violaceae	Dil
irregularis	Aceraceae	Ros
serotina	Hypericaceae	Dil
violae	Violaceae	Dil
autumnalis	Ulmaceae	Ham
tortilis	Betulaceae	Ham
urticae	Urticaceae	Ham
fastidiosa	Betulaceae Inq	Ham
interbracta	Betulaceae Inq	Ham
gleditchiae	Fabaceae	Ros
pustulans	Rosaceae	Ros
tiliae	Tiliaceae	Dil
traili	Ranunculaceae	Ran
papaveris	Papaveraceae	Ran
ribis	Saxifragaceae	Ros
acrophila	Oleaceae	Ros
angelicae	Apiaceae	Ros
bistortae	Polygonaceae	Car
brassicae	Brassicaceae	Dil
engstfeldi	Rosaceae	Ros
epilobii	Onagraceae	Ros
harrisoni	Rosaceae	Ros
oxyacanthae	Rosaceae	Ros
populeti	Salicaceae	Dil
saxifragae	Saxifragaceae	Ros
sisymbrii	Brassicaceae	Dil
tetensi	Saxifragaceae	Ros
thomasiana	Tiliaceae	Dil
ulmaria	Rosaceae	Ros

Figure 3. CAFCA cladogram for 54 *Dasineura* species showing host plant families and subclass affiliations. Arrangement of the gall midge species within the polytomies is alphabetical. Ast = Asteridae; Car = Caryophyllidae; Com = Commelinidae; Dil = Dilleniidae; Ham = Hamamelidae; Inq. = inquiline; Ran = Ranaunculidae; Ros = Rosidae.

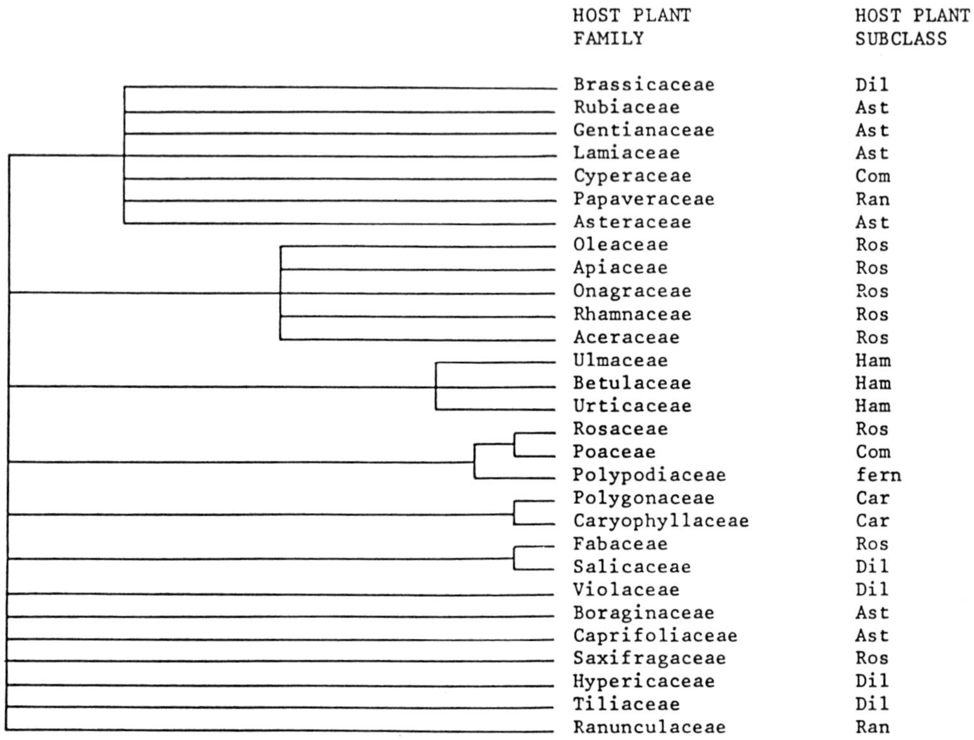

Figure 4. CAFCA cladogram for plant families, as based on the gall midge cladogram of Fig. 3. See Fig. 3 for abbreviations of the plant subclasses.

contrast with 214 steps using HENNIG-86. The CAFCA cladogram is much less resolved, hence the larger number of steps. It consists of a polytomy of 24 clades, 10 of which contain more than one species (Fig. 3), and has the families of the subclass Asteridae placed in one clade. All species associated with the Poaceae are also in a single clade. The sister-group for the midges on grasses is *D. filicina*, on ferns. Furthermore, all gall-inducing species on the subclass Hamamelidae form one clade, as do the two inquilines on *Betula*. Therefore, at least some groups of *Dasineura* might have coradiated with their host plants, at a taxonomically high level. However, there is no connection with the age of the plant taxa, as indicated by pollen analysis (Muller, 1981). The Asteridae families, for example, range in age from 50 million (Caprifoliaceae) to 10 million (Lamiaceae) years. The grasses evolved in the Paleocene, about 60 million years ago. Some of the species clusters that emerged from the PCA are members of the same clade according to the CAFCA cladogram. This includes the inquilines on *Betula* and the species associated with *Viola*, Rubiaceae, Boraginaceae, Poaceae and partially, Fabaceae.

If we construct a phylogeny for the plant families based on the CAFCA analysis for the midges, we get a single 75 step (CI = 0.87) cladogram (Fig. 4).

Six clades of a 13 clade polytomy contain more than one plant family. Some of these clades are indeed consistent with current angiosperm classification. One clade consists of all superordines of the Rosidae, the Rosanae excepted. Another clade consists of all of the Hamamelidae. There is one clade for the Caryophyllidae. The clade for the Asteridae, however, also harbors the Brassicaceae, Papaveraceae and Cyperaceae, whereas the Boraginaceae are placed in a separate clade.

As expected, the pattern of phylogenetic relationships within *Dasineura* is very complex at, and above, the family level for host plants. There are some indications of congruence in the radiations of the associates. However, there are indications of conflicting patterns as well.

3.2 **Patterns within Host Plant Families**

3.2.1 **Rosaceae**

This plant family harbors by far the largest number of *Dasineura* species; we used 12 in our analysis. Within the Rosaceae, the distribution of these 12 species is heterogeneous. Five species occur on *Filipendula* and two on *Crataegus* (*D. tortrix* is accompanied by an inquiline species, *D. sodalis*).

The species that are associated with Rosaceae are labeled in the PCA scattergrams (Fig. 5) for the semaphoronts. Three clusters of midge species are evident on the scattergrams for all semaphoronts. There is coherence among all species of *Filipendula* and those on Maloideae. Furthermore, the morphological similarity of *D. plicatrix* (*Rubus*, Rosoideae) and *D. tortrix* (*Prunus*, Prunoideae) is apparent. Within these clusters among the scattergrams, the ordination of species differs, depending on which semaphoront was used in the analysis. In the scattergrams for males and larvae, *D. pustulans* shows an eccentric position, whereas among females, the position of *D. ulmaria* is eccentric. The inquiline species *D. socalis* is close to the *D. plicatrix*/*D. tortrix* pair in the scattergrams for both males and females; in the scattergram for larvae, however, this species is situated in the multivariate space that is spanned by the midges on Maloideae. In the scattergram for males, *Dasineura potentillae* is close to the species on *Filipendula*, but on the scattergram for females it is closer to the cluster involving *D. plicatrix*/*D. tortrix*/*D. sodalis*.

Fig. 6 presents a comparison of (1) an intuitive cladogram, (2) a dendrogram based on single-linkage clustering of pairwise Euclidean distances from the PCA of linked quantitative data, and (3) a cladogram derived using HENNIG-86. The intuitive cladogram is mainly based upon a transformation series of the mediobasal lobes of the male postabdomen (qualitative character 10, Table 2). Reconstruction (using CAFCA) of the character states on the root of this intuitive cladogram indicates that *D. potentillae*, with its thick, rectangular mediobasal lobes, represents the local out-group. The species on *Filipendula* (*D. filipendula* excepted; adults unavailable) form a monophyletic group because of their solitary galls. Within the right hand cluster, the branches for the *D. oxyacanthae*/*D. pyri* pair, *D. sodalis* and the *D. plicatrix*/*D. tortrix* pair were constructed using their overall similarity because qualitative arguments were absent. This intuitive cladogram has 46 steps (CI = 0.59).

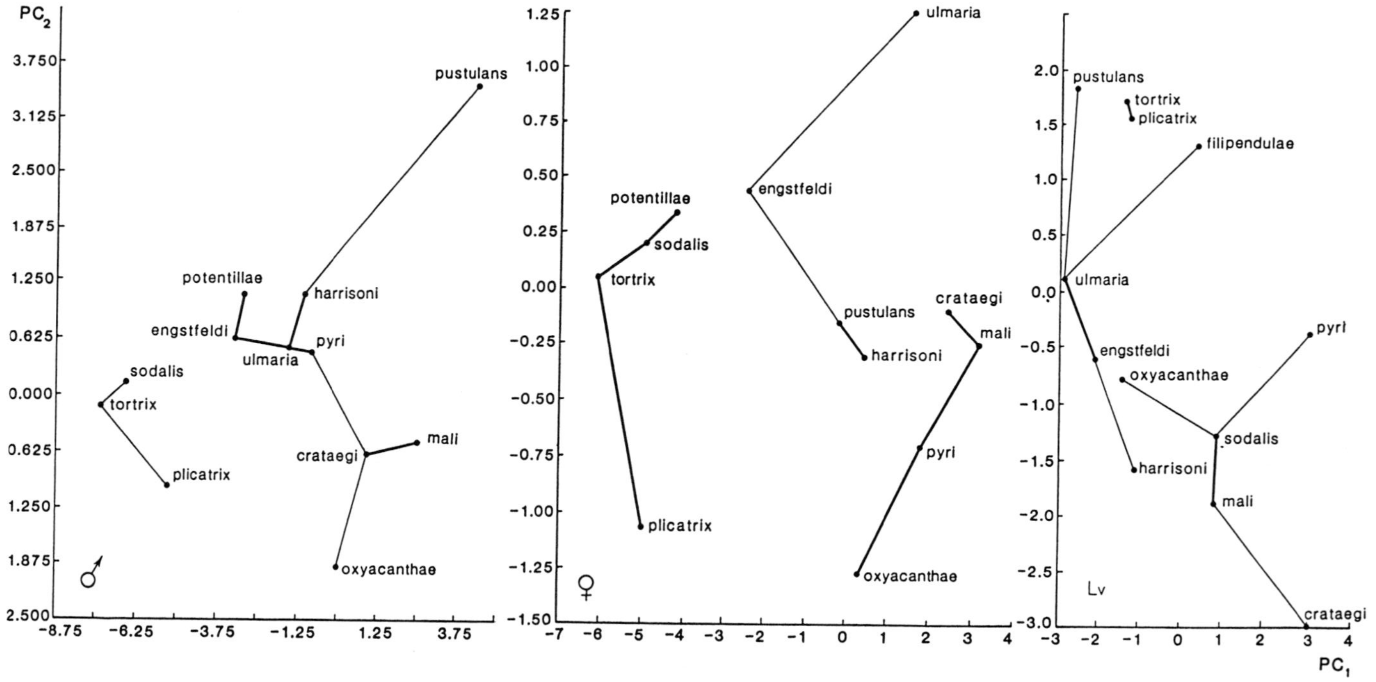

Figure 5. Comparison of PCA axes 1 versus 2 for *Dasineura* species associated with Rosaceae. Male (left), female (center) and larval (right) semaphoronts. Short [< 2] and long [> 2] Euclidean distances between species are indicated by thick and thin lines, respectively.

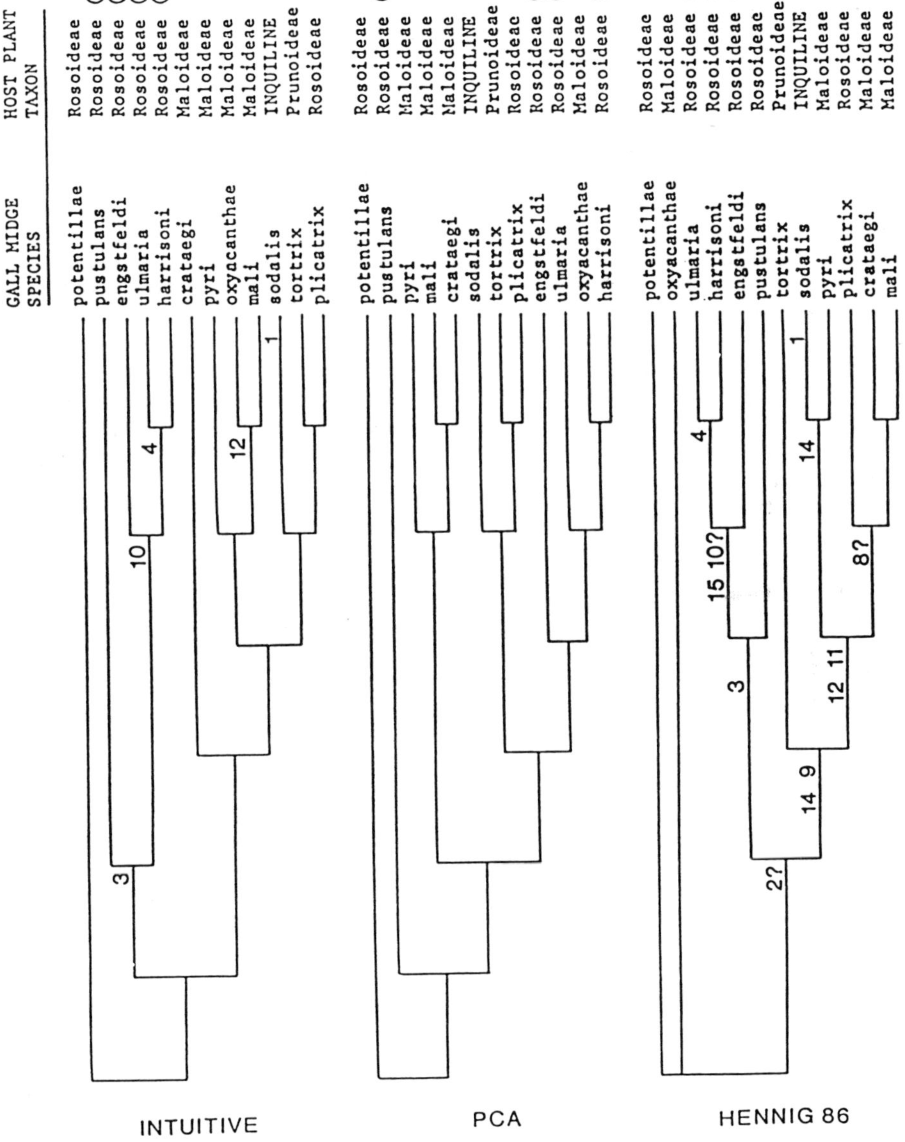

Figure 6. Comparison of phylogeny reconstructions for *Dasineura* species associated with Rosaceae. The numbers on stems represent synapomorphies for respective quantitative characters, as listed in Tables 3 and 4. Question marks indicate putative regressive evolutions in transformation series. Fil. = *Filipendula*. See text for further explanation.

The PCA dendrogram in Fig. 6 is based on overall 'gestalt' for quantitative characters. It is one step longer (47 stems) than the intuitive cladogram. Moreover, the species on *Filipendula* form a polyphyletic group. This dendrogram should be rejected if the most current concept of Rosaceae phylogeny prevails (Engler, 1964).

The CAFCA analysis derived two 45 step cladograms (CI = 0.27), both with *D. potentillae* as out-group for *Dasineura* on Rosaceae. The species on *Filipendula* do not form a monophyletic group in either of these cladograms. Using HENNIG-86, with *D. potentillae* as the root, 72 cladograms were found with a length of 43 steps (CI = 0.69). The species on *Filipendula* constitute a monophyletic group in only four of these cladograms. An ordination showing *D. ulmaria* close to *D. pustulans* as depicted in two of these four cladograms is unlikely because of the putative plesiomorphic condition of the mediobasal lobes when compared with *D. potentillae* as the out-group. The two remaining cladograms differ with respect to the transformation series of the shape of the gonostyli. One cladogram (Fig. 6, HENNIG-86) is preferable because *D. mali* and *D. cartaegi* share curved, blunt gonostyli which is a putative synapomorphy compared with the constricted gonostyli of the local out-group, *D. plicatrix*. It is three steps shorter (43 steps) than the intuitive cladogram. The species on *Filipendula* are monophyletic on this cladogram. The species on Maloideae, though, are polyphyletic. It is remarkable that the HENNIG-86 analysis indicates that "vegetative galls" (character 2) is a synapomorphy for all species except *D. oxyacanthae* and *D. potentillae*; we consider this trait to actually be a homoplasy. Although the flower galls induced by these species are uncommon in *Dasineura*, we suspect that flower galls may have evolved in parallel, separately. *Dasineura plicatrix*, *D. tortrix* and *D. sodalis* do not form a monophyletic group, which makes the origin of this triplet from an ancestor living on Maloideae unlikely. The first dichotomy in *Dasineura* on Rosaceae involves a specialization in the attacked tissue of the host. The second dichotomy deals with a radiation on *Filipendula*. The third dichotomy contrasts *D. tortrix* (on Prunoideae) with a group of species on Maloideae. Within the latter group, two species, *D. plicatrix* and *D. sodalis*, have shifted to Rosoideae and an inquiline trophic type, respectively. We are uncertain whether biological arguments involving host plants, such as a monophyletic radiation on Maloideae, effectively overrule the parsimony criteria used to reconstruct the HENNIG-86 cladogram shown.

3.2.2 **Radiation on Monocots, Gymnosperms and Ferns**

Only a few species of *Dasineura* were able to colonize monocots, gymnosperms and ferns. We analyzed the phylogenetic relationships for nine species on grasses and sedges, but also comment on *D. filicina* and *D. pteridicola* on *Pteridium*, *D. abietiperda* on *Picea*, and *D. asparagi* on *Asparagus*.

Dasineura filicina and *D. pteridicola* appear as a closely linked pair in the PCA scattergrams for adults (Fig. 7). Both species have low scores on the second principal axis for males and females, but high scores on that same axis for larvae. Their pairwise Euclidean distances are short for males [1.90] and females [1.53]; however, they are high for larvae [5.77]. A difference in the larval character "spatula shape" also makes a close affinity doubtful.

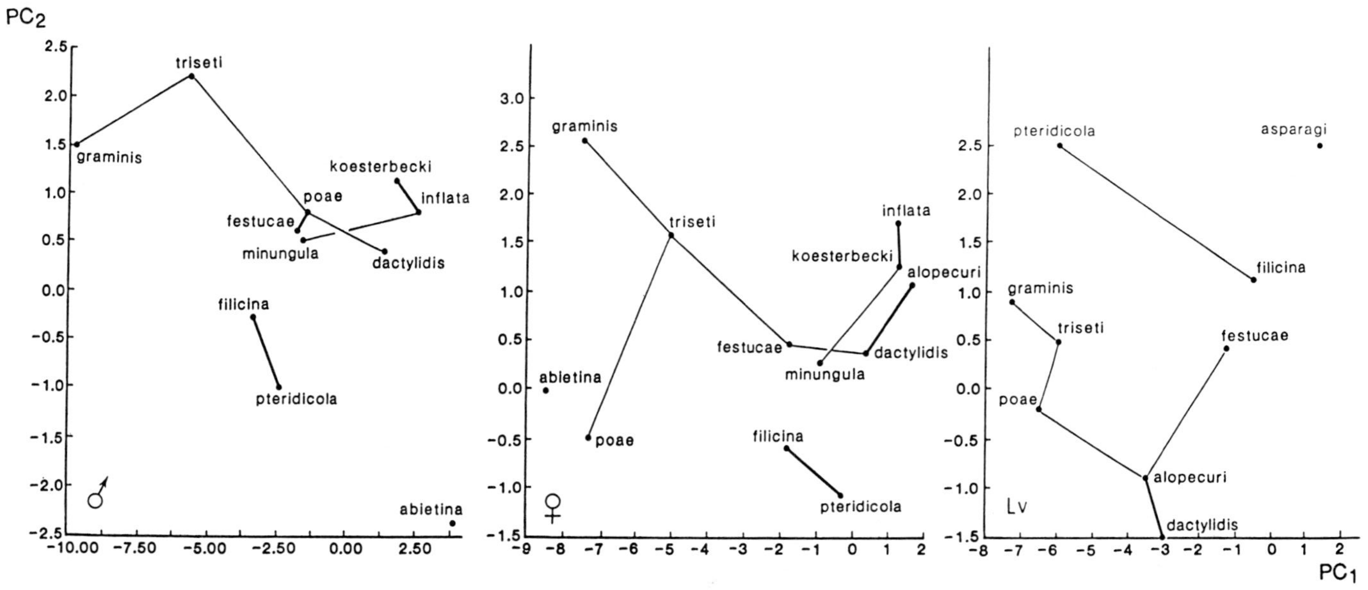

Figure 7. Comparison of PCA axes 1 versus 2 for *Dasineura* species associated with monocots, gymnosperms and ferns. Male (left), female (center) and larval (right) semaphoronts. Short [<2] and long [>2] Euclidean distances between species are indicated by thick and thin lines, respectively. See Fig. 5 for further explanation.

Consequently, the similarity shown by the males and females may be plesiomorphic, or possibly homoplastic. *Dasineura asparagi* (larvae only) and *D. abietiperda* (adults only) have eccentric positions in their respective scattergrams. Although impressions of similarity based upon 'gestalt' suggest a close affinity between these species and others on monocots or ferns, we believe these are false.

The three species on sedges, *D. koesterbecki*, *D. inflata* and *D. minungula*, are mutually close. The multivariate space they that span in the scattergrams overlaps with that of *D. alopecuri*, *D. dactylidis* and *D. festucae* on grasses. *Dasineura inflata* and *D. minungula* both pupate in their galls and have relatively short stalks of the male antennal segments. These characters are synapomorphies based upon comparison with *D. koesterbecki* as an out-group and consequently indicate a close relationship for both species. A synapomorphy for all species on sedges is absent.

The interpretation of the qualitative and quantitative character sets of species on grasses is problematic because the transformation series are conflicting for the various character sets and semaphoronts. The shape of the mediobasal lobes of the postabdomen, and the length of the stalks of the flagellomeres, are discriminating characters for males. A transformation series of the progressive reduction of stalk length is reflected on the first principal axis. We constructed an intuitive cladogram that is based on this transformation series (Fig. 8). In this 17-step dendrogram, we accepted a regressive evolution of the mediobasal lobes. It supports, though, the transformation series of the female cercus. *Dasineura graminis* and *D. triseti* share conical cerci; a condition absent in all other *Dasineura*. *Dasineura festucae*, *D. graminis* and *D. dactylidis* have similar spatula size and shape, and fused anal papillae. *Dasineura dactylidis* has large anal papillae on separate mamelons. *Dasineura alopecuri*, for which males were unavailable, has small anal papillae and is unique in the shape and size of the spatula. This indicates that the species is a representative of a local out-group for *Dasineura* on Poaceae.

The intuitive cladogram equals one of 10 cladograms found by CAFCA. Seven of the 10 generated CAFCA cladograms were rejected because they did not include *D. festucae*, *D. graminis* and *D. triseti* as a monophyletic group. Another was rejected because it included *D. poae* and *D. dactylidis* in a monophyletic group without possessing a synapomorphy for that group. The remaining cladogram was rejected because it showed *D. poae* to be an out-group, in conflict with a hypothesized transformation series for the ovipositor and larval anal papillae. Using HENNIG-86, two cladograms were found (16 steps, CI = 0.93), one of which is congruent with the intuitive cladogram. Based upon this congruency, we believe it is safe to conclude that the phylogenetic history of *Dasineura* on grasses had regressive transformations for the width of the larval spatula (character 5), the shape of the mediobasal lobe (character 10), and the gonostyli of the male postabdomen (character 12). The dendrogram based on linked quantitative characters and derived from the PCA in Fig. 8, is considerable less parsimonious (22 steps) than the shortest cladograms. Moreover, because it completely disagrees with the intuitive cladogram based upon qualitative data, and with the cladograms found using CAFCA and HENNIG-86, we must reject it.

If a cladogram is constructed for the species of *Dasineura* on both Poaceae and Cyperaceae, no synapomorphy can be found uniting the species on

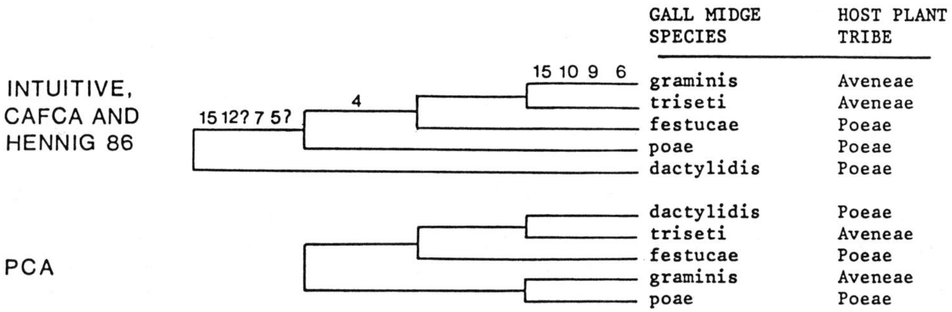

Figure 8. Comparison of phylogeny reconstructions for *Dasineura* species associated with Poaceae. See Fig. 6 for further explanation.

Cyperaceae. Nevertheless, if the species on Cyperaceae are considered monophyletic, quantitative and qualitative characters both indicate that *D. koesterbecki* is the sister-group of the remaining two species. The species-triplet, *D. festucae*, *D. graminis* and *D. triseti*, on Poaceae, is a not monophyletic on any generated cladogram that is based on data for the adult semaphoront only, because the transformation series for some adult characters become distorted by the inclusion of Cyperaceae species. Therefore, present evidence indicates that the midges on Poaceae and Cyperaceae cannot be regarded as sister-groups.

The host plants of all midges on grasses belong to the subfamily Pooideae. *Festuca*, *Dactylis* and *Poa* belong to the tribe Poeae. *Agrostis* and *Trisetum* are in the tribe Aveneae; therefore, *Dasineura graminis* on *Avena*, and *D. triseti* on *Trisetum*, share an apomorphy for relationship of their host plants. The host plant of *D. koesterbecki* is *Carex acutiformis* (subgenus *Carex*). The host plants of *D. inflata* and *D. paniculata*, *C. otrubae* and *C. paniculata*, respectively, belong to the subgenus *Vignea*.

4 Conclusions

In this chapter we compared a phenetic and a phylogenetic approach for analysis of the species of a large gall-inducing midge genus. Because of the supposed pervasive effect of the endophytophagous specialization of these insects on their radiation, we analyzed whether a classification of associated host plants might be used for phylogenetic reconstruction of the group of gall-inducers using reciprocal illumination.

In our study, the PCA-based dendrograms that were generated were not corroborated by numerical cladistic trees based on either compatibility or parsimony, using qualitative characters. The attempts at phylogenetic reconstruction, based on a principal components analysis of continuous characters, failed because objective criteria to determine the polarity of character transformation were lacking. We are unable to conclude whether species' with eccentric positions in the generated multivariate space, representing "divergent" shapes, or species' with positions at the grand centroid of PCA hyperspace, representing "average" shapes, should be regarded as derived. This necessarily hampers proper rooting of any generated network.

This does not, however, invalidate the use of this quantitative technique. Many situations observed here, where groupings of species in the multivariate space of PCA appear to reflect host plant affiliations, would only have a coevolutionary meaning if apomorphic information were used to generate the apparent spatial relationships. Examples include: the pairs and triplets of midge species that are associated with members of a same host plant family, the clusters of midges associated with *Filipendula* in Rosaceae or those associated with grasses. Within *Filipendula*, however, a diffuse evolutionary pattern is present because the midges did not speciate with the evolutionary radiation (speciation) of *Filipendula*, but did so in connection with resource exploitation; this suggests a resource-differentiating mode of midge phylogeny. Different *Dasineura* induce distinct types of galls on only one *Filipendula* species, *F. ulmaria*. Although *Filipendula* has more species than considered here, and they are all associated with *Dasineura*, these midge species, or at least some of their semaphoronts, were unavailable for the analyses.

Interpreting the results of multivariate methods that are based upon quantitative characters can be problematic. Cladistic treatments based upon coded, qualitative traits, however, also have shortcomings, even though intraspecific variation was absent in the qualitative characters that we used here. Too few qualitative characters, relative to the many midge species analyzed, were available for our data set to allow reliable cladistic analyses using either CAFCA or HENNIG-86. It was impossible to select a particular cladogram from the sometimes hundreds that are generated, using objective criteria other than parsimony. Moreover, we were often unable to judge whether a qualitative character represented a synapomorphy or homoplasy. There are indications that sometimes various character states are the result of simple switches at the genome level and these switches occur randomly, without apparent canalization by natural selection.

A striking example of such a homoplasy is the shape of the cercus of the female ovipositor. A remarkable type of an acute, striate cercus is rare, but occurs in *D. glyciphylli*, *D. rossi*, *D. autumnalis*, *D. ribis* and *D. papaveris*. This type of cerci has the most distinctive shape within *Dasineura* and would be considered very unlikely to represent a homoplasy. Therefore, in an attempt to break up *Dasineura* in monophyletic entities, one might erect a new genus for the species sharing this "synapomorphic" ovipositor. The same type of ovipositor, however, also occurs in *Wachtliella rosarum*, *Rabdophaga marginemtorguens*, *Jaapiella medicaginis* and *Macrolabis pavida*, indicating convergence across generic bounds, and suggesting that the use of this character state as a synapomorphy would be unwise. Alternatively, one might believe these non-*Dasineura* species could be erroneously classified, and actually should be in *Dasineura*, preserving the validity of this trait as a synapomorphy; several other pertinent characters indicate, however, that at least the latter three species do not belong in *Dasineura*.

More homoplastic examples are found in the shape of the larval spatula, implying that transformation series for characters emerging from cladistic analyses must be interpreted with great caution. Given that the phylogenetic reconstructions, based on qualitative and quantitative characters, that were generated here are (almost) equally parsimonious, but conflicting, no a priori argument can be formulated about the superiority of either approach. A caveat, however, is that 'gestalt' concepts of evolutionary relationships, in the sense of overall similarity, can be very misleading because it does not discriminate apomorphic from plesiomorphic similarity. In both the examples that we analyzed for Rosaceae and grasses, the PCA dendrograms were less parsimonious than the cladograms. Moreover, their hierarchies could be rejected because combinations of synapomorphies neatly fit our sequential evolution hypotheses (i.e., radiation on *Filipendula* within Rosaceae, and on grasses within Aveneae).

Nevertheless, because phylogenetic arguments emerge from a more sound theoretical base than phenetics, we feel they should play the prevailing role in determining past evolutionary scenarios. Although a phylogenetic classification for *Dasineura* is still far from complete, a combined approach using qualitative and quantitative techniques should shed light on the evolutionary relationships of these insects, and on the role that host plant associations play in unravelling their phylogeny.

5 Acknowledgments

For the loan of specimens in their care, we thank K.M. Harris, (British Museum [Natural History], London), W. Nijveldt, (Zoological Museum, Amsterdam), H. Stelter (Post Broderstorf, Germany) and E. Sylvén, (Museum of Natural History, Stockholm). For their critical and helpful reviews of the manuscript we thank R. Gagné (U.S. National Museum of Natural History, Washington, D.C.), H. Stelter and E. Sylvén. A substantial part of the material was measured by S. Kofman, and the drawings were prepared by J. Herzberg; both are gratefully acknowledged.

6 References

Archie, J.W., 1985. Methods for coding variable morphological features for numeric taxonomic analysis. *Systematic Zoology*, 34: 326-345.

Brooks, D.R., 1981. Hennig's parasitological method. A proposed solution. *Systematic Zoology*, 30: 229-249.

Buhr, H., 1964-1965. Bestimmungstabellen der Gallen (Zoo- und Phytocecidien) an Pflanzen Mittel- und Nordeuropas. Fischer, Jena, 1572 pp.

Dahlgren, R. and Bremer, K., 1985. Major clades of the angiosperms. *Cladistics*, 1: 349-368.

Engler, A., 1964. Syllabus der Pflanzenfamilien. II, (Twelfth Edition). Borntraeger, Berlin, 666 pp.

Farris, J.S., 1988. HENNIG, Hennig86 Reference, (Version 1.5). J.S. Farris, Department of Ecology and Evolution, University of New York at Stony Brook, Stoney Brook, New York.

Futuyma, D.J. and Slatkin, M. (Editors), 1983. Coevolution. Sinauer Associates Inc., Sunderland, Massachusetts, 555 pp.

Gagné, R.J., 1981. Cecidomyiidae. In: J.F. McAlpine, B.V. Peterson, G.E. Shewell, H.J. Teskey, J.R. Vockeroth and D.M. Wood (Editors), Manual of

Nearctic Diptera, Volume 1. Research Branch, Agriculture Canada, Monograph 27, pp. 257-292.
Gagné, R.J., 1989. The Plant-Feeding Gall Midges of North America. Cornell University Press, Ithaca, New York, 356 pp.
Hennig, W., 1966. Phylogenetic Systematics. University of Illinois Press, Chicago, 263 pp.
Jermy, T., 1976. Insect-host plant evolution: co-evolution or sequential evolution? *Symbolae Biologicae Hungaricae*, 16: 109-113.
Mani, M.S., 1973. Plant Galls of India. Macmillan India, Madras, India, 354 pp.
Muller, J., 1981. Fossil pollen records of extant angiosperms. *Botanical Review*, 47: 1-142.
Nitecki, M.H. (Editor), 1983. Coevolution. University of Chicago Press, Chicago, 392 pp.
Pimentel, R.A., 1979. Morphometrics, the Multivariate Analysis of Biological Data. Kendal/Hunt Publishing Company, Dubuque, Iowa, 276 pp.
Pimentel, R.A. and Riggins, R., 1987. The nature of cladistic data. *Cladistics*, 3: 201-209.
Roskam, J.C., 1977. Biosystematics of insects living in female birch catkins. I. Gall midges of the genus *Semudobia* Kieffer (Diptera, Cecidomyiidae). *Tijdschrift voor Entomologie*, 120: 153-197.
Roskam, J.C., 1985. Evolutionary patterns in gall midge-host plant associations (Diptera, Cecidomyiidae). *Tijdschrift voor Entomologie*, 128: 193-213.
Roskam, J.C., 1992. Evolution of the gall-inducing guild. In: J.D. Shorthouse and O. Rohfritsch (Editors), Biology of Insect-Induced Galls. Oxford University Press, Oxford, U.K., pp. 36-49.
Simon, C., 1983. A new coding procedure for morphometric data with an example from periodical cicada wing veins. In: J. Felsenstein (Editor), Numerical Taxonomy, NATO Advanced Study Institute, Series G, Number 1. Springer-Verlag, New York, pp. 378-382.
Skuhravá, M., 1986. Cecidomyiidae. In: A. Soós and L. Papp (Editors), Catalogue of Palaearctic Diptera, 4. Elsevier Science Publishers, Amsterdam, pp. 72-297.
Sylvén, E., 1975. Study on relationship between habits and external structures in Oligotrophidi larvae (Diptera, Cecidomyiidae). *Zoologica Scripta*, 4: 55-92.
Sylvén, E. and Carlbäcker, U., 1981. Morphometric studies on Oligotrophini adults (Diptera: Cecidomyiidae) including an attempt to correct for allometric deviations. *Entomologica Scandinavica, Supplement*, 15: 185-210.
Takhtajan, A.L., 1980. Outlines of the classification of flowering plants (Magnoliophyta). *Botanical Review*, 46: 225-359.
Wiley, E.O., 1987. Historical ecology and coevolution. In: P. Hovenkamp (Editor), Systematics and Evolution: a Matter of Diversity. Utrecht University Press, Utrecht, Netherlands, pp. 331-341.
Yukawa, J., 1971. A revision of the Japanese gall midges (Diptera, Cecidomyiidae). *Memoirs of the Faculty of Agriculture, Kagoshima University*, 8: 1-203.
Zandee, M., 1987. A computing environment for cladistic analysis. In: P. Hovenkamp (Editor), Systematics and Evolution: a Matter of Diversity.

Utrecht University Press, Utrecht, Netherlands, pp. 111-140.
Zandee, M., 1989. CAFCA: a Collection of APL Functions for Cladistic Analysis, User Manual, (PC-Version 1.9.6). Institute for Theoretical Biology, Leiden, Netherlands.
Zandee, M. and Geesink, R., 1987. Phylogenetics and legumes: a desire for the impossible? In: C.H. Stirton (Editor), Advances in Legume Systematics, 3. Royal Botanical Gardens, Kew, U.K., pp. 131-167.
Zandee, M. and Roos, M.C., 1987. Component compatibility in historical biogeography. *Cladistics*, 3: 305-332.

PRONOTAL SHAPE: A SOURCE OF CONFUSION OR PANACEA IN SYSTEMATIC STUDIES OF TREEHOPPERS (HOMOPTERA: MEMBRACIDAE)?

T.K. Wood[1] and J.D. Pesek[2]

[1] *Department of Entomology and Applied Ecology, University of Delaware, Newark, Delaware, USA 19711;* [2] *Department of Food and Resource Economics, University of Delaware, Newark, Delaware 19711, U.S.A.*

Abstract

Treehoppers are easily distinguished at the family level by their swollen pronotum, but their identification and taxonomic separation below the tribal level is often difficult because frequently pronotal shape is the only attribute that characterizes their genera. Within the tribe Smiliini, we explore the potential use of pronotal shape, in profile, to distinguish genera and deduce a phylogenetic network. We digitized the pronotal outline of members of the tribe, using 20 spaced nodal points along each of four pronotal contours. Cubic splines were used to calculate distances between the analogous nodal points among the contours, and the angles of the resulting segments were determined. The data were size scaled and subjected to discriminant function analysis, after which the centroid values for the taxa were networked using a Euclidean distance phylogenetic networking algorithm. The resulting networks, as phylogenetic hypotheses, produce credible evidence for the evolution of host plant utilization by the treehoppers, and for their historical biogeography.

1 Introduction

Shape of an individual or a characteristic structure is often one the most diagnostic features that confers membership in a taxon, but it is also the most elusive feature to measure. The elegant shape transformations by D'Arcy Thompson (1961) of fish, or Raup (1962, 1966) for snail shells, provides convincing evidence not only of the importance of shape analysis in evolutionary biology (Futuyma, 1986) but also its potential in systematic studies. Bookstein et al. (1985) have formalized this approach to shape by using trusses that can be transformed through a biorthogonal grid system. At present, the analysis can be used to transform pairs of taxa, but it has not been developed to the point where it can be applied to a large array of taxa.

Systematists, in general, do not deal with shape directly in this manner, but rather look for suites of homologous, discrete multistate characters that can be polarized or characterize variations in shape as a multistate condition (Wiley, 1981); such discrete state characters may then be used in a cladistic analysis. After cladistic analysis, the result can be a cladogram in which the array of taxa visually resemble a shape transformation series. Such results are not surprising, because morphological, multistate characters in monophyletic lineages, in theory, are an external measure of shared and derived developmental processes that are controlled by genetic loci. How these characters are spatially arranged is dictated by intraspecific allometric relations, although interspecific differences may reflect shared or derived patterns of allometry (Futuyma, 1986).

One of the confusing aspects of allometric patterns, and that of size, is that although there is a strong genetic component, the phenotype potentially can be influenced by environmental factors (Atchley and Rutledge, 1980). Without techniques or direct experimental evidence that permit the elimination of environmental influences on size measures, they can provide misleading results in systematic studies (Humphries et al. 1981). Here we consider how to deal with groups of organisms that can be distinguished by the overall shape and size of a particular structure, but which, nevertheless, appear to have a dearth of characters that can be categorized into discrete, multistate conditions.

2 Pronotum

The Membracidae are easily distinguished from other families of auchenorrhynchus Homoptera by the dorsal expansion of their pronotum. The variety and complexity of pronotal shapes and associated color patterns are as visually striking as the spectacular color patterns of some butterflies and birds. To naturalists, such as Poulton (1903), pronotal shapes and color patterns suggested crypsis, mimicry and aposematic functions. At present, there is no experimental evidence to support either cryptic or mimic functions in any treehopper species.

Experimental evidence for aposematic coloration exists for adults of several species (Wood, 1975a, 1977). These species change color as they mature and lose their defensive chemical qualities, but in others the shape and hardness of the pronotum provide a physical defense (Wood, 1975a). Perforations in the pronotum of one species provides an easily detached segment that could provide the basis for an escape from some types of predators (Mann, 1912; Strumpel, 1983).

Early workers (Funkhouser, 1951) thought the pronotum had no physiological function and was simply a hollow dorsal extension of the prothorax. At least in one species, the presence of trachea, nerves and neurosecretory tissue suggests that it is far from physiologically inert (Wood

1975b). The surface architecture of many species is covered with articulating setae which protrude over deep conical pits that penetrate to the inner wall of the pronotum (Wood and Morris, 1974). Although direct evidence is lacking, the dorsal pronotal expansions may increase the surface area for the reception of sensory information (Wood and Morris, 1974), or act as a dispersal site for pheromones. Sexual dimorphism is common in the Membracidae, raising the possibility that pronotal shape and color may be involved in the location and recognition of mates. Males, after locating a female, spend a considerable amount of time on the side of the female pronotum during precopulatory pairing (Wood, 1974, 1976; Wood et al. ,1984; Wood and Dowell, 1985).

Although the exact functions of the pronotum are at present conjecture, it is clear that it is structurally complex. As might be expected, the pronotum has played a central role in the taxonomy and classification of the group. Early workers (Funkhouser, 1951; Goding, 1926, 1928a, 1928b, 1929; Van Duzee, 1908) used the shape of the pronotum, in addition to other characters, to define subfamilies and tribes. They clearly saw shape transformations but had no way to measure or compare shape, except through written descriptions or line drawings. The end result was confusion at all levels of classification. With the advent of Hennig (1966), modern workers have tended to avoid pronotal shape in an attempt to find more definitive characters that would permit a cladistic analysis.

The only pronotal trait used in higher classification in the Membracoidea, is whether the scutellum is covered. Strumpel (1972) used this character to divide the Membracidae into two subfamilies, the Centrotinae and the Membracinae. Although not defined in a quantitative sense, Strumpel (1972) recognized that within each subfamily there appeared to be a transformation series of pronotal shapes. In contrast, Deitz (1975) used the same pronotal trait as one of a suite of characters to divide the New World Membracoidea into seven subfamilies and 27 tribes. Although he recognizes that pronotal shape may be useful at some levels of classification (L. Deitz, personal communication), he defines subfamilies and tribes without the vagueness associated with written descriptions of shape.

Even though subfamilies and tribes can be defined on the basis of other characters, most generic and specific definitions depend a great deal on pronotal shape. Early monographs relied heavily on elaborate written descriptions and line drawings to describe shape (e.g., Ball, 1932). Caldwell (1949) was the first to attempt a classification of the Ceresini using male genitalic characters. Kopp and Yonke (1979), in the most recent revision of this group, relied heavily on male genitalia, but also used a number of pronotal characters. Although genitalia are potentially very useful characters, they are only helpful in identifying one sex, or in distinguishing complexes of species.

In spite of recent work, diagnostic features for species are often body measurements, such as length, or a series of outline drawings of the pronotum. Regional faunal monographs that provide keys which attempt to minimize the use of pronotal shape (Kopp and Yonke, 1973a, 1973b, 1973c, 1974; Dennis, 1952, 1964, 1965), ultimately have to depend on line drawings, size and color patterns for accurate identification. In general, this works rather well for a given region because of the limited array of species present. However, this approach has limited value in generic revisions and phylogenetic investigations. The quest to find a suite of discrete characters that define genera and species is laudable, but may be unattainable because divergence at these levels may be the result of selection on aspects of pronotal shape that are difficult to define. We argue here that to ignore pronotal shape is to ignore a major and important aspect of the genome that has undergone transformation in the same way as a suite of discrete characters. The major difficulty with shape is that it is difficult to measure and translate into a set of discrete, multistate conditions that are operational for traditional cladistic analyses.

Because many species descriptions are based on pronotal shape and size, we have been examining techniques to measure these variables. The overall objective of this work is to develop a technique that: (1) provides rapid, accurate, defendable identifications and classifications; (2) can be used to examine intraspecific morphological variation that is related to host plants and geography; and (3) can provide a basis to infer phylogenetic relationships. In this chapter, we report a technique that we have developed to meet these objectives, and we explore its use in defining groups of species in the tribe Smiliini.

3 Tribe Smiliini

Approximately 46% of North American treehoppers are in the Smiliini, and a majority of the genera are restricted to oaks (*Quercus*). Historically, this tribe, as delineated by Deitz (1975), was divided into two tribes, the Telamonini and Smiliini. This early division was based upon size, the shape of the pronotum, and the presence or absence of a cross vein on the hind wing. Unfortunately, the cross vein character is unreliable because Ball (1932), Deitz (1975) and TKW (unpublished data) have found bilaterally differing specimens with the vein present on one wing but not the other. At the generic and specific levels, Ball (1932), Woodruff (1919, 1920, 1924) Dennis (1952, 1964, 1965), and Kopp and Yonke (1973a, 1973b, 1973c, 1974), all failed to find consistent, discrete characters, and without exception, all relied on the shape of the pronotum, color pattern and size measurements for diagnostic features.

At present, genera are characterized by several attributes of the pronotum and size: the presence or absence of a dorsal crest, the origin of the dorsal crest in relation to the humeral hump, the location of the crest's highest point relative to the humeral hump, the presence or absence and location of various surface depressions; descriptive features of the posterior process; and length of the pronotum or body (Fig. 1.). With practice, good illustrations and available types, species identifications are possible, but not particularly reliable among the 22 genera and 200+ species in this tribe of membracids. Both identifications and rigorous defense of apparent pronotal shape transformations between and within genera would be considerably easier and more dependable if sufficient tools were developed to that reflect the ability of the specialist's "gestalt."

4 Generic Assignments

We selected what appeared to be a typical female specimen from each of 168 described and undescribed species within the Smiliini. Each specimen was assigned initially to one of 22 recognized genera (Metcalf and Wade, 1965; Deitz, 1975). Because the generic limits within the tribe are rather vague, we

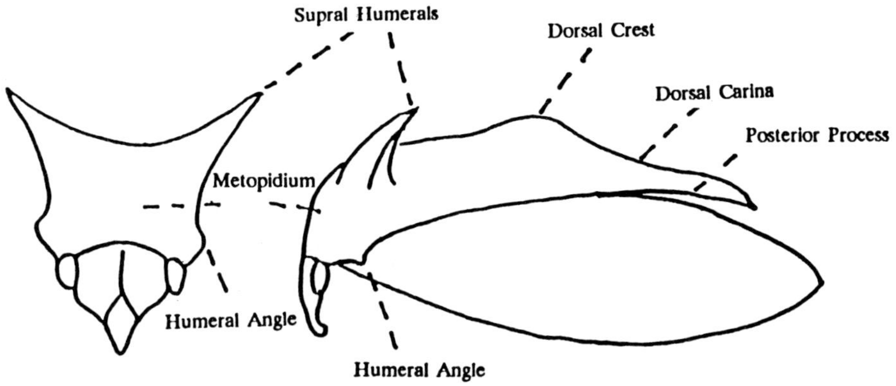

Figure 1. Topographical features of the pronotum that have been historically used to define genera and species of membracids.

combined some genera and reassigned some species accordingly. These reassignments were made on the basis of TKW's interpretation of present generic limits. After reassignments were made, this reduced the number of genera or groups to 16 (Appendix 1).

5 General Morphometric Considerations

With the advent of digitizing equipment, a number of workers with other taxa have experimented with complex equations to define shape. Fourier series have been used to describe the shape of snails (Ferson et al., 1985), bryozoans (Anstey and Delmet, 1973), and diatoms (Younker and Ehrlich, 1977; Stoermer and Ladewski, 1982; Ehrlich et al., 1983). This approach has been criticized (Bookstein et al., 1982, 1985) on the grounds that the coefficients used to characterize shape have no homology and are of little value in understanding morphological transformations in systematic studies. These views are not shared by Ehrlich et al. (1983). Our initial approach was to use coefficients from Legendre polynomials to describe the contours of the pronotum. These coefficients provided very good data reconstruction, but failed dismally to reveal distinguishable groups, in either principal components analysis or discriminant analysis, simply because the coefficients, as Bookstein et al. (1985) pointed out, are neither analogous or homologous among species.

An approach that Bookstein et al. (1985) advocate is the use of homologous landmarks, construction of truss networks and biorthogonal grids for the analysis of morphological transformations. This approach seems to work rather well with organisms whose contours or outlines are marked with homologous landmarks, such as fins. The problem with the membracid pronotum is that the only landmarks are its beginning and end, with no

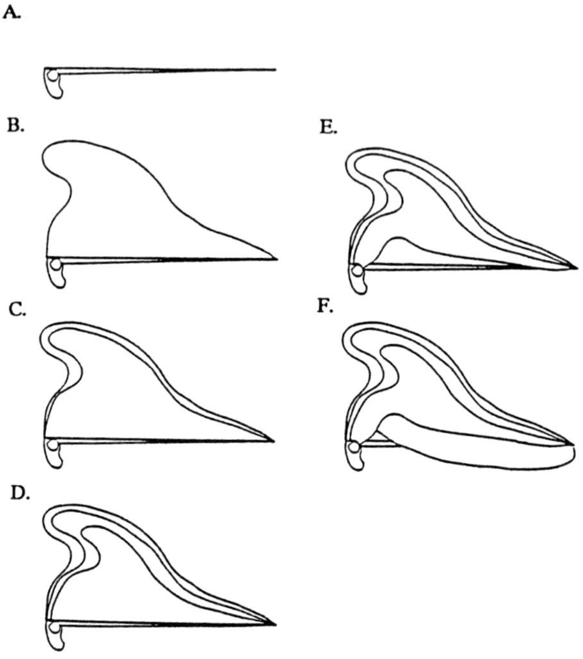

Figure 2. Diagrammatic representation of the pronotum and the contours traced. (A) The dorsal expansion of the pronotum removed, leaving the mid-line and lateral edge. (B) The dorsal expansion along the mid-line (contour 1). (C) The second contour begins lateral to the mid-line, following changes in the surface topography to the tip of the posterior process. (D) The third contour begins above the eye, following the surface topography to the tip of the posterior process. (E) The fourth contour begins at the outside of the eye, continuing along the lateral edge of the pronotum to the tip of the posterior process. (F) Representation of all contours relative to the attachment of the wing.

identifiable, homologous landmarks between them.

Others, such as Lohmann (1983) and Ray (in press) working with leaf shape, identify contours that extend between landmarks. Ray divides these "homologous contours" into line segments of equal length and uses the angles between segments in principal components analysis. Although, initially, we were unaware of Lohmann's or Ray's work, the technique we developed is similar to their analyses of outlines, but also incorporates the spirit of Bookstein's truss analysis.

Our procedure can be envisioned as follows. If the dorsal portion of the pronotum is excised and viewed laterally, it resembles a triangle (Fig. 2A). From this triangular base, a series of lateral contours rise dorsally; these contours extend from above the head and run ultimately to the posterior, terminating at the same point at the tip of the posterior process. Four points above the head were selected (Figs. 2B-2E) at progressively greater lateral distances from the mid-line: (1) the point at the mid-line (Fig. 2B) where the dorsal carinae rises from the head [contour 1]; (2) the point where the dorsal carina is separated from the humeral hump (Fig. 2C) [contour 2]; (3) the point above the inner-most corner of the eye (Fig. 2D) [contour 3]; and (4) the point defined by the outer-most edge of the eye and the pronotum (Fig. 2E) [contour 4].

6 Digitizing

Specimens were aligned under a Wilde 420 macroskop, with the longitudinal profile of the pronotum in focus along the mid-lateral line. This could be clearly defined as the point where the dorsal carinae rises from above the head and passes along the dorsal surface of the pronotum to the tip of the posterior process. Specimens were then tilted to bring the lateral humeral angle into focus. When this image was transmitted, via a television camera, to the computer screen, it was possible to discern not only silhouette outlines, but also surface topography consisting of bulges and depressions that extended laterally from the mid-line of the specimen.

With the use of a digitizing pad (connected to a Zeiss Videoplan II) set to record x- and y-coordinates every 0.1 mm and a mouse, traces of the contours were made from each of the four landmarks above the head to the tip of the posterior process (Figs. 2B-2E). The Zeiss Videoplan II was directly wired to an IBM PC computer, so that when a given trace was completed, the x- and y-coordinates were captured in a form that could be transferred directly to a IBM (3090 300E) mainframe computer for analysis. Before a trace was made, the following data were entered into the file for each specimen: contour number, sex, species code, individual code, and generic code. These data were carried along with all subsequent data manipulations and statistical analyses. Because the treehoppers in this tribe vary greatly in length and height of the pronotum, magnifications on the macroskop were changed so that the image filled the screen on the Zeiss Videoplan II; when magnifications were changed, the digitizing pad was rescaled. This assured that for each contour trace there would be at least 85 x- and y-coordinates.

7 Measurement of Shape and Size Components

To analyze size and shape successfully, a Fortran program (Appendix 2) was written for the IBM mainframe which, in sequence:

(A) Established a hypothetical reference line between the first point in contour 1 and the tip of the posterior process to provide a common perspective. The original, digitized x- and y-coordinates were translated, so that the first point on contour 1, which began above the head at the mid-line, was considered to be the origin, with $x,y = 0,0$.

(B) Rotated the new coordinates, so that reference line became the horizontal axis (the tip of the posterior process then had a y-coordinate of 0).

(C) Scaled all coordinates, so that all pronota, regardless of species, had reference line lengths of 1; thus, the new coordinates became for the first point on contour 1 ($x,y = 0,0$) and the last ($x,y = 1,0$). The remaining three contours were also similarly scaled.

(D) Translated contours 2 through 4, so that their coordinates for the tip of the posterior process coincided with that for contour 1.

(E) Calculated the arclength from the first point of each contour to successive, adjusted x- and y-coordinates along that same contour. After this was done, the arclength for each contour was normalized (the arclength between points was divided by total arclength) so that the distance from the first point to the last was 1.

(F) Calculated separately, for the x and y-coordinates of each of the four contours, a cubic spline that was relative to the normalized arclength. Using these splines, the x and y-coordinates of points at 0, 5, 10, 15, 20, 25, 30 . . . 100% intervals along each contour (as measured by normalized arclength) were calculated. The resulting 21 nodes divided each contour into 20 line segments, with the terminal node at the tip of the posterior process common among all four contours.

(G) Calculated the (two-dimensional) distance between each of 20 nodes (excluding the terminal node) in contour 1 and analogous nodes in contour 2. Similar, analogous, internodal distances were calculated between contours 2 and 3, and 3 and 4.

(H) Calculated, for each contour, the angle between a reference line and each generated internodal segments. These angles were output to a SAS program, which used them to determine the angles between adjacent internodal segments along each contour.

(I) Calculated, using the unscaled, digitized coordinates, the actual arclength (in mm) of each of the four contours, plus the length (in mm) from the first point of contour 1 to its last point, at the tip of the posterior process.

(J) Calculated, using the unscaled, digitized coordinates, pronotal area (in mm^2) between contours 1 and 4 by using Green's theorem and the trapezoidal rule.

8 **Analysis of Essential Shape and Size Components**

Strumpel (1972) was the first to attempt an analysis of the relationship between pronotal area and body length. He examined seven species in the genus *Membracis* and found that they formed an allometric series. Because the Smiliini (current limits) was previously divided on the basis of size and pronotal shape by early workers into the Telamonini and Smiliini, we evaluated whether pronotal area (square root), length and the perimeter could be used to partition the tribe. Although there is a tendency for the Telamonini to be larger in all three of these measurements (Fig. 3), there is considerable overlap with the Smiliini and no clear separation between these taxa using these variables. Thus, raw size (length) and general shape measurements, such as pronotal area and perimeter, are insufficient to discriminate among subtribes.

We used canonical discriminant analysis (SAS Institute Inc., 1985) to determine which shape and size measurements, or combinations thereof, if any, could be used to discriminate groups of species. The analysis reduces the number of variables that summarize between-class (group) variation by deriving linear combinations of variables that have the highest multiple correlation with groups. Canonical correlations are normalized so that within-group variance of the canonical variable is one (unity). The second correlation is obtained by finding linear combinations that are uncorrelated with the first canonical variable. Subsequent variables are extracted until the

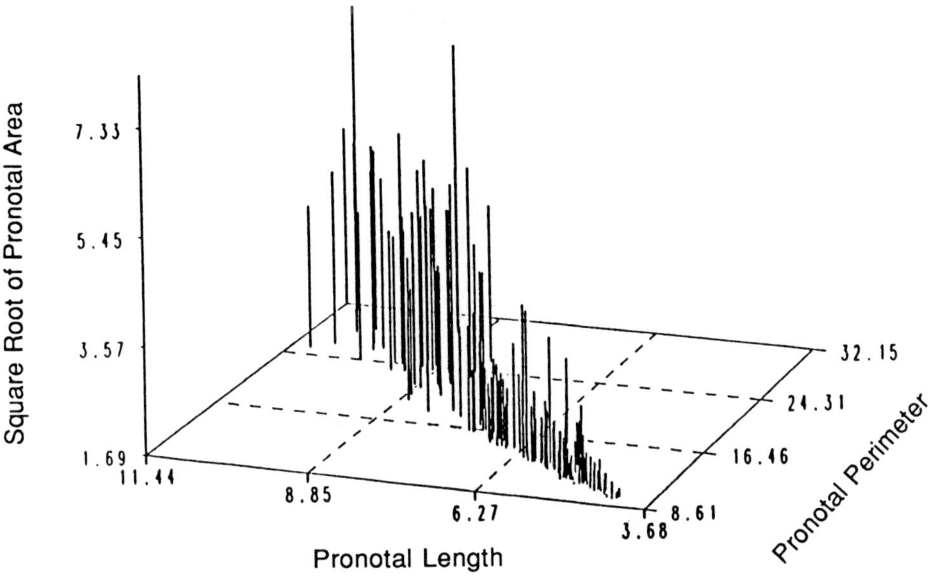

Figure 3. Three dimensional plot of the pronotal area (square root), length and perimeters to determine if the Smiliini can be divided into the Telamonini and Smiliini.

number of new variables equals the number of original variables or one less than the number of groups [*n* - 1] (SAS Institute, 1985).

We approached the analysis in two ways, using separate analyses on two parallel sets of variables, and sequentially increasing the number of variables within each series as it was analyzed. For the first set of variables, we used measurements that included components of both shape and size (e.g, area [= square root], contour length[s], pronotal length), plus variables that were size-independent (e.g., the angles and distances between spline nodes among contours from the normalized and scaled pronota). The measurements of variables with shape and size variance were transformed using logarithms (base 10), as suggested by Bookstein et al. (1985) to make them comparable to the dimensionless measurements of the scaled, normalized pronota. In the second set of variables, we attempted to directly measure shape, independent of size, by using only the angles and distances between spline nodes among the contours on normalized, scaled pronota.

Our first analysis was conducted using 23 variables; these were: area, pronotal length, length of contour 1, and the 20 angles from contour 1 (Fig. 4A). The results showed that discrimination was rather poor whether size measurements were included (Fig. 4A') or not (Fig. 4A"). When the angles from, and length of, contour 4 were added (Fig. 4B), the results showed a tendency to partition the data plots into two major groups, whether size measurements were included (Fig. 4B') or not (Fig. 4B"). In both the latter cases, the three-dimensional plots of various combinations of canonical variates seemed to separate some genera, and two major, but rather diffuse,

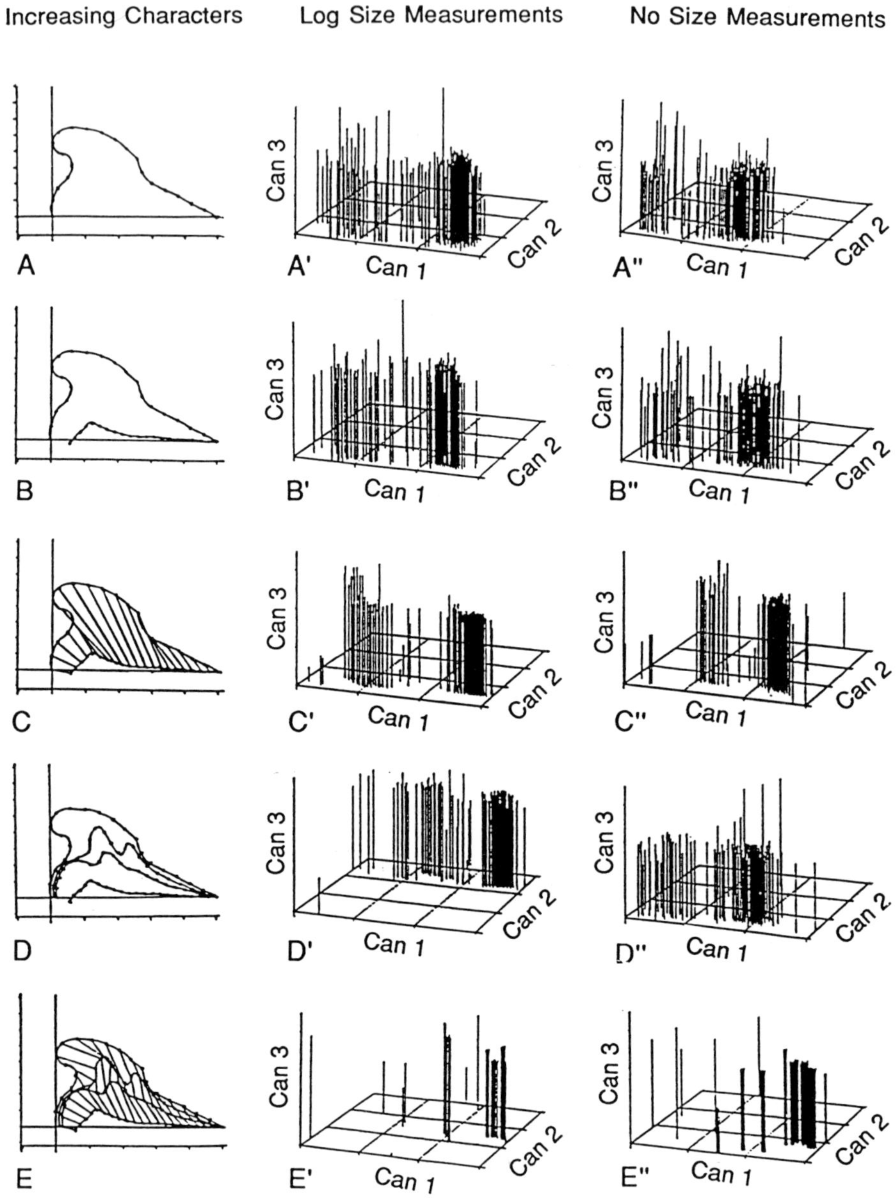

Figure 4. The effect of progressively using more variables on the discrimination among the sixteen groups in this study. The left column (A-E) diagrammatically represents an increasing number of angles and distances between analogous pronotal spline nodes. The middle column (A'-E') represents the degree of group discrimination when log square root area and the appropriate contour lengths are included with those represented in the left column. The right column (A"-E") represents discrimination when size measurements are excluded.

"clouds" of taxa within the tribe. Thus, simple measures of the pronotal outline at the mid-line or the outer boundary, combined with size measurements, were insufficient to describe and define shape differences.

We then modified our program to calculate the distance from the spline nodes on contour 1 to analogous nodes on contour 4; this increased the maximum number of measurements (variables) analyzed to 64 when size variables were included (Fig. 4C). With the inclusion of size variables (Fig. 4C'), relatively tight clouds resulted which represented the Telamonini and Smiliini; within the Telamonini cloud, even a few smaller groups appear as relatively tight clouds. When the size measurements were excluded (Fig. 4C"), there was slightly better discrimination. The distance measures between analogous spline nodes represented the degree of expansion and contraction between the two contours that were not reflected by any of the other measurements.

We expanded the number of measurements to 86, which included area, pronotal length, the lengths of contours 1 through 4, and all the angles between the line segments along each of four contours (Fig. 4D). When size measurements (Fig. 4D') where included in these analyses, two major clouds, and some relatively tight groupings of species that represent genera, were evident. This pattern was also evident when size measurements were excluded (Fig. 4D").

As a last step, we calculated the distances between the analogous spline nodes on contour pairs 1 and 2, 2 and 3, and 3 and 4 (Fig. 4E). These measurements were added to those described above to make a total of 146 variables for each individual, when size was included. With size variables included (Fig. 4E'), all 16 species groups form tight clouds when the first three canonical variables are plotted against each other. The analysis reduces the 146 variables, so that 92% of the between-group variation occurs in the first six canonical variables. When the six size measurements are excluded (Fig. 4E") only 13 of the 16 groups can be delineated on the same plot by the first three canonical variables. However, when the scale is expanded to delete outlying groups, 15 of the 16 groups were separated.

These analyses indicate that within the Smiliini, as currently defined, there appear to be three major subdivisions. When we plot only the Telamonini (Fig. 5A) (but adding *Smilia*), there appear to be two divisions. One group (*Hemicardiacus*, *Archasia*, *Smilia*, *Antianthe*, and *Telamonanthe*) have high, rounded, foliaceous pronota (Fig. 6: lineage group 2), while the other group (*Telamona*, *Heliria*, *Glossonotus*, *Thelia*) have elevated pronota that are more triangular (Fig. 6: lineage groups 1 and 3). The third division (Fig. 5C) reflects, in part, the older concept of the Smiliini (which originally included *Smilia* but not *Carynota*), although here it also includes: *Atymna*, *Grandolobus*, *Godingia*, *Cyrtolobus*, *Xantholobus*, *Ophiderma*, and *Carynota*. As a major grouping, these genera resemble each other in shape and size (Fig. 6: lineage groups 4 and 5).

The inclusion of size measurements in the analyses had relatively little effect in discriminating the groups in what are defined here as the Telamonini (Figs. 5A, 5B). The genera of that division are morphologically distinct and both analyses reflect differences in shape. The necessity of including size measurements is clearly evident in the old Smiliini (Figs. 5C, 5D). The exclusion of size produces discernable, but relatively diffuse, clusters of species in contrast to those produced when size is included. Therefore, when overall

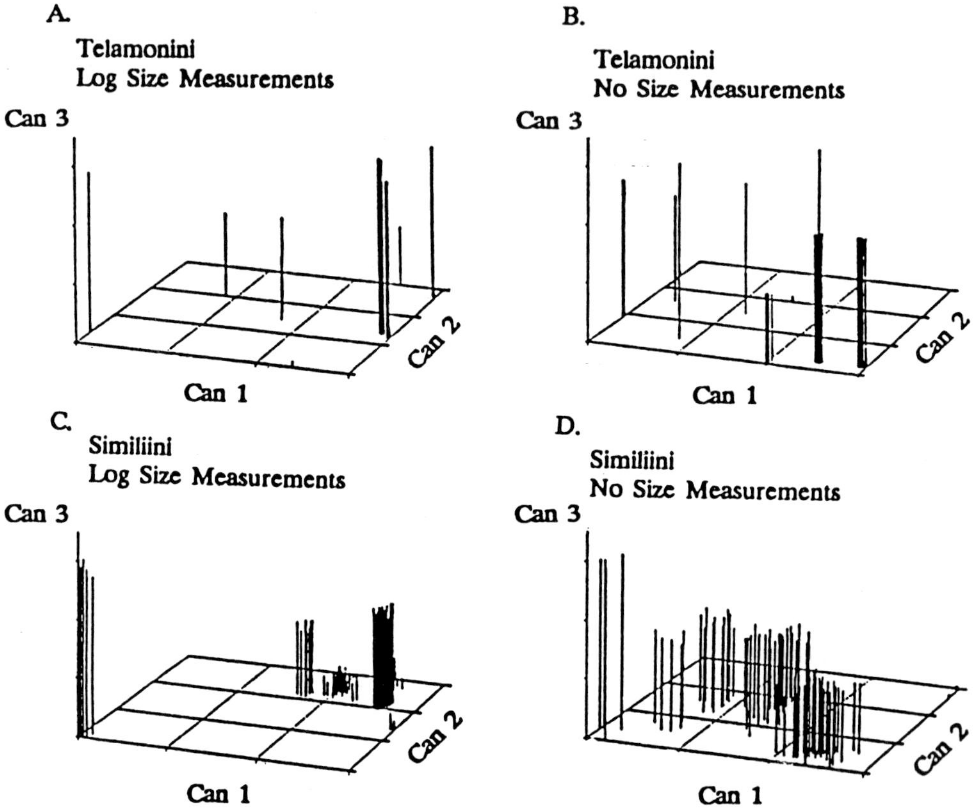

Figure 5. Expanded three dimensional plots of Fig. 4 E'-E" to show two major divisions of the Smiliini. (A) and (B) Groups that have been historically assigned to the Telamonini (but here includes *Smilia*). (C) and (D) Groups that have been historically assigned to the Smiliini (but here includes *Carynota* and *Tropidarnis*).

shape is similar, size variables become essential for group discrimination.

9 Phylogenetic Inferences

The canonical discriminant analysis generated a square Mahalanobis' distance matrix that is free of correlations among quantitative measurements (Felsenstein, 1988). Several Mahalanobis' distance matrices that resulted from our canonical analyses of the 16 groups were down-loaded to an IBM AT personal computer, and served as input data for the Fitch distance matrix program (options: P = 2, branch lengths unconstrained) in PHYLIP (version 2.7) (Felsenstein, 1984).

Our first phylogenetic analysis (sensu Sorensen, 1987, this volume; Sorensen and Foottit, this volume) was conducted to determine if the shape

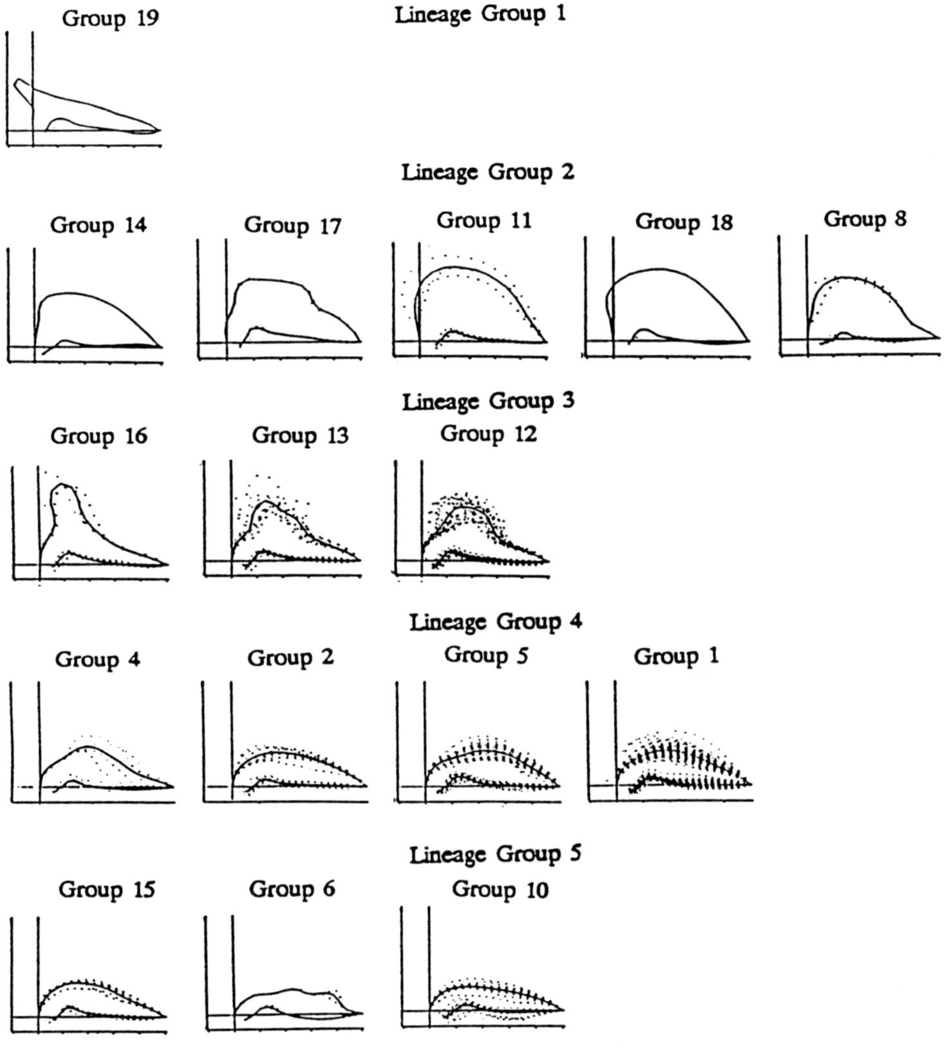

Figure 6. The average shape of the array of species in each group. The shape is designated by solid lines along contours 1 and 4. Dots represent the scatter at analogous spline nodes of species within each group. The five lineage groups were defined from dendrograms in Figs. 7 and 8, and are used to examine allometric relationships in Fig. 9.

variables of angles and distances between spline nodes (140 measurements) on scaled, normalized individuals would provide a possible transformation array for our 16 groups. The program produced and examined 414 trees; these had a sum of squares of 1.12 and an average standard deviation of 6.88%. After examining the lengths between nodes, we rooted the best tree as shown in Fig.

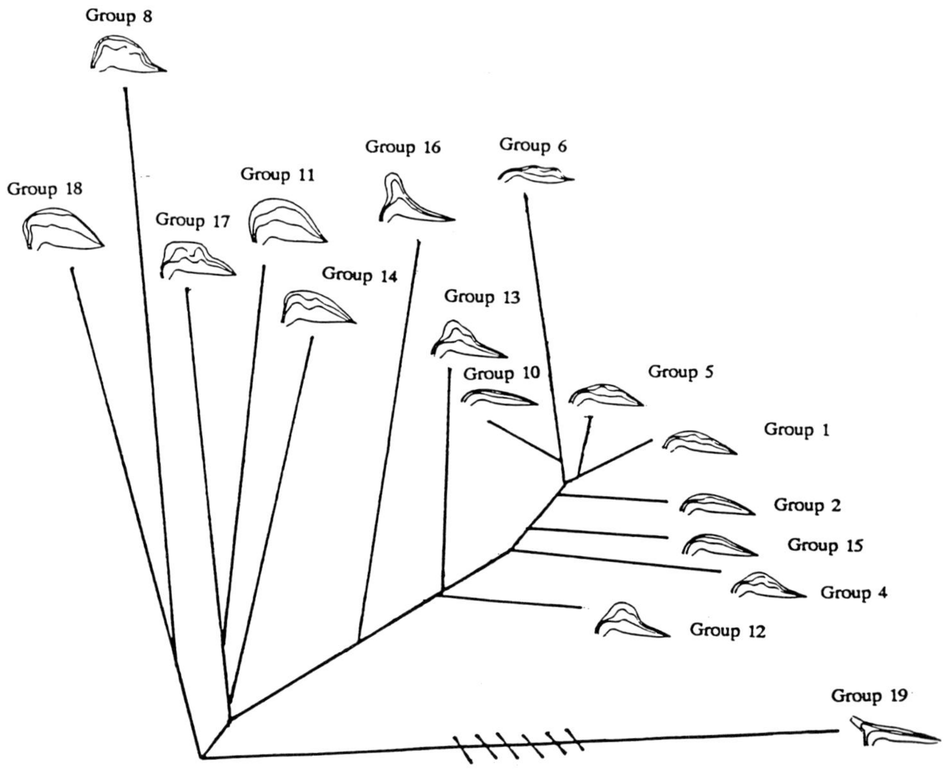

Figure 7. A Fitch-Margoliash tree generated from Mahalanobis' distances resulting from canonical discriminate analysis of 140 pronotal angles and distances, but *excluding* size variables.

7. In general, this network provided a credible transformation array for shapes, wherein the most distinctive pronotal shapes are located at the base of the tree. This general transformation array shows an initial divergence of forms with a frontal horn, from those with dorsally elevated pronota. The apex of the tree (Fig. 7) also shows that the transformation for the remainder of the groups is from high, foliaceous pronota to low, conical pronota.

A subsequent phylogenetic analysis included the six size measurements; PHYLIP produced 382 trees with a sum of squares of 1.21 and an average standard deviation of 7.13%. We rooted the best tree (Fig. 8) after examining the lengths between nodes. This size-included tree also provided, at least visually, a credible transformation series of pronotal shapes very similar to that on the size-excluded tree (Fig. 7). This tree separates groups with high foliaceous rounded pronota as an independent lineage. The remaining groups show a progressive transformation where the elevation of the pronotum is shifted posteriorly, from anteriodorsad of the head to behind it; this occurs as the elevation is decreased until the pronotum becomes conical (Fig. 8).

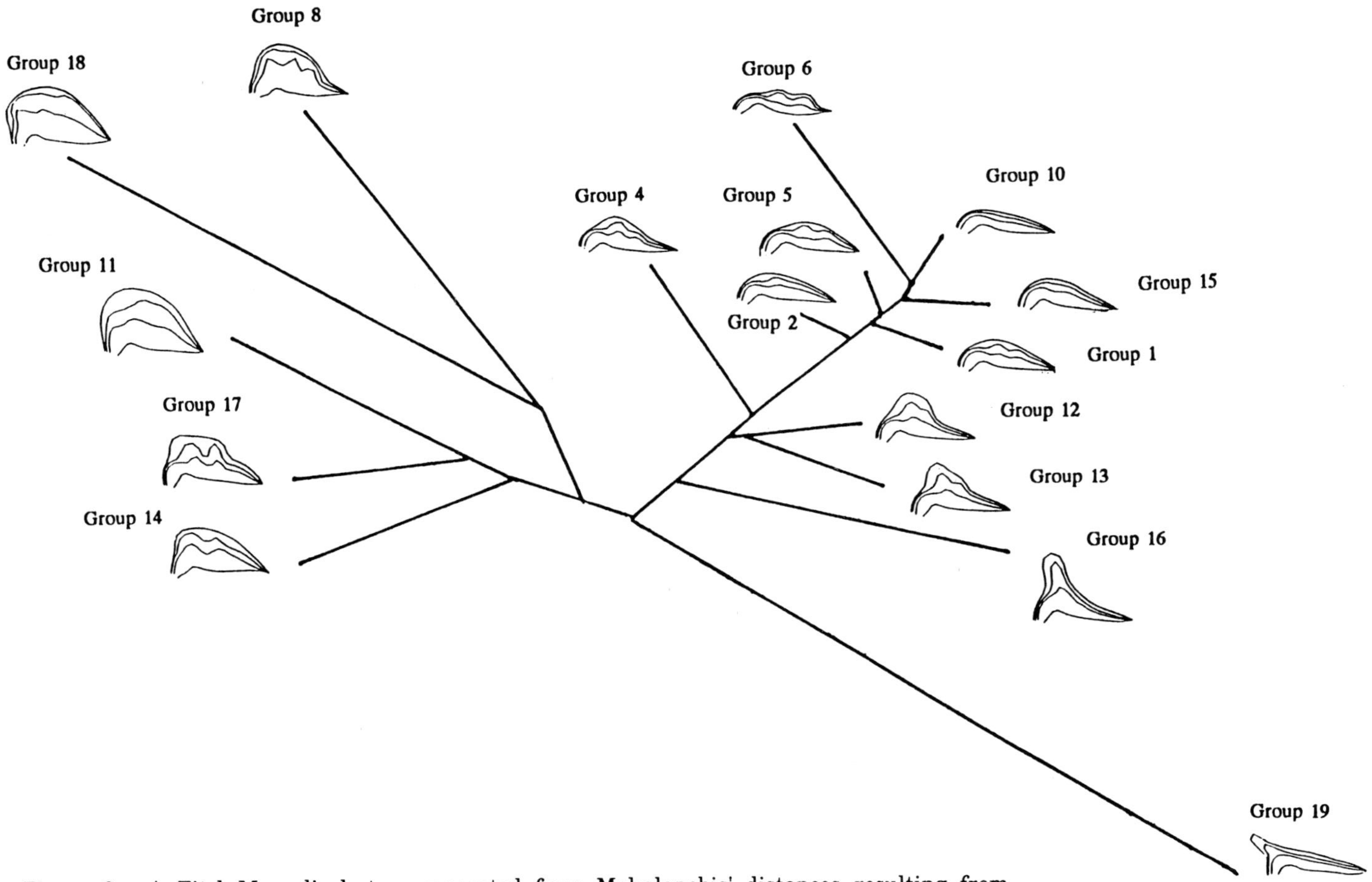

Figure 8. A Fitch-Margoliash tree generated from Mahalanobis' distances resulting from canonical discriminate analysis of 140 pronotal angles and distances, but *including* size variables.

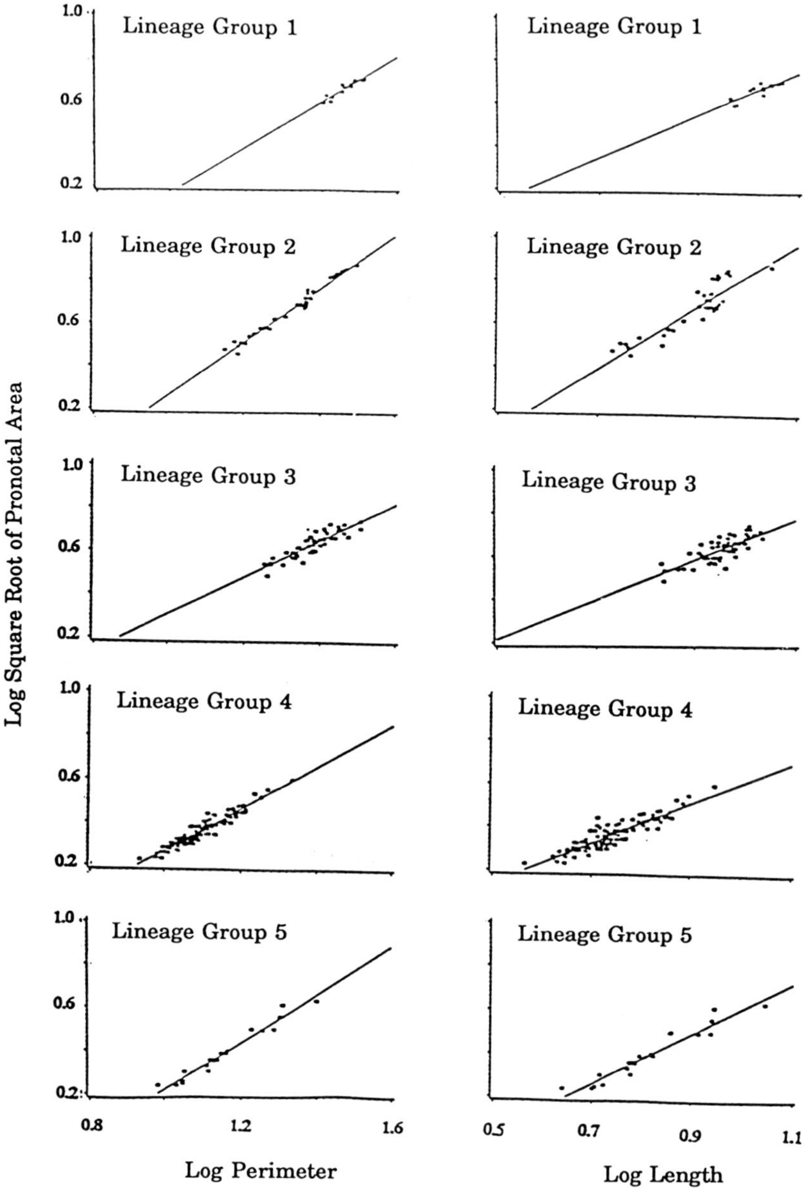

Figure 9. Regression of the log square root of pronotal area against the log perimeter (left column) and the log length (right column) for each of five lineage groups defined in Fig. 6. Vertical axes in all figures = Log Square Root of Pronotal Area; horizontal axes of left column figures = Log Perimeter; horizontal axes of right column figures = Log Length. Comparison of slopes and intercepts are in Tables 1 and 2.

10 Associated Correlates with a Size Related Phylogeny

Some authors have made a considerable effort to develop techniques that remove general-size from morphometric analyses to examine shape variation (Humphries et al., 1981). They imply that general-size is a confounding variable for systematic studies because of potential environmental effects. However, in many groups of organisms, major shifts in body size are associated with evolutionary divergence leading to radiation; therefore, size should not be ignored in phylogenetic studies.

Treehopper pronota show two components of size, vertical and horizontal. To demonstrate the within-group variation in the vertical axis, we calculated the average for contours 1 and 4 for each group (using scaled, normalized pronota) and then plotted the range of variation in that group (Fig. 6) along the mean contour. Although there is variation around the two contours within the groups, there is a consistent pattern between groups.

To determine the importance of allometric relations in defining groups, we took the original 16 groups and divided them into five lineage groups, based on their position on the tree in Fig. 8; these are defined in Fig. 6. Because lineage groups 1 and 2 contained few individuals, additional specimens were measured. The relationship between the square root of area and the perimeter incorporated not only a measure of size but also shape. We transformed (log base 10) each of these variables and performed a regression analysis for each of the five groups. Although there is some overlap among groups along the regression lines, which are presented in Fig. 9, three of the groups are characterized by high area/perimeter relationships, and two by low area/perimeter relationships; the regression values for the lines in Fig. 9 are given in Table 1. When the slopes of the lines were compared (Table 2), lineage group 1, although morphologically the most distinct, did not differ from any of the lineage groups. In contrast, the remaining lineage groups all differed from each other, indicating differences in allometry. On the tree that included size, this group diverged at the base as a separate lineage. The remainder of the groups on that tree have similar, but different, slopes; this suggests that they share common developmental pathways. An almost identical pattern emerged for pronotal area and length (Fig. 9) The only difference is that the slope for lineage group 3 did not differ from groups 4 and 5 (Table 3). This suggests that major changes in pronotal length are associated with gradual, but substantial, changes in shape.

There are two alternative interpretations of these allometric relationships. Strumpel (1972) interpreted similar regressions as an evolutionary tendency to increase the area of the pronotum in *Membracis*. There are, however, several arguments against this interpretation for the Smiliini. First, Fig. 7, which is based on scaled, normalized pronota and, therefore, contains no actual size related measurements, represents a progression from larger to smaller values for pronotal areas and lengths. Thus, the analysis objectively polarized these shapes toward decreasing pronotal area and length. Second, the number of points on the respective regression lines (Fig. 9) is a relative measure of extant species (with the exception of lineage groups 1 and 2); it seems clear that with a major reduction in pronotal area, there is a dramatic increase in the number of species. This is consistent with the concept that major reductions in body size may have permitted radiation within the tribe. Third, a reduction in pronotal area should be a developmentally and energetically more realistic

Table 1.
Values for regressions of lineage groups shown in Fig. 9.

Regression Values for Log $\sqrt{\text{Pronotal Area}}$ to *Log Perimeter*
(*Left column*, Fig. 9)

Lineage Group	log $\sqrt{\text{area}}$	*F*	R^2	*n*
1	-0.89 + 1.05 log (perimeter)	89.40	0.88	13
2	-0.98 + 1.24 log (perimeter)	1377.06	0.98	41
3	-0.53 + 0.84 log (perimeter)	188.98	0.77	57
4	-0.70 + 0.97 log (perimeter)	1069.47	0.92	100
5	-0.87 + 1.09 log (perimeter)	443.64	0.96	19

Regression Values for Log $\sqrt{\text{Pronotal Area}}$ to *Log Length*
(*Right column*, Fig. 9)

Lineage Group	log $\sqrt{\text{area}}$	*F*	R^2	*n*
1	-0.36 + 1.00 log (length)	42.34	0.78	13
2	-0.61 + 1.44 log (length)	160.37	0.80	41
3	-0.25 + 0.92 log (length)	87.38	0.61	57
4	-0.33 + 0.94 log (length)	463.00	0.82	100
5	-0.54 + 1.15 log (length)	263.12	0.94	19

evolutionary response for groups that feed upon nutritionally difficult host plants, such as *Quercus*. Smaller body sizes could facilitate the shorten developmental periods that are characteristic of these oak-inhabiting groups, by allocating energy to earlier reproduction.

The life history of oak-inhabiting groups is, in general, confined to the early spring flush of foliation; the majority of adults in such groups appear to die by midsummer, when oak leaves are fully tanned (T.K. Wood, personal observation) and become nutritionally poor (Feeny, 1970). With the exception of one species, all of the groups at the apex of the generated tree are restricted to *Quercus*. The overall selection pressure that is imposed by host plants appears to be for rapid adult maturation, and would favor reductions in body size. Thus, our interpretation of the polarity of the regressions is that evolution occurs with a decrease in pronotal area, so that the most derived species within each group are those with low area/perimeter relationships.

11 Discussion

The sensitivity of pronotal shape to environmental effects is perhaps the most important potential factor limiting its use to infer phylogenies. Although

Table 2.
Comparison of intercepts and slopes from regression of pronotal log square root area against log length (Fig. 9) for each of the five lineage groups defined in Fig. 6. Results of the analysis of covariance (F = 1342.9, P = 0.0001, R^2 = 0.98, n = 234), test for heterogeneity of intercepts (F = 8.7, P = 0.0001) and slope (F = 11.7, P = 0.0001) suggested differences among the five groups. Paired comparisons below suggest a pattern of allometry that is associated with the dendrogram in Fig's. 7 and 8.

NS = not significant; * = $P < 0.05$; ** = $P < 0.009$; *** = $P < 0.0009$.

		Comparison to Lineage Group				
Lineage Group	Comparison for	1	2	3	4	5
1	Intercept	--	NS	NS	NS	NS
	Slope	--	NS	NS	NS	NS
2	Intercept		--	***	***	NS
	Slope		--	***	***	**
3	Intercept			--	*	***
	Slope			--	*	*
4	Intercept				--	**
	Slope				--	*
5	Intercept					--
	Slope					--

few people would suggest that phenotypic shape does not have a genetic component, many would argue that it could be influenced by nutritional differences associated with different host plant species, or by other environmental variables, and that it should be, therefore, of dubious value in systematics. Wood and Datz (unpublished data) have completed a series of experiments designed to determine the sensitivity of pronotal shape to environmental influences. When photoperiod and temperature were held constant but the species of host plant was varied, six species of the *Enchenopa binotata* complex retained their characteristic pronotal shape, although inappropriate host species in some cases had an effect on body size. Thus, pronotal shape, as measured by the techniques presented here, appears to be phenotypically stable, and should serve as a indicator of the genotype.

We believe the phylogenetic tree that incorporates both shape and size, provides the best tentative transformation series for pronotal shape within the tribe. This tree reflects at least two evolutionary patterns. In the first, a shift occurs from highly elevated pronota at the basal portion of the tree, toward lower, rounded, conical pronota at the tree's terminal branches. In the second, pronotal area and length decrease with increased evolutionary distance from the tree's basal node. These general patterns of morphological evolution are reflected within each divergent group, suggesting that once the groups separated, selection favored similar trends within each lineage. The tree is consistent with the notion that the most derived groups are morphologically the most similar, suggesting convergence and less time for differentiation. These derived groups are extremely rich in species compared

Table 3.
Comparison of intercepts and slopes from regression of pronotal log square root area against log length (Fig. 9) for each of the five lineage groups defined in Fig. 6. Results of the analysis of covariance (F = 498.82, P = 0.0001, R^2 = 0.95, n = 234), test for heterogeneity of intercepts (F = 4.97, P = 0.0007) and slope (F = 7.98, P = 0.0001) suggested differences among the five groups. Paired comparisons below suggest a pattern of allometry that is associated with the dendrogram in Fig's. 7 and 8.

NS = not significant; * = P < 0.05; ** = P < 0.009; *** = P < 0.0009.

		Comparison to Lineage Group				
Lineage Group	Comparison for	1	2	3	4	5
1	Intercept	--	NS	NS	NS	NS
	Slope	--	NS	NS	NS	NS
2	Intercept		--	***	***	NS
	Slope		--	***	***	**
3	Intercept			--	NS	*
	Slope			--	NS	NS
4	Intercept				--	**
	Slope				--	*
5	Intercept					--
	Slope					--

to the more primitive (basal) groups which are either monotypic, or contain less than six species (Table 4). This suggests a recent radiation among derived, morphologically similar groups, and either extinction or developmental constraints to divergence among plesiomorphic (basal) groups that are more morphologically distinct.

Congruent with the above trends are patterns of host plant utilization (Table 4). Although it is impossible to identify the ancestral host plant for this tribe, the inferred phylogeny suggests a credible candidate. *Thelia*, which is the first group to diverge at the base of the tree, is unique because of the presence of a frontal horn and because one species (*Thelia bimaculata*) is the only member of the tribe that is restricted to a leguminous host plant (*Robinia pseudoacacia*) (Funkhouser, 1915). A major portion of the African (Capener, 1962), Indian (Ananthasubramanian and Ananthakrishnan, 1975), Central American (Wood, 1984) and South American (Richter, 1942) treehopper faunas are either restricted to, or utilize, host plants in the Fabaceae.

Old World membracids are restricted to the Centrotinae and related subfamilies, which are considered to be the most plesiomorphic of the Membracidae (Strumpel, 1972; Deitz, 1975). The majority of the African fauna appears to be restricted to legumes (Capener, 1962). With only a few minor exceptions, in one subfamily, there is no overlap between the Old and New World treehopper faunas (Strumpel, 1972; Deitz, 1975). This suggests that association with leguminous host species may have preceded the separation of the continents.

Table 4.
Lineage group composition: host plants and distributions. Grp = assigned lineage group, # Sp = number of described species in genus, Q = *Quercus* as host plant, O = other non-*Quercus* host plant, U = unknown host plant, NW = Pacific Northwest, CA = California, RM = Rockie Mountains, SW = Arizona and New Mexico, MX = Mexico, CSA = Central and South America, E = eastern North America.

			Host Plants			Geographic Region						
Grp	Genus	# Sp	Q	O	U	NW	CA	RM	SW	MX	CSA	E
1	*Thelia*	2	0	2	0	0	0	0	0	0	0	2
2	*Hemicardiacus*	1	1	0	0	0	0	0	0	1	0	0
2	*Archasia*	3	3	0	0	0	0	1	1	0	0	3
2	*Telamonanthe*	2	2	0	0	0	1	2	2	2	0	0
2	*Antianthe*	6	0	2	4	0	1	0	1	4	6	0
2	*Smilia*	3	2	0	1	0	0	0	0	1	0	2
3	*Telonaca*	2	2	0	0	0	1	0	0	0	0	1
3	*Glossonotus*	5	4	1	0	1	0	1	0	0	0	5
3	*Helonica*	1	1	0	0	0	0	0	0	1	0	1
3	*Heliria*	12	5	5	2	1	0	6	2	3	0	10
3	*Palonica*	6	2	3	1	0	0	3	1	1	1	3
3	*Telmona*	29	15	6	8	2	3	6	2	3	4	22
4	*Grandolobus*	5	5	0	0	0	0	0	1	3	1	0
4	*Atymna*	12	9	1	2	0	0	1	2	2	2	6
4	*Ashmeadea*	1	1	0	0	0	0	0	1	1	0	0
4	*Xantholobus*	12	12	0	0	0	0	3	6	0	0	6
5	*Godingia*	1	1	0	0	0	0	0	0	1	0	0
5	*Tropidarnis*	1	1	0	0	0	0	0	1	1	0	0
5	*Carynota*	4	1	3	0	0	0	0	0	0	0	4
5	*Ophiderma*	16	16	0	0	0	0	1	4	0	2	9

If we assume that a legume was the ancestral host of the Smiliini, the tree suggests that the colonization of *Quercus* led to extensive radiation. If the basal node (Fig. 8) represents the colonization of *Quercus*, lineage group 2, with high foliaceous pronota, diverged and one member (*Antianthe*) moved to *Solanum* (Wood, 1984). *Antianthe* is unique in the tribe not only in the host plants that it uses but also in its presocial, multivoltine life history.

The progression in the main lineage is toward restriction to *Quercus*. The first groups to diverge are primarily on *Quercus*, but a number of species are restricted to other deciduous host trees in several different families (Table 4). Because taxonomic revisions have not been done for these groups, we cannot determine whether the utilization of oak is a derived condition for them. Many of the species, which occur on other genera of hosts, in these groups tend to have lower, rounded pronota which, according to our tree, is a derived condition. If the species using hosts other than oaks are derived, it suggests that the primitive positions of these groups reflects a sufficient time interval to permit the colonization of new host plant genera.

Of the remaining seven terminal groups, six are restricted to *Quercus* (one *Atymna* species is found on *Castanae dentata* [Fagaceae]). The single exceptional group (14) has a species on *Quercus* and the remaining three are on other deciduous trees. Its position on the tree suggests that it is the oldest of the three terminal groups, which is consistent with longer time intervals for the colonization of new host genera.

The last point in support of the hypothesized phylogeny relates to the broad geographic distributions of members of this tribe and of *Quercus*. *Quercus* has a worldwide distribution, but with the exception of several records in Pakistan (I. Ahmad, personal communication), it has not apparently accumulated the extensive treehopper fauna that is characteristic of North America. During the Tertiary, the North American *Quercus* flora became regionally differentiated and isolated from that of Asia (Axelrod, 1983). Major centers of oak diversity were established in the central highlands of Mexico, California, Arizona and the eastern United States. The greatest diversity of species is in central Mexico, followed sequentially by the eastern U.S., California, and Arizona (Trelease, 1924).

Treehoppers in the Smiliini are also regionally differentiated (Table 4): of the 164 described species (Metcalf and Wade, 1965), 104 are found in the eastern U.S., 32 are in Mexico, 31 are in Arizona, 27 are in the Rocky Mountains, 16 are in Central and South America, and 7 are in California. At first glance, this suggests that many regions may be under sampled. This does not appear to be the case, however, because TKW has collected extensively in each region during the last five years and examined many insect collections, finding little evidence to indicate a large number of species remain undiscovered in any of these regions.

The similarities between the distributions of the treehoppers and *Quercus* suggest that both have been biogeographically affected by the same geological events. Outlined below is our speculation of a possible historical reconstruction for the Smiliini.

(1) Because no other members of the subfamily Smiliinae occur in the Old World (Asia or Africa), differentiation of the tribe had to have been within the last 65 million years, after the separation of South America.

(2) Possible relict species in some genera in South America (Table 4) support this time of divergence, and suggest that the tribe moved into North America within the last five million years, when these areas in the Americas became connected.

(3) The pervasive use of leguminose host plants by primitive subfamilies in all major zoogeographic regions suggests that members of this plant family served as the ancestral hosts in North America.

(4) Treehoppers, such as *Thelia bimaculata* (restricted to the eastern U.S., leguminose hosts and the most morphologically differentiated), appear to be relicts of the early treehopper fauna.

(5) The lack of an extensive treehopper fauna on *Quercus*, an extremely old genus (Axelrod, 1983), in any other region of the world suggests that historically, it is a recently colonized host.

(6) The pan-United States distribution of a few species in primitive (basal) groups of the tribe (those with representatives in California and the Pacific Northwest) suggests their broad distribution prior to the last glaciation.

(7) The colonization of *Quercus* may have occurred prior to the last glaciation, because all of the California Smiliini, except *Antianthe* (which may

have been introduced, [R.J. Gill, personal communication]), occur on this host genus.

(8) Because *Cyrtolobus* has a single representative in California, all genera showing divergence nodes on the phylogeny below *Cyrtolobus* must have also diverged by the last glaciation.

(9) The effect of the last glaciation was to restrict the eastern deciduous forest (Webb, 1981; Davis, 1983), and the treehoppers associated with these hosts, to the southern portions of the U.S.

(10) These glaciers appear to have created disjunctions in treehopper and *Quercus* distributions across the U.S.

(11) Such geographic isolation during glaciation may have promoted much of the species richness found in the more derived (terminal) taxa in this tribe.

(12) With the retreat of the ice shield, beginning 18,000 years ago, oaks dispersed throughout the Eastern U.S. in a roughly northeast direction (Davis, 1983; Webb, 1981).

(13) The treehoppers followed the northeasterly dispersal of the oaks, which reached their northern limit within the last 6,000 years.

(14) With the gradual aridification of mid-Texas and northern Mexico during the last 5,000 years, complete regional separation between these faunas was achieved.

12 Future Considerations

The technique, as presented here, is being used to understand geographic variation, sexual dimorphism, and the underlying genetics of pronotal shape, not only in the Smiliini but in other groups. When inventories of tropical forests are made, collecting and maintenance of specimens will be critical limitations, but without faster techniques for the identification and analysis of data, it will be several generations before we know what has been found and what it means. The technique presented here has promise, at least for treehoppers, for coping with the logistical problems that will face systematists.

We are beginning to explore the following possible avenue for handling specimens from massive faunal surveys. First, establishing calibration data sets for each level of classification within the family. Second, as described in this chapter, digitizing specimens and associating that data with relevant collection information. Third, the input of this digitized data directly into a series of programs that allow identification within seconds. Fourth, to set the probabilities of a correct match (identification) to a member taxon of the training data set high enough, so that potentially undescribed species can be immediately recognized as such and brought to a specialist's attention. Once new species are identified, their associated digitized information can be quickly added to the calibration data to reduce the interval between the recognition of a new taxa (description, publication) and its use in further identifications. This approach will provide an almost instant access to large data sets, not only for systematists, but also ecologists and biogeographers. Only through an approach to systematics that is similar to this procedure will it be possible to know what is there (or was), but more importantly to understand the relationships between taxa, trophic levels and geographic regions.

13 Acknowledgements

We thank K. Kruluts and M. Hardin for their help in the early portions of this work, G. Pratt for commenting on various drafts of the manuscript, and T. Green for general assistance in the preparation of figures. This work was supported by National Science Foundation Grant BSR-8400245. It is published as a miscellaneous paper of the Delaware Agricultural Experiment Station; contribution number 604 of the Department of Entomology and Applied Ecology, University of Delaware, Newark; and contribution number 128 of the Ecology Program, School of Life and Health Sciences, University of Delaware, Newark.

14 References

Ananthasubramanian, K.S. and Ananthakrishnan, T.N., 1975. Taxonomic, biological and ecological studies on some Indian membracids. Part II. *Records of the Zoological Survey of India*, 68: 305-340.

Anstey, R.L. and Delmet, D.A., 1973. Fourier analysis of zooecial shapes in fossil tubular bryozoans. *Bulletin of the Geological Society of America*, 84: 1753-1764.

Atchley, W.R. and Rutledge, J.J., 1980. Genetic components of size and shape. I. Dynamics of components of phenotype variability and covariability during ontogeny of the Laboratory Rat. *Evolution*, 34: 1161-1173.

Axelrod, D.I., 1983. Biogeography of oaks in the arcto-Tertiary province. *Annals of the Missouri Botanical Garden*, 70: 629-657.

Ball, E.D., 1932. A monographic revision of the treehoppers of the tribe Telamonini of North America. *Entomologica Americana*, 12: 1-69.

Bookstein, F.L., Chernoff, B., Elder, R.L., Humphries, J.M., Smith, G.R. and Strauss, R.E., 1985. Morphometrics in Evolutionary Biology. Special Publication, 15. Academy of Natural Sciences, Philadelphia, 277 pp.

Bookstein, F.L., Strauss, R.E., Chernoff, B., Elder, R.L. and Smith, G.R., 1982. A comment upon the uses of Fourier methods in systematics. *Systematic Zoology*, 31: 85-92.

Caldwell, J.S., 1949. A generic revision of the treehoppers of the tribe Ceresini in America, north of Mexico, based on a study of the male genitalia. *Proceedings of the United States National Museum*, 98: 491-521.

Capener, A L., 1962. The taxonomy of the African Membracidae. Part 1. The Oxyrhachinae. *Republic of South Africa, Department of Agriculture Technical Services, Entomology Memoirs, 6*, 164 pp.

Davis, M.B., 1983. Quaternary history of deciduous forests of eastern North America and Europe. *Annals of the Missouri Botanical Garden*, 70: 550-563.

Deitz, L.L., 1975. Classification of the higher categories of the New World treehoppers (Homoptera: Membracidae). *North Carolina State University, Agriculture Experiment Station, Technical Bulletin, 225*, 177 pp.

Dennis, C.J., 1952. The Membracidae of Wisconsin. *Transactions of the Wisconsin Academy of Science and Letters*, 41: 129-152.

Dennis, C.J., 1964. Membracidae of southeastern Oklahoma. *Proceedings of the Oklahoma Academy of Science*, 44: 47-51.

Dennis, C.J., 1965. Oklahoma treehoppers (Homoptera: Membracidae). *Proceedings of the Oklahoma Academy of Science*, 45: 50-64.

Ehrlich, R., Pharr, R.B. Jr. and Healy-Williams, N., 1983. Comments on the validity of Fourier descriptors in systematics: a reply to Bookstein et al. *Systematic Zoology*, 32: 202-206.

Feeny, P.P., 1970. Seasonal changes in oak leaf tannins and nutrients as a cause of spring feeding by winter moth caterpillars. *Ecology*, 51: 565-581.

Felsenstein, J., 1984. PHYLIP, Phylogeny Inference Package. J. Felsenstein, Department of Genetics, University of Washington, Seattle.

Felsenstein, J., 1988. Phylogenies and quantitative characters. *Annual Review of Ecology and Systematics*, 19: 445-471.

Ferson, S., Rohlf, F.J. and Koehn, R.K., 1985. Measuring shape variation of two-dimensional outlines. *Systematic Zoology*, 34: 59-68.

Funkhouser, W.D., 1915. Life history of *Thelia bimaculata* Fabr. (Membracidae). *Annals of the Entomological Society of America*, 8: 140-151.

Funkhouser, W.D., 1951. Homoptera Fam. Membracidae. *Genera Insectorum*, 208: 1-383.

Futuyma, D.J., 1986. Evolutionary Biology. Sinauer Associates Inc., Sunderland, Massachusetts, 600 pp.

Goding, F.W., 1926. Classification of the Membracidae of America. *Journal of the New York Entomological Society*, 34: 295-317.

Goding, F.W., 1928a. The Membracidae of South America and the Antilles. II. Subfamily Centrotinae. *Journal of the New York Entomological Society*, 35: 391-408.

Goding, F.W., 1928b. Membracidae of South America and Antilles. III. Subfamily Membracinae. *Journal of the New York Entomological Society*, 36: 201-234.

Goding, F.W., 1929. The Membracidae of South America and the Antilles. IV. Subfamilies Hoplophorioninae, Darninae, Smiliinae, Tragopinae (Homoptera). *Transactions of the American Entomological Society*, 55: 197-330.

Hennig, W., 1966. Phylogenetic Systematics. University of Illinois Press, Chicago, 262 pp.

Humphries, J.M., Bookstein, F.L., Chernoff, B., Smith, G.R., Elder, R.L. and Poss, S.G., 1981. Multivariate discrimination by shape in relation to size. *Systematic Zoology*, 30: 291-308.

Kopp, D.D. and Yonke, T.R., 1973a. The treehoppers of Missouri: part 1. Subfamilies Centrotinae, Hoplophorioninae, and Membracinae (Homoptera: Membracidae). *Journal of the Kansas Entomological Society*, 46: 42-64.

Kopp, D.D. and Yonke, T.R., 1973b. The treehoppers of Missouri: part 2. Subfamily Smiliinae; Tribes Acutalini, Ceresini, and Polyglyptini (Homoptera: Membracidae). *Journal of the Kansas Entomological Society*, 46: 233-276.

Kopp, D.D. and Yonke, T.R., 1973c. The treehoppers of Missouri: part 3. Subfamily Smiliinae; Tribe Smiliini. *Journal of the Kansas Entomological Society*, 46: 375-421.

Kopp, D.D. and Yonke, T.R., 1974. The treehoppers of Missouri: part 4. Subfamily Smiliinae; Tribe Telamonini (Homoptera: Membracidae). *Journal of the Kansas Entomological Society*, 47: 80-130.

Kopp, D.D. and Yonke, T.R., 1979. A taxonomic review of the tribe Ceresini. *Miscellaneous Publications of the Entomological Society of America*, 11 (2): 1-98.

Lohmann, G.P., 1983. Eigenshape analysis of microfossils: a general morphometric procedure for describing changes in shape. *Mathematical Geology*, 15: 659-672.

Mann, W.M., 1912. A protective adaptation in a Brazilian membracid. *Psyche*, 19: 145-147.

Metcalf, Z.P. and Wade, V., 1965. General Catalogue of the Homoptera. Membracoidea. Sec. I and II. North Carolina State University, Raleigh, 1552 pp.

Poulton, E.B., 1903. Suggestions as to the meaning of the shapes and colours of the Membracidae, in the struggle for existence. In: G.B. Buckton. Monograph of the Membracidae. Lovell Reeve Company, London, pp. 273-285.

Raup, D.W., 1962. Computer as aid in describing form in gastropod shells. *Science*, 138: 150-152.

Raup, D.W, 1966. Geometric analysis of shell coiling: general problems. *Journal of Paleontology*, 40: 1178-1190.

Ray, T.S., (in press). Landmark eigenshape analysis: homologous contours. *American Journal of Botany*.

Richter, L., 1942. Catalogo de los membracidos de Colombia. *Revista de la Academia Colombiana de Ciencias Exactas, Fisicas y Naturales, Correspondiente de la Espanola*, 5: 405-409.

SAS Institute Inc., 1985. SAS User's Guide: Statistics, (Version 5). SAS Institute Inc., Cary, North Carolina, 429 pp.

Sorensen, J.T., 1987. Multivariate statistical approach to deduction of phylogeny within *Essigiella* [sic] (Aphididae: Lachninae). In: J. Holman, J. Pelikan, A.G.F. Dixon and L. Weismann (Editors), Population Structure, Genetics and Taxonomy of Aphids and Thysanoptera, Proceedings of International Symposia, held at Smolenice, Czechoslovakia, September 9-14, 1985. SPB Academic Publishing, The Hague, pp. 243-260.

Sorensen, J.T., (this volume). The use of discriminant function analysis for estimation of phylogeny: partitioning, perspective and problems.

Sorensen, J.T. and Foottit, R.G., (this volume). The evolutionary quantitative genetic rationales for use of ordination analyses in systematics: phylogenetic implications.

Stoermer, E.F. and Ladewski, T.B., 1982. Quantitative analysis of shape variation in type and modern populations of *Gomphoneis herculena*. *Beihefte zur Nova Hedwigia*, 73: 347-386.

Strumpel, H., 1972. Beitrag zur phylogenie der Membracidae Rafinesque. *Zoologische Jahrbuecher Abteilung fuer Systematik, Oekologie und Geographie der Tiere*, 99: 323-407.

Strumpel, H., 1983. Beobachtungen zur autotomie des pronotums bei Anchistrotus-Arten (Homoptera, Membracidae). *Entomologische Mitteilungen Aus Dem Zoologischen Museum Hamburg*, 7 (119): 391-396.

Thompson, D.W., 1961. On Growth and Form, (Abridged Edition: J.T. Bonner, Editor). Cambridge University Press, Cambridge, U.K., 346pp.

Trelease, W., 1924. The American oaks. *Memoirs of the United States National Academy of Science*, 20: 1-255.

Van Duzee, E.P., 1908. Studies in North American Membracidae. *Bulletin of the Buffalo Society of Natural Sciences*, 9: 29-127.

Webb, T. III, 1981. The past 11,000 years of vegetational change in eastern North America. *Bioscience*, 31: 501-506.

Wiley, E.O., 1981. Phylogenetics: The Theory and Practice of Phylogenetic Systematics. John Wiley and Sons, New York, 439 pp.

Wood, T.K., 1974. Aggregating behavior of *Umbonia crassicornis* (Homoptera: Membracidae). *Canadian Entomologist*, 106: 169-173.

Wood, T.K., 1975a. Defense in presocial membracids (Homoptera: Membracidae). *Canadian Entomologist*, 107: 1227-31.

Wood, T.K., 1975b. Studies on the function of the membracid pronotum (Homoptera). II. Histology. *Proceedings of the Entomological Society of Washington*, 77: 78-82.

Wood, T.K., 1976. Biology and presocial behavior of *Platycotis vittata* F. (Homoptera: Membracidae). *Annals of the Entomological Society of America*, 69: 807-11.

Wood, T.K., 1977. Defense in *Umbonia crassicornis*: the role of the pronotum and adult aggregations. *Annals of the Entomological Society of America*, 70: 524-528.

Wood, T.K., 1984. Life history patterns of tropical membracids (Homoptera: Membracidae). *Sociobiology*, 8: 299-344.

Wood, T.K. and Dowell, R., 1985. Reproductive behavior and dispersal in *Umbonia crassicornis*. *Florida Entomologist*, 68: 151-158.

Wood, T.K., Guttman, S.I. and Taylor, M., 1984. Mating behavior of *Platycotis vittata*. *American Midland Naturalist*, 112: 305-313.

Wood, T.K. and Morris, G.K., 1974. Studies on the function of the membracid pronotum (Homoptera). 1. Occurrence and distribution of articulated hairs. *Canadian Entomologist*, 106: 143-148.

Woodruff, L.B., 1919. A review of our local species of the Membracidae genus *Ophiderma* Fairm. (Hemip. - Homop.). *Journal of the New York Entomological Society*, 27: 249-260.

Woodruff, L.B., 1920. Further notes on the membracid genus *Ophiderma* Fairm. (Hemip. - Homop.). *Journal of the New York Entomological Society*, 28: 212-214.

Woodruff, L.B., 1924. Critical observations in the membracid genus *Cyrtolobus* Goding (Hemip. - Homop.). *Journal of the New York Entomological Society*, 32: 1-62.

Younker, J.L. and Ehrlich, R., 1977. Fourier biometrics: harmonic amplitudes as multivariate shape descriptors. *Systematic Zoology*, 26: 336-342.

15 Appendices

Appendix 1.
List of 168 species in the tribe Smiliini (as defined by Deitz, 1975) and their assignment to the 16 groups referred to in this paper. Generic and specific names are those used in the Metcalf and Wade catalogue (1965). Specific names are tentative, because synonymies have not been worked out. Unlisted species names are specimens collected by TKW that cannot, with some degree of assurance, be matched with an existing species description or are known to be new species. Code numbers are those that we are presently using to designate morphospecies, while one of us (TKW) revises the genera. Area = the square root of pronotal area (mm), length = the distance at the mid line of the pronotum between the first point above the head to the tip of the posterior process (mm), contour 1 = is the length of the mid line outline (mm).

Group	Code	Genus	Species	Area	Length	Contour 1
1	1	*Cyrtolobus*	*vau*	2.28	5.33	6.33
1	2	*Cyrtolobus*	*puritanus*	1.94	4.66	5.48
1	4	*Cyrtolobus*	*parvulus*	2.03	4.91	5.77
1	5	*Cyrtolobus*	*maculifrontis*	2.10	5.12	6.01
1	6	*Cyrtolobus*	*gratiosus*	2.01	4.89	5.80
1	8	*Cyrtolobus*	*pictus*	2.04	5.13	5.82
1	10	*Cyrtolobus*		2.33	5.55	6.70
1	11	*Cyrtolobus*	*frigidus*	1.79	4.22	5.16
1	12	*Cyrtolobus*	*togatus*	1.82	4.44	5.27
1	14	*Aeshmeadea*	*carinata*	2.70	6.51	7.67
1	15	*Cyrtolobus*	*vanduzeei*	2.60	5.71	7.06
1	16	*Cyrtolobus*	*woodruffi*	2.56	5.40	7.18
1	17	*Cyrtolobus*		2.06	5.28	6.08
1	18	*Cyrtolobus*		2.97	7.23	8.46
1	20	*Cyrtolobus*		2.22	5.78	6.70
1	21	*Cyrtolobus*		1.97	5.25	5.88
1	27	*Cyrtolobus*		2.39	5.37	6.38
1	28	*Cyrtolobus*		2.20	5.65	6.48
1	29	*Cyrtolobus*		2.02	4.60	5.54
1	32	*Cyrtolobus*	*limus*	2.19	4.86	5.88
1	34	*Cyrtolobus*		2.06	5.12	6.03
1	45	*Cyrtolobus*	*inermis*	2.16	5.22	6.19
1	46	*Cyrtolobus*	*pallidifrontis*	2.08	5.40	6.30
1	47	*Cyrtolobus*	*fuliginosus*	2.72	6.08	7.90
1	48	*Cyrtolobus*	*fuscipennis*	2.21	5.40	6.39
1	49	*Cyrtolobus*	*griseus*	2.40	5.68	6.69
1	50	*Cyrtolobus*	*gramatanus*	2.05	4.92	5.77
1	52	*Cyrtolobus*	*rufulus*	2.29	5.69	6.52
1	53	*Cyrtolobus*		2.00	4.68	5.61
1	56	*Cyrtolobus*	*fenestratus*	2.52	6.07	7.57
1	57	*Cyrtolobus*	*clarus*	2.65	6.25	8.26
1	58	*Cyrtolobus*	*discoidalis*	2.40	5.80	7.07
1	64	*Cyrtolobus*		2.41	5.33	6.64
1	70	*Cyrtolobus*		1.93	4.53	5.37
1	72	*Cyrtolobus*		2.67	6.92	8.08
1	73	*Cyrtolobus*		2.33	5.59	6.99
1	74	*Cyrtolobus*		2.33	5.44	6.64
1	76	*Cyrtolobus*		2.11	5.49	6.44
1	79	*Cyrtolobus*		1.86	4.94	5.63

1	81	*Cyrtolobus*	*funkhouseri*	2.43	5.59	7.07
1	82	*Cyrtolobus*	*flavolatus*	2.12	5.31	6.30
1	83	*Cyrtolobus*	*ovatus*	2.76	5.16	7.36
1	86	*Cyrtolobus*	*arcuatus*	2.92	6.39	8.28
1	87	*Cyrtolobus*	*accuminatus*	2.81	7.08	8.74
1	88	*Cyrtolobus*	*auroreus*	2.45	5.81	7.07
1	90	*Cyrtolobus*		2.39	5.91	7.11
1	92	*Cyrtolobus*		2.78	6.78	8.28
1	93	*Cyrtolobus*		2.56	5.64	6.97
1	94	*Cyrtolobus*		2.80	7.24	8.39
1	96	*Cyrtolobus*		3.27	7.68	9.81
1	97	*Cyrtolobus*		3.02	6.91	8.65
1	98	*Cyrtolobus*		1.73	4.28	4.90
1	99	*Cyrtolobus*		1.81	4.63	5.37
1	100	*Cyrtolobus*		2.70	5.78	7.37
1	129	*Cyrtolobus*		1.92	4.32	5.32
1	130	*Cyrtolobus*		2.22	5.89	6.76
1	132	*Cyrtolobus*	*celsus*	2.39	4.99	6.38
1	134	*Cyrtolobus*		2.48	5.20	7.16
1	135	*Cyrtolobus*		2.10	4.69	5.79
1	139	*Cyrtolobus*		1.99	5.39	6.27
1	140	*Cyrtolobus*		2.19	5.50	6.58
2	9	*Atymna*		2.79	7.32	8.50
2	42	*Atymna*	*helena*	2.45	6.10	7.12
2	55	*Atymna*	*querci*	1.96	4.73	5.46
2	60	*Atymna*	*simplex*	2.87	7.13	8.79
2	65	*Atyminia*	*elongata*	2.50	6.84	7.78
2	67	*Atymna*		2.11	4.88	5.98
2	80	*Atymna*	*castanae*	2.87	6.28	7.94
2	89	*Cyrtolobus*	*dixianus*	2.67	6.42	7.83
2	156	*Cyrtolobus*	*maxeini*	2.39	5.71	6.81
2	166	*Atymna*		2.26	5.15	6.63
4	13	*Grandolobus*	*grandis*	2.99	6.81	8.25
4	22	*Grandolobus*		2.75	6.74	8.16
4	23	*Grandolobus*		2.21	5.86	7.34
4	59	*Cyrtolobus*	*tuberosus*	3.39	7.47	9.42
4	71	*Cyrtolobus*	*cristiferus*	3.96	8.89	12.16
5	41	*Xantholobus*	*nigrocinctus*	2.48	6.50	7.48
5	44	*Xantholobus*	*muticus*	2.80	6.65	8.11
5	54	*Xantholobus*		3.55	7.89	10.24
5	68	*Xantholobus*	*lateralis*	2.57	6.39	7.69
5	102	*Xantholobus*		2.12	4.64	5.95
5	103	*Xantholobus*		1.98	5.14	5.90
5	104	*Xantholobus*		2.75	6.84	8.06
5	105	*Xantholobus*		2.79	6.86	8.14
5	106	*Xantholobus*		1.96	4.48	5.64
5	107	*Xantholobus*		2.14	4.80	5.99
5	108	*Xantholobus*		1.83	4.20	5.25
5	110	*Xantholobus*		1.89	4.49	5.51
5	111	*Xantholobus*		2.00	4.95	5.68
5	114	*Xantholobus*		1.69	3.68	4.62
5	124	*Xantholobus*		2.18	5.34	6.46
5	125	*Xantholobus*		2.89	6.69	8.11
5	152	*Xantholobus*		1.70	4.43	5.13

5	153	*Xantholobus*		2.08	5.60	6.33
5	154	*Xantholobus*		1.97	4.49	5.64
5	155	*Xantholobus*		1.93	4.89	5.70
5	161	*Xantholobus*		2.40	5.10	6.53
5	162	*Xantholobus*		2.41	4.99	6.39
5	163	*Xantholobus*		2.53	5.14	6.99
5	165	*Xantholobus*		2.08	5.27	6.08
6	91	*Godingia*		2.29	6.04	7.23
6	128	*Godingia*	*guerreroensis*	3.14	8.22	9.35
8	84	*Smilia*	*camelus*	4.79	8.39	13.27
8	85	*Smilia*	*fasciata*	4.13	7.51	11.21
10	159	*Ophiderma*	*flavicephala*	2.28	5.95	6.81
10	167	*Ophiderma*	*pallida*	2.00	5.20	5.76
10	168	*Ophiderma*	*definita*	1.73	5.01	5.51
10	169	*Ophiderma*	*evelyna*	2.27	6.12	6.97
10	170	*Ophiderma*	*pubescens*	2.28	6.12	6.88
10	173	*Ophiderma*		2.02	6.00	6.65
10	175	*Ophiderma*	*flava*	2.47	6.67	7.50
10	177	*Ophiderma*	*salamandra*	2.52	6.60	7.54
10	178	*Ophiderma*	*grisea*	2.14	5.92	6.59
10	179	*Ophiderma*		1.76	5.07	5.78
10	180	*Ophiderma*		1.80	5.28	5.59
10	181	*Ophiderma*		1.73	4.37	5.04
10	182	*Ophiderma*		2.45	6.25	7.43
11	212	*Archasia*	*belfragei*	5.15	8.46	13.97
11	213	*Archasia*	*auriculata*	7.00	9.19	19.20
11	214	*Archasia*	*pallida*	5.43	8.22	14.30
12	158	*Telamona*	*monticola*	4.44	9.25	14.19
12	183	*Palonica*	*pyramidata*	4.39	9.12	14.54
12	184	*Palonica*	*portola*	4.48	9.13	15.32
12	189	*Telamona*	*maculata*	5.03	9.35	15.32
12	190	*Telamona*	*compacta*	3.44	7.35	10.08
12	191	*Telamona*	*dubium*	3.68	8.75	11.78
12	192	*Telamona*	*collina*	4.33	9.48	14.20
12	193	*Telamona*	*tristis*	3.66	6.80	11.51
12	194	*Telamona*	*concava*	4.62	8.61	14.19
12	196	*Telamona*	*reclivata*	3.88	8.32	11.58
12	197	*Telamona*	*salvini*	3.48	7.55	10.31
12	198	*Telamona*	*spreta*	5.15	10.71	16.49
12	199	*Telamona*	*ampelopsis*	4.77	10.01	15.42
12	200	*Telamona*	*extrema*	4.89	8.93	14.56
12	201	*Telamona*	*gibbera*	3.74	8.56	12.60
12	202	*Telamona*	*barbara*	4.23	9.09	14.18
12	203	*Telamona*	*vestita*	4.42	8.88	13.24
12	204	*Telamona*	*decorada*	3.89	8.46	12.32
12	205	*Telamona*	*unicolor*	5.27	10.24	15.62
12	206	*Telamona*	*lugubrus*	3.80	8.93	12.12
12	207	*Telamona*		4.80	10.81	14.52
12	218	*Telamona*		3.43	7.90	10.06
12	230	*Telamona*		4.36	10.13	13.33
12	239	*Telamona*		3.04	6.87	10.73
13	185	*Heliria*		4.19	9.30	13.59
13	209	*Telonaca*	*ramona*	4.59	8.90	14.11
13	215	*Helonaca*	*excelsa*	5.27	10.10	18.21

13	224	*Heliria*	*scalaris*	4.49	8.10	16.86
13	225	*Heliria*	*cristata*	4.98	10.46	17.25
13	226	*Heliria*	*praelata*	3.96	9.59	14.02
13	227	*Heliria*	*gemma*	4.93	10.14	15.06
13	228	*Heliria*	*sinuata*	4.46	9.29	13.13
13	229	*Heliria*	*strombergi*	4.18	9.51	12.94
14	211	*Antianthe*	*expansa*	4.10	7.48	11.18
15	186	*Tropidarnis*		4.25	11.10	13.39
15	187			4.08	8.84	11.14
15	195	*Telamona*	*wescotti*	3.15	8.70	10.09
15	216	*Carynota*	*mera*	3.61	8.75	10.84
15	217	*Carynota*	*marmorata*	3.16	7.25	8.99
16	210	*Telonaca*	*alta*	4.95	9.54	17.62
16	219	*Glossonotus*	*crategei*	3.41	6.89	12.51
16	220	*Glossonotus*	*univittatus*	4.01	8.21	15.08
16	221	*Glossonotus*	*accuminatus*	4.62	9.28	19.87
17	222	*Telamonanthe*	*pulchella*	3.43	6.07	9.72
17	223	*Telamonanthe*	*reilyi*	3.20	5.71	9.39
18	188	*Hemicardiacus*		7.33	11.18	19.43
19	208	*Thelia*	*bimaculata*	4.92	11.44	19.83

Appendix 2.
Fortran program for calculating pronotal shape of treehoppers.

```
C
C     From digitized data this program computes a grid of
C     analogous points using spline interpolation. Using
C     this grid, the program computes distances, angles,
C     lengths and area for use in canonical discriminant
C     analysis and discriminant function analysis.
C
C     The program conforms to the full ANSI 1977 Fortran
C     standard. However, the subroutine DCSINT which does
C     the cubic spline interpolation is in the IBM
C     Engineering and Scientific Library.
C
C     1. Declarations.
C
      INTEGER NMAX, NVALS, MAXC, NVM1
      DOUBLE PRECISION ONE, ZERO
      PARAMETER (NMAX=400, NVALS=21, MAXC=4*NMAX, NVM1=NVALS-1,
     *      ONE=1.0D0, ZERO=0.0D0)
      CHARACTER CODE*15, CONTUR*1 REST*29, REST2*29
      INTEGER I, J, INIT, MCOUNT
      DOUBLE PRECISION PI
      DOUBLE PRECISION DNVALS
      DOUBLE PRECISION XSMC, YSMC
      DOUBLE PRECISION XB, YB, SLOPE, SLOPE2, COSTH, SINTH, DMC
      DOUBLE PRECISION DIMAG, DSQRT, DATAN, DABS
      DOUBLE PRECISION CRENDX, CRENDY
      DOUBLE PRECISION X(NMAX), Y(NMAX), XS(NMAX), YS(NMAX)
```

```
      DOUBLE PRECISION T(NVALS), C(MAXC), S(NMAX)
      DOUBLE PRECISION XH(NVALS), YH(NVALS)
      DOUBLE PRECISION XC1(NVM1), XC2(NVM1), XC3(NVM1)
      DOUBLE PRECISION YC1(NVM1), YC2(NVM1), YC3(NVM1)
      DOUBLE PRECISION D12(NVM1), D23(NVM1), D34(NVM1)
      DOUBLE PRECISION XMIN, XMAX, XDIF, SCALE, WIDTH
      DOUBLE PRECISION ANGLE(NVALS,4)
      DOUBLE PRECISION LC(4), TLC
      DOUBLE PRECISION AREA, AREA1, AREA2
C
C     First executable statement
C
C     Compute pi for later use
C
      PI=4.0D0*DATAN(1.0D0)
C
C     Compute values at which spline interpolation is done.
C
      DNVALS=NVALS
      DO 700 I=1,NVALS
      T(I)=(DFLOAT(I)-ONE)/(DNVALS-ONE)
700   CONTINUE
C
C     2. Read data.
C
C     (The input starts with an identification line. The first
C     column indicates the  contour after which is information
C     identifying the specimen.  After that the digitized
C     coordinates follow one point to a line.  Coordinates
C     greater than or equal to 99 signal the end of the
C     contour.  Contours 2, 3, and 4 are recorded the same
C     way except that there is no information after column
C     1 in the identification line.)
C
 900  READ (30,1,END=9999) CONTUR, REST
 1    FORMAT(A1,A29)
C
      IF(CONTUR.EQ.'1')REST2=REST
      I=1
151   READ (30,*) X(I), Y(I)
C
C     Check for end of contour
C
      IF (X(I) .GE. 99.0)  GO TO  150
C
C     Check data for duplicate points. (Duplicates are ignored).
C
      IF ( I .GT. 1 ) THEN
            IF  ( DABS(X(I)-X(I-l) ).LT. l.0D-03.AND.
     *              DABS(Y(I)-Y(I-l)).LT. l.0D-03 ) GO TO 151
      END IF
      MCOUNT=I
       I=I+1
      GO TO 151
```

```
150     CONTINUE
C
C       3. Translate, rotate and scale coordinates
C
        IF (CONTUR .EQ. 111) THEN
                XDIF=DSQRT( (X(MCOUNT)-X(1) )**2 + (Y(MCOUNT)-Y(1))**2 )
                XB=X(l)
                YB=Y(l)
                SLOPE=(Y(MCOUNT)-Y(1))/(X(MCOUNT)-X(1) )
                SLOPE2=SLOPE*SLOPE
                COSTH=ONE/DSQRT(ONE+SLOPE2)
                SINTH=SLOPE*COSTH
        ENDIF
        DO 701 I=1, MCOUNT
                XS(I) = SCALE*(COSTH*(X(I)-XB)+SINTH*(Y(I)-YB))
                YS(I) = SCALE*(-SINTH*(X(I)-XB)+COSTH*(Y(I)-YB))
701     CONTINUE
C
C       4. Translate contours 2, 3 & 4 so tip of posterior process
C       coincides with that of contour 1.
C
                IF (CONTUR .EQ. '1') THEN
                        CRENDX=XS(MCOUNT)
                        CRENDY=YS(MCOUNT)
                ELSE
                        XSMC=XS(MCOUNT)
                        YSMC=YS(MCOUNT)
                        DO 702 I=1, MCOUNT
                                XS(I)=XS(I)+ CRENDX-XSMC
                                YS(I)=YS(I)+ CRENDY-YSMC
702                     CONTINUE
                END IF
C
C       5. A. Compute contribution of line integral for area from
C       contours 1 and 4. (By Green's theorem the area
C       computation is reduced to numerical integration of
C       a line integral).
C
        IF(CONTUR.EQ.'1')THEN
                AREA1=ZERO
                DO 703 I=2, MCOUNT
                        AREA1=AREA1-.5D0*(XS(I)+XS(I-1))*(YS(I)-YS(I-l))
703             CONTINUE
        END IF
        IF(CONTUR.EQ.'4')THEN
                AREA2=ZERO
                DO 704 I=2, MCOUNT
                        AREA2=AREA2+.5D0*(XS(I)+XS(I-l))*(YS(I)-YS(I-l))
704             CONTINUE
        END IF
C
C       6. Compute arclength and normalize.
C
        S(1)=ZERO
```

```
      DO 705 I=2, MCOUNT
              S(I)=S(I-1)
     *        +DSQRT( ( XS(I) - XS(I-1))**2 + (YS(I)-YS(I-1))**2 )
705   CONTINUE
      TLC=XDIF*S(MCOUNT)
      DO 706 I=1, MCOUNT
              S(I)=S(I)/S(MCOUNT)
706   CONTINUE
C
C     7. Spline Interpolation.
C
      INIT=0
      CALL DCSINT(S, XS, C, MCOUNT, INIT, T, XH, NVALS)
      INIT=0
      CALL DCSINT(S, YS, C, MCOUNT, INIT, T, YH, NVALS)
C
C     8. Compute distances.
C
      IF(CONTUR.EQ.'1' ) THEN
              DO 707 I=1, NVM1
                      XC1(I)=XH(I)
                      YC1(I)=YH(I)
707           CONTINUE
      ELSE IF(CONTUR.EQ.'2' ) THEN
              DO 708 I=1, NVM1
                      XC2(I)=XH(I)
                      YC2(I)=YH(I)
708           CONTINUE
      ELSE IF(CONTUR.EQ.'3' ) THEN
              DO 709 I=1, NVM1
                      XC3(I)=XH(I)
                      YC3(I)=YH(I)
709           CONTINUE
      ELSE IF(CONTUR.EQ.'4')THEN
              WIDTH=XDIF*DSQRT((XC1(1)-XH(1))**2+(YC1(1)-YH(1))**2)
              DO 710 I=1, NVM1
                      D12(I)=DSQRT((XC1(I)-XC2(I))**2+(YC1(I)-YC2(I))**2)
                      D23(I)=DSQRT((XC2(I)-XC3(I))**2+(YC2(I)-YC3(I))**2)
                      D34(I)=DSQRT((XC3(I)-XH(I))**2+(YC3(I)-YH(I))**2)
710           CONTINUE
      ENDIF
C
C     9. Compute angles of segments with the reference segment.
C
C     (Note: We chose to compute the angle by forming a complex
C     number whose real and imaginary parts are the differences
C     of the X-coordinates and of the Y-coordinates of the
C     endpoints respectively of the segment. The angle (in
C     radians) is then the imaginary part of the logarithm
C     of this number.)
C
      IF(CONTUR.EQ.'1' ) THEN
              DO 711 I=1, NVM1
              LC(1)=TLC
```

```
          ANGLE(I,1)=180.0D0/PI*
     *         DIMAG(CDLOG(DCMPLX(XH(I+l)-XH(I),YH(I+1)-YH(I))))
711       CONTINUE
      ELSE IF(CONTUR.EQ.'2' ) THEN
          LC(2)=TLC
          DO 712 I=1, NVM1
          ANGLE(I,2)=180.0D0/PI*
     *         DIMAG(CDLOG(DCMPLX(XH(I+1)-XH(I),YH(I+1)-YH(I))))
712       CONTINUE
      ELSE IF(CONTUR.EQ.'3' ) THEN
          DO 713 I=1, NVM1
          LC(3)=TLC
          ANGLE(I,3)=180.0D0/PI*
     *         DIMAG(CDLOG(DCMPLX(XH(I+1)-XH(I),YH(I+1)-YH(I))))
713       CONTINUE
      ELSE IF(CONTUR.EQ.'4')THEN
          DO 714 I=1, NVM1
          LC(4)=TLC
          ANGLE(I,4)=180.0D0/PI*
     *         DIMAG(CDLOG(DCMPLX(XH(I+l)-XH(I),YH(I+l)-YH(I))))
714       CONTINUE
C
C     5. B. Complete computation of area by adding contribution
C     of gap between starts of contours 1 and 4. Then
C     multiply by the length of the reference segment
C     squared to get square millimeters.
C
          AREA=AREA1+AREA2-.5DO*(XC1(1)+XH(1))*(YC1(1)-YH(1))
          AREA=XDIF*XDIF*AREA
C
C     10. Write results to a file.
C
C     Identification.
C
          WRITE (31,2) REST2
2         FORMAT(A29)
C
C     Length, distance from beginning of contour 1 to that of
C     contour 4 and area.
C
          WRITE (31,*) XDIF, WIDTH, AREA
          DO 715 I=1, NVM1
C
C     Distances.
C
          WRITE (31,*) D12(I), D23(I), D34(I)
715       CONTINUE
C
C     Lengths of contours.
C
          WRITE (31,*) (LC(J),J=1,4)
C
C     Angles.
C
```

```
          DO 716 I=1, NVM1
          WRITE (31,*) (ANGLE(I,J),J=1,4)
 716      CONTINUE
      END IF
      GO TO 900
C
C     11. End of computation.
C
 9999 CONTINUE
      STOP
      END
```

INDEX